Introduction to Analysis with Complex Numbers

Introduction to Analysis with Complex Numbers

Irena Swanson
Purdue University, USA

World Scientific

EW JERSEY · LONDON · SINGAPORE · BEIJING · SHANGHAI · HONG KONG · TAIPEI · CHENNAI · TOKYO

Published by

World Scientific Publishing Co. Pte. Ltd.

5 Toh Tuck Link, Singapore 596224

USA office: 27 Warren Street, Suite 401-402, Hackensack, NJ 07601

UK office: 57 Shelton Street, Covent Garden, London WC2H 9HE

Library of Congress Control Number: 2021000399

British Library Cataloguing-in-Publication Data
A catalogue record for this book is available from the British Library.

INTRODUCTION TO ANALYSIS WITH COMPLEX NUMBERS

ISBN 978-981-122-585-7 (hardcover)
ISBN 978-981-122-769-1 (paperback)
ISBN 978-981-122-586-4 (ebook for institutions)
ISBN 978-981-122-587-1 (ebook for individuals)

For any available supplementary material, please visit
https://www.worldscientific.com/worldscibooks/10.1142/11983#t=suppl

Desk Editor: Soh Jing Wen

Contents

Preface

These notes were written expressly for Mathematics 112 at Reed College, with first use in the spring of 2013. The title of the course is "Introduction to Analysis". The prerequisite is calculus. Recently used textbooks have been Steven R. Lay's "Analysis, With an Introduction to Proof" (Prentice Hall, Inc., Englewood Cliffs, NJ, 1986, 4th edition), and Ray Mayer's in-house notes "Introduction to Analysis" (2006, available at http://www.reed.edu/~mayer/math112.html/index.html). Ray Mayer's notes strongly influenced the coverage in this book.

In Math 112 at Reed College, students learn to write proofs while at the same time they learn about binary operations, orders, fields, ordered fields, complete fields, complex numbers, sequences, and series. We also review limits, continuity, differentiation, and integration. My aim for these notes is to constitute a self-contained book that covers the standard topics of a course in introductory analysis, that handles complex-valued functions, sequences, and series, that has enough examples and exercises, that is rigorous, and is accessible to undergraduates. I maintain two versions of these notes, one in which the natural, rational and real numbers are constructed and the Least upper bound theorem is proved for the ordered field of real numbers, and one version in which the Least upper bound property is assumed for the ordered field of real numbers. You are reading the longer, former version.

Chapter 1 is about how we do mathematics: basic logic, proof methods, and Pascal's triangle for practicing proofs. Chapter 2 introduces foundational concepts: sets, Cartesian products, relations, functions, binary operations, fields, ordered fields, Archimedean property for the set of real numbers. In Chapters 1 and 2 we assume knowledge of high school mathematics so that we do not practice abstract concepts and methods in a vacuum. Chapter 3 through Section 3.8 takes a step back: we "forget" most previously learned mathematics, and we use the newly learned abstract tools to construct natural numbers, integers, rational numbers, real numbers, with all the arithmetic and order. I do not teach these constructions in great detail; my aim is to give a sense of them and to practice abstract logical thinking. The remaining sections in Chapter 3 are new material for most students: the field of complex numbers, and some topology. I cover the last section of Chapter 3 very lightly. Subsequent chapters cover standard material for introduction to analysis: limits, continuity, differentiation, integration, sequences, series, ending with the development of the power series $\sum_{k=0}^{\infty} \frac{x^k}{k!}$, the exponential and the trigonometric functions. Since students have seen limits, continuity, differentiation and integration before, I go through chapters 4 through 7 quickly. I will slow down for sequences and series (the last three chapters).

An effort is made throughout to use only what had been proved. For this reason, the chapters on differentiation and integration do not have the usual palette of trigonometric and exponential examples of other books. The final chapter makes up for it and works out much trigonometry in great detail and depth.

I acknowledge and thank the support from the Dean of Faculty of Reed College to fund exercise and proofreading support in the summer of 2012 for Maddie Brandt, Munyo Frey-Edwards, and Kelsey Houston-Edwards. I also thank the following people for their valuable feedback: Mark Angeles, Josie Baker, Marcus Bamberger, Anji Bodony, Zachary Campbell, Nick Chaiyachakorn, Safia Chettih, Laura Dallago, Andrew Erlanger, Joel Franklin, Darij Grinberg, Rohr Hautala, Palak Jain, Ya Jiang, Albyn Jones, Willow Kelleigh, Mason Kennedy, Christopher Keane, Michael Keppler, Ryan Kobler, Jay Kruer, Oleks Lushchyk, Molly Maguire, Benjamin Morrison, Samuel Olson, Kyle Ormsby, Angélica Osorno, Shannon Pearson, David Perkinson, Jeremy Rachels, Ezra Schwartz, Jacob Sharkansky, Marika Swanberg, Simon Swanson, Matyas Szabo, Ruth Valsquier, Xingyi Wang, Emerson Webb, Livia Xu, Qiaoyu Yang, Dean Young, Eric Zhang, Jialun Zhao, and two anonymous reviewers. If you have further comments or corrections, please send them to `irena@purdue.edu`.

The briefest overview, motivation, notation

What are the meanings of the following:

$5 + 6$

$7 \cdot 9$

$8 - 4$

$4/5$

$\sqrt{2}$

$4 - 8$

$1/3 = 0.333\ldots \qquad 1 = 3 \cdot 1/3 = 0.999\ldots$

$(a \cdot b)^2 = a^2 \cdot b^2$

$(a + b) \cdot (c + d) = ac + ad + bc + bd$

$(a + b) \cdot (a - b) = a^2 - b^2$

$(a + b)^2 = a^2 + 2ab + b^2$

$\sqrt{a} \cdot \sqrt{b} = \sqrt{ab}$ (for which a, b?)

What is going on: $\sqrt{-4} \cdot \sqrt{-9} = \sqrt{(-4)(-9)} = \sqrt{36} = 6$,

$$\sqrt{-4} \cdot \sqrt{-9} = 2i \cdot 3i = -6$$

You **know** all of the above except possibly the complex numbers in the last two rows, where obviously something went wrong. We will not resolve this last issue until later in the semester, but the point for now is that we do need to reason carefully.

The main goal of this class is to learn to reason carefully, *rigorously*. Since one cannot reason in a vacuum, we will (but of course) be learning a lot of mathematics as well: sets, logic, various number systems, fields, the field of real numbers, the field of complex numbers, sequences, series, some calculus, and that $e^{ix} = \cos x + i \sin x$.

We will make it all rigorous, i.e., we will be doing proofs. A *proof* is a sequence of steps that logically follow from previously accepted knowledge.

But no matter what you do, never divide by 0. For further wise advice, turn to Appendix A.

[NOTATIONAL CONVENTION: TEXT BETWEEN SQUARE BRACKETS IN THIS FONT SHOULD BE READ AS A POSSIBLE REASONING GOING ON IN THE BACKGROUND IN YOUR HEAD, AND NOT AS PART OF FORMAL WRITING.]

†**1.** Exercises with a dagger are invoked later in the text.

***2.** Exercises with a star are more difficult.

Chapter 1

How we will do mathematics

1.1 Statements and proof methods

Definition 1.1.1. *A* **statement** *is a reasonably grammatical and unambiguous sentence that can be declared either true or false.*

Why do we specify "reasonably grammatical"? We do not disqualify a statement just because of poor grammar, nevertheless, we strive to use correct grammar and to express the meaning clearly. And what do we mean by **true** or **false**? For our purposes, a statement is **false** if there is at least one counterexample to it, and a statement is **true** if it has been proved so, or if we **assume** it to be true.

Examples and non-examples 1.1.2.

   (i) The sum of 1 and 2 equals 3. (This is a true statement.)

   (ii) Seventeen. (This is not a statement.)

   (iii) Seventeen is the seventh prime number. (This is a true statement.)

   (iv) Is x positive? (This is not a statement.)

   (v) $1 = 2$.* (This is a false statement.)

   (vi) For every real number $\epsilon > 0$ there exists a real number $\delta > 0$ such that for all x, if $0 < |x - a| < \delta$ then x is in the domain of f and $|f(x) - L| < \epsilon$. (This is a statement, and it is (a part of) the definition of the limit of a (special) function f at a being L. Out of context, this statement is neither true or false, but we can prove it or assume it for various functions f.)

---

* This statement can also be written in plain English as "One equals two." In mathematics it is acceptable to use symbolic notation to some extent, but keep in mind that too many symbols can make a sentence hard to read. In general we avoid starting sentences with a symbol. In particular, do not make the following sentence. "=" is a verb. Instead make a sentence such as the following one. Note that "=" is a verb.

(vii) Every even number greater than 4 can be written as a sum of two odd primes. (This statement is known as **Goldbach's conjecture**. No counterexample is known, and no proof has been devised, so it is currently not known if it is true or false.)

These examples show that not all statements have a definitive truth value. What makes them statements is that after possibly arbitrarily assigning them truth values, different consequences follow. For example, if we assume that (vi) above is true, then the graph of f near a is close to the graph of the constant function L. If instead we assume that (vi) above is false, then the graph of f near a has infinitely many values at some vertical distance away from L no matter how much we zoom in at a. With this in mind, even "I am good" is a statement: if I am good, then I get a cookie, but if I am not good, then you get the cookie. On the other hand, if "Hello" were to be true or false, I would not be able to make any further deductions about the world or my next action, so that "Hello" is not a statement, but only a sentence.

A useful tool for manipulating statements is a **truth table**: it is a table in which the first few columns may set up a situation, and the subsequent columns record truth values of statements applying in those particular situations. Here are two examples of truth tables, where "T" of course stands for "true" and "F" for "false":

f	constant	continuous	differentiable everywhere
$f(x) = x^2$	F	T	T
$f(x) = |x|$	F	T	F
$f(x) = 7$	T	T	T

x	y	$xy > 0$	$xy \leq 0$	$xy < 0$
$x > 0$	$y > 0$	T	F	F
$x > 0$	$y \leq 0$	F	T	F
$x < 0$	$y > 0$	F	T	T
$x < 0$	$y \leq 0$	F	F	F

Note that in the second row of the last table, in the exceptional case $y = 0$, the statement $xy < 0$ is false, but in "the majority" of the cases in that row $xy < 0$ is true. The one counterexample is enough to declare $xy < 0$ not true, i.e., false.

Statements can be manipulated just like numbers and variables can be manipulated, and rather than adding or multiplying statements, we **connect** them (by compounding the sentences in grammatical ways) with connectors such as "not", "and", "or", and so on.

Statement connecting:

(1) **Negation** of a statement P is a statement whose truth values are exactly opposite from the truth values of P (under any specific circumstance). The negation of P is denoted "not P" (or "$\neg P$"). Some simple examples: the negation of "$A = B$" is "$A \neq B$"; the negation of "$A \leq B$" is "$A > B$"; the negation of "I am here" is "I am not here" or "It is not the case that I am here".

Now go back to the last truth table. Note that in the last line, the truth values of "$xy > 0$" and "$xy \leq 0$" are both false. But one should think that "$xy > 0$" and "$xy \leq 0$" are negations of each other! So what is going on, why are the two truth values not opposites of each other? The problem is of course that the circumstances $x < 0$ and $y \leq 0$ are not specific enough. The statement "$xy > 0$" is under these circumstances false precisely when $y = 0$, but then "$xy \leq 0$" is true. Similarly, the statement "$xy \leq 0$" is under the given circumstances false precisely when $y < 0$, but then "$xy > 0$" is true. Thus, once we make the conditions specific enough, then the truth values of "$xy > 0$" and "$xy \leq 0$" are opposite, so that the two statements are indeed negations of each other.

(2) **Conjunction** of statements P and Q is a statement that is true precisely when both P and Q are true, and it is false otherwise. It is denoted "P and Q" or "$P \wedge Q$". We can record this in a truth table as follows:

P	Q	P and Q
T	T	T
T	F	F
F	T	F
F	F	F

For example, P and (not P) is always false, and P and P simplifies to P.

(3) **Disjunction** of statements P and Q is a statement that is false precisely when both P and Q are false, and it is true otherwise. We denote it as "P or Q" or as "$P \vee Q$". In other words, as long as either P or Q is true, then P or Q is true. In plain language, unfortunately, we use "or" in two different ways: "You may take cream or sugar" says you may take cream or sugar or both, just like in the proper logical way, but "Tonight we will go to the movies or to the baseball game" implies that we will either go to the movies or to the baseball game but we will not do both. The latter connection of two sentences is in logic called **exclusive or**, often denoted **xor**. Even "either-or" does not disambiguate between "or" and "xor". The truth table for the two disjunctions is:

P	Q	P or Q	P xor Q
T	T	T	F
T	F	T	T
F	T	T	T
F	F	F	F

(4) **Implication** or a **conditional statement** is a statement of the form "P implies Q," or variants thereof, such as all of the following:

> P implies Q.
> If P then Q.
> P is a sufficient condition for Q.
> P only if Q.
> Q if P.
> Q provided P.
> Q given P.
> Q whenever P.
> Q is a necessary condition for P.

P is called the **antecedent** and Q the **consequent**. A symbolic abbreviation is "$P \Rightarrow Q$."

An implication is true when a true conclusion follows a true assumption, or whenever the assumption is false. In other words, $P \Rightarrow Q$ is false exactly when P is true and Q is false.

P	Q	$P \Rightarrow Q$
T	T	T
T	F	F
F	T	T
F	F	T

It may be counterintuitive that a false antecedent always makes the implication true. Bertrand Russell once lectured on this and claimed that if $1 = 2$ then he (Bertrand Russell) was the pope. An audience member challenged him to prove it. So Russell reasoned somewhat like this: "If I am the pope, then the consequent is true. If the consequent is false, then I am not the pope. But if I am not the pope, then the pope and I are two different people. By assumption $1 = 2$, so we two people are one, so I am the pope. Thus no matter what, I am the pope." Furthermore, if $1 = 2$, then Bertrand Russell is also not the pope. Namely, if he is not the pope, the consequent is true, but if he is the pope, then the pope and he are one, and since one equals two, then the pope and he are two people, so Russell cannot be the pope.

A further discussion about why false antecedent makes the implication true is in the next discussion (5).

Unfortunately, the implication statement is not used consistently in informal spoken language. For example, your grandmother may say: "You may have ice cream if you eat your broccoli" when she means "You may have ice cream only if you eat your broccoli." Be nice to your grandmother and eat that broccoli even if she does not express herself precisely because you know precisely what she means. But in mathematics you do have to express yourself precisely! (Well, read the next paragraph.)

Even in mathematics some shortcuts in precise expressions are acceptable. Here is an example. The statements "An object x has property P if something-or-other holds" and "An object x has property P if and only if something-or-other holds" (see (5) below for "if and only if") in general have different truth values and the proof of the second is longer. However, the **definition** of what it means for an object to have property P in terms of something-or-other is usually phrased with "if", but "if and only if" is meant. For example, the following is standard: "**Definition:** A positive integer strictly bigger than 1 is **prime** if whenever it can be written as a product of two positive integers, one of the two factors must be 1." The given definition, if read logically precisely, since it said nothing about numbers such as $4 = 2 \cdot 2$, would allow us to call 4 prime. However, it is an understood shortcut that **only** the numbers with the stated property are called prime.

(5) **Equivalence** or the **logical biconditional** of P and Q stands for the compound statement $(P \Rightarrow Q)$ and $(Q \Rightarrow P)$. It is abbreviated "$P \Leftrightarrow Q$" or "P iff Q", and is true precisely when P and Q have the same truth values.

For example, for real numbers x and y, the statement "$x \leq y + 1$" is equivalent to "$x - 1 \leq y$." Another example: "$2x = 4x^2$" is equivalent to "$x = 2x^2$," but it is **not** equivalent to "$1 = 2x$." (Say why!)

We now backtrack on the truth values of $P \Rightarrow Q$. We can certainly fill in some parts without qualms, leaving some unknown truth values x and y:

P	Q	$P \Rightarrow Q$	$Q \Rightarrow P$	$P \Leftrightarrow Q$
T	T	T	T	T
T	F	F	x	F
F	T	x	F	F
F	F	y	y	T

Since the last column above is the conjunction of the previous two, the last line forces the value of y to be T. If x equals F, then the truth values of $P \Rightarrow Q$ are the same as the truth values of $P \Leftrightarrow Q$, which would say that the statements $P \Rightarrow Q$ and $P \Leftrightarrow Q$ are logically the same. But this cannot be: "If $r > 0$ then $r \geq 0$" is true whereas "If $r \geq 0$ then $r > 0$" is false. So this may convince you that the truth values for the third and the fifth column have to be distinct, and this is only possible if x is T.

Here is the truth table for all the connectives so far:

P	Q	$\text{not } P$	$P \text{ and } Q$	$P \text{ or } Q$	$P \text{ xor } Q$	$P \Rightarrow Q$	$P \Leftrightarrow Q$
T	T	F	T	T	F	T	T
T	F	F	F	T	T	F	F
F	T	T	F	T	T	T	F
F	F	T	F	F	F	T	T

One can form more elaborate truth tables if we start not with two statements P and Q but with three or more. Examples of logically compounding P, Q, and R are: $P \text{ and } Q \text{ and } R$, $(P \text{ and } Q) \Rightarrow Q$, etc. For manipulating three statements, we would fill a total of 8 rows of truth values, for four statements there would be 16 rows, and so on.

(6) **Proof** of P is a series of steps (in statement form) that establish beyond doubt that P is true under all circumstances, weather conditions, political regimes, time of day... . The logical reasoning that goes into mathematical proofs is called **deductive reasoning**. Whereas both guessing and intuition can help you find the next step in your mathematical proof, only the logical parts are trusted and get written down. Proofs are a mathematician's most important tool; the book contains many examples, and the next few pages give some examples and ideas of what proofs are.

Ends of proofs are usually marked by $\square$, $\blacksquare$, $//$, or QED (for "quod erat demonstrandum", which is Latin for "that which was to be proved"). It is a good idea to mark the completions of proofs especially when they are long or with many parts and steps — that helps the readers know that nothing else is to be added.

The most trivial proofs simply invoke a definition or axiom, such as "An even integer is of the form 2 times an integer," "An odd integer is of the form 1 plus 2 times an integer," or, "A positive integer is **prime** if whenever it can be written as a product of two positive integers, one of the two factors is 1."

Another type of proof consists of filling in a truth table. For example, $P \text{ or } (\text{not } P)$ is always true, no matter what the truth value of P is, and this can be easily verified with the truth table:

P	not P	P or not P
T	F	T
F	T	T

A formula using logical statements that is always true is called a **tautology**. So P or not P is a tautology. Here is another example of tautology: $((P \Rightarrow Q) \text{ and } P) \Rightarrow Q$, and it is proved below with the truth table:

P Q	$P \Rightarrow Q$	$(P \Rightarrow Q) \text{ and } P$	$((P \Rightarrow Q) \text{ and } P) \Rightarrow Q$
T T	T	T	T
T F	F	F	T
F T	T	F	T
F F	T	F	T

This particular tautology is called **modus ponens**, and its most famous example is the following:

> Every man is mortal. (If X is a man, then X is mortal.)
> Socrates is human.
> Therefore, Socrates is mortal.

Here is a more mathematical example of modus ponens:

> Every differentiable function is continuous.
> f is differentiable.
> Therefore, f is continuous.

Another tautology is **modus tollens**: $((P \Rightarrow Q) \text{ and } (\text{not } Q)) \Rightarrow (\text{not } P)$. To prove it, one constructs a truth table as before for modus ponens — it is a common proof technique to invoke the similarity principle with previous work that allows one to not carry out all the steps, as I just did. However, whenever you invoke the proof-similarity principle, you better be convinced in your mind that the similar proof indeed does the job; if you have any doubts, show all work instead! In this case, I am sure that the truth table does the job, but if you are seeing this for the first time, you may want to do the actual truth table explicitly to get a better grasp on these concepts.

Here is a mathematical example of modus tollens:

Every differentiable function is continuous.

f is not continuous.

Therefore, f is not differentiable.

Here is another example on more familiar ground:

If you are in Oregon, then you are in the USA.

You are not in the USA.

Therefore, you are not in Oregon.

Some proofs can be pictorial/graphical. Here we prove with this method that for any real numbers x and y, $|x| < y$ if and only if $-y < x < y$. (We will see many uses of **absolute values**.) Proof: [FOR A BICONDITIONAL $P \Leftrightarrow Q$ WE NEED TO PROVE $P \Rightarrow Q$ AND $Q \Rightarrow P$.] The assumption $|x| < y$ implies that y must be positive, and the assumption $-y < x < y$ implies that $-y < y$, which also says that y must be positive. So, with either assumption, we can draw the following part of the real number line:

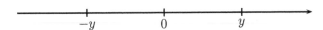

Now, by drawing, the real numbers x with $|x| < y$ are precisely those real numbers x with $-y < x < y$. A fancier way of saying this is that $|x| < y$ if and only if $-y < x < y$. □

Similarly, for all real numbers x and y, $|x| \leq y$ if and only if $-y \leq x \leq y$. (Here, the word "similarly" is a clue that I am invoking the proof-similarity principle, and a reader who wants to practice proofs or is not convinced should at this point work through a proof by mimicking the steps in the previous one.)

Some (or actually most) proofs invoke previous results without re-doing the previous work. In this way we prove the **triangle inequality**, which asserts that for all real numbers x and y, $|x \pm y| \leq |x| + |y|$. (By the way, we will use the triangle inequality intensely, so understand it well.) Proof: Note that always $-|x| \leq x \leq |x|$, $-|y| \leq \pm y \leq |y|$. Since the sum of smaller numbers is always less than or equal to the sum of larger numbers, we then get that $-|x| - |y| \leq x \pm y \leq |x| + |y|$. But $-|x| - |y| = -(|x| + |y|)$, so that $-(|x| + |y|) \le x \pm y \le |x| + |y|$. But then by the previous result, $|x \pm y| \le |x| + |y|$.   □

Most proofs require a combination of methods. Here we prove that whenever x is a real number with $|x - 5| < 4$, then $|x^3 - 3x| < 900$. Proof: The following is standard formatting that you should adopt: first write down the left side of the desired inequality ($|x^3 - 3x|$), then start manipulating it algebraically, in intermediate steps add a **clever** 0 here and there, multiply by a **clever** 1 here and there, rewrite, simplify, make it less than or equal to something else, and so on, every step should be either obvious or justified on the right, until at the end you get the quantity on the right (900):

$|x^3 - 3x| \le |x^3| + |3x|$ (by the triangle inequality)

$= |x|^3 + 3|x|$

$= |x - 5 + 5|^3 + 3|x - 5 + 5|$ (by adding a clever 0)

$\le (|x - 5| + 5)^3 + 3(|x - 5| + 5)$

(by the triangle inequality and since $a \le b$

implies that $a^3 \le b^3$)

$\le (4 + 5)^3 + 3(4 + 5)$ (since by assumption $|x - 5| < 4$)

$= 9^3 + 3 \cdot 9$

$= 9(9^2 + 3)$

$< 900.$

Here is a pictorial proof establishing the basis of trigonometry and the definition of slope as rise over run: namely that $\frac{B}{A} = \frac{b}{a}$.

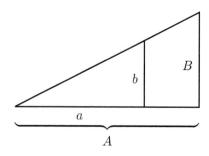

Proof: The areas of the big and small triangles are $\frac{1}{2}AB$ and $\frac{1}{2}ab$, and the area of the difference is base times the average height, i.e., it is $(A - a)\frac{b+B}{2}$. Thus

$$\frac{1}{2}AB = \frac{1}{2}ab + (A - a)\frac{b + B}{2}.$$

By multiplying through by 2 we get that $AB = ab + (A - a)(b + B) = ab + Ab + AB - ab - aB$, so that, after cancellations, $Ab = aB$. Then, after dividing through by aA we get $\frac{B}{A} = \frac{b}{a}$. □

Another common method of proof of a statement P is **proof by contradiction**: you assume that not P is true and by using only correct logical and mathematical steps you derive some nonsense (contradiction). This then says that not P is false, so that not (not P) = P must be true. Beware: proofs by contradiction are in general not considered *elegant*, nevertheless, they can be very powerful.

(Due to Pythagoras.) $\sqrt{2}$ is not a rational number. (A number is **rational** if it is a ratio of two whole integers, where the denominator is not zero.) *Proof by contradiction:* Suppose that $\sqrt{2}$ is rational. This means that $\sqrt{2} = \frac{a}{b}$ for some whole numbers a and b with b non-zero. Let d be the greatest common divisor of a and b. Write $a = a_0 d$ and $b = b_0 d$ for some integers $a_0, b_0 \neq 0$. Then $\sqrt{2} = \frac{a}{b} = \frac{a_0}{b_0}$, so $b_0\sqrt{2} = a_0$, and so by squaring both sides we get that $2b_0^2 = a_0^2$. So $a_0^2 = 2b_0^2$ is an even number, which by Exercise 1.1.12 means that a_0 is an even number. Write $a_0 = 2a_1$ for some integer a_1. Then $4a_1^2 = a_0^2 = 2b_0^2$, so that $b_0^2 = 2a_1^2$ is even, whence again b_0 is even. But then 2 divides both a_0 and b_0, so that $2d$ divides both a and b, which contradicts the assumption that d was the greatest common divisor of a and b. Thus it is not the case that $\sqrt{2}$ is rational, so it must be irrational. □

Exercises for Section 1.1

1.1.1. Determine and justify the truth value of the following statements.

i) 3 is odd or 5 is even.

ii) If n is even, then $3n$ is prime.

iii) If $3n$ is even, then n is prime.

iv) If n is prime, then $3n$ is odd.

v) If $3n$ is prime, then n is odd.

vi) $(P \text{ and } Q) \Rightarrow P$.

1.1.2. Sometimes statements are not written precisely enough. For example, "It is not the case that 3 is prime and 5 is even" may be saying "not (3 is prime and 5 is even)," or it may be saying "(not (3 is prime)) and (5 is even)." The first option is true and the second is false.

Similarly analyze several possible interpretations of the following ambiguous sentences:

i) If 6 is prime then 7 is even or 5 is odd.

ii) It is not the case that 3 is prime or if 6 is prime then 7 is even or 5 is odd.

General advice: Write precisely; aim to not be misunderstood.

1.1.3. Add the following columns to the truth table in (5): P and not Q, (not P) and (not Q), (not P) or (not Q), $P \Rightarrow$ not Q, (not P) $\Rightarrow$ not Q. Are any of the new columns negations of the columns in the truth table (5) or of each other?

1.1.4. Suppose that $P \Rightarrow Q$ is true and Q is false. Prove that P is false.

1.1.5. Prove that $P \Rightarrow Q$ is equivalent to (not Q) $\Rightarrow$ (not P).

1.1.6. Simplify the following statements:

i) $(P \text{ and } P) \text{ or } P$.

ii) $P \Rightarrow P$.

iii) $(P \text{ and } Q) \text{ or } (P \text{ or } Q)$.

1.1.7. Prove with truth tables that the following statements are true.

i) $(P \Leftrightarrow Q) \Leftrightarrow [(P \Rightarrow Q) \text{ and } (Q \Rightarrow P)]$.

ii) $(P \Rightarrow Q) \Leftrightarrow (Q \text{ or not } P)$.

iii) $(P \text{ and } Q) \Leftrightarrow [P \text{ and } (P \Rightarrow Q)]$.

iv) $[P \Rightarrow (Q \text{ or } R)] \Leftrightarrow [(P \text{ and not } Q) \Rightarrow R]$.

1.1.8. Assume that P or Q is true and that $R \Rightarrow Q$ is false. Determine with proof the truth values of P, Q, R, or explain if there is not enough information.

1.1.9. Assume that $(P \text{ and } Q) \Rightarrow R$ is false. Determine with proof the truth values of P, Q, R, or explain if there is not enough information.

1.1.10. Suppose that x is any real number such that $|x + 2| < 3$. Prove that $|x^3 - 3x| < 200$.

1.1.11. Suppose that x is any real number such that $|x - 1| < 5$. Find with proof a positive constant B such that for all such x, $|3x^4 - x| < B$.

1.1.12. (Odd-even integers)

i) Prove that the sum of two odd integers is an even integer.

ii) Prove that the product of two integers is odd if and only if the two integers are both odd.

iii) Suppose that the product of two integers is odd. Prove that the sum of those two integers is even.

iv) Suppose that the sum of the squares of two integers is odd. Prove that one of the two integers is even and the other is odd.

v) Prove that the product of two consecutive integers is even. Prove that the product of three consecutive integers is an integer multiple of 6.

vi) Prove that the sum of two consecutive integers is odd. Prove that the sum of three consecutive integers is an integer multiple of 3.

1.1.13. (The quadratic formula) Let a, b, c be real numbers with a non-zero. Prove (with algebra) that all solutions x of the quadratic equation $ax^2 + bx + c = 0$ are of the form

$$x = \frac{-b \pm \sqrt{b^2 - 4ac}}{2a}.$$

1.1.14. Prove that $\sqrt{3}$ is not a rational number. (Remark: It is harder to prove that π and e are not rational.)

1.1.15. For which integers n is $\sqrt{n}$ not a rational number?

1.1.16. (Cf. Exercise 10.2.3.) Draw a unit circle and a line segment from the center to the circle. Any real number x uniquely determines a point P on the circle at angle x radians from the line. Draw the line from that point that is perpendicular to the first line. The length of this perpendicular line is called $\sin(x)$, and the distance from the intersection of the two perpendicular lines to the center of the circle is called $\cos(x)$. This is our definition of cos and sin.

Consider the following picture inside the circle of radius 1:

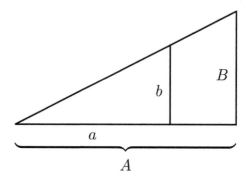

i) Label the line segments of lengths $\sin(x), \sin(y), \cos(y)$.
ii) Use ratio geometry (from page 10) to assert that the smallest vertical line in the bottom triangle has length $\sin(x)\cos(y)$.
iii) Use trigonometry and ratio geometry to assert that the vertical line in the top triangle has length $\sin(y)\cos(x)$.
iv) Prove that $\sin(x+y) = \sin(x)\cos(y) + \sin(y)\cos(x)$.

1.1.17. Use an illustration similar to the one in the previous exercise to prove $\cos(x+y) = \cos(x)\cos(y) - \sin(x)\sin(y)$.

1.1.18. Assuming that the area of the circle of radius r is πr^2, convince yourself with proportionality argument that the area of the region below, where x is measured in radians, is $\frac{1}{2}xr^2$.

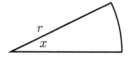

† **1.1.19.** (Invoked in Theorem 10.2.5.) Let x be a small positive real number. Consider the following picture with a circular segment of radius 1 and two right triangles:

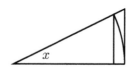

i) Assert that the area of the small triangle is strictly smaller than the area of the wedge, which in turn is strictly smaller than the area of the big triangle.

ii) Using the previous part and ratio geometry (from page 10), prove that $\frac{1}{2}\sin(x)\cos(x) < \frac{1}{2}x < \frac{1}{2}\tan x$.

iii) Using the previous part, prove that $0 < \cos(x) < \frac{x}{\sin x} < \frac{1}{\cos x}$.

*1.1.20. (Logic circuits) Logic circuits are simple circuits which take as inputs logical values of true and false (or 1 and 0) and give a single output. Logic circuits are composed of logic gates. Each logic gate stands for a logical connective you are familiar with — it could be and, or, or not (more complex logic circuits incorporate more). The shapes for logical **and, or, not** are as follows:

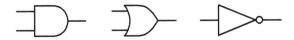

Given inputs, each of these logic gates outputs values equal to the values in the associated truth table. For instance, an "and" gate only outputs "on" if both of the wires leading into it are "on". From these three logic gates we can build many others. For example, the following circuit is equivalent to xor.

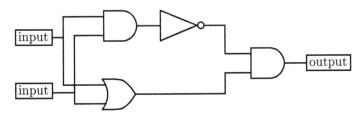

A circuit that computes xor. The output on the right is on when exactly one of the inputs on the left is on.

Make logic circuits that complete the following tasks. (It may be helpful to make logic tables for each one.)

i) xor in a different way than the circuit above.

ii) "P implies Q" implication.

iii) xor for three inputs.

iv) Is a 3-digit binary number greater than 2?

v) Is a 4-digit binary string a palindrome?

1.2   Statements with quantifiers

"The number x equals 1" is true for some x and false for some x. For determining a statement's veracity we possibly need a further qualification. We can use the **universal quantifier** "for all", "for every", or the **existential quantifier** "there exists", "for some". The statement above could be modified to one of the following:

(1) "There exists a real number x such that $x = 1$."

(2) "For all real numbers x, $x = 1$," which is logically the same as "For every real number x, $x = 1$," and the same as "Every real number equals 1."

Certainly the first statement is true and the second is false.

For shorthand we abbreviate "for all" with the symbol $\forall$, and "there exists" with $\exists$. These abbreviations come in handy when we manipulate logical statements. The general forms of abbreviated statements with quantifiers are:

"$\forall x$ with a certain specification, $P(x)$ holds" $=$ "$\forall x \, P(x)$"

"$\exists x$ with a certain specification, $P(x)$ holds" $=$ "$\exists x \, P(x)$"

where P is some property that can be applied to objects x in question. The forms on the left have an explicit specifications on the scope of the x, and in the forms on the right the scope of the x is implicit.

Warning: For ease of readability it may be better to write out full words rather than symbolic abbreviations.

We read the displayed statements above as "for all x, P of x [holds/is true]" and "there exists x such that P of x [holds/is true]", respectively. The part "such that" only appears with the existential quantifier as a grammar filler but without any logical meaning; it can be replaced with "for which", and can sometimes be shortened further. For example: "There exists a function f such that for all real numbers x, $f(x) = f(-x)$" can be rewritten with equal meaning as "There exists a function f that is defined for all real numbers and is even," or even shorter as "There exists an even function." (No "such that" appears in the last two versions.)

Read the following symbolic statement (it defines the limit of the function f at a to be L; see Definition 4.1.1):

$$\forall \epsilon > 0 \; \exists \delta > 0 \; \forall x, 0 < |x - a| < \delta \Rightarrow |f(x) - L| < \epsilon.$$

When are the truth values and the negations of statements with quantifiers? We first write a truth table with all the possible situations with regards to P in the first column, and other columns give the truth values of the quantifier statements:

a possible situation	$\forall x\, P(x)$	$\forall x$ not $P(x)$	$\exists x$ not $P(x)$	$\exists x P(x)$
there are no x of the specified type	T vacuously	T vacuously	F	F
there are x of the specified type, P true for all	T	F	F	T
there are x of the specified type, P false for all	F	T	T	F
there are x of the specified type, P true for some, false for some	F	F	T	T

A "for all" statement is true precisely when without exception all x with the given description have the property P, and a "there exists" statement is true precisely when at least one x with the given description satisfies property P. One proves a "for all" statement by determining that each x with the given description has the property P, and one proves a "there exists" statement by producing one specimen x with the given description and then proving that that specimen has property P. If there are no x with the given specification, then any property holds for those no-things x **vacuously**. For example, any positive real number that is strictly smaller than -1 is also zero, equal to 15, greater than 20, product of distinct prime integers, and any other fine property you can think of.

Notice that among the columns with truth values, one and three have opposite values, and two and four have opposite values. This proves the following:

Theorem 1.2.1. *The negation of "$\forall x\, P(x)$" is "$\exists x$ not $P(x)$." The negation of "$\exists x\, P(x)$" is "$\forall x$ not $P(x)$."*

Thus "$\forall x\, P(x)$" is false if there is even one tiny tiniest example to the contrary. "Every prime number is odd" is false because 2 is an even prime number. "Every whole number divisible by 3 is divisible by 2" is false because 3 is divisible by 3 and is not divisible by 2.

Remark 1.2.2. The statement "For all whole numbers x between $1/3$ and $2/3$, x^2 is irrational" is true vacuously. Another reason why "For all whole numbers x between $1/3$ and $2/3$, x^2 is irrational" is true is that its negation, "There exists a whole number x between $1/3$ and $2/3$ for which x^2 is rational", is false because there is no whole number between $1/3$ and $2/3$: since the negation is false, we get yet more motivation to declare the original statement true.

Exercises for Section 1.2

1.2.1. Show that the statements are false by providing counterexamples.
  i) No number is its own square.
  ii) All numbers divisible by 7 are odd.
  iii) The square root of all real numbers is greater than 0.
  iv) For every real number x, $x^4 > 0$.

1.2.2. Determine the truth value of each statement. Justify your answer.
  i) For all real numbers a, b, $(a + b)^2 = a^2 + b^2$.
  ii) For all real numbers a, b, $(a + b)^2 = a^2 + 2ab + b^2$.
  iii) For all real numbers $x < 5$, $x^2 > 16$.
  iv) There exists a real number $x < 5$ such that $x^2 < 25$.
  v) There exists a real number x such that $x^2 = -4$.
  vi) There exists a real number x such that $x^3 = -8$.
  vii) For every real number x there exists a positive integer n such that $x^n > 0$.
  viii) For every real number x and every integer n, $|x| < x^n$.

ix) For every integer m there exists an integer n such that $m + n$ is even.

x) There exists an integer m such that for all integers n, $m + n$ is even.

xi) For every integer n, $n^2 - n$ is even.

xii) Every list of 5 consecutive integers has one element that is a multiple of 5.

xiii) Every odd number is a multiple of 3.

1.2.3. Explain why the following statements have the same truth values:

i) There exists x such that there exists y such that P holds for the pair (x, y).

ii) There exists y such that there exists x such that P holds for the pair (x, y).

iii) There exists a pair (x, y) such that P holds for the pair (x, y).

1.2.4. Explain why the following statements have the same truth values:

i) $\forall x > 0 \,\forall y > 0, xy > 0$.

ii) $\forall y > 0 \,\forall x > 0, xy > 0$.

iii) $\forall (x, y)$, $x, y > 0$ implies $xy > 0$.

1.2.5. (Contrast with the switching of quantifiers in the previous two exercises.) Explain why the following two statements do not have the same truth values:

i) For every $x > 0$ there exists $y > 0$ such that $xy = 1$.

ii) There exists $y > 0$ such that for every $x > 0$, $xy = 1$.

1.2.6. Rewrite the following statements using quantifiers:

i) 7 is prime.

ii) There are infinitely many prime numbers.

iii) Everybody loves Raymond.

iv) Spring break is always in March.

1.2.7. Let "xLy" represent the statement that x loves y. Rewrite the following statements symbolically: "Everybody loves somebody," "Somebody loves everybody," "Somebody is loved by everybody," "Everybody is loved by somebody." At least one statement should be of the form "$\forall x \,\exists y, xLy$". Compare its truth value with that of "$\exists x \,\forall y, xLy$".

1.2.8. Find a property P of real numbers x, y, z such that "$\forall y \,\exists x \,\forall z, P(x, y, z)$" and "$\forall z \,\exists x \,\forall y, P(x, y, z)$" have different truth values.

1.2.9. Suppose that it is true that there exists x of some kind with property P. Can we conclude that all x of that kind have property P? (A mathematician and a few other jokers are on a train and see a cow through the window. One of them generalizes: "All cows in this state are brown," but the mathematician corrects: "This state has a cow whose one side is brown.")

1.3   More proof methods

When statements are compound, they can be harder to prove. Fortunately, proofs can be broken down into simpler statements. An essential chart of this breaking down is in the chart on the next page.

Example 1.3.1. Integers 2 and 3 are prime, i.e., 2 is a prime integer and 3 is a prime integer.

Proof. Let m and n be whole numbers strictly greater than 1. If $m \cdot n = 2$, then $1 < m, n \leq 2$, so $m = n = 2$, but $2 \cdot 2$ is not equal to 2. Thus 2 cannot be written as a product of two positive numbers different from 1, so 2 is a prime number. If instead $m \cdot n = 3$, then $1 < m, n \leq 3$. Then all combinations of products are $2 \cdot 2, 2 \cdot 3, 3 \cdot 2, 3 \cdot 3$, none of which is 3. Thus 3 is a prime number.   □

Example 1.3.2. A positive prime number is either odd or it equals 2. (Often the term "prime" implicitly assumes positivity, but -2 can be thought of as a prime number as well.)

Proof. Let p be a positive prime number. Suppose that p is not odd. Then p must be even. Thus $p = 2 \cdot q$ for some positive whole number q. Since p is a prime, it follows that $q = 1$, so that $p = 2$.   □

Example 1.3.3. If an integer is a multiple of 2 and a multiple of 3, then it is a multiple of 6. (Implicit here is that the factors are integers.)

Proof. Let n be an integer that is a multiple of 2 and of 3. Write $n = 2 \cdot p$ and $n = 3 \cdot q$ for some integers p and q. Then $2 \cdot p = 3 \cdot q$ is even, which forces that q must be even. Hence $q = 2 \cdot r$ for some integer r, so that $n = 3 \cdot q = 3 \cdot 2 \cdot r = 6 \cdot r$. Thus n is a multiple of 6.   □

Statement	How to prove it
P (via contradiction).	Suppose not P. Establish some nonsense that makes not P impossible so that P must hold.
P and Q.	Prove P. Prove Q.
P or Q.	Suppose that P is false. Then prove Q. Alternatively: Suppose that Q is false and then prove P. (It may even be the case that P is always true. Then simply prove P. Or simply prove Q.)
If P then Q.	Suppose that P is true. Then prove Q. Contrapositively: Suppose that Q is false. Prove that P is false.
$P \Leftrightarrow Q$.	Prove $P \Rightarrow Q$. Prove $Q \Rightarrow P$.
For all x of a specified type, property P holds for x.	Let x be arbitrary of the specified type. Prove that property P holds for x. (Possibly break up into a few subcases.)
There exists x of a specified type such that property P holds for x.	Find/construct an x of the specified type. Prove that property P holds for x. Alternatively, invoke a theorem guaranteeing that such x exists.
An element x of a specified type with property P is unique.	Suppose that x and x' are both of specified type and satisfy property P. Prove that $x = x'$. Alternatively, show that x is the only solution to an equation, or the only element on a list, or ... .
x with property P is unique.	Suppose that x and y have property P. Prove that $x = y$.

Example 1.3.4. For all real numbers x, $x^2 = (-x)^2$.

Proof. Let x be an arbitrary real number. Then $(-x)^2 = (-x)(-x) = (-1)x(-1)x = (-1)(-1)x \cdot x = 1 \cdot x^2 = x^2$. $\square$

Example 1.3.5. There exists a real number x such that $x^3 - 3x = 2$.

Proof. Observe that 2 is a real number and that $2^3 - 3 \cdot 2 = 2$. Thus $x = 2$ satisfies the conditions. $\square$

Example 1.3.6. There exists a real number x such that $x^3 - x = 1$.

Proof. Observe that $f(x) = x^3 - x$ is a continuous function. Since 1 is strictly between $f(0) = 0$ and $f(2) = 6$, by the Intermediate value theorem (Theorem 5.3.1 in this book) [INVOKING A THEOREM RATHER THAN CONSTRUCTING x, AS OPPOSED TO IN THE PREVIOUS EXAMPLE]* there exists a real number x strictly between 0 and 2 such that $f(x) = 1$. $\square$

Example 1.3.7. (Mixture of methods) For every real number x strictly between 0 and 1 there exists a positive real number y such that $\frac{1}{x} + \frac{1}{y} = \frac{1}{xy}$.

Proof. [WE HAVE TO PROVE THAT FOR ALL x AS SPECIFIED SOME PROPERTY HOLDS.] Let x be in $(0, 1)$. [FOR THIS x WE HAVE TO FIND y ...] Set $y = 1 - x$. [WAS THIS A LUCKY FIND? NO MATTER HOW WE GOT INSPIRED TO DETERMINE THIS y, WE NOW VERIFY THAT THE STATED PROPERTIES HOLD FOR x AND y.] Since x is strictly smaller than 1, it follows that y is positive. Thus also xy is positive, and $y + x = 1$. After dividing the last equation by the positive number xy we get that $\frac{1}{x} + \frac{1}{y} = \frac{y+x}{xy} = \frac{1}{xy}$. $\square$

Furthermore, y in the example above has no choice but to be $1 - x$.

Example 1.3.8. (The Fundamental theorem of arithmetic) Any positive integer $n > 1$ can be written as $p_1^{a_1} p_2^{a_2} \cdots p_k^{a_k}$ for some positive

---

* Recall that this font in brackets indicates the reasoning that should go on in the background in your head; these statements are not part of a proof.

prime integers $p_1 < \cdots < p_k$ and some positive integers $a_1, \ldots, a_k$. (Another standard part of the Fundamental theorem of arithmetic is that the p_i and the a_i are unique, but we do not prove that. Once you are comfortable with proofs you can prove that part yourself.)

Proof. (This proof is harder; it is fine to skip it.) Suppose for contradiction that the conclusion fails for some positive integer n. Then on the list $2, 3, 4, \ldots, n$ let m be the smallest integer for which the conclusion fails. If m is a prime, take $k = 1$ and $p_1 = m$, $a_1 = 1$, and so the conclusion does not fail. Thus m cannot be a prime number, and so $m = m_1 m_2$ for some positive integers m_1, m_2 strictly bigger than 1. Necessarily $2 \leq m_1, m_2 < m$. By the choice of m, the conclusion is true for m_1 and m_2. Write $m_1 = p_1^{a_1} p_2^{a_2} \cdots p_k^{a_k}$ and $m_2 = q_1^{b_1} q_2^{b_2} \cdots q_l^{b_l}$ for some positive prime integers $p_1 < \cdots < p_k$, $q_1 < \cdots < q_k$ and some positive integers $a_1, \ldots, a_k, b_1, \ldots, b_l$. Thus $m = p_1^{a_1} p_2^{a_2} \cdots p_k^{a_k} q_1^{b_1} q_2^{b_2} \cdots q_l^{b_l}$ is a product of positive prime numbers, and after sorting and merging the p_i and q_j, the conclusion follows also for m. But we assumed that the conclusion fails for m, which yields the desired contradiction. Hence the conclusion does not fail for any positive integer. $\square$

Example 1.3.9. Any positive rational number can be written as $\frac{a}{b}$, where a and b are positive whole numbers and where in any prime factorizations of a and b as in the previous example, the prime factors for a are distinct from the prime factors for b.

Proof. [WE HAVE TO PROVE THAT FOR ALL ...] Let x be a(n arbitrary) positive rational number. [WE REWRITE THE MEANING OF THIS IN A MORE CONCRETE AND USABLE FORM NEXT.] Thus $x = \frac{a}{b}$ for some whole numbers a, b. If a is negative, since x is positive necessarily b has to be negative. But then $-a, -b$ are positive numbers, and $x = \frac{-a}{-b}$. Thus by possibly replacing a, b with $-a, -b$ we may assume that a, b are positive. [A REWRITING TRICK.] There may be many different pairs of a, b, and we choose a pair for which a is the smallest of all possibilities. [A CHOOSING TRICK. BUT DOES THE SMALLEST a EXIST?] Such a does exist because in a collection of given positive integers there is always a smallest one. Suppose that a and b have a (positive) prime factor p in common. Write $a = a_0 p$ and $b = b_0 p$ for some positive whole numbers a_0, b_0. Then $x = \frac{a_0}{b_0}$, and since $0 < a_0 < a$, this contradicts the choice of the pair a, b. Thus a and b could not have had a prime factor in common.   □

Exercises for Section 1.3

1.3.1. Prove that every whole number is either odd or even.

1.3.2. Prove that the successor of any odd integer is even.

1.3.3. Prove that if n is an even integer, then either n is a multiple of 4 or $n/2$ is odd.

1.3.4. Prove that if $0 < x < 1$, then $x^2 < x < \sqrt{x}$.

1.3.5. Prove that if $1 < x$, then $\sqrt{x} < x < x^2$.

1.3.6. Prove that there exists a real number x such that $x^2 + \frac{5}{12}x = \frac{1}{6}$.

1.3.7. Find at least three functions f such that for all real numbers x, $f(x^2) = x^2$.

1.3.8. Prove that $f(x) = x$ is the unique function that is defined for all real numbers and that has the property that for all x, $f(x^3) = x^3$.

1.3.9. Prove that there exists a real number y such that for all real numbers x, $xy = y$.

1.3.10. Prove that there exists a real number y such that for all real numbers x, $xy = x$.

1.3.11. Prove that for every real number x there exists a real number y such that $x + y = 0$.

1.3.12. Prove that there exists no real number y such that for all real numbers x, $x + y = 0$.

1.3.13. Prove that for every real number y there exists a real number x such that $x + y \neq 0$.

1.4   Logical negation

In order to be able to prove statements effectively, we often have to suppose the negation of a part, say for proving statements with "or" and for proofs by contradiction. Work and think through the following negations:

Statement	Negation
not P	P
P and Q	$(\text{not } P)$ or $(\text{not } Q)$
P or Q	$(\text{not } P)$ and $(\text{not } Q)$
$P \Rightarrow Q$	P and $(\text{not } Q)$
$P \Leftrightarrow Q$	$P \Leftrightarrow (\text{not } Q) = (\text{not } P) \Leftrightarrow Q$
For all x of the specified type, property P holds for x.	There exists x of the specified type such that P is false for x.
There exists x of the specified type such that property P holds for x.	For all x of the specified type, P is false for x.

Warning: The negation of a conditional statement is not another conditional statement! Practice this one!

Example 1.4.1. There are infinitely many (positive) prime numbers. *Proof by contradiction:* (Due to Euclid.) Suppose that there are only finitely many prime numbers. Then we can enumerate them all: $p_1, p_2, \ldots, p_n$. Let $a = (p_1 p_2 \cdots p_n) + 1$. Since we know that $2, 3, 5$ are primes, necessarily $n \geq 3$ and so $a > 1$. By the Fundamental theorem of arithmetic (Example 1.3.8), a has a prime factor p. Since $p_1, p_2, \ldots, p_n$ are all the primes, necessarily $p = p_i$ for some i. But then $p = p_i$ divides a and $p_1 p_2 \cdots p_n$, whence it divides $1 = a - (p_1 p_2 \cdots p_n)$, which is a contradiction. So it is not the case that there are only finitely many prime numbers, so there must be infinitely many. $\qquad\square$

Another proof by contradiction: (Due to R. Meštrović, *American Mathematical Monthly* **124** (2017), page 562.) Suppose that there are only finitely many prime numbers. Then we can enumerate them all: $p_1 = 2, p_2 = 3, \ldots, p_n$. The positive integer $p_2 p_3 \cdots p_n - 2$ has no odd prime factors, and since it is odd, it must be equal to 1. Hence $p_2 p_3 \cdots p_n = 3$, which is false since $p_2 = 3$, $p_3 = 5$, and so on. $\square$

Exercises for Section 1.4

1.4.1. Prove that the following pairs of statements are negations of each other:

    i) $P \Leftrightarrow Q$.

    $(P \text{ and } Q) \text{ or } (\text{ not } P \text{ and not } Q)$.

    ii) $P \text{ and } (P \Rightarrow Q)$.

    not P or not Q.

    iii) $(P \Rightarrow \text{ not } Q) \text{ and } (R \Rightarrow Q)$.

    $(P \text{ and } Q) \text{ or } (R \text{ and not } Q)$.

    iv) $P \text{ and } (Q \text{ or not } R)$.

    $P \Rightarrow (\text{ not } Q \text{ and } R)$.

1.4.2. Why are "f is continuous at all points" and "f is not continuous at 3" not negations of each other?

1.4.3. Why are "Some continuous functions are differentiable" and "All differentiable functions are continuous" not negations of each other?

1.4.4. Why is "$P \Rightarrow \text{ not } Q$" not the negation of "$P \Rightarrow Q$"?

1.4.5. Negate the following statements:

    i) The function f is continuous at 5.

    ii) If $x > y$ then $x > z$.

    iii) For every $\epsilon > 0$, there exists $\delta > 0$ such that $f(\delta) = \epsilon$.

    iv) For every $\epsilon > 0$, there exists $\delta > 0$ such that for all x, $f(x \cdot \delta) = \epsilon x$.

    v) For every $\epsilon > 0$, there exists $\delta > 0$ such that for all x, $0 < |x - a| < \delta$ implies $|f(x) - L| < \epsilon$.

1.4.6. Prove that the following are false:

    i) For every real number x there exists a real number y such that $xy = 1$.

    ii) $5^n + 2$ is prime for all non-negative integers n.

    iii) For every real number x, if $x^2 > 4$ then $x > 2$.

    iv) For every real number x, $\sqrt{x^2} = x$.

    v) The cube of every real number is positive.

    vi) The square of every real number is positive.

1.5   Summation

There are many reasons for not writing out the sum of the first hundred numbers in full length: it would be too long, it would not be any clearer, and we might start doubting the intelligence of the writer. Instead we express such sums with the **summation sign** Σ:

$$\sum_{k=1}^{100} k \quad \text{or} \quad \sum_{n=1}^{100} n.$$

The counters k and n above are dummy variables, they vary from 1 to 100, as indicated below and above the summation sign. We could use any other name in place of k or n. In general, if f is a function defined at $m, m+1, m+2, \ldots, n$, we use the summation shortening as follows:

$$\sum_{k=m}^{n} f(k) = f(m) + f(m+1) + f(m+2) + \cdots + f(n).$$

This is one example where good notation saves effort and often clarifies the concept. For typographical reasons, to prevent lines jamming into each other, we also write this as $\sum_{k=m}^{n} f(k)$.

Now is a good time to discuss polynomials. A **polynomial function** is a function of the form $f(x) = a_0 + a_1 x + \cdots + a_n x^n$ for some non-negative integer n and some numbers $a_0, a_1, \ldots, a_n$. We call $a_0 + a_1 x + \cdots + a_n x^n$ a polynomial, and if a_n is non-zero, we say that the polynomial has **degree** n. (More on the degrees of polynomial functions is in Example 1.6.5, Exercise 2.6.15.) It is convenient to write this polynomial with the shorthand notation

$$f(x) = a_0 + a_1 x + \cdots + a_n x^n = \sum_{k=0}^{n} a_k x^k.$$

Here, of course, x^0 stands for 1. When we evaluate f at 0, we get $a_0 = a_0 + a_1 \cdot 0 + \cdots + a_n \cdot 0^n = \sum_{k=0}^{n} a_k 0^k$, and we deduce that **notationally** 0^0 stands for 1 here.

Remark 1.5.1. 0^0 could possibly be thought of also as $\lim_{x \to 0^+} 0^x$, which is surely equal to 0. But then one can wonder whether 0^0 equals 0 or 1 or to something else entirely? Well, it turns out that 0^0 is not equal to that zero limit — you surely know of other functions f for which $\lim_{x \to c} f(x)$ exists but the limit is not equal to $f(c)$. (Check out also Exercise 7.6.9.)

Examples 1.5.2.

(1) $\displaystyle\sum_{k=1}^{5} 2 = 2 + 2 + 2 + 2 + 2 = 10.$

(2) $\displaystyle\sum_{k=1}^{5} k = 1 + 2 + 3 + 4 + 5 = 15.$

(3) $\displaystyle\sum_{k=1}^{4} k^2 = 1^2 + 2^2 + 3^2 + 4^2 = 30.$

(4) $\displaystyle\sum_{k=10}^{12} \cos(k\pi) = \cos(10\pi) + \cos(11\pi) + \cos(12\pi) = 1 - 1 + 1 = 1.$

(5) $\displaystyle\sum_{k=-1}^{2} (4k^3) = -4+0+4+4\cdot 8 = 32 = 4(-1+0+1+1\cdot 8) = 4\sum_{k=-1}^{2} k^3.$

(6) $\displaystyle\sum_{k=1}^{n} 3 = 3$ added to itself n times $= 3n.$

(7) $\displaystyle\sum_{k=a}^{b} 2 = 2$ added to itself $b - a + 1$ times $= 2(b - a + 1).$

We can even deal with **empty sums** such as $\sum_{k=1}^{0} a_k$: here the index starts at $k = 1$ and keeps increasing and we stop at $k = 0$, but there are no such indices k. What could possibly be the meaning of such an empty sum? Note that

$$\sum_{k=1}^{4} a_k = \sum_{k=1}^{2} a_k + \sum_{k=3}^{4} a_k = \sum_{k=1}^{1} a_k + \sum_{k=2}^{4} a_k = \sum_{k=1}^{0} a_k + \sum_{k=1}^{4} a_k,$$

or explicitly written out:

$$a_1 + a_2 + a_3 + a_4 = (a_1 + a_2) + (a_3 + a_4)$$
$$= (a_1) + (a_2 + a_3 + a_4)$$
$$= () + (a_1 + a_2 + a_3 + a_4),$$

from which we deduce that this empty sum must be 0. Similarly, every empty sum equals 0.

Similarly we can shorten products with the **product sign Π**:

$$\prod_{k=m}^{n} f(k) = f(m) \cdot f(m+1) \cdot f(m+2) \cdot \; \cdots \; \cdot f(n).$$

In particular, for all non-negative integers n, the product $\prod_{k=1}^{n} k$ is used often and is abbreviated as $n! = \prod_{k=1}^{n} k$. See Exercise 1.5.6 for the fact that $0! = 1$.

Exercises for Section 1.5

1.5.1. Compute $\sum_{k=0}^{4}(2k+1)$, $\sum_{k=0}^{4}(k^2+2)$.

1.5.2. Determine all non-negative integers n for which $\sum_{k=0}^{n} k = \sum_{k=0}^{n} n$.

1.5.3. Prove that

i) $c \sum_{k=m}^{n} f(k) = \sum_{k=m}^{n} cf(k).$

ii) $\sum_{k=m}^{n} f(k) + \sum_{k=m}^{n} g(k) = \sum_{k=m}^{n} (f(k) + g(k)).$

1.5.4. Prove that $\sum_{k=1}^{0} k = \frac{0 \cdot (-1)}{2}$.

1.5.5. Prove that for all integers $m \le n$, $\sum_{k=1}^{m-1} f(k) + \sum_{k=m}^{n} f(k) = \sum_{k=1}^{n} f(k).$

1.5.6. Prove that the empty product equals 1. In particular, we can declare that $0! = 1$. This turns out to be very helpful notationally.

1.5.7. Prove:

i) $\displaystyle\prod_{k=1}^{5} 2 = 32.$

ii) $\displaystyle\prod_{k=1}^{5} k = 120.$

1.6 Proofs by (mathematical) induction

So far we have learned a few proof methods. There is another type of proofs that deserves special mention, and this is **proof by (mathematical) induction**, sometimes referred to as the **principle of mathematical induction**. This method can be used when one wants to prove that a property P holds for all integers n greater than or equal to an integer n_0. Typically, n_0 is either 0 or 1, but it can be any integer, even a negative one.

Induction is a two-step procedure:

(1) **Base case:** Prove that P holds for n_0.

(2) **Inductive step:** Let $n > n_0$. Assume that P holds for all integers $n_0, n_0 + 1, n_0 + 2, \ldots, n - 1$. Prove that P holds for n.

Why does induction succeed in proving that P holds for all $n \geq n_0$? By the base case we know that P holds for n_0. The inductive step then proves that P also holds for $n_0 + 1$. So then we know that the property holds for n_0 and $n_0 + 1$, whence the inductive step implies that it also holds for $n_0 + 2$. So then the property holds for n_0, $n_0 + 1$ and $n_0 + 2$, whence the inductive step implies that it also holds for $n_0 + 3$. This establishes that the property holds for n_0, $n_0 + 1$, $n_0 + 2$, and $n_0 + 3$, so that by inductive step it also holds for $n_0 + 4$. We keep going. For any integer $n > n_0$, in $n - n_0$ step we similarly establish that the inductive step holds for n_0, $n_0 + 1$, $n_0 + 2, \ldots, n_0 + (n - n_0) = n$. Thus for any integer $n \geq n_0$, we eventually prove that P holds for it.

The same method can be phrased with a slightly different two-step process, with the same result, and the same name:

(1) **Base case:** Prove that P holds for n_0.

(2) **Inductive step:** Let $n > n_0$. Assume that P holds for integer $n - 1$. Prove that P holds for n.

Similar reasoning as in the previous case also shows that this induction principle succeeds in proving that P holds for all $n \geq n_0$.

Example 1.6.1. Prove the equality $\sum_{k=1}^{n} k = \frac{n(n+1)}{2}$ for all $n \geq 1$.

Proof. Base case $n = 1$: The left side of the equation is $\sum_{k=1}^{1} k$ which equals 1. The right side is $\frac{1(1+1)}{2}$ which also equals 1. This verifies the base case.

Inductive step: Let $n > 1$ and we assume that the equality holds for $n - 1$. [WE WANT TO PROVE THE EQUALITY FOR n. WE START WITH THE EXPRESSION ON THE LEFT (MESSIER) SIDE OF THE DESIRED AND NOT-YET-PROVED EQUATION FOR n AND MANIPULATE THE EXPRESSION UNTIL IT RESEMBLES THE DESIRED RIGHT SIDE.] Then

$$
\begin{aligned}
\sum_{k=1}^{n} k &= \left(\sum_{k=1}^{n-1} k \right) + n \\
&= \frac{(n-1)(n-1+1)}{2} + n \quad \text{(by induction assumption for } n-1) \\
&= \frac{n^2 - n}{2} + \frac{2n}{2} \quad \text{(by algebra)} \\
&= \frac{n^2 + n}{2} \\
&= \frac{n(n+1)}{2},
\end{aligned}
$$

as was to be proved. $\qquad\qquad\qquad\qquad\qquad\qquad\qquad\qquad\qquad\qquad\square$

We can even prove the equality $\sum_{k=1}^{n} k = \frac{n(n+1)}{2}$ for all $n \geq 0$. Since we have already proved this equality for all $n \geq 1$, it remains to prove it for $n = 0$. The left side $\sum_{k=1}^{0} k$ is an empty sum and hence 0, and the right side is $\frac{0(0+1)}{2}$, which is also 0.

Example 1.6.2. Prove the equality $\sum_{k=1}^{n} k(k+1)(k+2)(k+3) = \frac{n(n+1)(n+2)(n+3)(n+4)}{5}$ for all $n \geq 1$.

Proof. Base case $n = 1$: $\sum_{k=1}^{1} k(k+1)(k+2)(k+3) = 1(1+1)(1+2)(1+3) = 1 \cdot 2 \cdot 3 \cdot 4 = \frac{1 \cdot 2 \cdot 3 \cdot 4 \cdot 5}{5} = \frac{1(1+1)(1+2)(1+3)(1+4)}{5}$, which verifies the base case.

Inductive step: Let $n > 1$ and we assume that the equality holds for $n - 1$. [WE WANT TO PROVE THE EQUALITY FOR n. WE START WITH THE EXPRESSION ON THE LEFT SIDE OF THE DESIRED AND NOT-YET-PROVED

EQUATION FOR n (THE MESSIER OF THE TWO) AND MANIPULATE THE EXPRESSION UNTIL IT RESEMBLES THE DESIRED RIGHT SIDE.] Then

$$\sum_{k=1}^{n} k(k+1)(k+2)(k+3)$$

$$= \left(\sum_{k=1}^{n-1} k(k+1)(k+2)(k+3)\right) + n(n+1)(n+2)(n+3)$$

$$= \frac{(n-1)(n-1+1)(n-1+2)(n-1+3)(n-1+4)}{5}$$

$$+ n(n+1)(n+2)(n+3) \quad \text{(by induction assumption)}$$

$$= \frac{(n-1)n(n+1)(n+2)(n+3)}{5} + \frac{5n(n+1)(n+2)(n+3)}{5}$$

$$= \frac{n(n+1)(n+2)(n+3)}{5}(n-1+5) \quad \text{(by factoring)}$$

$$= \frac{n(n+1)(n+2)(n+3)(n+4)}{5},$$

as was to be proved. □

Example 1.6.3. Assuming that the derivative of x is 1 and the product rule for derivatives, prove that for all $n \geq 1$, $\frac{d}{dx}(x^n) = nx^{n-1}$. (We introduce derivatives formally in Section 6.1.)

Proof. We start the induction at $n = 1$. By calculus we know that the derivative of x^1 is $1 = 1 \cdot x^0 = 1 \cdot x^{1-1}$, so equality holds in this case.

Inductive step: Suppose that equality holds for $1, 2, \ldots, n-1$. Then

$$\frac{d}{dx}(x^n) = \frac{d}{dx}(x \cdot x^{n-1})$$

$$= \frac{d}{dx}(x) \cdot x^{n-1} + x\frac{d}{dx}(x^{n-1})$$

$$\text{(by the product rule of differentiation)}$$

$$= 1 \cdot x^{n-1} + (n-1)x \cdot x^{n-2}$$

$$\text{(by induction assumption for 1 and } n-1)$$

$$= x^{n-1} + (n-1)x^{n-1}$$

$$= nx^{n-1}. \qquad \square$$

The following result will be needed many times, so remember it well.

Example 1.6.4. For any number x and any integer $n \geq 1$,

$$(1 - x)(1 + x + x^2 + x^3 + \cdots + x^n) = 1 - x^{n+1}.$$

Proof. When $n = 1$,

$$(1 - x)(1 + x + x^2 + x^3 + \cdots + x^n) = (1 - x)(1 + x) = 1 - x^2 = 1 - x^{n+1},$$

which proves the base case. Now suppose that equality holds for some integer $n - 1 \geq 1$. Then

$$(1 - x)(1 + x + x^2 + \cdots + x^{n-1} + x^n)$$
$$= (1 - x)\left((1 + x + x^2 + \cdots + x^{n-1}) + x^n\right)$$
$$= (1 - x)(1 + x + x^2 + \cdots + x^{n-1}) + (1 - x)x^n$$
$$= 1 - x^n + x^n - x^{n+1} \text{ (by induction assumption and algebra)}$$
$$= 1 - x^{n+1},$$

which proves the inductive step. □

Example 1.6.5. (Euclidean algorithm) Let $f(x) = a_0 + a_1 x + \cdots + a_n x^n$ for some numbers $a_0, a_1, \ldots, a_n$ and with $a_n \neq 0$, and let $g(x) = b_0 + b_1 x + \cdots + b_m x^m$ for some numbers $b_0, b_1, \ldots, b_m$ and with $b_m \neq 0$. Suppose that $m, n \geq 1$. Then there exist polynomials $q(x)$ and $r(x)$ such that $f(x) = q(x) \cdot g(x) + r(x)$ and such that the degree of $r(x)$ is strictly smaller than m.

Proof. We keep $g(x)$ fixed and we prove by induction on the degree n of $f(x)$ that the claim holds for all polynomials $f(x)$. If $n < m$, then we are done with $q(x) = 0$ and $r(x) = f(x)$. If $n = m$, then we set $q(x) = \frac{a_n}{b_n}$ and (necessarily)

$$r(x) = f(x) - \frac{a_m}{b_m} g(x)$$
$$= a_0 + a_1 x + \cdots + a_m x^m - \frac{a_m}{b_m} (b_0 + b_1 x + \cdots + b_m x^m)$$
$$= (a_0 - \frac{a_m}{b_m} b_0) + (a_1 - \frac{a_m}{b_m} b_1)x + \cdots + (a_{m-1} - \frac{a_m}{b_m} b_{m-1})x^{m-1},$$

which has degree strictly smaller than m. These are the base cases.

Now suppose that $n > m$. Set $h(x) = a_1 + a_2 x + a_3 x^2 + \cdots + a_n x^{n-1}$. By induction on n, there exist polynomials $q_1(x)$ and $r_1(x)$ such that $h(x) = q_1(x) \cdot g(x) + r_1(x)$ and such that the degree of $r_1(x)$ is strictly smaller than m. Then $xh(x) = xq_1(x) \cdot g(x) + xr_1(x)$. Since the degree of $xr_1(x)$ is at most m, by the second base case there exist polynomials $q_2(x)$ and $r_2(x)$ such that $xr_1(x) = q_2(x)g(x) + r_2(x)$ and such that the degree of $r_2(x)$ is strictly smaller than m. Now set $q(x) = xq_1(x) + q_2(x)$ and $r(x) = r_2(x) + a_0$. Then the degree of $r(x)$ is strictly smaller than m, and

$$
\begin{aligned}
q(x)g(x) + r(x) &= xq_1(x)g(x) + q_2(x)g(x) + r_2(x) + a_0 \\
&= xq_1(x)g(x) + xr_1(x) + a_0 \\
&= xh(x) + a_0 \\
&= f(x).
\end{aligned}
$$
$\square$

Remark 1.6.6. A common usage of the Euclidean algorithm is in finding the greatest common divisor of two polynomials. A polynomial $d(x)$ divides $f(x)$ and $g(x)$ exactly when it divides $g(x)$ and $r(x) = f(x) - q(x) \cdot g(x)$. It is easier to find factors of polynomials of smaller degree. As an example, let $f(x) = x^4 + 4x^3 + 6x^2 + 4x + 1$ and $g(x) = x^3 + 2x^2 + x$. The first step of the Euclidean algorithm gives

$$
r(x) = f(x) - (x+2)g(x) = x^2 + 2x + 1.
$$

So to find the greatest common divisor of $f(x)$ and $g(x)$ it suffices to find the greatest common divisor of $g(x)$ and $r(x)$. The Euclidean algorithm on these two gives $r_1(x) = g(x) - xr(x) = 0$, so that to find the greatest common divisor of $f(x)$ and $g(x)$ it suffices to find the greatest common divisor of $r(x)$ and 0. But the latter is clearly $r(x)$. In fact, $f(x) = (x+1)^4$ and $g(x) = x(x+1)^2$.

Example 1.6.7. For all positive integers n, $\sqrt[n]{n} < 2$.

Proof. Base case: $n = 1$, so $\sqrt[n]{n} = 1 < 2$.

Inductive step: Suppose that n is an integer with $n \geq 2$ and that $\sqrt[n-1]{n-1} < 2$. This means that $n - 1 < 2^{n-1}$. Hence $n < 2^{n-1} + 1 < 2^{n-1} + 2^{n-1} = 2 \cdot 2^{n-1} = 2^n$, so that $\sqrt[n]{n} < 2$. $\square$

Remark 1.6.8. There are two other equivalent formulations of mathematical induction for proving a property P for all integers $n \geq n_0$:

Mathematical induction, version III:
(1) **Base case:** Prove that P holds for n_0.
(2) **Inductive step:** Let $n \geq n_0$. Assume that P holds for all integers $n_0, n_0 + 1, n_0 + 2, \ldots, n$. Prove that P holds for $n + 1$.

Mathematical induction, version IV:
(1) **Base case:** Prove that P holds for n_0.
(2) **Inductive step:** Let $n \geq n_0$. Assume that P holds for integer n. Prove that P holds for $n + 1$.

Convince yourself that these two versions of the workings of mathematical induction differ from the original two versions only in notation.

Exercises for Section 1.6: Prove the following properties for n ≥ 1 by induction.

1.6.1. $\displaystyle\sum_{k=1}^{n} k^2 = \frac{n(n + 1)(2n + 1)}{6}$.

1.6.2. $\displaystyle\sum_{k=1}^{n} k^3 = \left(\frac{n(n + 1)}{2}\right)^2$.

1.6.3. The sum of the first n odd positive integers is n^2.

1.6.4. $\displaystyle\sum_{k=1}^{n}(2k - 1) = n^2$.

1.6.5. (Triangle inequality) For all positive integers n and for all real numbers $a_1, \ldots, a_n$, $|a_1 + a_2 + \cdots + a_n| \leq |a_1| + |a_2| + \cdots + |a_n|$. (Hint: there may be more than one base case. Why is that?)

1.6.6. $\displaystyle\sum_{k=1}^{n}(3k^2 - k) = n^2(n + 1)$.

1.6.7. $1 \cdot 2 + 2 \cdot 3 + 3 \cdot 4 + \cdots + n(n + 1) = \frac{1}{3}n(n + 1)(n + 2)$.

1.6.8. $7^n + 2$ is a multiple of 3.

1.6.9. $3^{n-1} < (n + 1)!$.

1.6.10. $\dfrac{1}{\sqrt{1}} + \dfrac{1}{\sqrt{2}} + \dfrac{1}{\sqrt{3}} + \cdots + \dfrac{1}{\sqrt{n}} \geq \sqrt{n}$.

1.6.11. $\dfrac{1}{1^2} + \dfrac{1}{2^2} + \dfrac{1}{3^2} + \cdots + \dfrac{1}{n^2} \le 2 - \dfrac{1}{n}$.

1.6.12. Let $a_1 = 2$, and for $n \ge 2$, $a_n = 3a_{n-1}$. Formulate and prove a theorem giving a_n in terms of n (no dependence on other a_i).

1.6.13. 8 divides $5^n + 2 \cdot 3^{n-1} + 1$.

1.6.14. $1(1!) + 2(2!) + 3(3!) + \cdots + n(n!) = (n+1)! - 1$.

1.6.15. $2^{n-1} \le n!$.

1.6.16. $\displaystyle\prod_{k=2}^{n} \left(1 - \dfrac{1}{k}\right) = \dfrac{1}{n}$.

1.6.17. $\displaystyle\prod_{k=2}^{n} \left(1 - \dfrac{1}{k^2}\right) = \dfrac{n+1}{2n}$.

1.6.18. $\displaystyle\sum_{k=0}^{n} 2^k(k+1) = 2^{n+1}n + 1$.

1.6.19. $\sum_{k=1}^{2^n} \frac{1}{k} \ge \frac{n+2}{2}$.

†1.6.20. (Invoked in Example 9.1.9.) $\sum_{k=1}^{2^n-1} \frac{1}{k^2} \le \sum_{k=0}^{n} \frac{1}{2^k}$.

† 1.6.21. (Invoked in the proof of Theorem 9.4.1.) For all numbers x, y, $x^n - y^n = (x - y)\left(x^{n-1} + x^{n-2}y + x^{n-3}y^2 + \cdots + y^{n-1}\right)$, i.e., $x^n - y^n = (x - y)\sum_{k=0}^{n-1} x^{n-1-k}y^k$.

† 1.6.22. (Invoked in the Ratio tests Theorems 8.6.6 and 9.2.3.) Let r be a positive real number.

　i) Suppose that for all positive integers $n \ge n_0$, $a_{n+1} < ra_n$. Prove that for all positive integers $n > n_0$, $a_n < r^{n-n_0}a_{n_0}$.

　ii) Suppose that $a_{n+1} \,\square\, ra_n$, where $\square$ is one of $\le, >, \ge$. Prove that $a_n \,\square\, r^{n-n_0} a_{n_0}$.

1.6.23. Let $A_n = 1^2 + 2^2 + 3^2 + \cdots + (2n - 1)^2$ and $B_n = 1^2 + 3^2 + 5^2 + \cdots + (2n-1)^2$. Discover formulas for A_n and B_n, and prove them (by using algebra and previous problems, and possibly not with induction).

1.6.24. (From the *American Mathematical Monthly* **123** (2016), page 87, by K. Gaitanas) Prove that for every $n \ge 2$, $\sum_{k=1}^{n-1} \frac{k}{(k+1)!} = 1 - \frac{1}{n!}$.

1.6.25. Let $A_n = \frac{1}{1\cdot 2} + \frac{1}{2\cdot 3} + \frac{1}{3\cdot 4} + \cdots + \frac{1}{n(n+1)}$. Discover a formula for A_n and prove it.

1.6.26. How many handshakes happen at a gathering of n people if everybody shakes everybody else's hands exactly once?

1.6.27. Find with proof an integer n_0 such that $n^2 < 2^n$ for all integers $n \geq n_0$.

1.6.28. Find with proof an integer n_0 such that $2^n < n!$ for all integers $n \geq n_0$.

1.6.29. For any positive integer n and any real number x define $S_n = 1 + x + x^2 + \cdots + x^n$.

i) Prove (with easy algebra) that for any integer $n \geq 2$, $xS_{n-1} + 1 = S_n$ and $S_n(1 - x) = 1 - x^{n+1}$.

ii) Prove that if $x = 1$, then $S_n = n + 1$.

iii) Prove that if $x \neq 1$, then $S_n = \frac{1 - x^{n+1}}{1 - x}$. Compare with the proof by induction in Example 1.6.4.

1.6.30. (Fibonacci numbers) Let $s_1 = 1$, $s_2 = 1$, and for all $n \geq 2$, let $s_{n+1} = s_n + s_{n-1}$. This sequence starts with $1, 1, 2, 3, 5, 8, 13, 21, 34, \ldots$. (Many parts below are taken from the book *Fibonacci Numbers* by N. N. Vorob'ev, published by Blaisdell Publishing Company, 1961, translated from the Russian by Halina Moss; there is a new edition of the book with author's last name written as Vorobiev, published by Springer Basel AG, 2002, translated from the Russian by Mircea Martin.)

i) Fibonacci numbers are sometimes "motivated" as follows. You get the rare gift of a pair of newborn Fibonacci rabbits. Fibonacci rabbits are the type of rabbits who never die and each month starting in their second month produce another pair of rabbits. At the beginning of months one and two you have exactly that 1 pair of rabbits. In the second month, that pair gives you another pair of rabbits, so at the beginning of the third month you have 2 pairs of rabbits. In the third month, the original pair produces another pair of rabbits, so that at the beginning of the fourth month, you have 3 pairs of rabbits. Justify why the number of rabbits at the beginning of the nth month is s_n.

ii) Prove that for all $n \geq 1$, $s_n = \frac{1}{\sqrt{5}} \left(\frac{1+\sqrt{5}}{2} \right)^n - \frac{1}{\sqrt{5}} \left(\frac{1-\sqrt{5}}{2} \right)^n$. (It may seem **amazing** that these expressions with square roots of 5 always yield positive integers.) Note that the base case requires proving this for $n = 1$ and $n = 2$, and that the inductive step uses knowing the property for the previous two integers.

iii) Prove that $s_1 + s_3 + s_5 + \cdots + s_{2n-1} = s_{2n}$.

iv) Prove that $s_2 + s_4 + s_6 + \cdots + s_{2n} = s_{2n+1} - 1$.

v) Prove that $s_1 + s_2 + s_3 + \cdots + s_n = s_{n+2} - 1$.

vi) Prove that $s_1 - s_2 + s_3 - s_4 + \cdots + s_{2n-1} - s_{2n} = 1 - s_{2n-1}$.

vii) Prove that $s_1 - s_2 + s_3 - s_4 + \cdots + s_{2n-1} - s_{2n} + s_{2n+1} = s_{2n} + 1$.

viii) Prove that $s_1 - s_2 + s_3 - s_4 + \cdots + (-1)^{n+1} s_n = (-1)^{n+1} s_{n-1} + 1$.

ix) Prove that for all $n \geq 3$, $s_n > \left(\frac{1+\sqrt{5}}{2}\right)^{n-2}$.

x) Prove that for all $n \geq 1$, $s_1^2 + s_2^2 + \cdots + s_n^2 = s_n s_{n+1}$.

xi) Prove that $s_{n+1} s_{n-1} - s_n^2 = (-1)^n$.

xii) Prove that $s_1 s_2 + s_2 s_3 + \cdots + s_{2n-1} s_{2n} = s_{2n}^2$.

xiii) Prove that $s_1 s_2 + s_2 s_3 + \cdots + s_{2n} s_{2n+1} = s_{2n+1}^2 - 1$.

xiv) Prove that $n s_1 + (n-1) s_2 + (n-2) s_3 + \cdots + 2 s_{n-1} + s_n = s_{n+4} - (n+3)$.

xv) Prove that for all $n \geq 1$ and all $k \geq 2$, $s_{n+k} = s_k s_{n+1} + s_{k-1} s_n$.

xvi) Prove that for all $n, k \geq 1$, s_{kn} is a multiple of s_n. (Use the previous part.)

xvii) Prove that $s_{2n+1} = s_{n+1}^2 + s_n^2$.

xviii) Prove that $s_{2n} = s_{n+1}^2 - s_{n-1}^2$.

xix) Prove that $s_{3n} = s_{n+1}^3 + s_n^3 - s_{n-1}^3$.

xx) Prove that $s_{n+1} = \left(\frac{1+\sqrt{5}}{2}\right) s_n + \left(\frac{1-\sqrt{5}}{2}\right)^n$.

xxi) Prove that $s_3 + s_6 + s_9 + \cdots + s_{3n} = \frac{s_{3n+2}-1}{2}$. (Use the previous part.)

xxii) Prove that $s_1^3 + s_2^3 + s_3^3 + \cdots + s_n^3 = \frac{s_{3n+2} + (-1)^{n+1} 6 s_{n-1} + 5}{10}$.

xxiii) Prove that $\left| s_n - \frac{(1+\sqrt{5})^n}{2^n \sqrt{5}} \right| < \frac{1}{2}$.

xxiv)* If you know a bit of number theory, prove that for all positive integers m, n, the greatest common divisor of s_m and s_n is $s_{\gcd(m,n)}$.

xxv)* Prove that s_n is even if and only if n is a multiple of 3.

Prove that s_n is divisible by 3 if and only if n is a multiple of 4.

Prove that s_n is divisible by 4 if and only if n is a multiple of 6.

Prove that s_n is divisible by 5 if and only if n is a multiple of 5.

Prove that s_n is divisible by 7 if and only if n is a multiple of 8.

Prove that there are no Fibonacci numbers that have the remainder of 4 when divided by 8.

Prove that there are no odd Fibonacci numbers that are divisible by 17.

xxvi) If you know matrices, prove that for all integers $n \geq 2$, $\begin{bmatrix} 1 & 1 \\ 1 & 0 \end{bmatrix}^n = \begin{bmatrix} s_{n+1} & s_n \\ s_n & s_{n-1} \end{bmatrix}$.

1.6.31. (Via the grapevine, based on ideas of Art Benjamin, Harvey Mudd College, and Dan Velleman, Amherst College) A tromino is a plane figure composed of three squares in L-shape:

i) Prove that for every positive integer n, any $2^n \times 2^n$ square grid with exactly one of the squares removed can be tiled with trominoes.

ii) Prove that for every positive integer n, $4^n - 1$ is an integer multiple of 3.

1.6.32. Pick a vertex V in a triangle. Draw n distinct lines from V to the opposite edge of the triangle. If $n = 1$, you get the original triangle and two smaller triangles, for a total of three triangles. Determine the number of distinct triangles obtained in this way with arbitrary n.

1.6.33. (**Spiral of Theodorus**) Draw a triangle with vertices at $(0,0)$, $(1,0)$, $(1,1)$. The hypotenuse has length $\sqrt{2}$.

i) One of the vertices of the hypotenuse is at $(0,0)$. At the other vertex of the hypotenuse, draw an edge of length 1 at the right angle away from the first triangle. Make a triangle from the old hypotenuse and this new edge. What is the length of the hypotenuse of the new triangle?

ii) Repeat the previous step twice.

iii) Prove that one can draw $\sqrt{n}$ for every positive integer n.

1.6.34. (**Tower of Hanoi**) There are 3 pegs on a board. On one peg, there are n disks, stacked from largest to smallest. The task is to move all of the disks from one peg to a different peg, given the following constraints: you may only move one disk at a time, and you may only place a smaller peg on a larger one (never a larger one on a smaller one). Let S_n be the least number of moves to complete the task for n disks.

i) If $n = 1$, then what is the least number of moves it takes to complete the task? What if there are 2 disks? Repeat for 3, 4, 5 disks.

ii) Make a recursive formula (defining S_n based on S_{n-1}) for this S_n.

Then, make a guess for a non-recursive formula for S_n (defining S_n based on n without invoking S_{n-1}). Prove your guess using induction and the recursive formula that you wrote.

1.6.35. What is wrong with the following "proof by induction"? I will prove that $5^n + 1$ is a multiple of 4. Assume that this is true for $n - 1$. Then we can write $5^{n-1} + 1 = 4m$ for some integer m. Multiply this equation through by 5 to get that

$$5^n + 5 = 20m,$$

whence $5^n + 1 = 4(5m - 1)$. As $5m - 1$ is an integer, this proves that $5^n + 1$ is a multiple of 4.   □

1.6.36. What is wrong with the following "proof by induction" besides the fact that the conclusion is false for many n? I will prove that all horses are of the same color. This is the same as saying that for any integer $n \geq 1$ and any set of n horses, all the horses belonging to the set have the same color. If $n = 1$, of course this only horse is the same color as itself, so the base case is proved. Now let $n > 1$. If we remove one horse from this set, the remaining $n - 1$ horses in the set are all of the same color by the induction assumption. Now bring that one horse back into the set and remove another horse. Then again all of these horses are of the same color, so the horse that was removed first is the same color as all the rest of them.   □

1.7   Pascal's triangle

Pascal's triangle is very useful, so read this section with the exercises.

The following is rows 0 through 8 of **Pascal's triangle**, and the pattern is obvious for continuation into further rows:

row 0:	.	.	.	.	.	.	.	.	1						
row 1:	.	.	.	.	.	.	.	1		1					
row 2:	.	.	.	.	.	.	1		2		1				
row 3:	.	.	.	.	.	1		3		3		1			
row 4:	.	.	.	.	1		4		6		4		1		
row 5:	.	.	.	1		5		10		10		5		1	
row 6:	.	.	1		6		15		20		15		6		1
row 7:	.	1		7		21		35		35		21		7	1

Note that the leftmost and rightmost numbers in each row are all 1, and each of the other numbers is the sum of the two numbers nearest to it in the row above it. We number the slanted columns from left to right starting from 0: the 0th slanted column consists of all 1s, the 1st slanted column consists of consecutive numbers $1, 2, 3, 4, \ldots$, the 2nd slanted column consists of consecutive numbers $1, 3, 6, 10, \ldots$, and so on for the subsequent columns.

Let the entry in the nth row and kth column be denoted $\binom{n}{k}$. We read this as "**n choose k**". These are loaded words, however, and we will eventually justify them.

Pascal's triangle is defined so that for all $n \geq 1$ and all $k = 0, 1, \ldots, n - 1$,

$$\binom{n}{k} + \binom{n}{k+1} = \binom{n+1}{k+1}.$$

What would it take to compute $\binom{100}{5}$? It seems like we would need to write down rows 0 through 100 of Pascal's triangle, or actually a little less, only slanted columns 0 through 5 of these 101 rows. That is too much drudgery! We will instead be smart mathematicians and we will prove many properties of Pascal's triangle in general, including shortcuts for computing $\binom{100}{5}$. We will accomplish this through exercises, most of which can be proved by mathematical induction.

Exercises for Section 1.7

1.7.1. Prove that the sum of the entries in the nth row is 2^n.

1.7.2. Let k be an arbitrary non-negative integer. (This means that k is given to you not as a specific number but as an unknown integer.) Prove by induction on n that the sum of the entries in the slanted column k in rows $j \leq n$ is $\binom{n+1}{k+1}$. In other words, prove that $\binom{k}{k} + \binom{k+1}{k} + \binom{k+2}{k} + \cdots + \binom{n}{k} = \binom{n+1}{k+1}$, which is the same as proving that $\sum_{j=k}^{n} \binom{j}{k} = \binom{n+1}{k+1}$.

1.7.3. Prove that every integer $n \geq 0$ has the property that for all $k = 0, 1, 2, \ldots, n$, $\binom{n}{k} = \frac{n!}{k!(n-k)!}$.

1.7.4. Prove that $\binom{n}{k} = \frac{n(n-1)(n-2)\cdots(n-k+1)}{k!}$.

1.7.5. Compute $\binom{4}{2}$, $\binom{5}{2}$, $\binom{6}{2}$, $\binom{7}{2}$, $\binom{8}{2}$, $\binom{100}{2}$, $\binom{100}{3}$, $\binom{100}{4}$, $\binom{100}{5}$.

1.7.6. Prove that $\binom{n}{k}$ is the number of possible k-member teams in a club with exactly n members. For this reason $\binom{n}{k}$ is read **n choose k**.

†1.7.7. (Invoked in Theorem 2.9.2, Example 8.2.9, Theorem 6.2.3.) Prove that for all non-negative integers n,

$$(a+b)^n = \sum_{k=0}^{n} \binom{n}{k} a^k b^{n-k}.$$

(Since $a + b$ contains two summands, it is called a **binomial**, and the expansion of $(a+b)^n$ is called the **binomial expansion**, with coefficients $\binom{n}{i}$ being called by yet another name in this context: **binomial coefficients**.)

1.7.8. Express each of the following as $a + b\sqrt{2}$ for some integers a, b:
   i) $\sqrt{2} - 1$, $(\sqrt{2} - 1)^2$, $(\sqrt{2} - 1)^3$, $(\sqrt{2} - 1)^4$, $(\sqrt{2} - 1)^5$.
   ii) Write each of the five expressions in the previous part in the form $\sqrt{c} - \sqrt{d}$ for some positive integers c, d.
   iii)* Do you see a relation between c and d for each expression in the previous part? Is there a general rule? Can you prove it?

1.7.9. Prove that for all positive integers n, $\sum_{k=0}^{n}(-1)^k \binom{n}{k} = 0$. Compute $\sum_{k=0}^{n}(-1)^k \binom{n}{k}$ in case $n = 0$.

1.7.10. Prove that for any non-negative integer k,

$$\sum_{j=1}^{n} j(j+1)(j+2)\cdots(j+k) = \frac{n(n+1)(n+2)\cdots(n+k+1)}{k+1}.$$

(Hint: induction on n or instead use Exercises 1.7.2 and 1.7.4.)

1.7.11. Use Exercise 1.7.10 to get succinct simplifying formulas for

$$\sum_{j=1}^{n} j, \quad \sum_{j=1}^{n} j(j+1), \quad \sum_{j=1}^{n} j(j+1)(j+2), \quad \sum_{j=1}^{n} j(j+1)(j+2)(j+3).$$

   i) Note that $j^2 = j(j+1) - j$. Use the simplifications from above to prove that $\sum_{j=1}^{n} j^2 = \frac{n(n+1)(2n+1)}{6}$.
   ii) From $j^3 = j(j+1)(j+2) - 3j^2 - 2j = j(j+1)(j+2) - 3j(j+1) + j$ develop the formula for $\sum_{j=1}^{n} j^3$.
   iii) Mimic the previous work to develop the formula for $\sum_{j=1}^{n} j^4$.

1.7.12. Prove that for all non-negative integers n and all $k = 0, 1, \ldots n$, $\binom{n}{k} \leq \frac{n^k}{k!}$.

1.7.13. Fix a positive integer k. Prove that there exists a positive number C such that for all sufficiently large integers n, $Cn^k \leq \binom{n}{k}$.

1.7.14. Give reasons why we should have $\binom{n}{k} = 0$ for $n < k$ or if either k or n is negative.

1.7.15. Let d be a positive integer. This is about summing entries in Pascal's triangle along the dth northwest-southeast slanted column: Prove by induction on $n \geq 0$ that $\sum_{k=0}^{n} \binom{d+k}{k} = \binom{d+n+1}{n}$.

*1.7.16. Prove that for all non-negative integers n,

$$\left(\sum_{k=0}^{n} \binom{2n}{2k} 2^k\right)^2 - 1 = 2\left(\sum_{k=0}^{n-1} \binom{2n}{2k+1} 2^k\right)^2$$

and

$$\left(\sum_{k=0}^{n} \binom{2n+1}{2k} 2^k\right)^2 + 1 = 2\left(\sum_{k=0}^{n-1} \binom{2n+1}{2k+1} 2^k\right)^2.$$

For notation's sake you may want to label $E_n = \sum_{k=0}^{e_n} \binom{n}{2k} 2^k$ and $O_n = \sum_{k=0}^{o_n} \binom{n}{2k+1} 2^k$, where e_n, o_n are the largest integers such that $2e_n \leq n$ and $2o_n + 1 \leq n$. The claim is then that for all $n \geq 0$, $E_{2n}^2 - 1 = 2O_{2n}^2$ and $E_{2n+1}^2 + 1 = 2O_{2n+1}^2$. (Hint: use the definition of $\binom{n}{k}$ to rewrite E_n in terms of E_{n-1}, O_{n-1}. Proceed with induction.)

Chapter 2

Concepts with which we will do mathematics

In this chapter we introduce abstract structures and we examine them on familiar notions of numbers and functions from high school mathematics. The development of abstract tools enables us to analyze with common proofs rather than separately in particular all familiar and also many non-yet familiar number structures. This abstraction is akin to abstracting the concrete calculations $0^2, 0.1^2, 0.2^2, 0.25^2, 0.3^2, 1^2, 1.1^2, \ldots$ into the squaring function $f(x) = x^2$; while at first it may be hard to grasp x^2, by this point in your mathematical career you are able to manipulate this function in many ways with ease. By the end of this chapter you will similarly be at ease with the abstracted notions of a function or of a number field, and more.

2.1 Sets

What is a set? Don't we already have an idea of what a set is? The following informal definition relies on our intuitive idea of a set while making precise some notation and vocabulary of membership.

Definition 2.1.1. *A* **set** *is a collection of objects. These objects are called* **members** *or* **elements** *of that set. If m is a member of a set A, we write* $m \in A$*, and also say that A* **contains** *m. If m is not a member of a set A, we write* $m \notin A$*.*

The set of all polygons contains triangles, squares, rectangles, pentagons, and so on. The set of all polygons does not contain circles or disks. The set of all functions contains the trigonometric, logarithmic, exponential, constant functions, and so on.

Examples and notation 2.1.2.

(1) **Intervals** are sets:

$(0, 1)$ is the interval from 0 to 1 that does not include $0, 1$. From the context you should be able to distinguish between the interval $(0, 1)$ and a point $(0, 1)$ in the plane.

More generally, below we take real numbers a and b with $a < b$.

$(a, b]$ is the interval from a to b that includes b but not a.

$[a, b)$ is the interval from a to b that includes a but not b.

$[a, b]$ is the interval from a to b that includes a and b.

(a, ∞) is the interval of all real numbers strictly bigger than a.

$[a, \infty)$ is the interval of all real numbers bigger than or equal to a.

$(-\infty, b)$ is the interval of all real numbers strictly smaller than b.

$(-\infty, b]$ is the interval of all real numbers smaller than or equal to b.

$(-\infty, \infty)$ is the set of all real numbers.

$(0, \infty)$ is the set of all positive real numbers.

$[a, a]$ is the "interval" consisting of precisely one number a.

$(a, a), (a, a], [a, a)$ are the "interval" consisting of no numbers.

(2) The set with no elements is called **the empty set** and is denoted $\emptyset$ or $\{\}$.

The set $\{\emptyset\}$ is not empty because it contains the empty set.

(3) Descriptively we can say for example that a set A is the set of all prime numbers. The description may be more loaded with symbols: A is the set of all real numbers x with the property that $x^2 > 4x - 3$. (Use the quadratic formula to prove that A consists of the numbers that are strictly smaller than 1 or strictly larger than 3.)

(4) We can list the elements and surround them with curly braces to define a set:

(i) $\{1, 2, 3\}$ is the set consisting of precisely $1, 2$ and 3.

(ii) $\{1, 2, 3, 2\}$ is the set consisting of precisely $1, 2$ and 3. Thus $\{1, 2, 3, 2\} = \{1, 2, 3\}$.

(iii) $\{blue, \text{"}hello\text{"}, 5\}$ is the set consisting precisely of the color blue, of the word "hello", and of number 5.

(iv) $\{1, 2, \{1, 2\}\}$ is the set consisting of precisely of numbers 1 and 2 and of the set $\{1, 2\}$. This set has exactly three distinct elements, it is not the same as $\{1, 2\}$, and it is not the same as $\{\{1, 2\}\}$.

(5) When the list of elements is not small enough for reasonable explicit listing but the pattern of elements is clear, we can start the list and then add ", ..." when the pattern is clear:

(i) $\{1, 2, 3, \ldots, 10000\}$ is the set of all positive integers that are at most 10000.

(ii) $\{1, 4, 9, \ldots, 169\}$ is the set of the first 13 squares of integers.

(iii) $\{1, 2, 3, \ldots\}$ is the set consisting of all positive whole numbers. This set is often denoted by $\mathbb{N}$. To be absolutely clear, we will write $\mathbb{N}^+$ for $\{1, 2, 3, \ldots\}$ and $\mathbb{N}_0$ for $\{0, 1, 2, 3, \ldots\}$.

(6) Warning: $\{3, 5, 7, \ldots\}$ or $\{3, 5, 7, \ldots, 101\}$ could stand for the set of all odd primes (up to 101), or possibly for the set of all odd whole numbers strictly greater than 1 (up to 101). Avoid ambiguities: write more elements, or write an explicit description of the elements instead.

(7) The set of all whole numbers is written $\mathbb{Z}$, the set of all rational numbers is written $\mathbb{Q}$, the set of all real numbers is written $\mathbb{R}$, and the set of all complex numbers is written $\mathbb{C}$ (complex numbers are defined in Section 3.9, and until then do not worry when "$\mathbb{C}$" appears in the text). The set of all non-negative real numbers equals $[0, \infty)$, and we also write it as $\mathbb{R}_{\geq 0}$.

(8) We can define sets **propositionally**: if P is a property, then the set

$$\{x : P(x)\} \text{ or } \{x \in A : P(x)\}$$

consists of all x (or $x \in A$) for which P holds. Here are some explicit examples:

(i) $\{x \in \mathbb{R} : x^2 = x\}$, and this happens to be the set $\{0, 1\}$.

(ii) $\{x \in \mathbb{R} : x > 0 \text{ and } x < 1\}$, and this happens to be the interval $(0, 1)$.

(iii) $\mathbb{Q} = \{\frac{a}{b} : a, b \in \mathbb{Z} \text{ and } b \neq 0\}$.

(iv) The set of all positive prime numbers equals $\{n \in \mathbb{N} : n > 1$ and if $n = pq$ for some integers p, q then $|p| = 1$ or $|q| = 1\}$.

(v) Set $A = \{x : x$ is a positive integer that equals the sum of its proper factors$\}$. (Elements of A are called **perfect** numbers.) It is easy to verify that $1, 2, 3, 4, 5$ are not in A. But 6 has factors $1, 2, 3, 6$, and the sum of the factors other than 6 equals 6. Thus 6 is an element of A. You can verify that the three numbers $28, 496,$

and 8128 are also in A. (If you and your computer have a lot of time, write a program to verify that no other number smaller than 33 million is in A.)

(9) Proving that a property P holds for all integers $n \geq n_0$ is the same as saying that the set $A = \{n \in \mathbb{Z} : P \text{ holds for } n\}$ contains $\{n_0, n_0 + 1, n_0 + 2, \ldots\}$. By the principle of mathematical induction, P holds for all integers $n \geq n_0$ is the same as saying that $n_0 \in A$ and that $n - 1 \in A$ implies that $n \in A$.

Summary of example sets, and their notation

$\emptyset = \{\}$: the set with no elements.

$\{a, b, c\}$, $\{a, b, \ldots, z\}$.

$\{x : x \text{ can be written as a sum of three consecutive integers}\}$.

$\mathbb{N}$: the set of all **N**atural numbers. Depending on the book, this could be the set of all positive integers or it could be the set of all non-negative integers. The symbols below are unambiguous:

  $\mathbb{N}_0$: the set of all non-negative integers;

  $\mathbb{N}^+$: the set of all positive integers.

$\mathbb{Z}$: the set of all integers ("**Z**ahlen" in German).

$\mathbb{Q}$: the set of all rational numbers (**Q**uotients).

$\mathbb{R}$: the set of all **R**eal numbers.

$\mathbb{C}$: the set of all **C**omplex numbers (more about them starts in Section 3.9).

Just like numbers, functions, and logical statements, sets and their elements can also be related and combined in meaningful ways. The list below introduces quite a few new concepts that may be overwhelming at first, but in a few weeks you will be very comfortable with them.

Subsets: A set A is a **subset** of a set B if every element of A is an element of B. In that case we write $A \subseteq B$. For example, $\mathbb{N}^+ \subseteq \mathbb{N}_0 \subseteq \mathbb{Z} \subseteq \mathbb{Q} \subseteq \mathbb{R}$. The non-subset relation is expressed with the symbol $\not\subseteq$: $\mathbb{R} \not\subseteq \mathbb{N}$. Every set is a subset of itself, i.e., for every set A, $A \subseteq A$. The empty set is a subset of every set, i.e., for every set A, $\emptyset \subseteq A$. If A is a subset of B and A is not equal to B (so B contains at least one element that is not in A), then we say that A is a **proper subset** of B, and we write $A \subsetneq B$. For example, $\mathbb{N}^+ \subsetneq \mathbb{N}_0 \subsetneq \mathbb{Z} \subsetneq \mathbb{Q} \subsetneq \mathbb{R}$.

Equality: Two sets are **equal** if they consist of exactly the same elements. In other words, $A = B$ if and only if $A \subseteq B$ and $B \subseteq A$.

Intersection: The **intersection** of sets A and B is the set of all objects that are in A and in B:

$$A \cap B = \{x : x \in A \text{ and } x \in B\}.$$

When $A \cap B = \emptyset$, we say that A and B are **disjoint**.

Union: The **union** of sets A and B is the set of all objects that are either in A or in B:

$$A \cup B = \{x : x \in A \text{ or } x \in B\}.$$

Intersections and unions of arbitrary families of sets: We have seen intersections and unions of two sets at a time. We can also take intersections and unions of three, four, five, and even infinitely many sets at a time. Verify the equalities below:

$$(A \cap B) \cap C = A \cap (B \cap C),$$
$$(A \cup B) \cup C = A \cup (B \cup C),$$
$$(A \cap B) \cap (C \cap D) = A \cap (B \cap C \cap D), \text{ etc.}$$
$$(A \cup B) \cup (C \cup D) = (A \cup (B \cup C)) \cup D, \text{ etc.}$$

(Verification of the first equality above: Let $x \in (A \cap B) \cap C$. This holds if and only if $x \in A \cap B$ and $x \in C$, which holds if and only if $x \in A, x \in B$ and $x \in C$, which in turn holds if and only if $x \in A$ and $x \in (B \cap C)$, i.e., if and only if $x \in A \cap (B \cap C)$.)

Thus having established that parentheses above are irrelevant, we simply write the four sets above as $A \cap B \cap C$, $A \cup B \cup C$, $A \cap B \cap C \cap D$, $A \cup B \cup C \cup D$, respectively.

More generally, given sets $A_1, A_2, \ldots, A_n$, we write

$$\bigcap_{k=1}^{n} A_k = A_1 \cap A_2 \cap \cdots \cap A_n = \{a : a \in A_k \text{ for all } k = 1, \ldots, n\},$$

$$\bigcup_{k=1}^{n} A_k = A_1 \cup A_2 \cup \cdots \cup A_n = \{a : \exists k = 1, \ldots, n \text{ such that } a \in A_k\}.$$

The k in the subscripts are referred to as **indices** of unions or intersections. The indices can be taken from arbitrary sets, even infinite ones, and we call such sets **index sets**. Notationally, if I is an index set, we write

$$\bigcap_{k \in I} A_k = \{a : a \in A_k \text{ for all } k \in I\},$$

$$\bigcup_{k \in I} A_k = \{a : \text{ there exists } k \in I \text{ such that } a \in A_k\}.$$

When $I = \{1, 2, \ldots, n\}$, then the intersections and unions in the two lines above are the same as those in the previous display.

When I is the empty index set, one can argue as for empty sums in Section 1.5 that

$$\bigcup_{k \in \emptyset} A_k = \emptyset,$$

i.e., the empty union is that set which when unioned with any other set returns that other set. The only set which satisfies this property is the empty set. Similarly, the empty intersection should be that set which when intersected with any other set returns that other set. However, this empty intersection depends on the context: when the allowed other sets vary over all subsets of a set X, then the empty intersection equals X. We return to this theme in Section 2.5.

Complement: The **complement** of A in B is

$$B \setminus A = \{b \in B : b \notin A\}.$$

Some authors write $B - A$, but that has another meaning as well: $B - A : \{b - a : b \in B \text{ and } a \in A\}$. Always try to use precise and unambiguous notation.

We often have an implicit or explicit **universal set** that contains all elements of our current interest. Perhaps we are talking only about real numbers, or perhaps we are talking about all functions defined on the interval $[0, 1]$ with values being real numbers. In that case, for any subset A of the universal set U, the **complement** of A is the complement of A in U, thus $U \setminus A$, and this is denoted as A^c.

Summary notation and vocabulary

$a \in A$: a is an element of a set A.

$A \subseteq B$: A is a subset of set B; every element of A is an element of B.

$A \subsetneq B$: $A \subseteq B$ and $A \neq B$; A is a **proper** subset of B.

$A = B$: $A \subseteq B$ and $B \subseteq A$.

$A \cap B$: the set of all elements that are in A and in B.

A and B are **disjoint**: $A \cap B = \emptyset$.

$A \cup B$: the set of all elements that are either in A or in B.

$A \setminus B$: the set of all elements of A that are not in B.

A^c: the set of all elements in the universal set that are not in A.

Example 2.1.3. We prove that $\mathbb{Z} = \{3m + 4n : m, n \in \mathbb{Z}\}$. Certainly for any integers m and n, $3m + 4n$ is also an integer, so that $\{3m + 4n : m, n \in \mathbb{Z}\} \subseteq \mathbb{Z}$. Now let $x \in \mathbb{Z}$. Then

$$x = 1 \cdot x = (4 - 3) \cdot x = 3(-x) + 4x,$$

so that $x \in \{3m + 4n : m, n \in \mathbb{Z}\}$, whence $\mathbb{Z} \subseteq \{3m + 4n : m, n \in \mathbb{Z}\}$. Since we already proved the other inclusion, the proof is done. $\square$

Example 2.1.4. We prove that $A = \{6m + 14n : m, n \in \mathbb{Z}\}$ equals the set B of all even integers. Certainly for any integers m and n, $6m + 14n$ is an even integer, so that $A \subseteq B$. Now let $x \in B$. Then x is even, so $x = 2n$ for some integer n. Write

$$x = 2n = (14 - 2 \cdot 6)n = 6(-2n) + 14n,$$

so that $x \in \{6m + 14n : m, n \in \mathbb{Z}\} = A$. Thus $B \subseteq A$. Together with the first part this implies that $A = B$. $\square$

Example 2.1.5. The complement in $\mathbb{Z}$ of the set of even integers is the set of odd integers. The complement in $\mathbb{Q}$ of the set of even integers contains many more elements than odd integers. For example, it contains $\frac{1}{2}, \frac{2}{3}, \ldots$.

Example 2.1.6. If $A \subseteq C$, then $C \setminus (C \setminus A) = A$. In other words, the complement of the complement of A is A.

Proof. Let $x \in A$. Then x is not in the complement $C \subseteq A$ of A in C. Since $x \in C$ and not in the subset $C \subseteq A$ of C, it follows that $x \in C \setminus (C \setminus A)$. This proves that $A \subseteq C \setminus (C \setminus A)$.

Now let $x \in C \setminus (C \setminus A)$. Then $x \in C$ and x is not in $C \setminus A$. Thus necessarily $x \in A$. This proves that $C \setminus (C \setminus A) \subseteq A$.   □

Example 2.1.7. Let A and B be subsets of C. Then $A \subseteq B$ if and only if $C \setminus B \subseteq C \setminus A$.

Proof. ($\Rightarrow$) Suppose that $A \subseteq B$. Let $x \in C \setminus B$. Then $x \in C$ and x is not in B. Since $A \subseteq B$, then necessarily x is not in B. This proves that $x \in C \subseteq B$. Since x was arbitrary in $C \setminus B$. it follows that $C \setminus B \subseteq C \setminus A$.

($\Leftarrow$) Suppose that $C \setminus B \subseteq C \setminus A$. Then by the previous paragraph, $C \setminus (C \setminus A) \subseteq C \setminus (C \setminus B)$, and so by Example 2.1.6, $A \subseteq B$.   □

Example 2.1.8. For each $i \in \mathbb{N}^+$, let $A_i = [i, \infty)$, $B_i = \{i, i+1, i+2, i+3, \ldots\}$, and $C_i = (-i, i)$. Think through the following:

$$\bigcap_{k \in \mathbb{N}} A_k = \emptyset, \qquad \bigcap_{k \in \mathbb{N}} B_k = \emptyset, \qquad \bigcap_{k \in \mathbb{N}} C_k = (-1, 1),$$

$$\bigcup_{k \in \mathbb{N}} A_k = [1, \infty), \qquad \bigcup_{k \in \mathbb{N}} B_k = \mathbb{N}^+, \qquad \bigcup_{k \in \mathbb{N}} C_k = \mathbb{R}.$$

Example 2.1.9. For each real number r, let $A_r = \{r\}$, $B_r = [0, |r|]$. Then

$$\bigcap_{r \in \mathbb{R}} A_r = \emptyset, \qquad \bigcap_{r \in \mathbb{R}} B_r = \{0\}, \qquad \bigcup_{r \in \mathbb{R}} A_r = \mathbb{R}, \qquad \bigcup_{r \in \mathbb{R}} B_r = [0, \infty).$$

Set operations can be represented with a **Venn diagram**, especially in the presence of a universal set U. Here is an example:

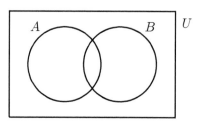

On this Venn diagram, sets are represented by the geometric regions: A is the set represented by the left circle, B is represented by the right circle, $A \cap B$ is the part of the two circles that is both in A and in B, $A \cup B$ is represented by the region that is either in A or in B, $A \setminus B$ is the left crescent after B is chopped out of A, etc. (There is no reason why the regions for sets A and B are drawn as circles, but this is traditional.)

Sometimes we draw a few (or all) elements into the diagram. For example, in

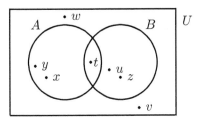

we read that $A = \{x, y, t\}$, $A \cap B = \{t\}$, etc.

Two disjoint sets A and B are represented by a Venn diagram as follows:

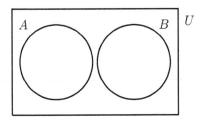

Theorem 2.1.10. *For all sets A, B, C, we have $A \cap (B \cup C) = (A \cap B) \cup (A \cap C)$.*

Consider the Venn diagram below.

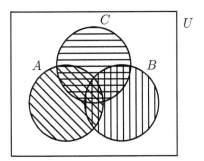

The region filled with either horizontal or vertical lines is $B \cup C$, A is the region filled with the southeast-northwest slanted lines, and so $A \cap (B \cup C)$ is the region that has simultaneously the southeast-northwest slanted lines and either horizontal or vertical lines. Also, $A \cap C$ is the region that has horizontal and southeast-northwest slanted lines, $A \cap B$ is the region that has vertical and southeast-northwest slanted lines, so that their union $(A \cap B) \cup (A \cap C)$ represents the total region of southeast-northwest slanted lines that either have horizontal or vertical cross lines as well, which is the same as the region for $A \cap (B \cup C)$.

We prove this also algebraically. We have to prove that $A \cap (B \cup C) \subseteq (A \cap B) \cup (A \cap C)$ and $(A \cap B) \cup (A \cap C) \subseteq A \cap (B \cup C)$.

Let x be arbitrary in $A \cap (B \cup C)$. This says that $x \in A$ and $x \in B \cup C$, and the latter says that either $x \in B$ or $x \in C$. But then either $x \in (A \cap B)$ or $x \in (A \cap C)$, so that $x \in (A \cap B) \cup (A \cap C)$. This proves one inclusion.

Now let x be arbitrary in $(A \cap B) \cup (A \cap C)$. This says that either $x \in A \cap B$ or $x \in A \cap C$, which in turn says that either x is in A and in B or else that x is in A and in C. In any case, $x \in A$, and either $x \in B$ or $x \in C$, so that $x \in A$ and $x \in B \cup C$, which finally says that $x \in A \cap (B \cup C)$. This proves the other inclusion.

Exercises for Section 2.1

2.1.1. Prove by induction on n that a set with n elements has exactly 2^n distinct subsets.

2.1.2. Prove that a set with n elements has $\binom{n}{i}$ subsets with exactly i elements.

2.1.3. Let $A, B \subseteq U$.

    i) Prove that there exist at most 16 distinct subsets of U obtained from A, B, U by intersections, unions, and complementation.

    ii) If A and B are disjoint, prove that there exist at most 8 such distinct subsets.

    iii) If $A = B$, prove that there exist at most 4 such distinct subsets.

    iv) If $A = B = U$, prove that there exist at most 2 such distinct subsets.

    v) If $A = B = U = \emptyset$, prove that there exists at most 1 such subset.

2.1.4. Assume that A and B are disjoint subsets of U. For each part below, draw a Venn diagram with A and B in U, and shade in the region described by the set: (i) $U \setminus B$, (ii) $A \cap B$, (iii) $U \setminus (U \setminus A)$, (iv) $(U \setminus A) \cap (U \setminus B)$, (v) $(U \cap B) \cup (U \setminus (B \cup A))$, (vi) $(B \cap U) \cup (B \setminus U)$, (vii) $(U \setminus (U \setminus A)) \cup B$, (viii) $(A \cup B) \cup (U \setminus A)$.

2.1.5. Prove the following:

   i) $\{x \in \mathbb{R} : x^2 = 3\} = \{\sqrt{3}, -\sqrt{3}\}$.

   ii) $\{x^3 : x \in \mathbb{R}\} = \mathbb{R}$.

   iii) $\{x^2 : x \in \mathbb{R}\} = \{x \in \mathbb{R} : x \geq 0\} = [0, \infty)$.

   iv) $\{2, 2, 5\} = \{2, 5\} = \{5, 2\}$.

   v) $\{x \geq 0 : x$ is an even prime number$\} = \{2\}$.

   vi) $\emptyset$ is a subset of every set. Elements of $\emptyset$ are green, smart, sticky, hairy, feathery, prime, whole, negative, positive,...

   vii) $\{x : x$ can be written as a sum of three consecutive integers$\} = \{3n : n \in \mathbb{Z}\}$.

   viii) If $A \subseteq B$, then $A \cap B = A$ and $A \cup B = B$.

2.1.6. Let $U = \{1, 2, 3, 4, 5, 6\}$, $A = \{1, 3, 5\}$, and $B = \{4, 5, 6\}$. Find the following sets:

   i) $(A \setminus B) \cup (B \setminus A)$.

   ii) $U \setminus (B \setminus A)$.

   iii) $U \cup (B \setminus A)$.

   iv) $U \setminus (A \cup B)$.

   v) $(U \cap A) \cup (U \cap B)$.

   vi) $A \setminus (A \setminus B)$.

   vii) $B \setminus (B \setminus A)$.

   viii) $\{A\} \cap \{B\}$.

2.1.7. Let $A, B, C \subseteq U$. Prove the following statements:

   i) $(A \cap C) \setminus B = (A \setminus B) \cap (C \setminus B)$.

   ii) $(A \setminus B) \cup (B \setminus A) = (A \cup B) \setminus (A \cap B)$.

   iii) $(A \cap B) \cup (U \setminus (A \cup B)) = (U \setminus (A \setminus B)) \setminus (B \setminus A)$.

   iv) $U \setminus (A \setminus B) = (U \setminus A) \cup B$.

   v) If $U = A \cap B$, then $A = B = U$.

2.1.8. Let $A, B \subseteq U$.

i) Prove that $(U \setminus A) \cap (U \setminus B) = U \setminus (A \cup B)$. (The intersection of the complements is the complement of the union.)

ii) Prove that $(U \setminus A) \cup (U \setminus B) = U \setminus (A \cap B)$. (The union of the complements is the complement of the intersection.)

2.1.9. Compute $\displaystyle\bigcap_{k \in \mathbb{N}^+} (-1/k, 1/k)$, $\displaystyle\bigcap_{k \in \mathbb{N}^+} [-1/k, 1/k]$, $\displaystyle\bigcap_{k \in \mathbb{N}^+} \{-1/k, 1/k\}$.

2.1.10. Compute $\displaystyle\bigcup_{k \in \mathbb{N}^+} (-1/k, 1/k)$, $\displaystyle\bigcup_{k \in \mathbb{N}^+} [-1/k, 1/k]$, $\displaystyle\bigcup_{k \in \mathbb{N}^+} \{-1/k, 1/k\}$.

2.2   Cartesian product

The set $\{a, b\}$ is the same as the set $\{b, a\}$, as any element of either set is also the element of the other set. Thus, the order of the listing of elements does not matter. But sometimes we want the order to matter. We can then simply make another new notation for **ordered pairs**, but in general it is not a good idea to be inventing many new notations and concepts; it is better if we can reuse and recycle old ones. We do this next:

Definition 2.2.1. *An* **ordered pair** (a, b) *is defined as the set* $\{\{a\}, \{a, b\}\}$.

So here we defined (a, b) in terms of already known constructions: (a, b) is the set one of whose elements is the set $\{a\}$ with exactly one element a, and the remaining element of (a, b) is the set $\{a, b\}$ that has exactly two elements a, b if $a \neq b$ and has exactly one element otherwise. Thus for example the familiar ordered pair $(2, 3)$ really stands for $\{\{2\}, \{2, 3\}\}$, $(3, 2)$ stands for $\{\{3\}, \{2, 3\}\}$, and $(2, 2)$ stands for $\{\{2\}, \{2, 2\}\} = \{\{2\}, \{2\}\} = \{\{2\}\}$.

Theorem 2.2.2. $(a, b) = (c, d)$ *if and only if* $a = c$ *and* $b = d$.

Proof. [Recall that $P \Leftrightarrow Q$ is the same as $P \Rightarrow Q$ and $P \Leftarrow Q$. Thus the proof consists of two parts.]

Proof of $\Rightarrow$: Suppose that $(a, b) = (c, d)$. Then by the definition of ordered pairs, $\{\{a\}, \{a, b\}\} = \{\{c\}, \{c, d\}\}$. If $a = b$, this says that $\{\{a\}\} = \{\{c\}, \{c, d\}\}$, so that $\{\{c\}, \{c, d\}\}$ has only one element, so that $\{c\} = \{c, d\}$, so that $c = d$. But then $\{\{a\}, \{a, b\}\} = \{\{c\}, \{c, d\}\}$ is saying that $\{\{a\}\} =$

$\{\{c\}\}$, so that $\{a\} = \{c\}$, so that $a = c$. Furthermore, $b = a = c = d$, which proves the consequent in case $a = b$. Now suppose that $a \neq b$. Then $\{\{a\}, \{a, b\}\} = \{\{c\}, \{c, d\}\}$ has two elements, and so $c \neq d$. Note that $\{a\}$ is an element of $\{\{a\}, \{a, b\}\}$, hence of $\{\{c\}, \{c, d\}\}$. Thus necessarily either $\{a\} = \{c\}$ or $\{a\} = \{c, d\}$. But $\{a\}$ has only one element and $\{c, d\}$ has two (since $c \neq d$), it follows that $\{a\} = \{c\}$, so that $a = c$. But then $\{a, b\} = \{c, d\}$, and since $a = c$, it follows that $b = d$. This proves the consequent in the remaining cases.

Proof of $\Leftarrow$: If $a = c$ and $b = d$, then $\{a\} = \{c\}$ and $\{a, b\} = \{c, d\}$, so that $\{\{a\}, \{a, b\}\} = \{\{c\}, \{c, d\}\}$. $\qquad\qquad\qquad\qquad\qquad\qquad\square$

Note that by our definition an ordered pair is a set of one or two sets.

Definition 2.2.3. *For any sets A and B, the* **Cartesian product $A \times B$** *of A and B is the set $\{(a, b) : a \in A$ and $b \in B\}$ of all ordered pairs where the first component varies over all elements of A and the second component varies over all elements of B.*

In general, one can think of $A \times B$ as the "rectangle" with A on the horizontal side and B on the perpendicular side.

Say, if A has 4 elements and B has 3 elements, then $A \times B$ is represented by the 12 points in the rectangle with base consisting of elements of A and height consisting of elements of B as follows:

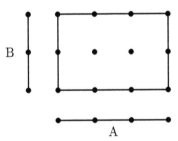

If instead A and B are intervals as above, then $A \times B$ is the indicated rectangle. When A and B extend infinitely far, then $A \times B$ is correspondingly a "large" rectangle: The familiar real plane is the Cartesian product $\mathbb{R} \times \mathbb{R}$.

(The three-dimensional space can be written as the Cartesian product $\mathbb{R} \times (\mathbb{R} \times \mathbb{R})$ or the Cartesian product $(\mathbb{R} \times \mathbb{R}) \times \mathbb{R}$. In the former case we write elements in the form $(a, (b, c))$, and in the latter case we write them in the form $((a, b), c)$. Those extra parentheses are there only for notation and to slow us down, they serve no better function, so by convention we write elements simply in the form (a, b, c).)

Exercises for Section 2.2

2.2.1. How many elements are in $\emptyset \times \mathbb{R}$?

2.2.2. Prove that $A \times (B \cup C) = (A \times B) \cup (A \times C)$.

2.2.3. Prove that $A \times (B \cap C) = (A \times B) \cap (A \times C)$.

2.2.4. Prove that $(A \cap C) \times (B \cap D) = (A \times B) \cap (C \times D)$.

2.2.5. Give examples of sets A, B, C, D showing that $(A \cup C) \times (B \cup D) \neq (A \times B) \cup (C \times D)$.

2.2.6. Let A have m elements and B have n elements. Prove that $A \times B$ has mn elements. Prove that $A \times B$ has 2^{mn} subsets.

2.3   Relations, equivalence relations

In this section we introduce relations in a formal way. Most relations that we eventually analyze are of familiar kind, such as $\leq$, $<$, is cousin, taller than, has same birth date, etc., but we get more structure with a formal approach.

Definition 2.3.1. *A **relation on A and B** is any subset of $A \times B$.*

*A **relation on A** is a relation on A and A, i.e., a subset of $A \times A$.*

We can give relation a name, such as R, and in place of "$(a, b) \in R$" we alternatively write "aRb". (We prefer to write "$3 \leq 5$" rather than "$(3, 5) \in \leq$".)

Examples 2.3.2.

(1) Some relations on $\mathbb{R}$ are $\leq, <, =, \geq, >$. We can write $(1, 2) \in \leq$, or more familiarly, $1 \leq 2$. As a subset of $\mathbb{R} \times \mathbb{R}$, $\leq$ consists of all points on or above the line $y = x$. This relation can be drawn (and read off) easily.

(2) Draw anything in $\mathbb{R} \times \mathbb{R}$. That defines a relation on $\mathbb{R}$ (which most likely cannot be expressed with a formula). The relation $R =$

$\{(a, b) : a, b \in \mathbb{R}$ and $a^2 < b + 1\}$ is drawn as the set of all points (x, y) above the parabola $y = x^2 - 1$.

(3) The following are all the possible relations on $A = \{1, 2\}$ and $B = \{a, b\}$:

$$\{(1, a), (1, b), (2, a), (2, b)\},$$
$$\{(1, b), (2, a), (2, b)\},$$
$$\{(1, a), (2, a), (2, b)\},$$
$$\{(1, a), (1, b), (2, b)\},$$
$$\{(1, a), (1, b), (2, a)\},$$
$$\{(1, a), (1, b)\},$$
$$\{(1, a), (2, a)\},$$
$$\{(1, a), (2, b)\},$$
$$\{(1, b), (2, a)\},$$
$$\{(1, b), (2, b)\},$$
$$\{(2, a), (2, b)\},$$
$$\{(1, a)\},$$
$$\{(1, b)\},$$
$$\{(2, a)\},$$
$$\{(2, b)\},$$
$$\{\}.$$

(4) A relation can be known by more than one name. Among the students in a typical college classroom, the relations cousin and aunt are identical, i.e., no one student is a cousin or aunt of another, so this relation equals $\{\}$, the empty relation.

Definition 2.3.3. *Let R be a relation on A.*

(1) *R is **reflexive** if for all $a \in A$, aRa.*

(2) *R is **symmetric** if for all $a, b \in A$, aRb implies bRa.*

(3) *R is **transitive** if for all $a, b, c \in A$, if aRb, bRc, then aRc.*

(4) *R is an **equivalence relation** if it is reflexive, symmetric and transitive.*

Examples 2.3.4.

(1) $\leq$ on $\mathbb{R}$ is reflexive and transitive but not symmetric.

(2) $<$ on $\mathbb{R}$ is transitive but not reflexive or symmetric.

(3) $=$ on any set A is reflexive, symmetric, and transitive.

(4) Being a cousin is symmetric and neither reflexive nor transitive.

(5) Being taller than is ...

(6) Let $A = \{u, v\}$. Any equivalence relation on A needs to contain (u, u) and (v, v) in order to achieve reflexivity. The relation

$$\{(u, u), (v, v)\}$$

is reflexive, symmetric and transitive. The relations

$$\{(u, u), (v, v), (u, v)\} \text{ and } \{(u, u), (v, v), (v, u)\}$$

are reflexive and transitive but not symmetric. The relation

$$\{(u, u), (v, v), (u, v), (v, u)\}$$

is an equivalence relation. Thus there are exactly two equivalence relations on A.

Definition 2.3.5. *Let R be an equivalence relation on a set A. For each $a \in A$, the set of all elements b of A such that aRb is called the* **equivalence class of** a. *We denote the equivalence class of a with the shorthand $[a]$.*

For example, if R is the equality relation, then the equivalence class of a is $\{a\}$. If $R = A \times A$, then the equivalence class of any a is A. If A is the set of all students in Math 112 this year, and aRb if students a and b are in the same section of Math 112, then $[a]$ is the set of all students that are in the same section as student a.

Theorem 2.3.6. *Let R be an equivalence relation on a set A. Two equivalence classes are either identical or they have no elements in common.*

Proof. Let $a, b \in A$, and suppose that their equivalence classes have an element in common. Call the element c.

We now prove that the equivalence class of a is a subset of the equivalence class of b. Let d be any element in the equivalence class of a. Then aRd, aRc and bRc imply by symmetry that dRa and cRb, so that by transitivity dRc. Then dRc, cRb and transitivity give dRb, so that by symmetry

bRd, which says that d is in the equivalence class of b. Thus the equivalence class of a is a subset of the equivalence class of b.

A symmetric proof shows that the equivalence class of b is a subset of the equivalence class of a, so that the two equivalence classes are identical.

$\square$

Remark 2.3.7. What this says is that whenever R is an equivalence relation on a set A, then every element of A is in a unique equivalence class. Thus A is the **disjoint union** of distinct equivalence classes. Conversely, if $A = \cup_{i \in I} A_i$ where the A_i are pairwise disjoint, define $R \subseteq A \times A$ as $(a, b) \in R$ precisely if a and b are elements of the same A_i. Then R is an equivalence relation: reflexivity and symmetry are obvious, and for transitivity, suppose that a and b are in the same A_i and b and c are in the same A_j. Since A_i and A_j have the element b in common, by the pairwise disjoint assumption necessarily $i = j$, so that a and c are both in A_i. Thus R is an equivalence relation.

Example 2.3.8. Let $A = \{1, 2, 3, 4, 5\}$. The writing of A as $\{1, 2\} \cup \{3, 4\} \cup \{5\}$ makes the following equivalence relation on A:

$$\{(1, 1), (1, 2), (2, 1), (2, 2), (3, 3), (3, 4), (4, 3), (4, 4), (5, 5)\}.$$

This means is that counting all the possible equivalence relations on A is the same as counting all the possible writings of A as unions of pairwise disjoint subsets. (In contrast, the number of all possible relations on a set A equals the number of subsets of $A \times A$.)

Important example 2.3.9. Let n be a positive integer. Let R be the relation on $\mathbb{Z}$ given by aRb if $a - b$ is a multiple of n. This relation is called **congruence modulo n**. It is reflexive because for every $a \in \mathbb{Z}$, $a - a = 0$ is an integer multiple of n. It is symmetric because for all $a, b \in \mathbb{Z}$, if aRb, then $a - b = x \cdot n$ for some integer x, and so $b - a = (-x) \cdot n$, and since $-x$ is an integer, this proves that bRa. Finally, this relation R is transitive: let $a, b, c \in \mathbb{Z}$, and suppose that aRb and bRc. This means that $a - b = x \cdot n$ and $b - c = y \cdot n$ for some integers x and y. Then $a - c = a + (-b + b) - c = (a - b) + (b - c) = x \cdot n + y \cdot n = (x + y) \cdot n$, and since $x + y$ is an integer, this proves that aRc. Thus R is an equivalence relation. If aRb for this relation R, we say that **a is congruent to b modulo n**, or that **a is congruent to b mod n**. (Normally in the literature this is written as $a \equiv b \bmod n$.) We denote the set of all equivalence classes with $\mathbb{Z}/n\mathbb{Z}$, and we read this as **"Z mod n Z"**. This set consists of $[0], [1], [2], \ldots, [n-1], [n] = [0], [n+1] = [1]$, etc., so that $\mathbb{Z}/n\mathbb{Z}$ has **at most** n equivalence classes. Since any two numbers among $0, 1, \ldots, n-1$ have difference strictly between 0 and n, it follows that this difference is not an integer multiple of n, so that $[0], [1], [2], \ldots, [n-1]$ are distinct. Thus $\mathbb{Z}/n\mathbb{Z}$ has **exactly** n equivalence classes. Two natural lists of representatives of equivalence classes are $0, 1, 2, \ldots, n-1$ and $1, 2, \ldots, n$. (But there are infinitely many other representatives as well.)

For example, modulo 12, the equivalence class of 1 is the set $\{1, 13, 25, 37, \ldots\} \cup \{-11, -23, -35, \ldots\}$, and the equivalence class of 12 is the set of all multiples of 12 (including 0).

In everyday life we use congruence modulo 12 (or sometimes 24) for hours, modulo 12 for months, modulo 7 for days of the week, modulo 4 for seasons of the year, modulo 3 for meals of the day ...

There are exactly two equivalence classes for the congruences modulo 2: one consists of all the even integers and the other of all the odd integers. There is exactly one equivalence class for the congruences modulo 1: all integers are congruent modulo 1 to each other. For the congruences modulo 0, each equivalence class consists of precisely one element.

Example 2.3.10. (Construction of $\mathbb{Z}$ from $\mathbb{N}_0$.) Consider the Cartesian product $\mathbb{N}_0 \times \mathbb{N}_0$. Elements are pairs of the form (a, b), with $a, b \in \mathbb{N}_0$. If $a, b, c, d \in \mathbb{N}_0$, we write $(a, b)R(c, d)$ if $a + d = b + c$. Thus R is a relation on $\mathbb{N}_0 \times \mathbb{N}_0$. (Certainly you are familiar with $\mathbb{Z}$, in which you may want to think of this relation simply saying that $(a, b)R(c, d)$ if $a - b = c - d$. The problem is that in $\mathbb{N}_0$ we may not be able to subtract b from a and still get an element from $\mathbb{N}_0$.)

(1) R is reflexive: because for all $(a, b) \in \mathbb{N}_0 \times \mathbb{N}_0$, by commutativity of addition, $a + b = b + a$, so that by definition of R, $(a, b)R(a, b)$.

(2) R is symmetric: if $(a, b)R(c, d)$, then by definition $a + d = b + c$, so that by commutativity of addition, $d + a = c + b$, and by symmetry of the $=$ relation, $c + b = d + a$. But by definition this says that $(c, d)R(a, b)$.

(3) R is transitive: if $(a, b)R(c, d)$ and $(c, d)R(e, f)$, then by definition $a + d = b + c$ and $c + f = d + e$. It follows that $(a + d) + (c + f) =$

$(b+c) + (d+e)$. By associativity and commutativity of addition, $(a+f) + (c+d) = (b+e) + (c+d)$, and by cancellation then $a + f = b + e$, which says that $(a,b)R(e,f)$.

Now we define a set $\mathbb{Z}$ to be the set of equivalence classes for this relation. Every element (a,b) of $\mathbb{N}_0 \times \mathbb{N}_0$ is in an equivalence class: if $a \geq b$, then $(a,b)R(a-b,0)$, and if $a < b$, then $(a,b)R(0,b-a)$. Thus for each $(a,b) \in \mathbb{N}_0 \times \mathbb{N}_0$, there is an element in the equivalence class of (a,b) of the form $(0,e)$ or $(e,0)$ for some $e \in \mathbb{N}_0$, and it is left for the reader to verify that this e is unique. In Definition 3.5.2 and beyond we identify the set of equivalence classes with the usual integers in $\mathbb{Z}$ as follows: the equivalence class of $(e,0)$ corresponds to the non-negative integer e and the equivalence class of $(0,e)$ corresponds to the non-positive integer $-e$. In Section 3.5 we develop the arithmetic on these equivalence classes (and hence on $\mathbb{Z}$) from the arithmetic on $\mathbb{N}_0$.

Exercises for Section 2.3

2.3.1. Let $A = \{a,b\}$ and $B = \{b,c,d\}$.

   i) How many elements are in the set $A \times A$?

   ii) How many elements are in the set $A \times B$?

   iii) How many elements are in the set $B \times B$?

   iv) How many relations are there on A? How many relations are there on B? (Recall Definition 2.3.1 for the definition of relation.)

   v) How many relations are there on A and B? How many relations are there on B and A?

   vi) How many relations are there on $A \cup B$ and $A \cap B$?

2.3.2. Let A have n elements and B have m elements. How many distinct relations on A and B are there? Prove.

2.3.3. In each part below, find a relation with the given properties. You may contrive a relation on a contrived set A.

   i) Reflexive, but not symmetric and not transitive.

   ii) Reflexive and symmetric, but not transitive.

   iii) Reflexive and transitive, but not symmetric.

   iv) Symmetric, but not reflexive and not transitive.

   v) Transitive, but not symmetric and not reflexive.

   vi) Transitive and symmetric, but not reflexive.

2.3.4. Let A be a set with 2 elements. Count all equivalence relations on A. Repeat first for A with 3 elements, then for A with 4 elements.

2.3.5. Let A be a set with n elements. Let R be an equivalence relation on A with fewest members. How many members does R have?

2.3.6. Let R be the relation on $\mathbb{R}$ given by aRb if $a - b$ is a rational number. Prove that R is an equivalence relation. Find at least three disjoint equivalence classes.

2.3.7. Let R be the relation on $\mathbb{R}$ given by aRb if $a - b$ is an integer.

i) Prove that R is an equivalence relation.

ii) Prove that for any $a \in \mathbb{R}$ there exists $b \in [0, 1)$ such that $[a] = [b]$.

2.3.8. Let R be a relation on $\mathbb{R} \times \mathbb{R}$ given by $(a, b)R(c, d)$ if and only if $a - c$ and $b - d$ are integers.

i) Prove that R is an equivalence relation.

ii) Prove that for any $(a, b) \in \mathbb{R} \times \mathbb{R}$ there exists $(c, d) \in [0, 1) \times [0, 1)$ such that $[(a, b)] = [(c, d)]$.

iii) Prove that the set of equivalence classes can be identified with $[0, 1) \times [0, 1)$.

iv) For fun: check out the video game Asteroids online for a demonstration of this equivalence relation. Do not get addicted to the game.

2.3.9. Let A be the set of all lines in the plane.

i) Prove that the relation "is parallel to" is an equivalence relation. Note that the equivalence class of a non-vertical line can be identified by the (same) slope of the lines in that class. Note that the vertical lines are in their own equivalence class.

ii) Prove that the relation "is perpendicular to" is not an equivalence relation.

†**2.3.10.** (Invoked in Section 3.6.) For $(a, b), (a', b') \in \mathbb{Z} \times (\mathbb{Z} \setminus \{0\})$, define $(a, b) \wr (a', b')$ if $a \cdot b' = a' \cdot b$.

i) Prove that $\wr$ is an equivalence relation. (Possibly mimic Example 2.3.10.)

ii) Describe the equivalence classes of $(0, 1)$, $(1, 1)$, $(2, 3)$?

iii) Find a natural identification between the equivalence classes and elements of $\mathbb{Q}$.

2.4   Functions

Here is the familiar definition of functions:

Definition 2.4.1. *Let A and B be sets. A **function from A to B** is a rule that assigns to each element of A a unique element of B. We express this with "$f : A \to B$ is a function." The set A is the **domain** of f and B is the **codomain** of f. The **range** or **image** of f is* $\mathrm{Image}(f) = \mathrm{Range}(f) = \{b \in B : b = f(a) \text{ for some } a \in A\}$.

But in the spirit of introducing few new notions, let's instead define functions with the concepts we already know. Convince yourself that the two definitions are the same:

Definition 2.4.2. *Let A and B be sets. A relation f on A and B is a **function** if for all $a \in A$ there exists $b \in B$ such that $(a, b) \in f$ and if for all $(a, b), (a, c) \in f$, $b = c$. In this case we say that A is the **domain** of f, B is the **codomain** of f, and we write $f : A \to B$. The **range** of f is* $\mathrm{Range}(f) = \{b \in B : \text{there exists } a \in A \text{ such that } (a, b) \in f\}$.

Note that this second formulation is also familiar: it gives us all elements of the **graph** of the function: $b = f(a)$ if and only if (a, b) is on the graph of f. We freely change between notations $f(a) = b$ and $(a, b) \in f$.

One should be aware that if f is a function, then $f(x)$ is an element of the range and is not by itself a function. (But often we speak loosely of $f(x)$ being a function, such as "x^2 is a function".)

To **specify** a function one needs to present its domain and its codomain, and to show what the function does to each element of the domain.

Examples 2.4.3.

(1) A function can be given with a **formula**.
For example, let $f : \mathbb{R} \to \mathbb{R}$ be given by $f(x) = \frac{1}{1+x^2}$. The range is $(0, 1]$: For all x, $1 + x^2 \geq 1$ with equality when $x = 0$. Thus $f(x) \in (0, 1]$. For any $y \in (0, 1]$, $1/y \geq 1$, so $1/y - 1 \geq 0$, so $x = \sqrt{\frac{1}{y} - 1}$ is a positive real number and $f(x) = y$. So indeed the range of f is $(0, 1]$.

(2) Here are formula definitions of two functions with domains $[0, \infty)$: $f(x) = \sqrt{x}$ and $g(x) = -\sqrt{x}$. Note, however, that $h(x) = \pm\sqrt{x}$ is NOT a function!

(3) There may be more than one formula for a function, each of which is applied to distinct elements of the domain. For example, define $f : \mathbb{N}^+ \to \mathbb{Z}$ by

$$f(n) = \begin{cases} \frac{n-1}{2}, & \text{if } n \text{ is odd;} \\ -\frac{n}{2}, & \text{if } n \text{ is even.} \end{cases}$$

(4) Let $f : \mathbb{N}^+ \to \mathbb{R}$ be given by the **description** that $f(n)$ equals the nth prime. By Euclid's theorem (proved on page 25) there are infinitely many primes so that f is indeed defined for all positive integers. We know that $f(1) = 2$, $f(2) = 3$, $f(3) = 5$, and with computer's help I get that $f(100) = 541$, $f(500) = 3571$. There is no algebraic formula for the nth prime.

(5) For any set A, the **identity function** $\text{id}_A : A \to A$ takes each x to itself.

(6) Let $b \in B$. A function $f : A \to B$ given by $f(a) = b$ for all a is called a **constant** function.

(7) The constant function $f : \mathbb{R} \to \mathbb{R}$ given by $f(x) = 1$ for all x is not the identity function.

(8) A function may be presented by a **table**. Here is an example.

x	$f(x)$
1	1
2	1
3	2

(9) A function may be presented in a pie chart, histogram, with words, in a weather map... .

(10) A function $f : \mathbb{N}^+ \to \mathbb{R}$ can be given recursively, such as the Fibonacci numbers $f(1) = 1$, $f(2) = 1$, and for all $n \geq 2$, $f(n+1) = f(n) + f(n-1)$. See Exercise 1.6.30 for more on these numbers.

(11) A function can be given by its **graph**:

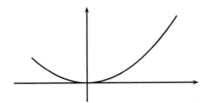

From this particular graph we surmise that $f(0) = 0$, but with the precision of the drawing and our eyesight it might be the case that $f(0) = 0.000000000004$. Without any further labels on the axes we cannot estimate the numerical values of f at other points. Typically the graph should be filled in with more information. The arrows on the graphs indicate increasing values.

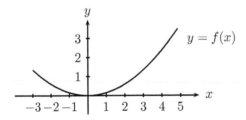

(12) A function may be presented with Venn diagrams and arrows:

Remark 2.4.4. Two functions are the same if they have the same domains, the same codomains, and if to each element of the domain they assign the same element of the codomain.

For example, $f : \mathbb{R} \to \mathbb{R}$ and $g : \mathbb{R} \to [0, \infty)$ given by $f(x) = x^2$ and $g(x) = x^2$ are not the same! On the other hand, the functions $h, k : \mathbb{R} \to \mathbb{R}$ given by $h(x) = |x|$ and $k(x) = \sqrt{x^2}$ are the same.

Notation 2.4.5. It is common to not specify the domain, in which case the domain is implicitly the largest possible subset of $\mathbb{R}$ on which the function is defined. For example, the domain of $f(x) = \frac{1}{x}$ is the set of all non-zero real numbers, and the domain of $f(x) = \sqrt{x}$ is the set of all non-negative real numbers. (After we introduce complex numbers the domain will implicitly be the largest possible subset of $\mathbb{C}$ on which the function is defined, see page 154.)

Sometimes instead we want to take smaller domains than possible:

Definition 2.4.6. *Let $f : A \to B$ be a function and let C be a subset of A. The function $g : C \to B$ defined by $g(c) = f(c)$ for all $c \in C \subseteq A$ is called* the restriction of f to C *or* f restricted to C. *It is commonly written as $g = f|_C$.*

For example, let $f : \mathbb{R} \to \mathbb{R}_{\geq 0}$ be given as $f(x) = \sqrt{x^2}$. Then f restricted to $\mathbb{R}_{\geq 0}$ equals the identity function and f restricted to $\{-2, 2\}$ equals a constant function.

Definition 2.4.7. *Let A, B, C, D be sets with $B \subseteq C$. If $f : A \to B$ and $g : C \to D$, then* the composition g composed with f *is a function* $\mathbf{g} \circ \mathbf{f} : A \to D$ *such that for all $a \in A$, $(g \circ f)(a) = g(f(a))$.*

If $f(x) = (x+1)^2$ and $g(x) = x^3 - 1$, then

$$(g \circ f)(x) = g(f(x)) = g((x+1)^2) = ((x+1)^2)^3 - 1 = (x+1)^6 - 1,$$

$$(f \circ g)(x) = f(g(x)) = f(x^3 - 1) = (x^3 - 1 + 1)^2 = x^6,$$

$$g(x)f(x) = (x^3 - 1)(x+1)^2,$$

and these three outcomes are distinct. For example, when plugging in $x = 1$, we get values $(g \circ f)(1) = 63$, $(f \circ g)(1) = 1$, $g(1)f(1) = 0$.

Remark 2.4.8. It is common to write $f^2 = f \circ f$, $f^3 = f^2 \circ f = f \circ f \circ f$, etc. Some exceptions to this notation are established in trigonometry for historical reasons: "$\sin^2(x)$" stands for "$(\sin x)^2$" and not for "$\sin(\sin x)$". If we discovered trigonometry today, then $\sin^2(x)$ would have the latter meaning.

Example 2.4.9. Let $f(x) = \sqrt{x}$, and $g(x) = x^2$. The domain of f is $\mathbb{R}_{\geq 0}$, and the domain of g is $\mathbb{R}$. It is possible to compose $g \circ f : \mathbb{R}_{\geq 0} \to \mathbb{R}$ to obtain $(g \circ f)(x) = x$, but when composing f and g in the other order, $(f \circ g)(x)$ is not always equal to x. Namely, for any negative number x, $(f \circ g)(x) = -x$.

We just demonstrated that in composing two functions, the order matters.

Definition 2.4.10. *A function* $f : A \to B$ *is* **injective** *(or* **one-to-one***) if for all* $a, a' \in A$, $f(a) = f(a')$ *implies* $a = a'$. *In other words,* f *is injective if whenever two things map via* f *to one object, then those two things are actually the same.*

A function $f : A \to B$ *is* **surjective** *(or* **onto***) if for all* $b \in B$, *there exists* $a \in A$ *such that* $f(a) = b$. *In other words,* f *is surjective if every element of the codomain is mapped onto via* f *by some element of the domain.*

A function $f : A \to B$ *is* **bijective** *if it is injective and surjective.*

For example, the identity function is always bijective. Constant functions are injective when the domain consists of one element, and are surjective when the codomain consists of one element. The function $f : \mathbb{R} \to \mathbb{R}$ given by $f(x) = x^2$ is not injective because $f(-1) = 1 = f(1)$. It is, however, surjective. The function $f : \mathbb{R}_{\geq 0} \to \mathbb{R}$ given by $f(x) = \sqrt{x}$ is injective because the square root function is strictly increasing (more about that in Section 2.9), but it is not surjective because -1 is not the square root of any non-negative real number. The function $f : \mathbb{R}_{\geq 0} \to \mathbb{R}_{\geq 0}$ given by $f(x) = \sqrt{x}$ is both injective and surjective, thus bijective.

The following are all the possible functions $f : \{1, 2\} \to \{1, 2\}$, and they are given in tabular form:

x	$f(x)$	x	$f(x)$	x	$f(x)$	x	$f(x)$
1	1	1	1	1	2	1	2
2	1	2	2	2	1	2	2

The first and the last are neither injective nor surjective, but the middle two are bijective.

The following are the eight possible functions $f : \{1, 2, 3\} \to \{1, 2\}$:

x	$f(x)$	x	$f(x)$	x	$f(x)$	x	$f(x)$	x	$f(x)$	x	$f(x)$	x	$f(x)$	x	$f(x)$
1	1	1	1	1	1	1	1	1	2	1	2	1	2	1	2
2	1	2	1	2	2	2	2	2	1	2	1	2	2	2	2
3	1	3	2	3	1	3	2	3	1	3	2	3	1	3	2

In this case no functions are injective, and all non-constant ones are surjective.

Theorem 2.4.11. *The composition of two injective functions is injective.*

Proof. Let $f : A \to B$ and $g : C \to D$ be injective functions, and suppose that $B \subseteq C$ so that we can compose the two functions.

If $(g \circ f)(a) = (g \circ f)(a')$, then $g(f(a)) = g(f(a'))$. Since g is injective, it follows that $f(a) = f(a')$, and since f is injective, it follows that $a = a'$. Thus $g \circ f$ is injective. $\square$

Theorem 2.4.12. *If $f : A \to B$ and $g : B \to C$ are surjective functions, so is $g \circ f : A \to C$. If f and g are both bijective, so is $g \circ f$.*

Proof. Let $c \in C$. Since g is surjective, there exists $b \in B$ such that $g(b) = c$. Since f is surjective, there exists $a \in A$ such that $f(a) = b$. Thus $(g \circ f)(a) = g(f(a)) = g(b) = c$, so that $g \circ f$ is surjective.

The last statement follows from the first part and Theorem 2.4.11. $\square$

Definition 2.4.13. *We have seen polynomial functions in Section 1.5: recall that for any subset $S \subseteq \mathbb{R}$, a function $f : S \to \mathbb{R}$ is a **polynomial function** if there exist a non-negative integer n and $c_0, c_1, \ldots, c_n \in \mathbb{R}$ such that for all $x \in S$, $f(x) = c_0 + c_1 x + c_2 x^2 + \cdots + c_n x^n$. A function $f : S \to \mathbb{R}$ is a **rational function** if there exist polynomial functions $f_1, f_2 : S \to \mathbb{R}$ such that for all $x \in S$, $f_2(x) \neq 0$ and $f(x) = f_1(x)/f_2(x)$.*

Similarly there are polynomial and rational functions if all occurrences of "$\mathbb{R}$" above are replaced by "$\mathbb{Q}$" or "$\mathbb{C}$".

Polynomial and rational functions are a workhorse of analysis. Below are some special properties. Further special properties of polynomial functions in greater generality appear in Exercise 2.6.13. (The reader has of course encountered trigonometric, exponential, and logarithmic functions, which are not polynomial or rational.)

Definition 2.4.14. *Let* $f : S \to \mathbb{R}$ *be a function. A* **zero** *or a* **root** *of* f *is any* $a \in S$ *such that* $f(a) = 0$.

(This meaning of "root" is different from the meaning of "root" in "square root", "cube root" or "100th root" of a number.)

Theorem 2.4.15. *If a polynomial function with real coeffiecients is not constant zero, then it has only finitely many roots. Specifically, if* $f(x) = c_0 + c_1 x + c_2 x^2 + \cdots + c_n x^n$, *then the number of roots is at most* n.

The domain of a rational function is the complement of a finite subset of $\mathbb{R}$.

(The same statement and proof work if the coefficients are complex numbers; complex numbers are introduced in Chapter 3.)

Proof. At least one of c_i is non-zero since $f(x)$ is not constant 0. By possibly renaming we may assume that $c_n \neq 0$, and by the non-constant assumption, $n \geq 1$. If $n = 1$, then f has only one root, namely $-c_0/c_1$. Suppose that $n \geq 2$.* In general, let a be any root of f. By the Euclidean algorithm (Example 1.6.5), there exist polynomial functions q and r such that

$$f(x) = q(x)(x - a) + r(x),$$

and the degree of r is strictly smaller than 1, i.e., $r(x)$ is a constant. But if we plug $x = a$ into both sides, since a is a root of f, we get that r is the constant zero function, so that

$$f(x) = q(x)(x - a).$$

---

* If $n = 2$, then by the **quadratic formula** the roots of $c_0 + c_1 x + c_2 x^2$ are $\frac{-c_1 \pm \sqrt{c_1^2 - 4c_0 c_2}}{2c_2}$. There are root formulas for cases $n = 3, 4$, but executing them is very time-consuming, they require familiarity with complex numbers, the solutions involve sums of cube roots with a few square roots thrown in for good measure, and furthermore it can be hard to identify that the ensuing long expression simplifies to a nice root such as 2. This, and the existence of computer capabilities, are the reasons that we do not teach such formulas. The formula for solutions of cubic polynomials was discovered by Niccolò Fontana Tartaglia (1500–1557) and for quartic ones by Lodovico Ferrari (1522–1565). Both formulas were popularized in a book by Gerolamo Cardano (1501–1576). There are no formulas for general polynomials of degree $n \geq 5$: not only do you and I do not know such a formula, but Niels Henrik Abel (1802–1829) and Évariste Galois (1811–1832) proved that no formulas exist using only radicals, sums, products, quotients.

We write $q(x) = b_0 + b_1 x + \cdots + b_m x^{m-1}$. By multiplying $q(x)(x-a)$ out we get that $m = n - 1$. By induction, $q(x)$ has at most $n - 1$ roots. If b is a root of f, then

$$0 = f(b) = q(b)(b-a),$$

so that either $q(b) = 0$ of $b - a = 0$. Thus b is either a root of q or $b = a$. Thus the roots of f are a and any of the at most $n - 1$ roots of q, so that f has at most n roots. This proves the first part.

A rational function is a quotient of two polynomial functions, and the rational function is defined everywhere except where the denominator is 0. By the first part this excludes only finitely many numbers.   □

This theorem is useful in that it assures us that the domain of a rational function contains infinitely many points, or even better, that the domain is all except finitely many real (or complex) numbers.

(You are aware that the trigonometric functions sine and cosine have infinitely many zeroes and that tangent and cotangent are not defined at infinitely many real numbers. This fact, together with the previous theorem, establishes that these four trigonometric functions are not polynomial or rational functions. For similar reasons the logarithmic functions are not polynomial or rational. We have to work harder to prove that the exponential functions are not polynomial or rational.)

Exercises for Section 2.4

2.4.1. Fix a positive integer n. Define $f : \mathbb{Z} \to \{0, 1, 2, 3, \ldots, n-1\}$ by $f(a)$ is the remainder after a is divided by n. In number theory instead of "$f(a) = b$" one says **a mod n** is b. We can also define a similar (but not the same!) function $g : \mathbb{Z} \to \mathbb{Z}/n\mathbb{Z}$ by $g(a) = [a]$. Graph the function f for $n = 2$ and $n = 3$.

2.4.2. Define the **floor function** $\lfloor \ \rfloor : \mathbb{R} \to \mathbb{R}$ to be the function such that for all $x \in \mathbb{R}$, $\lfloor x \rfloor$ is the largest integer that is less than or equal to x. The range is $\mathbb{Z}$. For example, $\lfloor \pi \rfloor = 3$, $\lfloor 1 \rfloor = 1$, $\lfloor -1.5 \rfloor = -2$. Graph this function.

2.4.3. The **ceiling function** $\lceil \ \rceil : \mathbb{R} \to \mathbb{R}$ is the function for which $\lceil x \rceil$ is the smallest integer that is greater than or equal to x. The range is $\mathbb{Z}$. For example, $\lceil \pi \rceil = 4$, $\lceil 1 \rceil = 1$, $\lceil -1.5 \rceil = -1$. Graph this function.

2.4.4. Let $A = [-a, a]$ or $A = (-a, a)$, and let B be a subset of $\mathbb{R}$ (or of $\mathbb{C}$). A function $f : A \to B$ is an **odd function** (resp. **even function**) if for all $x \in A$, $f(-x) = -f(x)$ (resp. $f(-x) = f(x)$). Let n be a positive integer, $c_0, c_1, \ldots, c_n$ real (or complex) numbers, and f the polynomial function $f(x) = c_0 + c_1 x + c_2 x^2 + \cdots + c_n x^n$.

   i) Suppose that for all even k, $c_k = 0$. Prove that f is an odd function.

   ii) Suppose that for all odd k, $c_k = 0$. Prove that f is an even function.

2.4.5. Construct a set A and functions $f, g : A \to A$ such that $f \circ g \neq g \circ f$.

2.4.6. Let $h : A \to B$, $g : B \to C$, and $f : C \to D$. Prove that $f \circ (g \circ h) = (f \circ g) \circ h$.

2.4.7. Let $f : \mathbb{R} \times \mathbb{R} \to \mathbb{R}$ be given by $f(x, y) = x$ and $g : \mathbb{R} \to \mathbb{R} \times \mathbb{R}$ by $g(x) = (x, 0)$.

   i) Compute the domain, codomain and algebraic formula for $f \circ g$. Repeat for $g \circ f$.

   ii) Which among $f, g, f \circ g, g \circ f$ are injective, surjective, bijective?

2.4.8. Let $A = \{1, 2\}$, $B = \{2, 4\}$, and $C = \{3, 6, 9\}$. Mark each of the following as a function $A \to B$, $B \to C$, or $C \to A$.

   i) $f = \{(2, 6), (4, 3)\}$.

   ii) $g = \{(9, 1), (6, 2), (3, 2)\}$.

   iii) $h = \{(1, 4), (2, 2)\}$.

2.4.9. For each of the following functions, state the domains, codomains, and how they are compositions of f, g, h from the previous exercise.

   i) $\{(2, 2), (4, 2)\}$.

   ii) $\{(9, 4), (3, 2), (6, 2)\}$.

   iii) $\{(2, 6), (1, 3)\}$.

2.4.10. Let A be a set with 3 elements and B be a set with 2 elements.

   i) Count all functions from A to B.

   ii) Count all injective functions from A to B.

   iii) Count all surjective functions from A to B. Repeat for functions from B to A.

   iv) Count all bijective functions from A to B.

   v) Count all functions from A to B that are injective but not surjective.

   vi) Count all functions from A to B that are surjective but not injective.

   vii) Count all functions from A to B that are neither surjective nor injective.

2.4.11. In each part below, find $f : \mathbb{R} \to \mathbb{R}$ with the specified condition.

i) f is a bijective function.

ii) f is neither injective nor surjective.

iii) f is injective, but not surjective.

iv) f is surjective, but not injective.

2.4.12. The **pigeonhole principle** states that if n items (such as pigeons) are put into m holes with $n > m$, then at least one hole has more than one item. Let A and B be sets with only finitely many elements.

i) Use the pigeonhole principle to demonstrate that if A has more elements than B, then $f : A \to B$ cannot be injective.

ii) Use the pigeonhole principle to demonstrate that if A has fewer elements than B, then $f : A \to B$ cannot be surjective.

iii) Use the pigeonhole principle to demonstrate that if A and B do not have the same number of elements, then $f : A \to B$ cannot be bijective.

2.4.13. Let A be a set with m elements and B a set with n elements. (To solve this problem, you may want to first examine the case $m = 1$, then $m = 2$, followed by $m = 3$, or possibly you have to start with small n, after which you will probably see a pattern. Once you have the correct pattern, the proof is straightforward.)

i) Count all functions from A to B.

ii) For which combinations of m, n are there injective functions from A to B?

iii) For m, n as in (ii), count all injective functions from A to B.

iv) For which combinations of m, n are there surjective functions from A to B?

v) For m, n as in (iv), count all surjective functions from A to B.

vi) For which combinations of m, n are there bijective functions from A to B?

vii) For m, n as in (vi), count all bijective functions from A to B.

2.4.14. Find an injective function $f : \mathbb{N}^+ \to \mathbb{N}^+$ that is not surjective. Find a surjective function $g : \mathbb{N}^+ \to \mathbb{N}^+$ that is not injective. Compare with parts (ii) and (iv) of the previous exercise. Does the pigeonhole principle apply?

2.4.15. Let $f : A \to B$ and $g : B \to C$ be functions.

i) Suppose that $g \circ f$ is injective. Prove that f is injective.

ii) Suppose that $g \circ f$ is surjective. Prove that g is surjective.

iii) Give an example of f, g such that $g \circ f$ is injective but g is not injective.

iv) Give an example of f, g such that $g \circ f$ is surjective but f is not surjective.

2.4.16. Find functions $f, g : \mathbb{R} \to \mathbb{R}$ such that f is surjective but not injective, g is injective but not surjective, and $f \circ g$ is bijective.

2.4.17. Prove that if f and g are both injective functions, then $f \circ g$ is also injective.

2.4.18. Prove that if f and g are both surjective functions, then $f \circ g$ is also surjective.

2.4.19. Prove that if f and g are both bijective functions, then $f \circ g$ is also bijective.

2.4.20. Suppose $f : A \to B$. We introduce the notation $f(S) = \{f(x) : x \in S\}$ for any subset S of A. Let C be an arbitrary subset of A.

i) Prove $f(A) \setminus f(C) \subseteq f(A \setminus C)$.

ii) With proof, what condition makes $f(A) \setminus f(C) = f(A \setminus C)$?

2.4.21. For the following polynomial and rational functions, determine the domains (largest sets on which the function is defined):

i) $f(x) = x^2 - \frac{1}{x^3 - x}$.

ii) $f(x) = \frac{x^4 - 2x^3 - 1}{x^6 + 2}$.

iii)* $f(x) = \frac{1}{x^5 - 17x^2 + \pi x - 4}$. (Gotcha! We do not know the roots of this denominator.)

2.4.22. Prove that $f : \mathbb{R} \to \mathbb{R}$ given by $f(x) = x^3 - 1$ is surjective and injective.

2.4.23. Prove that the composition of polynomial functions $f, g : \mathbb{R} \to \mathbb{R}$ is polynomial.

2.4.24. Prove that the composition of rational functions $f, g : \mathbb{R} \to \mathbb{R}$ is rational.

2.4.25. A function $f : A \to B$ is called **invertible** if there exists $g : B \to A$ such that $g \circ f = \mathrm{id}_A$ and $f \circ g = \mathrm{id}_B$.

    i) Suppose that $g, h : B \to A$ are such that $g \circ f = \mathrm{id}_A = h \circ f$ and $f \circ g = \mathrm{id}_B = f \circ g$. Prove that $g = h$. Such g is called the **inverse** of f.

    ii) Prove that a function is invertible if and only if it is bijective.

2.4.26. Let $f : \mathbb{R} \to \mathbb{R}$ be given by $f(x) = x^3 - 5$. Prove that f is invertible, find its inverse, and verify that the inverse is not a polynomial function.

2.4.27. (Lagrange interpolation) Let $c_1, \ldots, c_n$ be distinct real (or complex) numbers. For $j \in \{1, 2, \ldots, n\}$ set

$$g_j(x) = \frac{(x - c_1)(x - c_2) \cdots (x - c_{j-1})(x - c_{j+1}) \cdots (x - c_n)}{(c_j - c_1)(c_j - c_2) \cdots (c_j - c_{j-1})(c_j - c_{j+1}) \cdots (c_j - c_n)}.$$

    i) Prove that g_j is a polynomial function and that

$$g_j(c_i) = \begin{cases} 1, & \text{if } j = i; \\ 0, & \text{otherwise.} \end{cases}$$

    ii) Prove that for any real (or complex) numbers $d_1, \ldots, d_n$, there exists a polynomial g of degree at most $n - 1$ such that for all $i = 1, \ldots, n$, $g(c_i) = d_i$.

2.4.28. Let $f : \mathbb{N}^+ \to \mathbb{Z}$ and $g : \mathbb{N}^+ \to \mathbb{Q}^+$ be defined by

$$f(n) = \begin{cases} \frac{n-1}{2}, & \text{if } n \text{ is odd;} \\ -\frac{n}{2}, & \text{if } n \text{ is even,} \end{cases}$$

and for any positive integer with prime factorization $p_1^{e_1} \cdots p_k^{e_k}$ let $g(p_1^{e_1} \cdots p_k^{e_k}) = p_1^{f(e_1)} \cdots p_k^{f(e_k)}$. For example, the prime factorization of 1 has all e_i equal to 0, so $g(1) = 1$.

    i) Compute $f(1), f(2), f(3), f(4), f(5), f(6)$.

    ii) Compute $g(1), g(2), g(3), g(4), g(5), g(6)$.

    iii) Prove that f is bijective.

    iv) Assuming/knowing that prime factorization of positive integers is unique up to order, prove that g is bijective.

(This proof that $\mathbb{N}^+$ and $\mathbb{Q}^+$ have "the same infinite number" of elements is from the article *Counting the rationals* by Y. Sagher, in *American Mathematical Monthly* **96** (1989), page 823. Another proof of this size equality appears on page 315.)

2.4.29. When defining a function, one has to pay attention that it is **well-defined**. Read through examples below showing how a definition can go wrong. As a result, the purported definition of a function does not define a function at all.

  i) The codomain is not large enough, say for $f : \{1, 2\} \to \{1, 2\}$ with $f(x) = x^2$.

  ii) The function is not defined for all elements in the specified domain, say for $f : \mathbb{R} \to \mathbb{R}$ that is the inverse of the squaring function or for $f : \mathbb{R} \to \mathbb{R}$ that returns the multiplicative inverse.

  iii) The function does not return an unambiguous value. For example, let $f : \mathbb{Z}/12\mathbb{Z} \to \mathbb{Z}$ be defined by $f([a]) = a$. Since $[0] = [12]$, then $0 = f([0]) = f([12]) = 12$, but the integer 0 is not equal to the integer 12.

  iv) The values may not be known to exist as specified. For example, let $f : \{1\} \to \mathbb{Z}$ be defined as $f(1)$ is the smallest even integer greater than 4 that cannot be written as the sum of two primes. The reason why this function is not well-defined is that according to the Goldbach conjecture (see page 2) we do not know whether such a number exists.

  v) Let $f : \{1\} \to \mathbb{Z}$ be defined as $f(1)$ is the smallest prime number bigger than $10^{10^{10}}$. We know from page 25 that such a prime exists, so this f is well-defined as a function. However, we do not know its numerical value due to computational limitations, which makes this function useless for some purposes.

2.4.30. Let A be the set of all differentiable functions $f : \mathbb{R} \to \mathbb{R}$. Define a relation R on A by fRg if $f(0) = g(0)$.

  i) Prove that R is an equivalence relation.

  ii)* Let S be the set of all equivalence classes. Define $F : S \to \mathbb{R}$ by $F([f]) = f(0)$. Prove that F is a well-defined function, in the sense that if $[f] = [g]$, then $F([f]) = F([g])$. Also prove that F is bijective.

2.5   Binary operations

Definition 2.5.1. A **binary operation** *on a set A is a function* $\circ : A \times A \to A$. *An arbitrary element of* $A \times A$ *is an ordered pair* (a, b) *for* $a, b \in A$, *and when we plug that element into the function* $\circ$, *we should perhaps write* $\circ((a, b))$, *but normally one set of parentheses is removed, so we write* $\circ(a, b)$. *But depending on the exact form of* $\circ$, *we often traditionally write* $a \circ b$ *rather than* $\circ(a, b)$ *(see examples below).*

Binary operations that we have seen are:

$$+, -, \cdot, /, \times, \circ, \cap, \cup, \text{ or }, \text{ and }, \text{ xor}, \Longrightarrow, \Longleftrightarrow \ .$$

Examples and non-examples 2.5.2.
(1) $+, \cdot$ on $\mathbb{N}$, $\mathbb{Z}$, $\mathbb{Q}$, $\mathbb{R}$ (we of course write $a + b$ rather than $+(a, b)$ and $a \cdot b$ or even ab rather than $\cdot(a, b)$);
(2) Subtraction $-$ on $\mathbb{Z}$, $\mathbb{Q}$, $\mathbb{R}$ (but not on $\mathbb{N}$);
(3) Division $/$ on $\mathbb{Q} \setminus \{0\}$, $\mathbb{R} \setminus \{0\}$ (but not on $\mathbb{Q}$, $\mathbb{R}$, $\mathbb{N}^+$, $\mathbb{Z} \setminus \{0\}$);
(4) The additive inverse operation $-$ is not a binary operation on $\mathbb{R}$. (It is **unary**, i.e., it acts on one number at a time.)
(5) The multiplicative inverse operation $\_^{-1}$ is not a binary operation on $\mathbb{R} \setminus \{0\}$. (It is **unary**.)
(6) Let S be a set, and let A be the set of all functions $f : S \to S$. The composition (of functions) is a binary operation on A.
(7) $\cap, \cup, \setminus$ are binary operations on the set of subsets of a set S.
(8) and & or are binary operations on the set of logical statements.

Definition 2.5.3. *Let* $\circ$ *be a binary operation on a set* A. *An element* $e \in A$ *is an* **identity element** *for* $\circ$ *if for all* $a \in A$, $a \circ e = a = e \circ a$.

(In this chapter we use "e" for the identity element; this is a symbol that is unrelated to the base of the exponential function as in Definition 7.6.6.)

Examples 2.5.4. An identity (or is it *the* identity?) for $+$ on $\mathbb{Z}, \mathbb{Q}, \mathbb{N}_0, \mathbb{R}$ is 0. An identity for multiplication on $\mathbb{Z}, \mathbb{Q}, \mathbb{N}_0, \mathbb{R}$ is the number 1.

An identity for composition of functions from a set A to itself is the identity function id_A (the function with $id_A(x) = x$ for all $x \in A$).

Subtraction on $\mathbb{Z}$ does not have an identity element because if e were an identity element, then $1 - e = 1 = e - 1$, which says that $e = 0$ and $e = 2$, which is nonsense.

If S is a set and A is the collection of all subsets of S, then S is the identity element for $\cap$ and $\emptyset$ is the identity element for $\cup$. The binary operation $\setminus$ on A does not have an identity unless the only element of A is the empty set.

The theorem below resolves the problem of "an identity" versus "the identity".

Theorem 2.5.5. *Let $\circ$ be a binary operation on A. Suppose that e and f are both identities for $\circ$. Then $e = f$. In other words, if an identity exists for a binary operation, it is unique. Hence we talk about* **the** *identity for $\circ$.*

Proof. Since for all $a \in A$, $e \circ a = a$, we get in particular that $e \circ f = f$. Also, for every $a \in A$, $a \circ f = a$, hence $e \circ f = e$. Thus $e = e \circ f = f$. $\square$

Note: we used symmetry and transitivity of the equality relation.

Definition 2.5.6. *Let $\circ$ be a binary operation on A and suppose that e is its identity. Let x be an element of A. An* **inverse** *of x is an element $y \in A$ such that $x \circ y = e = y \circ x$. To emphasize what the operation is, we may also say that y is a $\circ$-***inverse** *of x (or see specific terms below).*

Examples and non-examples 2.5.7.
(1) Let $\circ = +$ on $\mathbb{Z}$. Then 0 is the identity element and every element has an **additive inverse**.
(2) Let $\circ = +$ on $\mathbb{N}_0$. Then 0 is the identity element and only 0 has an inverse.
(3) Let $\circ = \cdot$ on $\mathbb{Q}$. Then 1 is the identity element and every non-zero element has a **multiplicative inverse**.
(4) Let $\circ = \cdot$ on $\mathbb{Z}$. Then 1 is the identity element and only ± 1 have inverses.
(5) If S is a set and A is the collection of all subsets of S, then only S has an inverse for $\cap$, the inverse being S itself, and only $\emptyset$ has an inverse for $\cup$, the inverse being $\emptyset$.

(6) Here is a new binary operation $\circ$ on the set $S = \{a, b, c, d\}$ presented via its $\circ$-**table**:

$\circ$	a	b	c	d
a	a	b	c	d
b	b	c	d	a
c	c	d	a	b
d	d	a	b	c

Note that a is the identity element, the inverse of a is a, the inverse of b is d, the inverse of c is c, the inverse of d is b.

Definition 2.5.8. *A binary operation $\circ$ on A is* **associative** *if for all $a, b, c \in A$, $a \circ (b \circ c) = (a \circ b) \circ c$.*

Examples and non-examples 2.5.9.

(1) $+$ and $\cdot$ are associative.

(2) $-, /$ are not associative.

(3) Composition of functions is associative. (See Exercise 2.4.6.)

(4) $\cap, \cup$ are associative.

Theorem 2.5.10. *Let $\circ$ be an associative binary operation on A with identity e. If x has an inverse, that inverse is unique.*

Proof. Let y and z be inverses of x. Then

$$y = y \circ e \text{ (by property of identity)}$$
$$= y \circ (x \circ z) \text{ (since } z \text{ is an inverse of } x)$$
$$= (y \circ x) \circ z \text{ (since } \circ \text{ is associative)}$$
$$= e \circ z \text{ (since } y \text{ is an inverse of } x)$$
$$= z \text{ (by property of identity).}$$

Thus by the transitivity of equality, $y = z$. $\qquad\square$

Definition 2.5.11. *We say that x is* **invertible** *if x has an inverse. The (abstract) inverse is usually denoted x^{-1}.*

Be careful! What is the number 5^{-1} if $\circ$ equals $+$?

If ∘ is associative, we in the future omit parentheses in $a \circ b \circ c$ (or in $a \circ b \circ c \circ d$ etc.), as the order of the computation does not matter.

If ∘ is not associative, we need to keep parentheses! For example, in $\mathbb{Z}$, $a - b - c - d$ can have parentheses inserted in many different ways, and four different values can be obtained. You may want to find specific four integers a, b, c, d for which you get four distinct values with different placements of parentheses.

Notation 2.5.12. More generally, if ∘ is associative, we may and do omit parentheses in expressions such as $a_1 \circ a_2 \circ \cdots \circ a_n$, as the meaning is unambiguous. We abbreviate $a_1 \cdot a_2 \cdot \cdots \cdot a_n$ (in this order!) to $\prod_{k=1}^n a_k$. When ∘ is addition, we abbreviate $a_1 \circ a_2 \circ \cdots \circ a_n$ to $\sum_{k=1}^n a_k$. Examples were already worked out in Section 1.5.

Notation 2.5.13. Just like for functions in Remark 2.4.8, also for arbitrary associative binary operation ∘ we abbreviate $a \circ a$ with a^2, $(a \circ a) \circ a$ with a^3, etc., and in general for all positive integers n we write

$$a^n = a^{n-1} \circ a = a \circ a^{n-1}.$$

This notation is familiar also when ∘ equals multiplication: then 2^5 stands for 32. When ∘ is addition, then the abstract "2^5" stands for 10, but of course we prefer to not write 10 this way; instead we write it in the additive notation $2 + 2 + 2 + 2 + 2$, or briefly, $5 \cdot 2$. The empty product a^0 makes sense if the set has the identity, and in that case $a^0 = e$. (See Exercise 1.5.6 for the first occurrence of an empty product.)

If a has a multiplicative inverse, then a^{-1} is that inverse, and in that case if ∘ is also associative,

$$a^{-n} = a^{-(n-1)} \circ a^{-1}.$$

To prove this by induction on n, we multiply

$$(a^{-(n-1)} \circ a^{-1}) \circ a^n = (a^{-(n-1)} \circ a^{-1}) \circ (a \circ a^{n-1})$$
$$= a^{-(n-1)} \circ (a^{-1} \circ a) \circ a^{n-1}$$
$$= a^{-(n-1)} \circ a^{n-1} = e,$$

and similarly $a^n \circ (a^{-(n-1)} \circ a^{-1}) = e$, which proves that $a^{-n} = a^{-(n-1)} \circ a^{-1}$.

Theorem 2.5.14. *Let* $\circ$ *be an associative binary operation on* A. *Let* f, g *have inverses. Then* $g \circ f$ *also has an inverse, and the inverse is* $f^{-1} \circ g^{-1}$.

Proof. $(g \circ f) \circ f^{-1} \circ g^{-1} = g \circ (f \circ f^{-1}) \circ g^{-1} = g \circ e \circ g^{-1} = e$, and similarly $f^{-1} \circ g^{-1} \circ (g \circ f) = e$, so that $(g \circ f)^{-1} = f^{-1} \circ g^{-1}$. $\square$

In particular, if $\circ$ equals $+$, then for all $x, y \in A$,

$$-(x + y) = (-x) + (-y),$$

and if $\circ$ equals $\cdot$, then

$$(x \cdot y)^{-1} = y^{-1} \cdot x^{-1}.$$

Theorem 2.5.15. *If* x *is invertible, then its inverse is also invertible, and the inverse of the inverse is* x.

Proof. By definition of inverses of x, $x^{-1} \circ x = e = x \circ x^{-1}$, which also reads as "the inverse of x^{-1} is x." $\square$

Theorem 2.5.16. Cancellation. *Let* $\circ$ *be an associative binary operation on a set* A, *and* z *an invertible element in* A. *Then for all* $x, y \in A$,

$$x \circ z = y \circ z \Rightarrow x = y,$$
$$z \circ x = z \circ y \Rightarrow x = y.$$

Proof. We prove only the first implication. If $x \circ z = y \circ z$, then $(x \circ z) \circ z^{-1} = (y \circ z) \circ z^{-1}$, hence by associativity, $x \circ (z \circ z^{-1}) = y \circ (z \circ z^{-1})$. Thus by the definition of inverses and identities, $x = x \circ e = y \circ e = y$.

Another proof of the same fact is as follows:

$$x = x \circ e$$
$$= x \circ (z \circ z^{-1})$$
$$= (x \circ z) \circ z^{-1} \text{ (by associativity)}$$
$$= (y \circ z) \circ z^{-1}$$
$$= y \circ (z \circ z^{-1}) \text{ (by associativity)}$$
$$= y \circ e$$
$$= y.$$ $\square$

Definition 2.5.17. *A binary operation* $\circ$ *on* A *is* **commutative** *if for all* $a, b \in A$, $a \circ b = b \circ a$.

Examples and non-examples 2.5.18.

(1) $+, \cdot$ are commutative on all examples we have seen so far. (If you have seen matrices, you know that matrix multiplication is not commutative.)

(2) $\cap, \cup$ are commutative.

(3) Function composition is not commutative (cf. Example 2.4.9).

We end this section with an important example. The reader is familiar with manipulations below when $n = 12$ or $n = 24$ for hours of the day (we do not say "28 o'clock"), when $n = 7$ for days of the week, when $n = 3$ for standard meals of the day, etc.

Important example 2.5.19. Let n be a positive integer. Recall the definition of $\mathbb{Z}/n\mathbb{Z}$ from Example 2.3.9: elements are equivalence classes $[0], [1], [2], \ldots, [n-1]$. Define $+$ on $\mathbb{Z}/n\mathbb{Z}$ as follows: $[a] + [b] = [a+b]$. Well, first of all, is this even a function? Namely, we need to verify that whenever $[a] = [a']$ and $[b] = [b']$, then $[a+b] = [a'+b']$, which says that any choice of representatives gives the same final answer. Well, $a - a'$ and $b - b'$ are integer multiples of n, hence $(a+b) - (a'+b') = (a-a') + (b-b')$ is a sum of two multiples of n and hence also a multiple of n. Thus $[a+b] = [a'+b']$, which says that $+$ is indeed a binary operation on $\mathbb{Z}/n\mathbb{Z}$. It is straightforward to verify that $+$ on $\mathbb{Z}/n\mathbb{Z}$ is commutative and associative, the identity elements is $[0]$, and every element $[a]$ has an additive inverse $[-a] = [n - a]$.

Similarly, we can define $\cdot$ on $\mathbb{Z}/n\mathbb{Z}$ as follows: $[a] \cdot [b] = [a \cdot b]$. It is left to the reader that this is a binary operation that is commutative and associative, and the identity elements is $[1]$. The multiplication tables for $n = 2, 3, 4$ are below, where, for ease of notation, we abbreviate "$[a]$" with "a":

$\mathbb{Z}/2\mathbb{Z}$:

$\cdot$	0	1
0	0	0
1	0	1

$\mathbb{Z}/3\mathbb{Z}$:

$\cdot$	0	1	2
0	0	0	0
1	0	1	2
2	0	2	1

$\mathbb{Z}/4\mathbb{Z}$:

$\cdot$	0	1	2	3
0	0	0	0	0
1	0	1	2	3
2	0	2	0	2
3	0	3	2	1

Note that for multiplication in $\mathbb{Z}/4\mathbb{Z}$, $[3] = [-1]$ is the multiplicative inverse of itself, and $[2]$ has no multiplicative inverse.

Exercises for Section 2.5

2.5.1. Let A be the set of all bijective functions $f : \{1,2\} \to \{1,2\}$.

i) How many elements are in A?

ii) Prove that function composition $\circ$ is a binary operation on A.

iii) For all $f, g \in A$ compute $f \circ g$. (How many compositions is this?)

iv) What is the identity element?

v) Verify that every element of A is its own inverse.

vi) Verify that $\circ$ is commutative.

2.5.2. For any integer $n \geq 3$, let A be the set of all bijective functions $f : \{1,2,3,\ldots,n\} \to \{1,2,3,\ldots,n\}$.

i) How many elements are in A?

ii) Prove that function composition $\circ$ is a binary operation on A.

iii) Find $f, g \in A$ such that $f \circ g \neq g \circ f$.

2.5.3. Find a set A with a binary operation $\circ$ such that for some invertible $f, g \in A$, $(g \circ f)^{-1} \neq g^{-1} \circ f^{-1}$. (This is sometimes called the socks-and-shoes problem. Say why?)

2.5.4. Refer to Example 2.5.19: Write addition tables for $\mathbb{Z}/2\mathbb{Z}$, $\mathbb{Z}/3\mathbb{Z}$, $\mathbb{Z}/4\mathbb{Z}$.

2.5.5. Write multiplication tables for $\mathbb{Z}/5\mathbb{Z}, \mathbb{Z}/6\mathbb{Z}, \mathbb{Z}/7\mathbb{Z}$.

2.5.6. Determine all $[a] \in \mathbb{Z}/12\mathbb{Z}$ that have a multiplicative inverse.

*2.5.7. Determine all $[a] \in \mathbb{Z}/p\mathbb{Z}$ that have a multiplicative inverse if p is a prime number.

2.5.8. Here is an opportunity to practice induction. Let $f, g : A \to A$ be functions, with f invertible. Prove that $(f \circ g \circ f^{-1})^n = f \circ g^n \circ f^{-1}$. (Notation is from Remark 2.4.8.)

2.5.9. Consider the following binary operation $\circ$ on $\{a,b,c\}$:

$\circ$	a	b	c
a	a	b	c
b	b	a	a
c	c	a	a

i) Show that $\circ$ is commutative and has an identity element.

ii) Show that every element has an inverse and that inverses need not be unique.

iii) Prove that $\circ$ is not associative. (Hint: Theorem 2.5.10.)

*2.5.10. Let S be the set of all logical statements. We define a relation $\sim$ on S as follows: $P \sim Q$ if P and Q have the same truth values in all conditions. For example, "$1 = 1$" and "$2 = 2$" are related via $\sim$.

  i) Prove that $\sim$ is an equivalence relation on S. Let A be the set of all equivalence classes.

  ii) Verify that and, or and xor are naturally binary operations on A.

  iii) Find the identity elements, if they exist, for each of the three binary operations.

  iv) For each binary operation above with identity, which elements of A have inverses?

2.5.11. Define a binary operation $\oplus$ on $\mathbb{R}$ as $a \oplus b = a + b + 2$, where $+$ is the ordinary addition on $\mathbb{R}$. Prove that $\oplus$ is commutative and associative. Find the identity element of $\oplus$, and for each $x \in \mathbb{R}$, find its inverse.

2.5.12. Define a binary operation $\odot$ on $\mathbb{R}$ as $a \odot b = a + 2 \cdot a + 2 \cdot b + 2$, where $+$ and $\cdot$ are the ordinary addition and multiplication on $\mathbb{R}$. Prove that $\odot$ is commutative and associative. Find the identity element of $\odot$, and for each $x \in \mathbb{R} \setminus \{-2\}$, find its inverse.

2.6 Fields

The motivation for the abstract definition of fields below comes from the familiar properties of the set of all real numbers.

Definition 2.6.1. *A set F is a **field** if it has two binary operations on it, typically denoted $+$ and $\cdot$, and special elements $0, 1 \in F$, such that the following identities hold for all $m, n, p \in F$:*

  (1) (**Additive identity**) $m + 0 = m = 0 + m$.

  (2) (**Associativity of addition**) $m + (n + p) = (m + n) + p$.

  (3) (**Commutativity of addition**) $m + n = n + m$.

  (4) (**Multiplicative identity**) $m \cdot 1 = m = 1 \cdot m$.

  (5) (**Distributivity**) $m \cdot (n + p) = (m \cdot n) + (m \cdot p)$.

  (6) (**Associativity of multiplication**) $m \cdot (n \cdot p) = (m \cdot n) \cdot p$.

  (7) (**Commutativity of multiplication**) $m \cdot n = n \cdot m$.

(8) (**Existence of additive inverses**) *There exists $r \in F$ such that* $m + r = r + m = 0$.

(9) (**Existence of multiplicative inverses**) *If $m \neq 0$, there exists* $r \in F$ *such that* $m \cdot r = r \cdot m = 1$.

(10) $1 \neq 0$.

0 *is called the* **additive identity** *and* 1 *is called the* **multiplicative identity**.

The familiar $\mathbb{N}^+$, $\mathbb{N}_0$, $\mathbb{Z}$, $\mathbb{Q}$, $\mathbb{R}$ all have the familiar binary operations $+$ and $\cdot$ on them. Among these, $\mathbb{N}^+$ lacks the additive identity, but all others have the additive identity 0. In $\mathbb{N}_0$, all non-zero elements lack additive inverses, and in $\mathbb{Z}$, all non-zero elements other than 1 and -1 lack a multiplicative inverse. Thus $\mathbb{N}^+, \mathbb{N}_0$ and $\mathbb{Z}$ are not fields.

We take it for granted (until Chapter 3) that $\mathbb{Q}$ and $\mathbb{R}$ are fields. In Section 3.9 we construct a new field, the field of complex numbers. There are many other fields out there, such as the set of all real-valued rational functions with real coefficients. A few fields are developed in the exercises to this section.

Notation 2.6.2. By Section 2.5, we know that the additive and multiplicative identities and inverses are unique in a field. The additive inverse of m is denoted $-m$, and the multiplicative inverse of a non-zero m is denoted m^{-1}, or also $1/m$. The sum $n + (-m)$ of n and $-m$ is also written as $n - m$, and the product $n \cdot m^{-1}$ of n and m^{-1} is also written as n/m. The latter two operations are also called **subtraction** and **division**. The functions $-$ and $\_^{-1}$ are unary (see definition on page 78) with domains F and $F \setminus \{0\}$, respectively. By Theorem 2.5.15, $-(-m) = m$, and for any non-zero m, $\frac{1}{\frac{1}{m}} = (m^{-1})^{-1} = m$.

It is standard to omit "$\cdot$" when no confusion arises. Note that the expression $2 \cdot 222 + 4$ is different from $2\,222 + 4$, but $2 \cdot x + 4$ is the same as $2x + 4$.

Another bit of notation: $\cdot$ takes precedence over addition, so that "$(a \cdot b) + c$" can be written simply as "$a \cdot b + c$", or with the omission of the multiplication symbol, as "$ab + c$".

Theorem 2.6.3. (The other distributive property) *If F is a field, then for all $m, n, p \in F$, $(m + n)p = mp + np$.*

Proof. $(m + n)p = p(m + n)$ (by commutativity of multiplication)

$$= pm + pn \text{ (by distributivity (5))}$$

$$= mp + np \text{ (by commutativity of multiplication)}. \qquad \square$$

Theorem 2.6.4. *If F is a field, then for all $m \in F$, $m \cdot 0 = 0 = 0 \cdot m$.*

Proof. We use the trick of adding a clever zero.

$m \cdot 0 = m \cdot 0 + 0$ (since 0 is the additive identity).

$$= m \cdot 0 + (m \cdot 0 + (-(m \cdot 0))) \text{ (by the definition of additive inverses)}$$

$$= (m \cdot 0 + m \cdot 0) + (-(m \cdot 0)) \text{ (by associativity of addition)}$$

$$= m \cdot (0 + 0) + (-(m \cdot 0)) \text{ (by the distributive property)}$$

$$= m \cdot 0 + (-(m \cdot 0)) \text{ (since 0 is the additive identity)}$$

$$= 0 \text{ (by the definition of additive inverses)}.$$

Since multiplication is commutative, it also follows that $0 \cdot m = 0$. $\qquad \square$

We cannot divide by 0. Never divide by 0. For one thing, in an abstract field, dividing by 0 is simply not on the list of allowed operations, and for another, n/m always stands for that unique element of the field, which, when multiplied by m, yields n. In other words, $n = (n/m) \cdot m$. If m somehow – horribly – happened to be 0, then we would have $n = (n/0) \cdot 0$, and by Theorem 2.6.4, this product n would be 0. So, if we were to divide by 0, the only number that could possibly be divided by it is 0 itself. But — continuing the horrible detour — what should $0/0$ be? How about the muddle in the following usage of other axioms that seems to require also division of 1 by 0: $0/0 = 1 \cdot (0/0) = (1 \cdot 0)/0 = (0 \cdot 1)/0 = 0 \cdot (1/0)$. In any case, "division" by 0 is inconsistent, and not allowed. In a mathematics paper, **never** write "$\frac{x}{0}$" or "$x/0$".

At this stage of your mathematical life, you of course never write something like "$3/0$" (my eyes hurt seeing this!), but a common college mistake that is essentially division by 0 is cancelling x when solving an equation such as $x^2 = 3x$ to obtain only one solution $x = 3$. This cancellation was division by 0 when x was the other possible solution! **Avoid even hidden division by** 0, so that you can find all the solutions.

Exercises for Section 2.6

2.6.1. Verify that the set $\{0\}$ satisfies axioms (1)–(9) of fields, with 0 being the additive and the multiplicative identity. Obviously $\{0\}$ fails axiom (10).

2.6.2. Use the set-up in Example 2.3.9. Prove that $\mathbb{Z}/2\mathbb{Z}$ is a field. Note that in this field $[2] = [0]$, so $[2]$ does not have a multiplicative inverse. Also note that in this field, every number has a square and cube root.

2.6.3. Use the set-up in Example 2.3.9. Prove that $\mathbb{Z}/3\mathbb{Z}$ is a field. Note that in this field $[3] = [0]$, so $[3]$ does not have a multiplicative inverse. Note that in this field, $[2]$ is not the square of any number.

2.6.4. Use the set-up in Example 2.3.9, and let n be a positive integer strictly bigger than 1 that is not a prime integer. Prove that $\mathbb{Z}/n\mathbb{Z}$ is not a field.

*__2.6.5.__ Use the set-up in Example 2.3.9. Prove that $\mathbb{Z}/p\mathbb{Z}$ is a field for any prime integer p. Note that in $\mathbb{Z}/7\mathbb{Z}$, $[2]$ is the square of $[3]$ and of $[4]$.

2.6.6. Prove using only the axioms of fields that for any x in a field, $(-1)\cdot x$ is the additive inverse of x.

2.6.7. Let F be a field. Prove that for any $x \in F$, $-(-x) = x$. Prove that for any non-zero $x \in F$, $1/(1/x) = x$.

2.6.8. Let F be a field. Prove that for any $x, y \in F$, $(-x) \cdot y = -(x \cdot y) = x \cdot (-y)$. (Hint: Use the definition of additive inverses.)

2.6.9. Let F be a field. Prove that for any $x, y \in F$, $(-x) \cdot (-y) = x \cdot y$.

2.6.10. Let x be a non-zero element of a field F. Then $(-x)^{-1} = -(x^{-1})$.

2.6.11. Let A be a set and F a field. For any functions $f, g : A \to F$ we define new functions $f + g, f \cdot g : F \to G$ as $(f + g)(x) = f(x) + g(x)$ and $(f \cdot g)(x) = f(x) \cdot g(x)$. Here, the second $+$ and $\cdot$ are the binary operations on F, and the first $+$ and $\cdot$ are getting defined. Let S be the set of all functions from A to F.

   i) Prove that $+$ and $\cdot$ are binary operations on S.

   ii) If $A = F$, then S includes polynomial functions. Let T be the set of all polynomial functions from F to F. Prove that $+, \cdot$ and $\circ$ are binary operations on T.

2.6.12. Let F be a field and n a non-negative integer. We define **exponentiation by n** to be a function $f : F \to F$ given as $f(x) = x^n$, where $x^0 = 1$ for all x and where for positive n, $x^n = x \cdot x^{n-1}$. In this exponentiation, n is called the **exponent**, or **power**, and x is called the **base**.

  i) Prove by induction on n that exponentiation is a well-defined function.

  ii) We want to define exponentiation by negative integers as well. What is the largest subset D of F such that x^{-1} is defined for all $x \in D$. Is $D = F$? Why or why not?

  iii) Prover that for any integers m and n, if $x \in D$, then $(x^m)^n = (x^n)^m$.

2.6.13. (Euclidean algorithm over arbitrary fields) Let F be a field. Let $f(x) = a_0 + a_1 x + \cdots + a_n x^n$ and $g(x) = b_0 + b_1 x + \cdots + b_m x^m$ for some $a_0, a_1, \ldots, a_n, b_0, b_1, \ldots, b_m \in F$ and with $a_n b_m \neq 0$.

  i) Suppose that $m, n \geq 1$. Prove that there exist polynomials $q(x)$ and $r(x)$ such that $f(x) = q(x) \cdot g(x) + r(x)$ and such that the degree of $r(x)$ is strictly smaller than m.

  ii) Prove that there are at most n elements c in F such that $f(c) = 0$.

2.6.14. (Degree of a polynomial function) Let F be an infinite field. Let $f : F \to F$ be a polynomial function given by $f(x) = a_0 + a_1 x + \cdots + a_n x^n$ for some non-negative integer n and $a_1, a_2, \ldots, a_n \in F$.

  i) Suppose that f is the zero function. Prove that $a_0 = a_1 = \cdots = a_n = 0$. (Hint: Exercise 2.6.13.)

  ii) Prove that the coefficients $a_0, a_1, \ldots, a_n$ are uniquely determined. In particular, the degree of a polynomial function is uniquely determined.

2.6.15. (Degree of a polynomial function)

  i) Let $f, g : \mathbb{Z}/2\mathbb{Z} \to \mathbb{Z}/2\mathbb{Z}$ be defined by $f(x) = x^2$, $g(x) = x$. Show that f and g are an identical polynomial function given by polynomials of different degrees.

  ii) Find a non-zero polynomial $p(x)$ of degree 3 that equals the zero function on $\mathbb{Z}/2\mathbb{Z}$.

2.6.16. (An unusual field.) Let $\oplus$ and $\odot$ be binary operations on $\mathbb{R}$ as defined in Exercises 2.5.11 and 2.5.12. Prove that $\mathbb{R}$ is a field with these two binary operations.

2.7 Order on sets, ordered fields

If $<$ is a relation on a set S, we define relations $\leq, >, \geq$ on S by

$a \leq b$ means that $a < b$ or $a = b$.

$a > b$ means that $b < a$.

$a \geq b$ means that $b \leq a$.

Conversely, if $\leq$ is a relation on S, then we define $<$ on S by

$a < b$ if and only if $a \leq b$ and $a \neq b$,

which by before also defines $>, \geq$. Similarly, each of $>, \geq$ determines all four relations of this form. Thus one of these relations on a set S implies that we have all four relations naturally derived from the one. These relations impose the familiar notion of order.

We are familiar with these relations $<, \leq, >, \geq$ in $\mathbb{R}$. We can also use them in other contexts:

Examples 2.7.1.

(1) $<$ can be the relation "is a proper subset of" on a set S of all subsets of some universal set U. In this case, $\leq$ means "is a subset of", $>$ means "properly contains", and $\geq$ means "contains".

(2) If $<$ is the relation "has strictly fewer elements" on the set S of all subsets of the set $\{1, 2, 3, \ldots, 100\}$, then $\leq$ means "has fewer elements or is the same set" (rather than "has fewer or the same number of elements").

Definition 2.7.2. *Let $\leq$ be a relation on a set S.*

An element $b \in S$ is called an **upper bound** *(resp.* **lower bound***) of T (in S) if for all $t \in T$, $t \leq b$ (resp. $t \geq b$).**

A subset T of S is **bounded above** *(resp.* **bounded below***) (in S) if there exists an upper bound (resp. lower bound) of T in S.*

A set that is bounded above and below is called **bounded.**

---

* A sentence of the form "P is Q (resp. Q') if R (resp. R')" is shorthand for two sentences: "P is Q if R" and "P is Q' if R'".

An element $c \in S$ is called a **least upper bound**, or **supremum**, of T, if it is an upper bound of T, and if for all upper bounds b of T, $c \le b$. If $c \in T$, then c is also called **the maximum** of T.

An element $c \in S$ is called a **greatest lower bound**, or **infimum**, of T, if it is a lower bound of T, and if for all lower bounds b of T, $b \le c$. If $c \in T$, then c is also called **the minimum** of T.

The obvious standard abbreviations are: $\mathrm{lub}(T) = \sup(T)$, $\mathrm{glb}(T) = \inf(T)$, $\max(T)$, $\min(T)$, possibly without parentheses around T.

Examples 2.7.3.

(1) The set $\mathbb{N}_0$ has minimum 0. It is not bounded above, for any upper bound u would be strictly smaller than the positive integer $\lceil u \rceil + 1$ (the ceiling function), thus contradicting the assumption of upper bounds.

(2) The set $T = \{1/n : n \in \mathbb{N}^+\}$ has maximum 1, it is bounded below, the infimum is 0, and there is no minimum.

Proof: In long form, the set equals $\{1, 1/2, 1/3, 1/4, 1/5, \ldots\}$. From this re-writing it is clear that 1 is the maximum, that 0 is a lower bound and that 0 is not in the set, so 0 cannot be the minimum. Why is 0 the largest lower bound, i.e., why is 0 the infimum of T? Suppose that r is a positive real number. Set $n = \lceil \frac{1}{r} \rceil + 1$. Then n is a positive integer, and $n > \frac{1}{r}$. By cross multiplying we get that $r > \frac{1}{n}$, which proves that r is not a lower bound on T. Since r was arbitrary, this proves that no positive number is a lower bound on T, so that 0 is the greatest lower bound on T. $\square$

(3) The set $\{1/p : p \text{ is a positive prime number}\}$ has maximum $1/2$ and infimum 0. (There are infinitely many prime numbers; see the proof on page 25.)

(4) The sets $\{(-1)^n : n \in \mathbb{N}^+\}$ and $\{\sin(x) : x \in \mathbb{R}\}$ both have maximum 1 and minimum -1.

(5) The set of all positive rational numbers that are strictly smaller than π has infimum 0 and supremum π, but it has no minimum and no maximum.

(6) The set $\{e^x : x \in \mathbb{R}\}$ has no upper bound, it is bounded below with infimum 0 and no minimum.

(7) The empty subset has neither minimum nor maximum. Every element of S is vacuously an upper and a lower bound of the empty set.

(8) The set $\{x \in \mathbb{R} : -3 < x - 5 < 3\}$ has no minimum and maximum, but the infimum is 2 and the supremum is 8. The set $\{x \in \mathbb{R} : -3 \le x - 5 < 3\}$ has minimum 2, supremum 8, and no maximum. The set $\{x \in \mathbb{R} : -3 < x - 5 \le 3\}$ has maximum 8, infimum 2, and no minimum. The set $\{x \in \mathbb{R} : -3 \le x - 5 \le 3\}$ has minimum 2 and maximum 8.

(9) If $T = \{\{\}, \{1\}, \{2\}\}$, then the inclusion relation on T has minimum $\{\}$, and no upper bounds in T. If we think of T as a subset of the set S of all subsets of $\{1, 2\}$ (or of the set S of all subsets of $\mathbb{R}$), then T has supremum $\{1, 2\}$.

(10) If S is the set of all subsets of the set $\{1, 2, 3, \dots, 100\}$ and $<$ is the relation "has strictly fewer elements", then the empty set is the minimum and $\{1, 2, 3, \dots, 100\}$ is the maximum. If T is the subset of S consisting only of sets with at most two elements, then the minimum of T is the empty set, and there is no maximum or supremum in T. The $\binom{100}{2}$ elements $\{1, 2\}, \{1, 3\}, \dots, \{1, 100\}$, $\{2, 3\}, \{2, 4\}, \dots, \{99, 100\}$ are each greater than or equal to all elements of T and they are not strictly smaller than any other element of T.

In the sequel we restrict $<$ to relations that satisfy the trichotomy property:

Definition 2.7.4. *A relation $<$ on a set S satisfies the **trichotomy** if for all $s, t \in S$, exactly one of the following relations holds:*

$$s = t, \quad s < t, \quad t < s.$$

Examples 2.7.5.

(1) The familiar $<$ on $\mathbb{R}$ satisfies the trichotomy.

(2) If S is the set of all subsets of a universal set U, then the inclusion relation on S satisfies the trichotomy.

(3) If $S = \{\{\}, \{1\}, \{2\}\}$, then the inclusion relation on S does not satisfy the trichotomy.

Theorem 2.7.6. *Let $<$ on S satisfy the trichotomy. Then a supremum (resp. infimum) of a non-empty subset T of S, if it exists, is unique.*

Proof. Suppose that c, c' are suprema of T in S. Both c and c' are upper bounds on T, and since c is a least upper bound, necessarily $c' \leq c$. Similarly $c \leq c'$, so that by trichotomy $c = c'$. This proves that suprema are unique, and a similar proof shows that infima are unique. $\square$

Why did we assume that the subset T of S above be non-empty? By definition every element of S is an upper bound for $\emptyset$, so in particular if S has no minimum, then $\emptyset$ has no least upper bound.

Definition 2.7.7. *A set S with relation $\leq$ is **well-ordered** if $\inf(T) = \min(T)$ for every non-empty subset T of S. The element $\min(T)$ is called the **least element** of T.*

Examples 2.7.8.

(1) Any finite set with relation $\leq$ is well-ordered (simply check the finitely many pairings for which element is smaller).

(2) $\mathbb{Z}$ is not well-ordered as there is no smallest whole number.

(3) Similarly, the set of all positive rational numbers is not well-ordered.

(4) $\mathbb{N}_0$ is well-ordered because for any non-empty subset T of $\mathbb{N}_0$, by the fact that the set is not empty there exists an element $n \in T$, and after that one has to check which of the finitely many numbers $0, 1$ through n is the smallest one in T. Similarly, $\mathbb{N}^+$ is well-ordered. (Chapter 3 covers this more rigorously.)

Definition 2.7.9. *Let F be a set with a binary operation $+$, with (additive) identity $0 \in F$, and with a relation $<$ satisfying the trichotomy. Define $F^+ = \{x \in F : 0 < x\}$, and $F^- = \{x \in F : x < 0\}$. Elements of F^+ are called **positive**, and element of F^- are called **negative**. Elements of $F \setminus F^+$ are called **non-positive** and element of $F \setminus F^-$ are called **non-negative**.*

We define **intervals** in F to be sets of the following form, where $a, b \in F$ with $a < b$:

$$(a, b) = \{x \in F : a < x < b\},$$
$$(a, b] = \{x \in F : a < x \leq b\},$$

$$[a,b) = \{x \in F : a \leq x < b\},$$
$$[a,b] = \{x \in F : a \leq x \leq b\},$$
$$(a,\infty) = \{x \in F : a < x\},$$
$$[a,\infty) = \{x \in F : a \leq x\},$$
$$(-\infty,b) = \{x \in F : x < b\},$$
$$(-\infty,b] = \{x \in F : x \leq b\}.$$

Definition 2.7.10. *We say that a field F is an* **ordered** *field if it has a relation $<$ with the following properties:*

(1) *$<$ satisfies the trichotomy, i.e., for all $x, y \in F$, exactly one of the following is true:*

$$x < y, x = y, y < x.$$

(2) **(Transitivity of $<$)** *For all $x, y, z \in F$, if $x < y$ and $y < z$ then $x < z$.*

(3) **(Compatibility of $<$ with addition)** *For all $x, y, z \in F$, if $x < y$ then $x + z < y + z$.*

(4) **(Compatibility of $<$ with multiplication by positive elements)** *For all $x, y, z \in F$, if $x < y$ and $0 < z$ then $xz < yz$.*

A subset of an ordered field is called an **ordered set.**

Theorem 2.7.11. *Let F be an ordered set.*

(1) *For $x \in F$ with the additive inverse $-x \in F$, $x \in F^+$ if and only if $-x \in F^-$, and $x \in F^-$ if and only if $-x \in F^+$.*

(2) *$1 \in F^+$.*

(3) *For $x \in F$ with the multiplicative inverse $x^{-1} \in F$, $x \in F^+$ if and only if $x^{-1} \in F^+$, and $x \in F^-$ if and only if $x^{-1} \in F^-$.*

Proof. (1) $x \in F^+$ if and only if $0 < x$, and by compatibility of $<$ with addition this implies that $-x = 0 - x < x - x = 0$, so that $-x \in F^-$. The rest of (1) is equally easy.

(2) By assumption $1 \neq 0$. If $1 \notin F^+$, then by trichotomy $1 < 0$, and by (1), $0 < -1$. Thus by compatibility of $<$ with multiplication by positive

numbers, since -1 is supposedly positive, $0 = 0 \cdot (-1) < (-1) \cdot (-1)$. By Exercise 2.6.9, $(-1) \cdot (-1) = 1$, which by transitivity says that $0 < 1$. Since we also assumed that $1 < 0$, we get a contradiction to the trichotomy. So necessarily $1 \in F^+$.

(3) Suppose that $x \in F^+$. By trichotomy then exactly one of the following three inequalities holds: $x^{-1} < 0$, $x = 0$, $x^{-1} > 0$. Let $\bigcirc$ stand for the correct inequality (or equality). By compatibility of $<$ with multiplication by the positive number x, we then have $1 = x \cdot x^{-1} \bigcirc x \cdot 0 = 0$. By (2), the relation $\bigcirc$ must equal $>$, so that $x^{-1} > 0$.

If instead $x \in F^-$, then by (1), $-x \in F^+$, and by what we have proved of (3), $(-x)^{-1} \in F^+$. By Exercise 2.6.10 then $-x^{-1} = (-x)^{-1} \in F^+$, so that $x^{-1} \in F^-$ by (1). $\qquad\square$

Theorem 2.7.12. *Let F be an ordered field.*
  (1) For $x, y \in F^+$, $x + y$ is also in F^+.
  (2) For $x, y \in F^+$, $x \cdot y$ is also in F^+.
  (3) For $x, y \in F^-$, $x + y \in F^-$.
  (4) For $x, y \in F^-$, $x \cdot y \in F^+$.
  (5) For $x \in F^+$ and $y \in F^-$, $x \cdot y \in F^-$.

Proof. (1) By assumption, $0 < x$ and $0 < y$. Then by compatibility of $<$ with addition, $y = 0 + y < x + y$, and since $0 < y$, by transitivity of $<$, $0 < x + y$, i.e., $x + y \in F^+$.

(2) By assumption, $0 < x$ and $0 < y$. Then by compatibility of $<$ with multiplication by positive numbers, $0 = 0 \cdot y < x \cdot y$, which proves that $x \cdot y \in F^+$.

The proofs of the rest are similar. $\qquad\square$

Exercises for Section 2.7

2.7.1. Prove that if $<$ is transitive then $\leq$ is transitive.

2.7.2. Prove that $<$ (resp. $\leq$) is transitive if and only if $>$ (resp. $\geq$) is transitive.

2.7.3. (An exercise of this flavor is relevant in computing limits as in Section 4.1.) Under what conditions is the minimum of $\{0.2, \epsilon/7\}$ equal to 0.2, and when is the minimum $\epsilon/7$? Similarly determine $\min\{0.2, \epsilon/7, \epsilon^2/4\}$.

2.7.4. For each of the subsets of $\mathbb{R}$ below, determine its minimum, maximum, infimum, supremum, if applicable. Justify all answers.

i) $\{-1, 2, \pi, -7\}$.

ii) $\{(-1)^n/n : n \in \mathbb{N}^+\}$.

iii) The set of all positive prime numbers.

iv) $\{x \in \mathbb{R} : -1 < x < 5\}$.

v) $\{x \in \mathbb{R} : 2 \leq x < 5\}$.

vi) $\{x \in \mathbb{Q} : x^2 < 2\}$.

vii) $\{x \in \mathbb{R} : x^2 + x - 1 = 0\}$.

viii) $\{x \in \mathbb{Q} : x^2 + x - 1 = 0\}$.

ix) $\{n/(n+1) : n \in \mathbb{N}_0\}$.

2.7.5. Suppose that a subset T of an ordered field has a minimum (resp., maximum, infimum, supremum) b. Prove that the set $-T = \{-t : t \in T\}$ has a maximum (resp., minimum, supremum, infimum) $-b$.

2.7.6. Suppose that a subset T of positive elements of an ordered field has a minimum (resp., maximum, infimum, supremum) b. What can you say about the maximum (resp., minimum, supremum, infimum) of the set $\{1/t : t \in T\}$?

†**2.7.7.** (Invoked in Theorem 7.3.4.) Let S and T be subsets of an ordered field F. Let $S + T = \{s + t : s \in S \text{ and } t \in T\}$.

i) If S and T are bounded above, prove that $\sup(S + T) \leq \sup S + \sup T$.

ii) If S and T are bounded below, prove that $\inf(S+T) \geq \inf S + \inf T$.

2.7.8. Let F be an ordered field. Prove that $2, 3$ are positive (and so not zero).

2.7.9. Let F be a field and $x \in F$. Prove that $x^2 = 0$ if and only if $x = 0$.

2.7.10. Let F be an ordered field and $x \in F$. Let $x, y \in F$ be non-negative (resp. non-positive) such that $x + y = 0$. Prove that $x = y = 0$.

2.7.11. Let F be an ordered field. Suppose that $x \leq y$ and $p \leq q$. Prove that $x + p \leq y + q$. If in addition $x < y$ or $p < q$, prove that $x + p < y + q$.

2.7.12. Let F be an ordered field and $x, y \in F$. Prove that $x < y$ if and only if $0 < y - x$. Prove that $x \leq y$ if and only if $0 \leq y - x$.

2.7.13. Let F be an ordered field, and $x, y \in F^+$ with $x < y$. Prove that $1/y < 1/x$.

2.7.14. Let F be an ordered field. Suppose that $x < y$ and that x, y are non-zero. Does it follow that $1/y < 1/x$? Prove or give a counterexample.

† **2.7.15.** (In-betweenness in an ordered field) Let F be an ordered field. Let $x, y \in F$ with $x < y$. Prove that $x < (x + y)/2 < y$. (Why are we allowed to divide by 2?)

2.7.16. Find an ordered set without a minimum.

2.7.17. Let F be an ordered set. Prove that any non-empty finite subset S of F has a maximum and a minimum. Prove that for all $s \in S$, $\min(S) \le s \le \max(S)$.

2.7.18. Let $n > 1$ be an integer and $F = \mathbb{Z}/n\mathbb{Z}$. (This was defined in Example 2.3.9.) Prove that F is not an ordered set. In particular, using Exercise 2.6.5, for any prime integer p, $\mathbb{Z}/p\mathbb{Z}$ is a field that is not ordered.

2.8 What are the integers and the rational numbers?

So far we have taken it for granted that elements of $\mathbb{N}_0$ and $\mathbb{Z}$ are special and until Chapter 3 we take it for granted that $\mathbb{Q}$ and $\mathbb{R}$ are ordered fields. This section contains a formal definition of $\mathbb{N}_0$ as a subset of $\mathbb{R}$ with derivations of a few important properties. I recommend covering this section very lightly if at all. Construction of $\mathbb{N}_0$ independent of $\mathbb{R}$ is done formally in the next chapter, where we also derive all the properties from this section. Regardless of whether you read this section or not, you should be able to do all the exercises at the end.

Once we have a definition of the set $\mathbb{N}_0$ of non-negative integers, we define the set $\mathbb{Z}$ of all integers as $\mathbb{N}_0 \cup \{n \in \mathbb{R} : -n \in \mathbb{N}_0\}$ and the set $\mathbb{Q}$ of all rational numbers as $\{x \cdot y^{-1} : x \in \mathbb{Z}, y \in \mathbb{N}^+\}$. In this section we derive no special properties of $\mathbb{Z}$ or of $\mathbb{Q}$.

We accept that $\mathbb{R}$ is an ordered field (and we prove this formally in Theorem 3.8.2).

Definition 2.8.1. *A subset T of $\mathbb{R}$ is called **inductive** if $0 \in T$ and if for every element n of T, $n + 1$ is also in T.*

Examples of inductive sets are $\mathbb{R}$ and $\mathbb{R}^+ \cup \{0\}$.

Theorem 2.8.2. *There exists an inductive subset $\mathbb{N}_0$ of $\mathbb{R}$ that is a subset of every inductive subset of $\mathbb{R}$.*

(1) *$\mathbb{N}_0$ is the smallest inductive subset of $\mathbb{R}$ (in the sense that any inductive subset of $\mathbb{R}$ contains $\mathbb{N}_0$).*

(2) *If $m \in \mathbb{N}_0$ is non-zero, then $m = n + 1$ for some $n \in \mathbb{N}_0$.*

(3) *All elements of $\mathbb{N}_0 \setminus \{0\}$ are positive.*

Proof. The collection $\mathbf{S}$ of inductive subsets of $\mathbb{R}$ is a non-empty set because it contains $\mathbb{R}$. We define $\mathbb{N}_0$ as the intersection of all sets in $\mathbf{S}$. Since 0 is in every inductive set, then 0 is also in their intersection $\mathbb{N}_0$. If $n \in \mathbb{N}_0$, then n is in every inductive subset of $\mathbb{R}$, and so by the definition of inductive sets, $n+1$ is in every inductive subset of $\mathbb{R}$, and so it is in their intersection $\mathbb{N}_0$. This proves that $\mathbb{N}_0$ is an inductive set. By definition it is a subset of every inductive subset of $\mathbb{R}$. This proves (1).

Suppose that $m \in \mathbb{N}_0$ is not 0 and is not equal to $n + 1$ for any $n \in \mathbb{N}_0$. Let $T = \mathbb{N}_0 \setminus \{m\}$. Then $0 \in T$, and if $n \in T$, then $n + 1 \in T$, so that T is an inductive set. But $\mathbb{N}_0$ is a subset of every inductive set, so that $m \in \mathbb{N}_0$ would have to be in T, which is a contradiction. Thus every non-zero $m \in \mathbb{N}_0$ equals $n + 1$ for some $n \in \mathbb{N}_0$. This proves (2).

Let $T = \mathbb{N}_0 \setminus \mathbb{R}^-$. Then $0 \in T$. If $n \in T$, then by trichotomy, $n = 0$ or $n \in \mathbb{R}^+$, and hence $n + 1 \in \mathbb{R}^+ \cap T$. Hence T is inductive. Since $\mathbb{N}_0$ is contained in every inductive set, it follows that $\mathbb{N}_0 \subseteq T$, so that $\mathbb{N}_0$ contains no negative numbers. By trichotomy this proves (3).   □

Theorem 2.8.3. *Let $n \in \mathbb{N}_0$. There are no elements of $\mathbb{N}_0$ strictly between n and $n + 1$.*

Proof. Let T be the subset of $\mathbb{N}_0$ consisting of all n that satisfy the property that there are no elements of $\mathbb{N}$ strictly between n and $n+1$. We will prove that T is an inductive set.

Suppose that there exists $m \in \mathbb{N}_0$ strictly between 0 and $0 + 1 = 1$. Then by Theorem 2.8.2, $m = p + 1$ for some $p \in \mathbb{N}_0$. By compatibility of order with addition, $p = m - 1 < 1 - 1 = 0$, contradicting Theorem 2.8.2 which asserts that elements of $\mathbb{N}_0$ are non-negative. This proves that $0 \in T$.

Now suppose that $n \in T$. We want to prove that $n + 1 \in T$. Suppose for contradiction that there exists $m \in \mathbb{N}_0$ that is strictly between $n+1$ and $(n + 1) + 1$. Since $m > n + 1 \geq 1$, m is not zero, so that by Theorem 3.1.6,

$m = p + 1$ for some $p \in \mathbb{N}_0$. Then by compatibility of order with addition, p is strictly between n and $n + 1$, which contradicts the assumption that $n \in T$. Hence, necessarily there is no m with the stated property, so that $n + 1 \in T$.

This proves that T is an inductive subset of $\mathbb{N}_0$, and since $\mathbb{N}_0$ is contained in every inductive subset, the theorem is proved. □

Theorem 2.8.4. *If $n, m \in \mathbb{N}_0$ and $n \leq m$, then $m - n \in \mathbb{N}_0$.*

Proof. Let $T = \{n \in \mathbb{N}_0 : \text{if } m \in \mathbb{N}_0 \text{ and } n \leq m, \text{ then } m - n \in \mathbb{N}_0\}$. We will prove that T is an inductive set. Certainly $0 \in T$ as $m - 0 = m$. Now suppose that $n \in T$. We claim that $n+1 \in T$. Namely, let $m \in \mathbb{N}_0$ such that $n+1 \leq m$. Necessarily $1 \leq m$ so that by Theorem 3.1.6, $m = p+1$ for some $p \in \mathbb{N}_0$. By compatibility of $\leq$ with addition then $n \leq p$, and since $n \in T$, it follows that $p - n \in \mathbb{N}_0$. Hence $m - (n+1) = p+1 - (n+1) = p - n \in \mathbb{N}_0$. This proves that $n + 1 \in T$. Since n was arbitrary, this proves that T is an inductive subset of $\mathbb{N}_0$, and since $\mathbb{N}_0$ is a subset of every inductive set we have that $T = \mathbb{N}_0$. This proves the theorem. □

Theorem 2.8.5. (The well-ordering principle) $\mathbb{N}_0$ *is well-ordered (see Definition 2.7.7). In other words, every non-empty subset S of $\mathbb{N}_0$ has a **least element**, that is, S contains an element r such that for all $t \in S$, $r \leq t$.*

Proof. We will prove that the following set is inductive:

$$T = \{n \in \mathbb{N}_0 : \text{if } S \subseteq \mathbb{N}_0 \text{ and } n \in S, \text{ then } S \text{ has a least element}\}.$$

Note that $0 \in T$ because $0 \leq n$ for all $n \in \mathbb{N}_0$ and hence $0 \leq n$ for all $n \in S$.

Suppose that $n \in T$. We next prove that $n + 1 \in T$. Let $S \subseteq \mathbb{N}_0$ and $n + 1 \in S$. By assumption that $n \in T$, it follows that the set $S \cup \{n\}$ has a least element. Thus there exists $r \in S \cup \{n\}$ such that for all $t \in S$, $r \leq t$ and $r \leq n$. If $r \in S$, then we just showed that for all $t \in s$, $r \leq t$, so that S has a least element. Now suppose that $r \notin S$. So necessarily $r = n$, and this is not an element of S. Then we claim that $n + 1$ is the least element of S. Namely, let $t \in S$. We need to prove that $n + 1 \leq t$. Suppose for contradiction that $t < n + 1$. Since n is the least element of $S \cup \{n\}$, it follows that $n \leq t$. Thus we have $n \leq t < n + 1$, so that by Theorem 3.1.7, necessarily $n = t$. But then $n = t \in S$, which contradicts the assumption that $n \notin S$. Thus $n + 1 \leq t$, and since t was an arbitrary element of S, it follows that $n + 1$ is the least element of S. Thus in all cases, if $n + 1 \in S$, the set S has a least element. So $n + 1 \in T$. Thus T is an inductive set, and hence equal to $\mathbb{N}_0$. This means that every non-empty subset of $\mathbb{N}_0$ has a least element. $\square$

Exercises for Section 2.8

You should be able to do these problems without reading the section.

2.8.1. Let F be an ordered field and let $x \in F$ satisfy $x > 1$. Prove that for all positive integers n, $x^n > 1$ and $x^{n+1} > x$.

2.8.2. Let F be an ordered field and let $x \in F$ satisfy $0 < x < 1$. Prove that for all positive integers n, $0 < x^n < 1$ and $x^{n+1} < x$.

2.8.3. (**Bernoulli's inequality**) Prove that for all $x \in \mathbb{R}_{\geq 0}$ and all $n \in \mathbb{N}_0$, $(1 + x)^n \geq 1 + nx$.

2.8.4. Does the set $\{2^n/n : n \in \mathbb{N}^+\}$ have a lower (resp. upper bound)? Is it well-ordered? Justify. Repeat with $\{n/2^n : n \in \mathbb{N}^+\}$.

2.8.5. Prove that for all $n \in \mathbb{Z}$ there exist no integer strictly between n and $n + 1$. (Hint: if n is negative, then the interval $(n, n + 1)$ can be mirrored across 0 to the interval $(-n - 1, (-n - 1) + 1)$.)

2.8.6. Prove that for all $x \in \mathbb{R}$, the interval $(x, x + 1)$ can contain at most one integer.

2.9   Increasing and decreasing functions

Definition 2.9.1. *Let F, G be ordered sets (as in the previous section), and let $A \subseteq F$. A function $f : F \to G$ is **increasing** (resp. **decreasing**) on A if for all $x, y \in A$, $x < y$ implies that $f(x) \leq f(y)$ (resp. $f(x) \geq f(y)$). If furthermore $f(x) < f(y)$ (resp. $f(x) > f(y)$) for all $x < y$, then we say that f is **strictly increasing** (resp. **strictly decreasing**) on A. A function is* **(strictly) monotone** *if it is (strictly) increasing or (strictly) decreasing.*

Theorem 2.9.2. *Let n be a positive integer and F an ordered field. Then the function $f : F \to F$ defined by $f(x) = x^n$ when restricted to $F^+ \cup \{0\}$ is strictly increasing with the range in $F^+ \cup \{0\}$.*

Proof. Let $x, y \in F^+ \cup \{0\}$ with $x < y$. Then by Exercise 1.7.7,

$$
\begin{aligned}
f(y) - f(x) &= y^n - x^n \\
&= (x + (y - x))^n - x^n \\
&= \sum_{k=0}^{n} \binom{n}{k} x^k (y - x)^{n-k} - x^n \\
&= \sum_{k=0}^{n-1} \binom{n}{k} x^k (y - x)^{n-k} = (y - x)^n + \sum_{k=1}^{n-1} \binom{n}{k} x^k (y - x)^{n-k}.
\end{aligned}
$$

Since $y - x$ is positive, by Theorem 2.7.12, $(y - x)^n$ is also positive. Since in addition $x \geq 0$ and $\binom{n}{k}$ is a non-negative integer, then by the compatibilities of $>$, $\sum_{k=1}^{n-1} \binom{n}{k} x^k (y - x)^{n-k} \geq 0$. Thus $f(y) - f(x) > 0$. $\square$

Corollary 2.9.3. *Let n be a positive integer and F an ordered field. Suppose that $x, y \in F^+ \cup \{0\}$ have the property that $x^n < y^n$. Then $x < y$.*

Proof. If $x = y$, then $x^n = y^n$, which contradicts the assumption and trichotomy. If $x > y$, then by Theorem 2.9.2, $x^n > y^n$, which also contradicts the assumption. So by trichotomy $x < y$. $\square$

Theorem 2.9.4. *If F, G are ordered sets and $f : F \to G$ is strictly monotone, then f is injective, and there exists a strictly monotone function $g : \text{Range}(f) \to F$ such that for all $x \in F$, $(g \circ f)(x) = x$ and for all $y \in G$, $(f \circ g)(y) = y$. In other words, g is the inverse of the function $f : F \to \text{Range}(f)$.*
Furthermore, f is increasing if and only if g is increasing.

Proof. Let $y \in \text{Range}(f)$. Then $y = f(x)$ for some $x \in F$. If also $y = f(z)$ for some $z \in F$, since f is strictly monotone, $x = z$. So f is injective and x is unique. Thus we define $g : \text{Range}(f) \to F$ by $g(y) = x$. Then by definition for all $x \in F$, $g(f(x)) = x$ and for all $y \in \text{Range}(f)$, $f(g(y)) = y$. If f is increasing and $y_1, y_2 \in \text{Range}(f)$ such that $y_1 < y_2$, then $g(y_1) < g(y_2)$ for otherwise by the increasing property of f, $y_1 = f(g(y_1)) \geq f(g(y_2)) = y_2$, which is a contradiction. Thus if f is increasing, so is $g = f^{-1}$. Thus if $g = f^{-1}$ is increasing, so is $f = (f^{-1})^{-1}$. The same reasoning goes for the decreasing property. $\square$

If the exponentiation function in Corollary 2.9.3 with exponent n takes $F^+ \cup \{0\}$ onto $F^+ \cup \{0\}$, then by Theorem 2.9.4, we can define its inverse function $F^+ \cup \{0\} \to F^+ \cup \{0\}$. However, the function need not be surjective or have an inverse, witness $F = \mathbb{Q}$ and $n = 2$ as proved on page 11.

Remark 2.9.5. Let F be an ordered set and G an ordered field. Let $f, g : F \to G$ be functions. Below we need the definitions of the sum and product of functions (see Exercise 2.6.11).

(1) If f, g are both strictly increasing (resp. both decreasing), then $f + g$ is strictly increasing (resp. decreasing).

(2) If f, g are both strictly increasing (resp. both decreasing) and always take on positive values, then fg is strictly increasing (resp. decreasing).

(3) If f, g are both strictly increasing (resp. both decreasing) and always take on negative values, then fg is strictly decreasing (resp. increasing).

Proof of (3):   Let $x, y \in F$ with $x < y$. Suppose that f and g are both increasing functions, so that $f(x) < f(y) < 0$ and $g(x) < g(y) < 0$. Then $-f(y)$, $-g(x)$ are positive numbers, so by compatibility of $<$ with multiplication by positive numbers, $f(x)(-g(x)) < f(y)(-g(x))$ and $(-f(y))g(x) < (-f(y))g(y)$. By Exercise 2.6.8, this says that $-(f(x)g(x)) < -(f(y)g(x))$ and $-(f(y)g(x)) < -(f(y)g(y))$. By transitivity of $<$ then $-(f(x)g(x)) < -(f(y)g(y))$. By compatibility of $<$ with addition, by adding $f(x)g(x) + f(y)g(y)$ we get that $f(y)g(y) < f(x)g(x)$. With function notation, $(fg)(y) < (fg)(x)$, and since x and y were arbitrary, this says that fg is strictly decreasing. The proof in the case where both f and g are strictly decreasing is similar.   □

Exercises for Section 2.9

2.9.1. Let n be an odd positive integer and F an ordered field. Prove that the function $f : F \to F$ defined by $f(x) = x^n$ is strictly increasing.

2.9.2. Let n be an even positive integer and F an ordered field. Prove that the function $f : F^- \cup \{0\} \to F$ defined by $f(x) = x^n$ is strictly decreasing.

2.9.3. Let n be an odd positive integer and F an ordered field. Suppose that $a, b \in F$ and that $a^n < b^n$. Prove that $a < b$.

2.9.4. Let F be an ordered field, $a \in F^+$ and $f : F \to F$ defined by $f(x) = ax$. Prove that f is a strictly increasing function.

2.9.5. Let F be an ordered field, $a \in F^-$ and $f : F \to F$ defined by $f(x) = ax$. Prove that f is a strictly decreasing function.

2.9.6. Prove that the composition of (strictly) increasing functions is (strictly) increasing. Prove that the composition of (strictly) decreasing functions is (strictly) increasing.

2.9.7. Prove that the composition of a (strictly) increasing function with a (strictly) decreasing function, in any order, is (strictly) decreasing.

2.9.8. Suppose that $f : F \to G$ is strictly monotone.
    i) Let B be a subset of F, and define $g : B \to G$ by $g(x) = f(x)$. Prove that g is strictly monotone.
    ii) Define $h : B \to \text{Range}(g)$ by $h(x) = f(x)$. Prove that h is bijective.

2.9.9. Give an example of a non-decreasing function $f : \mathbb{R} \to \mathbb{R}$ that is not injective.

2.10 Absolute values

Definition 2.10.1. *Let F be an ordered field. The* **absolute value** *function* $| \ | : F \to F$ *is a function defined as*

$$|x| = \begin{cases} 0; & \text{if } x = 0; \\ x; & \text{if } x \in F^+; \\ -x; & \text{if } x \in F^-. \end{cases}$$

This defines the absolute value function on the ordered fields $\mathbb{Q}$ and $\mathbb{R}$. We think of $|x|$ as the distance of x from 0 on the real number line.

The following theorem lists the familiar properties of absolute values, and the reader may wish to prove them without reading the given proof.

Theorem 2.10.2. *Let F be an ordered field.*
    (1) For all $x \in F$, $|x| \geq 0$. Furthermore, $x \geq 0$ if and only if $x = |x|$; and $x \leq 0$ if and only if $x = -|x|$. (In particular, $|1| = 1$.)
    (2) For all $x \in F$, $|x| = |-x|$.
    (3) For all $x \in F$, $-|x| \leq x \leq |x|$.
    (4) For all $x, a \in F$, $|x| \leq a$ if and only if $-a \leq x \leq a$.
    (5) For all $x, a \in F$, $|x| < a$ if and only if $-a < x < a$.
    (6) For all $x, y \in F$, $|xy| = |x||y|$.

Proof. (1) is from the definition.

(2) Certainly $|0| = |-0| = 0$. Suppose that $x \in F^-$. Then $|x| = -x$ and $-x \in F^+$ so that $|-x| = -x$. Thus $|x| = -x = |-x|$. If $x \in F^+$, then $-x \in F^-$, so that by what we just proved, $|-x| = |-(-x)| = |x|$.

(3) If $x \geq 0$, then $|x| = x$, and if $x < 0$, then $x < 0 < |x|$. Thus by transitivity for all x, $x \leq |x|$. In particular, when applied to $-x$, this says that $-x \leq |-x| = |x|$, and by adding $x - |x|$ to both sides we get that $-|x| \leq x$.

(4) Suppose that $|x| \leq a$. Then by (3) and transitivity, $x \leq a$, and $-x \leq |-x| = |x| \leq a$, so that by transitivity and adding $x - a$ to both sides, $-a \leq x$.

(5) The proof of (5) is similar to that of (4).

(6) We may choose r and $s \in \{1, -1\}$ such that $rx \geq 0$ and $sy \geq 0$. Then

$$\begin{aligned}
|xy| &= |\pm(xy)| \text{ (by (2))} \\
&= |(rx)(sy)| \\
&= (rx)(sy) \text{ (by Theorem 2.7.12)} \\
&= |rx| \cdot |sy| \text{ (by the definition of } r, s) \\
&= |x||y| \text{ (by (2)).} \qquad \square
\end{aligned}$$

The last part of the theorem above shows that the absolute value works well with multiplication: the absolute value of the product is the product of absolute values. It is not the case that the absolute value of the sum of two numbers is always the sum of their absolute values. Instead we have triangle inequalities as in the theorem below. We use the standard notation "$\pm$" to mean that the result holds with either $+$ or $-$.

Theorem 2.10.3. *The following inequalities hold for an ordered field F.*

(1) **Triangle inequality:** *For all $x, y \in F$, $|x \pm y| \leq |x| + |y|$.*

(2) **Reverse triangle inequality:** *For all $x, y \in F$, $|x \pm y| \geq ||x| - |y|| = ||y| - |x||$.*

Proof. (1) By the first part of the previous theorem, $-|x| \leq x \leq |x|$ and $-|y| \leq y \leq |y|$. Thus

$-(|x| + |y|) = (-|x|) + (-|y|)$ (by Theorem 2.5.14)

$\leq x + (-|y|)$ (by compatibility of $\leq$ with addition)

$\leq x + y$ (by compatibility of $\leq$ with addition)

$\leq |x| + y$ (by compatibility of $\leq$ with addition)

$\leq |x| + |y|$ (by compatibility of $\leq$ with addition),

so that by transitivity, $-(|x| + |y|) \leq x + y \leq |x| + |y|$. Thus by part (4) of the previous theorem, $|x + y| \leq |x| + |y|$. It follows that $|x - y| = |x + (-y)| \leq |x| + |-y|$, and by the second part again this is equal to $|x| + |y|$.

(2) By (1), $|x| = |x \pm y - (\pm y)| \leq |x \pm y| + |y|$, so that $|x| - |y| \leq |x \pm y|$. Similarly, $|y| - |x| \leq |y \pm x|$. But by the second part of the previous theorem, $|y - x| = |-(y - x)| = |x - y|$ and $|y + x| = |x + y|$, so that $|x \pm y| \geq |x| - |y|$ and $|x \pm y| \geq |y| - |x| = -(|x| - |y|)$. Since $||x| - |y||$ is either $|x| - |y|$ or $|y| - |x|$, (2) follows. □

Observe that the proof of the reverse triangle inequality above used the triangle inequality and did not require referencing Theorem 2.10.2. This made the proof shorter.

Theorem 2.10.4. *Let F be an ordered field. Let $r \in F$.*

(1) If $r < \epsilon$ for all $\epsilon \in F^+$, then $r \leq 0$.

(2) If $r > -\epsilon$ for all $\epsilon \in F^+$, then $r \geq 0$.

(3) If $|r| < \epsilon$ for all $\epsilon \in F^+$, then $r = 0$.

Proof. Proof of (1): By Theorem 2.7.11, $1 \in F^+$, so that $0 < 1$, and by compatibility of $<$ with addition, $1 < 2$. Thus by transitivity of $<$, $0 < 2$, so that 2 is positive, and by Theorem 2.7.11 (3), $2^{-1} \in F^+$. If $r > 0$, by compatibility of $<$ with multiplication by positive numbers, $\epsilon = r/2$ is a positive number. By assumption, $r < \epsilon = r/2$. Again by compatibility of $<$ with multiplication, by multiplying through by $2r^{-1}$, we get that $2 < 1$, which contradicts the trichotomy (since we already established that $1 < 2$). Thus $r \notin F^+$, so that $r \leq 0$.

The proof of (2) is similar.

For (3), if $|r| < \epsilon$ for all $\epsilon \in F^+$, then $-\epsilon < r < \epsilon$. Then by (1) and (2), $0 \leq r \leq 0$. Since $F^+ \cap F^- = \emptyset$ by trichotomy, it follows that $r = 0$. □

Exercises for Section 2.10

2.10.1. Let F be an ordered field. Prove that the absolute value function on F is increasing on $F^+ \cup \{0\}$ and decreasing on $F^- \cup \{0\}$.

2.10.2. Let F be an ordered field.

i) Prove that for all $a, b \in F$, $\left| |a| + |b| \right| = |a| + |b|$.

ii) Prove that for all $a, b \in F^+$, $\left| |a| + |b| \right| = a + b$.

iii) Prove that for all $a, b \in F^-$, $\left| |a| + |b| \right| = -a - b$.

2.10.3. Let F be an ordered field, $a \in F$ and $r \in F^+$. Express the sets $\{x \in F : |x - a| < r\}$ and $\{x \in F : |x - a| \leq r\}$ in interval notation.

2.10.4. (Triangle inequality) Let F be an ordered field and $a_1, \ldots, a_n \in F$. Prove that

$$|a_1 + a_2 + \cdots + a_n| \leq |a_1| + |a_2| + \cdots + |a_n|.$$

2.10.5. (Reverse triangle inequality) Let F be an ordered field and $a_1, \ldots, a_n \in F$. Prove that

$$|a_1 + a_2 + \cdots + a_n| \geq |a_1| - |a_2| - \cdots - |a_n|.$$

Give an example of real numbers a_1, a_2, a_3 with $|a_1 + a_2 + a_3| \not\geq \left| |a_1| - |a_2| - |a_3| \right|$.

2.10.6. Give an example of a set S in $\mathbb{R}$ that is bounded above but $\{|s| : s \in S\}$ is not bounded above.

Chapter 3

Construction of the number systems

In Chapter 1 we laid the basic groundwork for how we do mathematics: how we reason logically, how we prove further facts from established truths, and we learned some notation. In Chapter 2 we introduced sets, and from those derived functions and binary operations that play a big role in mathematics. In this chapter, we use set theory to derive numbers and basic arithmetic: on $\mathbb{N}_0$ (non-negative whole numbers), $\mathbb{Z}$ (integers), $\mathbb{Q}$ (rational numbers), $\mathbb{R}$ (real numbers), $\mathbb{C}$ (complex numbers), in this order.

When I teach this course, I go through lightly the first eight sections of this very long chapter. Dedekind cuts give straightforward proofs, but if one wishes to avoid Dedekind cuts, then these theorems can (and should) be simply accepted as facts. Complex numbers play a central role in the rest of the book, so Sections 3.9 through 3.12 should be studied thoroughly. Sections 3.13 and 3.14 introduce topology, and they contain more information than strictly necessary for the rest of the book.

3.1 Inductive sets, a construction of natural numbers

Inductive sets present a new way of thinking about mathematical induction from Section 1.6.

We start with the most basic set: $\emptyset$. We can make the empty set be the unique element of a new set, like so: $\{\emptyset\}$. The sets $\emptyset$ and $\{\emptyset\}$ are distinct because the latter contains the empty set and the former contains no elements. With these two distinct sets we can form a new set with precisely these two elements: $\{\emptyset, \{\emptyset\}\}$. This set certainly differs from the empty set, and it also differs from $\{\emptyset\}$ because $\{\emptyset\}$ is not an element of $\{\emptyset\}$ (but it is a subset — think about this). So now we have three distinct sets, and we can form another set from these: $\{\emptyset, \{\emptyset\}, \{\emptyset, \{\emptyset\}\}\}$, giving us distinct sets

$$\emptyset, \quad \{\emptyset\}, \quad \{\emptyset, \{\emptyset\}\}, \quad \{\emptyset, \{\emptyset\}, \{\emptyset, \{\emptyset\}\}\}.$$

We could call these Zeno, Juan, Drew, Tricia, but more familiarly these sets can be called zero, one, two, three, and written 0, 1, 2, 3 for short. At this point, we are simply giving them names, with no assumption on any arithmetic properties. We could say "and so on", but that is not very rigorous, is it.

Definition 3.1.1. *For any set S, its* **successor** S^+ *is defined as* $S^+ = S \cup \{S\}$.

By our naming convention, 0 is not the successor of any set, $0^+ = 1$, $1^+ = 2$, $2^+ = 3$. We can also write $0^{++} = (0^+)^+ = 1^+ = 2$ and $0^{+++} = ((0^+)^+)^+ = (1^+)^+ = 2^+ = 3$. At this point I rely on your basic training for the **naming conventions** of $3^+ = 4$, $3^{++} = 5, \ldots$.

The definition of successors of successors of successors... is recursive. We can see informally that the set of all successive successors of 0 is infinite, in fact we hope that this set would be the familiar $\mathbb{N}_0$, but to get to that we would have to take a union of infinitely many sets that may not be elements of some universal set. Taking unions of infinitely many sets is subject to pitfalls, just like infinite sums have pitfalls.

Definition 3.1.2. *A set of sets J is called* **inductive** *if it satisfies two conditions:*

(1) $\emptyset \in J$.

(2) For any $n \in J$, the successor n^+ of n is also in J.

We accept as a given (we take it as an **axiom**) that there exists a set that contains $\emptyset$ and all its successors:

Axiom 3.1.3. *There exists an inductive set.*

An inductive set has to contain the familiar $0, 1, 2, 3, \ldots$, but possibly it can also contain the familiar numbers -1.3 or π, which are not part of the familiar (inductive set) $\mathbb{N}_0$.

Theorem 3.1.4. *Let J be an inductive set. Define $\mathbb{N}_0$ as the intersection of all inductive subsets of J. Then $\mathbb{N}_0$ is the smallest inductive subset of J in the sense that if T is any inductive subset of J, then $\mathbb{N}_0 \subseteq T$.*

Proof. Each inductive set contains $\emptyset$, hence their intersection contains $\emptyset$ as well, so that $\emptyset \in \mathbb{N}_0$. If n is in $\mathbb{N}_0$, then by definition of $\mathbb{N}_0$, this n is in all the inductive sets, hence n^+ is also in those same inductive sets, which says that n^+ is in the intersection. Thus $\mathbb{N}_0$ is an inductive set. Furthermore, the intersection of some sets is a subset of all of those, so that $\mathbb{N}_0$ is a subset of each inductive subset of J. □

Definition 3.1.5. (Definition of $\mathbb{N}_0$) *With definition of $\mathbb{N}_0$ as in Theorem 3.1.4, elements of $\mathbb{N}_0$ are called* **natural numbers.** *(Beware: in some books, this set is written as $\mathbb{N}$, whereas in some other books $\mathbb{N}$ stands for all non-zero natural numbers. In this book, I always write $\mathbb{N}_0$ for clarity.)*

Exercise 3.1.4 shows that $\mathbb{N}_0$ does not depend on the choice of J.

Some of the elements of $\mathbb{N}_0$ are: $\emptyset, \{\emptyset\}, \{\emptyset, \{\emptyset\}\}, \{\emptyset, \{\emptyset\}, \{\emptyset, \{\emptyset\}\}\}$, or written alternatively: $0, 1, 2, 3$. In fact, we prefer to write elements of $\mathbb{N}_0$ numerically, but for proofs we often need sets to resort to the notion of successors of sets.

Theorem 3.1.6. *0 is not the successor of any element of $\mathbb{N}_0$, and every element of $\mathbb{N}_0$ other than 0 is the successor of some element of $\mathbb{N}_0$.*

Proof. 0 stands for the empty set, which by definition cannot be the successor of any set.

Let $n \in \mathbb{N}_0 \setminus \{0\}$. If n is not the successor of any element in $\mathbb{N}_0$, then $\mathbb{N}_0 \setminus \{n\}$ is an inductive set which is strictly smaller than $\mathbb{N}_0$. But this contradicts Theorem 3.1.4. Thus there is no such n, which proves that every element of $\mathbb{N}_0 \setminus \{0\}$ is the successor of some element of $\mathbb{N}_0$. □

As the successor seems to act like "$+1$", this says that if π were to be in $\mathbb{N}_0$, then $\pi - 1$ would have to be in $\mathbb{N}_0$ as well. If we think of inclusions as "less than"s, the next theorem removes this possibility if we think of π as a number between 3 and 4.

Theorem 3.1.7. *Let $m, n \in \mathbb{N}_0$ be such that $n \subseteq m \subseteq n^+$. Then either $m = n$ or $m = n^+$.*

Proof. Suppose that $m \neq n^+$. As m is a subset of n^+, this means that some $x \in n^+ = n \cup \{n\}$ is not in m. But $n \subseteq m$, so every element of n is in m, which means that $x = n$, and necessarily $m = n$. $\qquad\square$

Recall mathematical induction from Section 1.6. The notion of inductive sets gives us a new perspective at induction and it enables us to eventually rigorously define the (usual) arithmetic. We will use the following new form of induction over and over:

Theorem 3.1.8. (Induction Theorem) *Let P be a property applicable to (some) elements of $\mathbb{N}_0$. Suppose that*

(1) $P(0)$ is true;

(2) For all $n \in \mathbb{N}_0$, $(P(n) \implies P(n^+))$ is true.

Then $P(n)$ is true for all $n \in \mathbb{N}_0$.

Proof. [WE ARE EXPECTED TO USE THE NEW CONCEPT OF INDUCTIVE SETS, SO WE PROBABLY NEED TO DEFINE A SET THAT TURNS OUT TO BE INDUCTIVE.] Let $T = \{n \in \mathbb{N}_0 : P(n) \text{ is true}\}$. [WE WILL PROVE THAT THIS T IS INDUCTIVE.] By assumption (1) we have that $0 \in T$. [SO THE FIRST PROPERTY OF INDUCTIVE SETS HOLDS FOR T.] If $n \in T$, then $P(n)$ is true, and since $(P(n) \implies P(n^+))$ is true, it follows that $P(n^+)$ is true. Thus $n^+ \in T$. This means that T is an inductive set. But T is a subset of $\mathbb{N}_0$ by definition, and $\mathbb{N}_0$ is the smallest inductive set by Theorem 3.1.4, which means that $\mathbb{N}_0 \subseteq T$, so that $P(n)$ is true for all $n \in \mathbb{N}_0$. $\qquad\square$

We illustrate below how the inductive theorem can be used. Many more examples are in the next three sections.

Theorem 3.1.9. *For all $n \in \mathbb{N}_0$, $0 \in n^+$.*

Proof. Set $T = \{n \in \mathbb{N}_0 : 0 \in n^+\}$. Since $0^+ = \{\emptyset\}$ contains $0 = \emptyset$, we have that $0 \in T$. Now assume $n \in T$. This means that $0 \in n^+ \subseteq n^+ \cup \{n^+\} = (n^+)^+$, so that $0 \in (n^+)^+$. Thus $n^+ \in T$. Hence the property holds for all $n \in \mathbb{N}_0$ by Theorem 3.1.8. $\qquad\square$

In more familiar language, this is saying that 0 is an element of (sets) $1, 2, 3$, etc.

I note that the following property is not true for arbitrary sets m and n, but it is true for sets that are elements of $\mathbb{N}_0$.

Theorem 3.1.10. *Let $m, n \in \mathbb{N}_0$. If $m \in n$ then $m \subseteq n$.*

Proof. The property in the conditional vacuously holds for $n = 0$ and all $m \in \mathbb{N}_0$. Suppose that $n \in \mathbb{N}_0$ has the property that for all $m \in \mathbb{N}_0$, if $m \in n$ then $m \subseteq n$. We want to prove that n^+ has the same property (as n). So let $m \in \mathbb{N}_0$, and suppose that $m \in n^+$. Since $n^+ = n \cup \{n\}$, either $m \in n$ or $m = n$. By the assumed property on n, this means that $m \subseteq n \subseteq n^+$. Thus by Theorem 3.1.8, the theorem is proved. □

Later, in Theorem 3.1.12, we prove more: that $m \in n$ if and only if $m \subsetneq n$.

Theorem 3.1.11. *For all $n \in \mathbb{N}_0$,*

$$n \notin n, \qquad n^+ \not\subseteq n, \qquad n^+ \notin n.$$

In particular, $n \neq n^+$.

Proof. Let T be the subset of $\mathbb{N}_0$ containing all n for which $n \notin n$, $n^+ \not\subseteq n$, and $n^+ \notin n$.

Since the empty set contains no elements, it follows immediately that $0 \notin 0$ and $0^+ \notin 0$. Furthermore, since $0^+ = \emptyset \cup \{\emptyset\} = \{\emptyset\}$, this set is not a subset of the empty set. Thus $0 \in T$.

Now suppose that $n \in T$.

Suppose (for contradiction) that $n^+ \in n^+$. By the definition of n^+, this says that $n^+ \in n \cup \{n\}$, so that either $n^+ \in n$ or $n^+ = n$. But both of these contradict the assumption that $n \in T$. Thus we have proved that $n^+ \notin n^+$.

Suppose (for contradiction) that $(n^+)^+ \subseteq n^+$. By the definition of $(n^+)^+$, this says that $n^+ \cup \{n^+\} \subseteq n^+$. In particular, $n^+ \in n^+ \cup \{n^+\} \subseteq n^+$, which contradicts the previous paragraph, and thus proves that $(n^+)^+ \not\subseteq n^+$.

Finally, suppose (for contradiction) that $(n^+)^+ \in n^+$. Since $n^+ = n \cup \{n\}$, this means that either $(n^+)^+ \in n$ or $(n^+)^+ = n$. By Theorem 3.1.10 both of these mean that $(n^+)^+ \subseteq n$, so that in particular the subset n^+ of $(n^+)^+$ is a subset of n, which contradicts the assumption that $n \in T$. □

Theorem 3.1.12. *For all $m, n \in \mathbb{N}_0$, $m \in n$ if and only if $m \subsetneq n$.*

Proof. Suppose that $m \in n$. By Theorem 3.1.10 we know that $m \subseteq n$, and by Theorem 3.1.11 we know that $m \neq n$. Thus if $m \in n$ then $m \subsetneq n$.

Set $T = \{n \in \mathbb{N}_0 : \text{for all } m \in \mathbb{N}_0, \text{ if } m \subsetneq n \text{ then } m \in n\}$. Note that $m \in \emptyset$ and $m \subsetneq \emptyset$ are both impossible, so that $n = \emptyset = 0 \in T$ vacuously. Now let $n \in T$, and let $m \in \mathbb{N}_0$ with $m \subsetneq n^+$. We want to prove that $m \in n^+$. If $n \in m$, then by the first paragraph of this proof, $n \subsetneq m$. As we also have $m \subsetneq n^+$, we get a contradiction to Theorem 3.1.7. So necessarily n is not an element of m. But then the assumption $m \subsetneq n^+ = n \cup \{n\}$ means that $m \subseteq n$. If $m = n$, then $m \in \{n\} \subseteq n^+$, and if $m \subsetneq n$, then since $n \in T$, this implies that $m \in n$. But n is a subset of n^+, so that $m \in n^+$.   □

Theorem 3.1.13. *For $m, n \in \mathbb{N}_0$, $m \subsetneq n$ if and only if $m^+ \subsetneq n^+$.*

Proof. Suppose that $m \subsetneq n$. By Theorem 3.1.12, this means that $m \in n$. Thus $m^+ = m \cup \{m\} \subset n$, and by Theorem 3.1.11, this is properly contained in n^+. This proves that $m^+ \subsetneq n^+$.

Suppose that $m^+ \subsetneq n^+$. This means that $m \cup \{m\} = m^+ \subsetneq n^+ = n \cup \{n\}$. If $n \subseteq m$, then by Theorem 3.1.7, $n = m$, which contradicts the assumption $m^+ \subsetneq n^+$. So n is not a subset of m. Thus the assumption $m \cup \{m\} \subsetneq n \cup \{n\}$ means that $m \cup \{m\} \subseteq n$, and thus $m \subseteq n$. But $m \in n$ by the same assumption but $m \notin m$ by Theorem 3.1.11, it follows that $m \subsetneq n$.   □

Theorem 3.1.14. *For all $m, n \in \mathbb{N}_0$, either $m \subseteq n$ or $n \subseteq m$.*

Proof. Let $T = \{m \in \mathbb{N}_0 : \text{for all } n \in \mathbb{N}_0, \text{ either } m \subseteq n \text{ or } n \subseteq m\}$. Certainly the empty set is a subset of any set, so that $0 \in T$.

Suppose that $m \in T$. We want to prove that $m^+ \in T$. Let n be arbitrary in $\mathbb{N}_0$. We have to prove that either $m^+ \subseteq n$ or $n \subseteq m^+$. This is certainly true if $n = 0$, so it suffices to prove this in case $n \neq 0$. Then by Theorem 3.1.6, $n = p^+$ for some $p \in \mathbb{N}_0$. Since $m \in T$, it follows that either $m \subseteq p$ or $p \subseteq m$, and by Theorem 3.1.13, this says that either $m^+ \subseteq p^+$ or $p^+ \subseteq m^+$. In other words, either $m^+ \subseteq n$ or $n \subseteq m^+$. Thus $m^+ \in T$, and Theorem 3.1.8 finishes the proof.   □

Exercises for Section 3.1

3.1.1. Prove that for every set S, $S \in S^+$ and $S \subseteq S^+$.

3.1.2. Let $T = \{\emptyset, \{\emptyset\}, \{\{\emptyset\}\}\}$. Prove that T is not the successor of any set S.

3.1.3. Prove that for all $n \in \mathbb{N}_0$, if $n \neq 0$ and $n \neq 1$, then $1 \in n$. (Hint: set $T = \{m \in \mathbb{N}_0 : 1 \in (m^+)^+\}$.)

3.1.4. Let J and K be inductive sets. Let N be the intersection of all inductive subsets of J, and let M be the intersection of all inductive subsets of K.

   i) Prove that $T = \{n \in N : n \in M\}$ is an inductive set.

   ii) Prove that $T = N = M$.

   iii) Prove that the definition of $\mathbb{N}_0$ is independent of the ambient inductive set J.

3.2   Arithmetic on $\mathbb{N}_0$

We apply the Induction Theorem (Theorem 3.1.8) to define addition and multiplication on $\mathbb{N}_0$. For example, for any $m \in \mathbb{N}_0$ we define the "adding m" function $A_m : \mathbb{N}_0 \to \mathbb{N}_0$ as follows:

$$A_m(0) = m; \qquad A_m(n^+) = A_m(n)^+ \text{ for } n \in \mathbb{N}_0.$$

The induction theorem says that $A_m(n)$ is defined for all m and n: $A_m(0)$ is in $\mathbb{N}_0$, and if $A_m(n) \in \mathbb{N}_0$, then $A_m(n)^+$ is in $\mathbb{N}_0$, hence $A_m(n^+) \in \mathbb{N}_0$, so that A_m is a function from $\mathbb{N}_0$ to $\mathbb{N}_0$. Similarly, we define "multiplication by m" as a function $M_m : \mathbb{N}_0 \to \mathbb{N}_0$ given by

$$M_m(0) = 0; \qquad M_m(n^+) = A_m(M_m(n)).$$

With this functions we define binary operations $+$ and $\cdot$ on $\mathbb{N}_0$:

$$m + n = A_m(n), \qquad m \cdot n = M_m(n).$$

Example 3.2.1. We will prove that $2 + 2 = 4$, and you can verify that $2 \cdot 2 = 4$.

$$2 + 2 = A_2(2)$$
$$= A_2(1^+)$$
$$= A_2(1)^+$$
$$= A_2(0^+)^+$$
$$= (A_2(0)^+)^+$$
$$= (2^+)^+$$
$$= 3^+$$
$$= 4.$$

Remark 3.2.2. For all $m, n \in \mathbb{N}_0$,

$$m + 1 = m^+, \quad m + (n^+) = (m + n)^+ \quad \text{and} \quad m \cdot (n^+) = m + (m \cdot n).$$

The last equality is simply rewriting the meanings of $\cdot$, M_m and A_m, the second equality follows by the definitions with $m + n^+ = A_m(n^+) = (A_m(n))^+ = (m + n)^+$, and when $n = 0$, this produces $m + 1 = m + 0^+ = (m + 0)^+ = (A_m(0))^+ = m^+$.

Theorem 3.2.3. *The following identities hold for all $m, n, p \in \mathbb{N}_0$:*
  (1) **(Additive identity)** $m + 0 = m = 0 + m$.
  (2) $m \cdot 0 = 0 = 0 \cdot m$.
  (3) **(Associativity of addition)** $m + (n + p) = (m + n) + p$.
  (4) **(Commutativity of addition)** $m + n = n + m$.
  (5) **(Multiplicative identity)** $m \cdot 1 = m = 1 \cdot m$.
  (6) **(Distributivity)** $m \cdot (n + p) = (m \cdot n) + (m \cdot p)$.
       (Distributivity) $(n + p) \cdot m = (n \cdot m) + (p \cdot m)$.
  (7) **(Associativity of multiplication)** $m \cdot (n \cdot p) = (m \cdot n) \cdot p$.
  (8) **(Commutativity of multiplication)** $m \cdot n = n \cdot m$.

Proof. (1) For all $m \in \mathbb{N}_0$, by definitions of $+$ and A_m, $m + 0 = A_m(0) = m$. It remains to prove that $0 + m = m$. Let $T = \{m \in \mathbb{N}_0 : 0 + m = m\}$. Since $0 + 0 = A_0(0) = 0$, it follows that $0 \in T$. If $m \in T$, then

$$0 + (m^+) = (0 + m)^+ \quad \text{(by Remark 3.2.2)}$$
$$= m^+ \quad \text{(as } m \in T\text{)},$$

so that $m^+ \in T$. Theorem 3.1.8 then says that for all $m \in \mathbb{N}_0$, $m + 0 = 0 = 0 + m$.

(2) By definition of $\cdot$, $m \cdot 0 = M_m(0) = 0$. Let $T = \{m \in \mathbb{N}_0 : 0 \cdot m = 0\}$. By $m \cdot 0 = 0$, $0 \in T$. If $m \in T$, then

$$
\begin{aligned}
0 \cdot (m^+) &= 0 + (0 \cdot m) \quad \text{(by Definition of $\cdot$)} \\
&= 0 + 0 \quad \text{(as $m \in T$)} \\
&= 0 \quad \text{(by (1))},
\end{aligned}
$$

so that $m^+ \in T$. Thus (2) is proved by Theorem 3.1.8.

(3) Fix $m, n \in \mathbb{N}_0$, and let $T = \{p \in \mathbb{N}_0 : m + (n + p) = (m + n) + p\}$. Since $m + (n + 0) = m + n = (m + n) + 0$, it follows that $0 \in T$. If $p \in T$, then

$$
\begin{aligned}
m + (n + (p^+)) &= m + ((n + p)^+) \quad \text{(by Remark 3.2.2)} \\
&= (m + (n + p))^+ \quad \text{(by Remark 3.2.2)} \\
&= ((m + n) + p)^+ \quad \text{(as $p \in T$)} \\
&= (m + n) + (p^+) \quad \text{(by Remark 3.2.2)},
\end{aligned}
$$

so that $p^+ \in T$, and so (3) holds by Theorem 3.1.8.

(4) You prove this. First prove that $n + 1 = 1 + n$ for all $n \in \mathbb{N}_0$.

(5) By definition of $\cdot$, $m \cdot 1 = m \cdot (0^+) = m + m \cdot 0$. In (2) we already proved that $m \cdot 0 = 0$, and in (1) we proved that $m + 0 = m$. Thus $m \cdot 1 = m \cdot 0 + m = 0 + m = m$. Let $T = \{m \in \mathbb{N}_0 : 1 \cdot m = m\}$. By (2), $0 \in T$. If $m \in T$, then

$$
\begin{aligned}
1 \cdot (m^+) &= 1 + (1 \cdot m) \quad \text{(by Definition of $\cdot$)} \\
&= 1 + m \quad \text{(as $m \in T$)}, \\
&= m + 1 \quad \text{(by (4))}, \\
&= m^+ \quad \text{(by Remark 3.2.2)},
\end{aligned}
$$

so that $m^+ \in T$, and Theorem 3.1.8 finishes the proof of (5).

(6) Fix $m, n \in \mathbb{N}_0$. Let $T = \{p \in \mathbb{N}_0 : m \cdot (n + p) = (m \cdot n) + (m \cdot p)\}$. By (2), $m \cdot 0 = 0$, and by (1), $(m \cdot n) + 0 = m \cdot n$, so that $0 \in T$. If $p \in T$, then

$$
\begin{aligned}
m \cdot (n + (p^+)) &= m \cdot ((n + p)^+) \quad \text{(by Remark 3.2.2)} \\
&= m + (m \cdot (n + p)) \quad \text{(by Definition of $\cdot$)} \\
&= m + ((m \cdot n) + (m \cdot p)) \quad \text{(as $p \in T$)}
\end{aligned}
$$

$$
\begin{aligned}
&= (m + (m \cdot n)) + (m \cdot p) \quad \text{(by (3))}\\
&= ((m \cdot n) + m) + (m \cdot p) \quad \text{(by (4))}\\
&= (m \cdot n) + (m + (m \cdot p)) \quad \text{(by (3))}\\
&= (m \cdot n) + (m \cdot (p^+)) \quad \text{(by Remark 3.2.2)},
\end{aligned}
$$

so that $p^+ \in T$. Thus the first part of (6) follows by Theorem 3.1.8.

Fix $m, n \in \mathbb{N}_0$. Let $T = \{p \in \mathbb{N}_0 : (m + n) \cdot p = (m \cdot p) + (n \cdot p)\}$. By (2) and (1), $(m + n) \cdot 0 = 0 + 0 = (m \cdot 0) + (n \cdot 0)$, so that $0 \in T$. If $p \in T$, then

$$
\begin{aligned}
(m + n) \cdot (p^+) &= M_{m+n}(p^+) \quad \text{(by Definition of } \cdot\text{)}\\
&= (m + n) + ((m + n) \cdot p) \quad \text{(by Definition of } M_{m+n}\text{)}\\
&= (m + n) + ((m \cdot p) + (n \cdot p)) \quad \text{(as } p \in T\text{)}\\
&= ((m + n) + (m \cdot p)) + (n \cdot p) \quad \text{(by (3))}\\
&= ((n + m) + (m \cdot p)) + (n \cdot p) \quad \text{(by (4))}\\
&= (n + (m + (m \cdot p))) + (n \cdot p) \quad \text{(by (3))}\\
&= (n + M_m(p^+)) + (n \cdot p) \quad \text{(by Definition of } M_m\text{)}\\
&= (M_m(p^+) + n) + (n \cdot p) \quad \text{(by (4))}\\
&= M_m(p^+) + (n + (n \cdot p)) \quad \text{(by (3))}\\
&= M_m(p^+) + (n + M_n(p^+)) \quad \text{(by Definition of } M_n\text{)}\\
&= (m \cdot p^+) + (n \cdot p^+) \quad \text{(by Definition of } \cdot\text{)},
\end{aligned}
$$

so that $p^+ \in T$, and so by Theorem 3.1.8, the second part of (6) holds for all $m, n, p \in \mathbb{N}_0$.

(7) You prove (7).

(8) Fix $m \in \mathbb{N}_0$. Let $T = \{n \in \mathbb{N}_0 : n \cdot m = m \cdot n \text{ for all } m \in \mathbb{N}_0\}$. By (2), $0 \in T$. If $n \in T$, then

$$
\begin{aligned}
m \cdot (n^+) &= m + m \cdot n \quad \text{(by Definition of } \cdot\text{)}\\
&= m + n \cdot m \quad \text{(since } n \in T\text{)}\\
&= 1 \cdot m + n \cdot m \quad \text{(by (5))}\\
&= (1 + n) \cdot m \quad \text{(by (6))}\\
&= (n + 1) \cdot m \quad \text{(by (4))}\\
&= n^+ \cdot m,
\end{aligned}
$$

so that $n^+ \in T$, which proves (8) by Theorem 3.1.8.   □

This gives us the familiar $\mathbb{N}_0$ with the familiar arithmetic properties. In particular, $\mathbb{N}_0$ has a binary operation $+$ with identity 0 and a binary operation $\cdot$ with identity 1. We already proved abstractly in Theorem 2.5.5 that such identities are unique.

Exercises for Section 3.2

3.2.1. Verify with the new definitions that $3 \cdot 2 = 6$ and that $2 \cdot 3 = 6$. Was one easier?

3.2.2. Prove (4) and (7) in Theorem 3.2.3.

3.2.3. Prove that for all $m, n \in \mathbb{N}_0$, $m^+ + n = n^+ + m$.

3.2.4. Prove that for all $m, n, p \in \mathbb{N}_0$, $(m^+) \cdot (p + n) = m \cdot (p + n) + (p + n)$.

3.3   Order on $\mathbb{N}_0$

Definition 3.3.1. *For $a, b \in \mathbb{N}_0$, we write $a \leq b$ to mean $a \subseteq b$. Relations $<, \geq, >$ are defined from $\leq$ as in Section 2.7. In particular, $a < b$ means that $a \subseteq b$ and $a \neq b$.*

Remark 3.3.2.

(1) Certainly $0 \leq n$ for all $n \in \mathbb{N}_0$ as 0 represents the empty set.

(2) $\leq$ is reflexive.

(3) $\leq$ is transitive, as set inclusion is transitive. It follows by Exercise 2.7.2 that $<, \geq, >$ are transitive.

(4) $\leq$ is not symmetric: $0 \leq 1$ but $1 \not\leq 0$ (because the empty set has no non-empty subsets).

(5) For all $m, n \in \mathbb{N}_0$, by definition of set equality, whenever $m \leq n$ and $n \leq m$, then $m = n$.

(6) By Theorem 3.1.12, $m \in n$ if and only if $m \subsetneq n$, which in the order notation says that $m \in n$ if and only if $m < n$.

(7) For all $n \in \mathbb{N}_0$, certainly $n \subseteq n^+$, so that $n \leq n^+$. Furthermore, by Theorem 3.1.11, $n < n^+$.

(8) By Theorem 3.1.13, $m < n$ if and only if $m^+ < n^+$.

(9) By Theorem 3.1.14, for all $m, n \in \mathbb{N}_0$, either $m \leq n$ or $n \leq m$.

Theorem 3.3.3. (Trichotomy on $\mathbb{N}_0$) *For all $m, n \in \mathbb{N}_0$, exactly one of the following conditions hold:*

(1) $m < n$,

(2) $n < m$,

(3) $m = n$.

Proof. By the last remark above, either $m \leq n$ or $n \leq m$, which means that one of the listed three conditions holds. If (1) and (2) hold or if (1) and (3) hold, then by transitivity of $<$, $m < m$, which contradicts the definition of $<$. If (2) and (3) hold, then by transitivity, $n < n$, which again gives a contradiction. So no two of the conditions can hold simultaneously, which proves the theorem. $\qquad\square$

Theorem 3.3.4. (The well-ordering principle) $\mathbb{N}_0$ *is well-ordered (see Definition 2.7.7). In other words, every non-empty subset S of $\mathbb{N}_0$ has a* **least element***, that is, an element r such that for all $t \in S$, $r \leq t$.*

Proof. We will prove that the following set is inductive:

$$T = \{n \in \mathbb{N}_0 : \text{ if } S \subseteq \mathbb{N}_0 \text{ and } n \in S, \text{ then } S \text{ has a least element}\}.$$

Note that $0 \in T$ because $0 \leq n$ for all $n \in \mathbb{N}_0$ (and hence $0 \leq n$ for all $n \in S$). Suppose that $n \in T$. We need to prove that $n^+ \in T$. Let $S \subseteq \mathbb{N}_0$ and $n^+ \in S$. By assumption that $n \in T$, it follows that the set $S \cup \{n\}$ has a least element. Thus there exists $r \in S \cup \{n\}$ such that for all $t \in S$, $r \leq t$ and $r \leq n$. If $r \in S$, then we just showed that for all $t \in s$, $r \leq t$, so that S has a least element. Now suppose that $r \notin S$. So necessarily $r = n$, and this is not an element of S. Then we claim that n^+ is the least element of S. Namely, let $t \in S$. We need to prove that $n^+ \leq t$. Suppose for contradiction that $t < n^+$. Since n is the least element of $S \cup \{n\}$, it follows that $n \leq t$. Thus we have $n \leq t < n^+$, so that by Theorem 3.1.7, necessarily $n = t$. But then $n = t \in S$, which contradicts the assumption that $n \notin S$. Thus $n^+ \leq t$, and since t was an arbitrary element of S, it follows that n^+ is the least element of S. Thus in all cases, if $n^+ \in S$, the set S has a least element. So $n^+ \in T$. Thus T is an inductive set, and hence equal to $\mathbb{N}_0$.

This means that every non-empty subset of $\mathbb{N}_0$ has a least element. $\qquad\square$

Now we relate order to arithmetic more directly:

Theorem 3.3.5. *If* $m, n \in \mathbb{N}_0$ *and* $m \le n$, *there exists* $r \in \mathbb{N}_0$ *such that* $m + r = n$.

Proof. Let $T = \{n \in \mathbb{N}_0 :$ for all $m \in \mathbb{N}_0$, if $m \le n$ then $\exists r \in \mathbb{N}_0$ such that $m + r = n\}$.

If $m \in \mathbb{N}_0$ and $m \le 0$, then since 0 stands for the empty set, necessarily $m = 0$, and hence we can set $r = 0$ to get $m + r = n$. This proves that $0 \in T$.

Suppose that $n \in T$. We want to prove that $n^+ \in T$. Let $m \in \mathbb{N}_0$ such that $m \le n^+$. If $m = 0$, set $r = n^+$; then $m + r = 0 + n^+ = n^+$. If instead $m > 0$, then by Theorem 3.1.6, $m = p^+$ for some $p \in \mathbb{N}_0$. Then by assumption $p^+ = m \le n^+$, and so by Exercise 3.3.1, $p \le n$. Since $n \in T$, there exists $r \in \mathbb{N}_0$ such that $p + r = n$. Hence $m + r = p^+ + r = (p + r)^+ = n^+$. Thus in all cases, whether $m = 0$ of $m > 0$, there exists $r \in \mathbb{N}_0$ such that $m + r = n^+$, so that $n^+ \in T$. Thus T is an inductive set, and since $\mathbb{N}_0$ is the smallest inductive set, it follows that $T = \mathbb{N}_0$. (Alternatively, we can invoke Theorem 3.1.8.) □

Exercises for Section 3.3

3.3.1. Let $m, n \in \mathbb{N}_0$. Prove that $m \le n$ if and only if $m^+ \le n^+$.

3.3.2. Prove that if $m < n$ and $n \le p$, then $m < p$. Verify also other standard combinations of $\le$ and $<$.

3.3.3. Let $m, n \in \mathbb{N}_0$ and $m < n$. Prove that there exists $r \in \mathbb{N}^+$ such that $m + r = n$.

3.3.4. (Compatibility of $\le$ and $<$ with addition) Let $m, n, p \in \mathbb{N}_0$. Prove that $m \ge n$ if and only if $m + p \ge n + p$ and that $m > n$ if and only if $m + p > n + p$.

3.3.5. (Compatibility of $\le$ with multiplication by non-zero numbers) Let $m, n, p \in \mathbb{N}_0$. Suppose that $m \le n$. Prove that $mp \le np$ for all p.

3.3.6. Suppose that $m, p, n \in \mathbb{N}_0$ and $m \le n$. Prove that $m + p < n + p^+$.

3.3.7. Prove that for all $n \in \mathbb{N}_0$, $n < 2^n$. (Recall Notation 2.5.13. Use inductive sets.)

3.3.8. Prove that for all $a, b \in \mathbb{N}^+$, $a \leq ab$ (and $b \leq ab$).

3.3.9. Let $a, b, c \in \mathbb{N}_0$. Is it true that $ab = c$ implies that $a \leq c$?

3.3.10. Let $m, n, r \in \mathbb{N}_0$ such that $m = n + r$. Prove that $m \leq n$.

3.4   Cancellation in $\mathbb{N}_0$

The familiar subtraction cannot be applied to arbitrary pairs of elements in $\mathbb{N}_0$: recall that elements of $\mathbb{N}_0$ are sets, and we have no way of identifying the familiar negative numbers with sets. However, we can cancel addition of equal terms, as we prove below in Theorem 3.4.2. We need an intermediate result:

Theorem 3.4.1.   *Let $m, n \in \mathbb{N}_0$. Then $m = n$ if and only if $m^+ = n^+$.*

Proof. By set inclusions and definition of order, $m = n$ is the same as saying $m \leq n$ and $n \leq m$. By Remark 3.3.2, this is the same as $m^+ \leq n^+$ and $n^+ \leq m^+$, and by set inclusions again this holds if and only if $m^+ = n^+$.

$\square$

Theorem 3.4.2. (Cancellation theorem) *Let $m, n, p \in \mathbb{N}_0$.*
  (1) $m + p = n + p$ if and only if $m = n$.
  (2) $m + n = 0$ if and only if $m = 0$ and $n = 0$.
  (3) $m \cdot n = 0$ if and only if $m = 0$ or $n = 0$.
  (4) If $p \neq 0$ and $p \cdot m = p \cdot n$, then $m = n$.

Proof. (1) Certainly if $m = n$, then $m + p = A_m(p) = A_n(p) = n + p$. Let $T = \{p \in \mathbb{N}_0 : \text{whenever } m + p = n + p \text{ for some } m, n \in \mathbb{N}_0, \text{ then } m = n\}$. By Theorem 3.2.3 (1), $0 \in T$. Suppose that $p \in T$. We will prove that p^+ is also in T. For this, suppose that for some $m, n \in \mathbb{N}_0$, $m + (p^+) = n + (p^+)$. By Remark 3.2.2 this says that $(m + p)^+ = (n + p)^+$. Lemma above (Theorem 3.4.2) then implies that $m + p = n + p$. This proves that $p^+ \in T$, so that T is an inductive set. This proves the additive cancellation.

(2) Certainly $0 + 0 = 0$. Now suppose that $m + n = 0$. This says that $A_m(n) = 0$. If $n \neq 0$, then $n = q^+$ for some $q \in \mathbb{N}_0$. Thus by the definition of addition, $A_m(n) = A_m(q^+) = (A_m(q))^+$, which cannot be the empty set. So necessarily $n = 0$. But addition is commutative, so similarly $m = 0$. This proves (2).

(3) In Theorem 3.2.3 (3) we proved that if $m = 0$ or $n = 0$, then $m \cdot n = 0$. Now suppose that $m \cdot n = 0$. If $n \neq 0$, then $n = p^+$ for some $p \in \mathbb{N}_0$ (by Theorem 3.1.6), hence $0 = m \cdot (p^+) = m \cdot p + m$. If $m \neq 0$, then m is a successor of some element k of $\mathbb{N}_0$. Thus $m \cdot p + m = m \cdot p + k^+ = m \cdot p + (k + 1) = (m \cdot p + k) + 1 = (m \cdot p + k)^+$ is the successor of some element of $\mathbb{N}_0$ as well, which means that $m \cdot p + m \neq 0$. So necessarily $m = 0$. We just proved that $n \neq 0$ implies $m = 0$, or in other words, we just proved that either $n = 0$ or $m = 0$. This proves (3).

(4) Fix $p \in \mathbb{N}_0 \setminus \{0\}$. Let

$$T = \{m \in \mathbb{N}_0 : \text{ whenever } p \cdot m = p \cdot n \text{ for some } n \in \mathbb{N}_0, \text{ then } m = n\}.$$

If $p \cdot 0 = p \cdot n$, then by (3), $0 = p \cdot n$ and $n = 0$, so that $0 \in T$. Suppose that $m \in T$ and that for some $n \in \mathbb{N}_0$, $p \cdot (m^+) = p \cdot n$. By (3), $n \neq 0$, so by Theorem 3.1.6, $n = r^+$ for some $r \in \mathbb{N}_0$. By Definition of $\cdot$, $p \cdot (m^+) = p \cdot m + p$, and $p \cdot n = p \cdot (r^+) = p \cdot r + p$, so that $p \cdot m + p = p \cdot r + p$. By (1), $p \cdot m = p \cdot r$, so that as $m \in T$, necessarily $m = r$. Then $m^+ = r^+ = n$, which proves that $m^+ \in T$. Thus $T = \mathbb{N}_0$ by Theorem 3.1.8. $\square$

Exercises for Section 3.4

3.4.1. Suppose that $m, p \in \mathbb{N}_0$ satisfy $m \cdot (p + m) = 0$. Prove that $m = 0$.

3.4.2. Suppose that $m, p, n \in \mathbb{N}_0$ satisfy $m + p = m \cdot (n^+)$. Prove that $p = m \cdot n$.

3.4.3. (**Compatibility of $<$ with multiplication by non-zero numbers**) Let $m, n, p \in \mathbb{N}_0$. Suppose that $m < n$. Prove that $mp \leq np$ for all p and that $mp < np$ for all $p \neq 0$.

3.5 Construction of $\mathbb{Z}$, arithmetic, and order on $\mathbb{Z}$

How are we supposed to think of the familiar -5? It does not seem to be possible to represent negative numbers with sets as we did for non-negative numbers. We go about it by using the rigorous construction of $\mathbb{N}_0$ and all the arithmetic on it. The following is a rehashing of Example 2.3.10.

Definition 3.5.1. *Consider the Cartesian product* $\mathbb{N}_0 \times \mathbb{N}_0$. *Elements are pairs of the form* (a, b), *with* $a, b \in \mathbb{N}_0$. *If* $a, b, a', b' \in \mathbb{N}_0$, *we write* $(a, b) \sim (a', b')$ *if* $a + b' = b + a'$.

In Example 2.3.10 it was proved that the relation $\sim$ is an **equivalence relation** on $\mathbb{N}_0 \times \mathbb{N}_0$. Thus we can talk about the **equivalence class** $[(a, b)]$ of (a, b).

Definition 3.5.2. *We define* $\mathbb{Z}$ *to be the set of all equivalence classes of elements of* $\mathbb{N}_0 \times \mathbb{N}_0$ *under the equivalence relation* $\sim$. *Elements of* $\mathbb{Z}$ *are called* **whole numbers,** *or* **integers.**

We have a natural inclusion of $\mathbb{N}_0$ into $\mathbb{Z}$ by **identifying** $n \in \mathbb{N}_0$ with $[(n, 0)]$ in $\mathbb{Z}$. Note that if $n, m \in \mathbb{N}_0$ are distinct, then so are $[(n, 0)]$ and $[(m, 0)]$, so that $\mathbb{N}_0$ is indeed identified with a natural subset of $\mathbb{Z}$. But $\mathbb{Z}$ is strictly larger: $[(0, 1)]$ is not equal to $[(m, 0)]$ for all $m \in \mathbb{N}_0$.

To define arithmetic on $\mathbb{Z}$, we first define arithmetic on $\mathbb{N}_0 \times \mathbb{N}_0$. Start with addition: for $a, b, c, d \in \mathbb{N}_0$, set

$$(a, b) + (c, d) = (a + c, b + d).$$

(There are three "+" here: the last two are the already well-studied addition in $\mathbb{N}_0$, and the first one is the one we are defining now on $\mathbb{N}_0 \times \mathbb{N}_0$. Yes, we are using the same name for two different operations; but that's ok: many families have a child named Pat, and the many Pats are not all equal.) Since + on $\mathbb{N}_0$ is binary, so is + on $\mathbb{N}_0 \times \mathbb{N}_0$. We show next that + is compatible with $\sim$:

Theorem 3.5.3. *Let* $a, a', b, b', c, c', d, d' \in \mathbb{N}_0$ *such that* $(a, b) \sim (a', b')$ *and* $(c, d) \sim (c', d')$. *Then* $(a + c, b + d) \sim (a' + c', b' + d')$.

Proof. By assumption, $a + b' = a' + b$ and $c + d' = c' + d$. Then by associativity and commutativity of + on $\mathbb{N}_0$, $(a+c)+(b'+d') = (a+b')+(c+d') = (a'+b)+(c'+d) = (a+c)+(b'+d')$, so that $(a+c, b+d) \sim (a'+c', b'+d')$. $\square$

This shows that + is **well-defined** on the equivalence classes of $\sim$, hence + makes sense on $\mathbb{Z}$:

$$[(a, b)] + [(c, d)] = [(a + c, b + d)].$$

For example, identifying $5, 6 \in \mathbb{N}_0$ with $[(5, 0)], [(6, 0)] \in \mathbb{Z}$, we get $5 + 6 = 11$ and $[(5, 0)] + [(6, 0)] = [(5 + 6, 0)] = [(11, 0)]$, as expected. Also, $[(5, 0)] + [(0, 6)] = [(5, 6)] = [(0, 1)]$.

Multiplication on $\mathbb{Z}$ is a little more complicated: if (a, b) and (c, d) are in $\mathbb{N}_0 \times \mathbb{N}_0$, then we declare

$$[(a, b)] \cdot [(c, d)] = [(a \cdot c + b \cdot d, a \cdot d + b \cdot c)].$$

We next prove that $\cdot$ is well defined on equivalence classes. Suppose that $[(a, b)] = [(a', b')]$ and $[(c, d)] = [(c', d')]$. This means that $a + b' = a' + b$ and $c + d' = c' + d$. By the established associativity, distributivity, and commutativity properties in $\mathbb{N}_0$, we then have that

$$(ac + bd + a'd' + b'c') + (a + b')d' + (a' + b)c' + b(c + d') + a(c' + d)$$
$$= ac + bd + a'd' + b'c' + ad' + b'd' + a'c' + bc' + bc + bd' + ac' + ad$$
$$= ad + bc + a'c' + b'd' + a'd' + bd' + ac' + b'c' + bc' + bd + ac + ad'$$
$$= (ad + bc + a'c' + b'd') + (a' + b)d' + (a + b')c' + b(c' + d) + a(c + d')$$
$$= (ad + bc + a'c' + b'd') + (a + b')d' + (a' + b)c' + b(c + d') + a(c' + d),$$

so by the cancellation Theorem 3.4.2 in $\mathbb{N}_0$, $ac + bd + a'd' + b'c' = ad + bc + a'c' + b'd'$. It follows that $[(ac + bd, ad + bc)] = [(a'c' + b'd', a'd' + b'c')]$, so that multiplication $\cdot$ on $\mathbb{Z}$ is well-defined.

Consequently, $[(5, 0)] \cdot [(6, 0)] = [(5 \cdot 6 + 0 \cdot 0, 5 \cdot 0 + 0 \cdot 6)] = [(30, 0)]$, as expected. Similarly compute $[(5, 0)] \cdot [(0, 6)]$.

We next prove the basic arithmetic laws on $\mathbb{Z}$.

Theorem 3.5.4. *The following identities hold for all* $m, n, p \in \mathbb{Z}$:

(1) **(Additive identity)** $m + [(0, 0)] = m = [(0, 0)] + m$.

(2) $m \cdot [(0, 0)] = [(0, 0)] = [(0, 0)] \cdot m$.

(3) **(Associativity of addition)** $m + (n + p) = (m + n) + p$.

(4) **(Commutativity of addition)** $m + n = n + m$.

(5) **(Multiplicative identity)** $m \cdot [(1, 0)] = m = [(1, 0)] \cdot m$.

(6) **(Distributivity)** $m \cdot (n + p) = (m \cdot n) + (m \cdot p)$.

(7) **(Associativity of multiplication)** $m \cdot (n \cdot p) = (m \cdot n) \cdot p$.

(8) **(Commutativity of multiplication)** $m \cdot n = n \cdot m$.

(9) **(Existence of additive inverses)** *There exists* $r \in \mathbb{Z}$ *such that* $m + r = r + m = [(0, 0)]$.

Proof. Let $a, b, c, d, e, f \in \mathbb{N}_0$ such that $m = [(a, b)]$, $n = [(c, d)]$ and $p = [(e, f)]$. With this, the proofs follow easily from the definitions of $+$ and $\cdot$. For example,

$$m + [(0, 0)] = [(a, b)] + [(0, 0)] = [(a + 0, b + 0)] = [(a, b)] = m,$$

and similarly $[(0, 0)] + m = m$. This proves (1).

Properties (2)–(4) are equally easy to prove.

To prove (5), we compute

$$m \cdot [(1, 0)] = [(a, b)] \cdot [(1, 0)]$$
$$= [(a \cdot 1 + b \cdot 0, a \cdot 0 + b \cdot 1)]$$
$$= [(a + 0, 0 + b)] = [(a, b)] = m,$$

and similarly $[(1, 0)] \cdot m = m$.

We check (6), and you explain each step:

$$(m \cdot (n + p)) = ([(a, b)] \cdot ([(c, d)] + [(e, f)]))$$
$$= [(a, b)] \cdot [(c + e, d + f)]$$
$$= [(a \cdot (c + e) + b \cdot (d + f), a \cdot (d + f) + b \cdot (c + e))]$$
$$= [(a \cdot c + a \cdot e + b \cdot d + b \cdot f, a \cdot d + a \cdot f + b \cdot c + b \cdot e)]$$

(why may we omit parentheses in addition?)

$$= [(a \cdot c + b \cdot d, a \cdot d + b \cdot c)] + [(a \cdot e + b \cdot f, a \cdot f + b \cdot e)]$$
$$= [(a, b)] \cdot [(c, d)] + [(a, b)] \cdot [(e, f)]$$
$$= m \cdot n + m \cdot p.$$

You verify (7) and (8).

(9) Set $r = [(b, a)]$. Then $m + r = [(a, b)] + [(b, a)] = [(a + b, b + a)] = [(a + b, a + b)] = [(0, 0)]$, and similarly $r + m = [(0, 0)]$.   □

Notation 3.5.5. The additive inverse of the equivalence class $[(a, b)]$ of (a, b) is the equivalence class $[(b, a)]$ of (b, a). The additive inverse of an element m is always denoted $-m$.

Notation 3.5.6. Furthermore, for any $a, b \in \mathbb{N}_0$, by trichotomy on $\mathbb{N}_0$ (Theorem 3.3.3), exactly one of the following three conditions is satisfied:

$$a < b, \quad a = b, \quad b < a.$$

In the first two cases, by Theorem 3.3.5, there exists $r \in \mathbb{N}_0$ such that $a + r = b$, and then $[(a, b)] = [(a, a + r)] = [(0, r)]$, and in the last case there exists $r \in \mathbb{N}_0$ such that $b + r = a$, and then $[(a, b)] = [(b + r, b)] = [(r, 0)]$. This proves that every element of $\mathbb{Z}$ can be written as the equivalence class of $[(a, b)]$ where a or b is 0.

Recall the identification of elements of $\mathbb{N}_0$ with elements in $\mathbb{Z}$: $n \in \mathbb{N}_0$ is identified with $[(n, 0)]$. Then with the last two paragraphs,

$$[(a, b)] = [(a, 0)] + [(0, b)] = a + (-b),$$

and it is standard shorthand to write this as $a - b$. Thus in the future we write elements $[(a, b)]$ in $\mathbb{Z}$ as $a - b$.

Theorem 3.5.7. *Let $m, n \in \mathbb{Z}$. Then $(-m) \cdot n = m \cdot (-n) = -(m \cdot n)$.*

Proof. Observe that

$$m \cdot (-n) + m \cdot n = m \cdot ((-n) + n) \text{ (by distributivity: Theorem 3.5.4 (6))}$$
$$= m \cdot 0 \text{ (by notation for additive inverses)}$$
$$= 0 \text{ (by Theorem 3.5.4 (2)).}$$

Similarly, $m \cdot n + m \cdot (-n) = 0$. Thus $m \cdot (-n)$ is an additive inverse of $m \cdot n$, and by uniqueness of inverses (Theorem 2.5.10), $m \cdot (-n)$ is the additive inverse of $m \cdot n$, i.e., $m \cdot (-n) = -(m \cdot n)$.

With that,

$$(-m) \cdot n = n \cdot (-m) \text{ (by commutativity of multiplication)}$$
$$= -(n \cdot m) \text{ (by the previous paragraph).} \qquad \square$$

Theorem 3.5.8. *Let $m, n \in \mathbb{Z}$. Then $m \cdot n = 0$ if and only if $m = 0$ or $n = 0$.*

Proof. We already proved that if $m = 0$ or $n = 0$, then $m \cdot n = 0$.

Now suppose that $m \cdot n = 0$. Write $m = [(a, b)]$ with $a, b \in \mathbb{N}_0$. By Notation 3.5.6 we may assume that either $a = 0$ or $b = 0$. Write $n = [(c, d)]$ for $c, d \in \mathbb{N}_0$.

If $a = 0$, then $[(0, 0)] = 0 = m \cdot n = [(0, b)] \cdot [(c, d)] = [(b \cdot d, b \cdot c)]$. This means (by definition of $\sim$) that $b \cdot d = b \cdot c$. If $b = 0$, then $m = [(0, 0)] = 0$, and if $b \neq 0$, then by Theorem 3.4.2, $d = c$, so that $n = [(c, d)] = [(0, 0)] = 0$.

Similarly, if $b = 0$, then either $m = 0$ or $n = 0$. But $a = 0$ or $b = 0$ covers all the possible cases for elements of $\mathbb{Z}$, so the theorem is proved. $\square$

Definition 3.5.9. (**Order on** $\mathbb{Z}$) *Define a relation $\leq$ on $\mathbb{Z}$ as follows:* $[(a, b)] \leq [(c, d)]$ *if $a + d \leq b + c$, where the latter $\leq$ is the already known relation on $\mathbb{N}_0$ (see Section 3.3). Similarly define $<, \geq, >$.*

It is left to the reader to verify the following:

Theorem 3.5.10.

(1) $\leq$ on $\mathbb{Z}$ is well-defined (this means that if one picks different representatives of the equivalence classes, $\leq$ for one pair holds if and only if it holds for the other pair).

(2) $\leq, \geq$ are reflexive and transitive but not symmetric.

(3) $<, >$ are transitive but not reflexive and not symmetric.

(4) If $m \in \mathbb{Z}$ and $0 \leq m$, then $m \in \mathbb{N}_0$.

(5) If $m \in \mathbb{Z}$ and $m \leq 0$, then $-m \in \mathbb{N}_0$.

Theorem 3.5.11. *The following properties hold on $\mathbb{Z}$:*

*(1) (**Trichotomy**) For any $m, n \in \mathbb{Z}$, exactly one of the following three conditions holds:*

i) $m = n$.

ii) $m < n$.

iii) $n < m$.

*(2) (**Transitivity**) For all $m, n, p \in \mathbb{Z}$, $m < n$ and $n < p$ imply $m < p$.*

*(3) (**Compatibility of $<$ with addition**) For all $m, n, p \in \mathbb{Z}$, $m < n$ implies $m + p < n + p$.*

*(4) (**Compatibility of $<$ with multiplication by positive numbers**) For all $m, n, p \in \mathbb{Z}$, $m < n$ and $0 < p$ imply $mp < np$.*

Proof. Write $m = [(a, b)]$, $n = [(c, d)]$, and $p = [(e, f)]$ for some a, b, c, d, e, $f \in \mathbb{N}_0$.

By Theorem 3.3.3, exactly one of the following three conditions holds:

$$a + d < b + c, \quad a + d = b + c, \quad b + c < a + d.$$

By definition, and since $<, =, >$ are compatible with the choice of representatives by Theorem 3.5.10, this says that exactly one of the following three conditions on $\mathbb{Z}$:

$$[(a,b)] < [(c,d)], \quad [(a,b)] = [(c,d)], \quad [(c,d)] < [(a,b)].$$

This proves (1).

Suppose that $m < n$ and $n < p$. This means that $a + d < b + c$ and $c + f < d + e$. Then by associativity and commutativity of addition in $\mathbb{N}_0$, and by transitivity of $<$ on $\mathbb{N}_0$, $(a + f) + (c + d) = (a + d) + (c + f) < (a + d) + (d + e) < (b + c) + (d + e) = (b + e) + (c + d)$, so that by Exercise 3.3.4, $a + f < b + e$. This says that $m < p$, which proves that $<$ is transitive. This proves (2).

The other properties follow similarly and are left to the reader. $\square$

With this, we can use the adjectives **positive, negative, non-positive, non-negative** from Definition 2.7.10 for elements of $\mathbb{Z}$, and in general we can use the properties developed in Sections 2.7 and 2.10, such as **absolute values, triangle inequality**, and **reverse triangle inequality**. We generalize Theorem 3.3.4 as follows:

Theorem 3.5.12. *Let S be a non-empty subset of $\mathbb{Z}$ that is bounded above (resp. below). Then S contains its largest (resp. least) element. (For definitions see Definition 2.7.2.)*

Proof. Let $s \in S$, and let $b \in \mathbb{Z}$ be an upper bound on S. Let $T = \{n \in \mathbb{N}_0 : b - n \in S\}$. Then T is a subset of $\mathbb{N}_0$. Note that $b - s \in \mathbb{N}_0$ and that $s = b - (b - s) \in S$, so that T is not empty. Thus by Theorem 3.3.4, T has a least element, call it t. Thus $b - t \in S$. We will prove that $b - t$ is an upper bound for S. Let $r \in S$. We need to prove that $b - t \geq r$. Since b is an upper bound on S and $r \in S$, we have that $b - r \in \mathbb{N}_0$. Since $r = b - (b - r) \in S$, we have that $b - r \in T$. But then by the choice of t, $t \leq b - r$, which proves that $r \leq b - t$, as desired. This proves the first part.

Now let $s \in S$, and let $b \in \mathbb{Z}$ be a lower bound on S. Let $T = \{n \in \mathbb{Z} : b + n \in S\}$. Since b is a lower bound, we conclude that T is a subset of $\mathbb{N}_0$. Note that $s - b \in \mathbb{N}_0$ and that $s = b + (s - b) \in S$, so that T is not empty. Thus by Theorem 3.3.4, T has a least element, call it t. Thus $b + t \in S$. We will prove that $b + t$ is a lower bound for S. Let $r \in S$. We need to prove that $b + t \leq r$. Since b is a lower bound on S and $r \in S$, we have that $r - b \in \mathbb{N}_0$. Since $r = b + (r - b) \in S$, we have that $r - b \in T$.

But then by the choice of t, $t \leq r - b$, which proves that $r \geq b + t$, as desired. This proves the second part, and hence the theorem.   □

Exercises for Section 3.5

3.5.1. Prove that for all $a, b, c \in \mathbb{N}_0$, $(a, b) \sim (a + c, b + c)$.

3.5.2. Suppose that $a, b, c, d \in \mathbb{N}_0$ such that $(a, b) \sim (c, d)$ and $a \geq c$. Prove that there exists $e \in \mathbb{N}_0$ such that $a = c + e$ and $b = d + e$.

3.5.3. Convince yourself that $\mathbb{Z}$ as defined in Definition 3.5.2 is the same as the familiar set of all integers.

3.5.4. **(Cancellation in $\mathbb{Z}$)** Let $m, n, p \in \mathbb{Z}$.
  i) If $m + p = n + p$, prove that $m = n$. (Hint: Theorem 2.5.16.)
  ii) If $p \neq 0$ and $m \cdot p = n \cdot p$, prove that $m = n$. (Note that Theorem 2.5.16 is inapplicable.)

3.5.5. Prove that if $(a, b) \sim (c, d)$ then $(a, c) \sim (b, d)$.

3.5.6. Let $a, b, c, d \in \mathbb{N}_0$. Suppose that $[(b, a)] < [(c, d)]$. Prove that $[(a, b)] > [(d, c)]$ and $[(a, b)] + [(c, d)] > [(0, 0)]$.

3.5.7. Prove (2), (3), (4), (7), (8) in Theorem 3.5.4.

3.5.8. Prove Theorem 3.5.10.

3.5.9. Prove (3) and (4) in Theorem 3.5.11.

3.6   Construction of the ordered field $\mathbb{Q}$

We next define the set $\mathbb{Q}$, arithmetic, and order on it, and we show that $\mathbb{Q}$ is a field. Elements of course correspond to the familiar rational numbers.

The following is a rehashing of the construction of $\mathbb{Q}$ from $\mathbb{Z}$ first given in Exercise 2.3.10. The relation $\wr$ on $\mathbb{Z} \times (\mathbb{Z} \setminus \{0\})$, given by

$$(m, n) \wr (m', n') \quad \text{if} \quad m \cdot n' = m' \cdot n,$$

is an equivalence relation. The reader may wish to go through the details of the proof, to check that only those properties of $\mathbb{Z}$ are used that we have proved in the previous section, and that no "well-known" knowledge creeps into the proof.

Remark 3.6.1. For any $a \in \mathbb{Z}$ and $b, c \in \mathbb{Z} \setminus \{0\}$, the equivalence class $[(ca, cb)]$ equals the equivalence class $[(a, b)]$. In particular, by possibly taking a negative c, we may always choose a representative of any equivalence class in $\mathbb{Z} \times \mathbb{N}^+$.

Definition 3.6.2. *We define the set of all equivalence classes of this relation to be* $\mathbb{Q}$. *The elements of* $\mathbb{Q}$ *are called* **rational numbers**.

We have a natural inclusion of $\mathbb{Z}$ into $\mathbb{Q}$ (and thus of $\mathbb{N}_0$ into $\mathbb{Q}$) by identifying $m \in \mathbb{Z}$ with $[(m, 1)]$. Note that if $m, n \in \mathbb{Z}$ are distinct, so are $[(m, 1)]$ and $[(n, 1)]$, so that $\mathbb{Z}$ is indeed identified with a natural subset of $\mathbb{Q}$. But $\mathbb{Q}$ is strictly larger: $[(1, 2)]$ is not equal to $[(m, 1)]$ for any $m \in \mathbb{Z}$. More about the familiar rational numbers is in Notation 3.6.5.

We declare the following binary operations $+$ and $\cdot$ on elements of $\mathbb{Z} \times (\mathbb{Z} \setminus \{0\})$:

$$(m, n) + (p, q) = (m \cdot q + p \cdot n, n \cdot q), \quad (m, n) \cdot (p, q) = (m \cdot p, n \cdot q).$$

(The $+$ and $\cdot$ inside the pair entries are the operations on $\mathbb{Z}$; the other $+$ and $\cdot$ are what we are defining here.) We show next that the new $+$ and $\cdot$ are compatible with $\wr$:

Theorem 3.6.3. *Let* $m, m', n, n', p, p', q, q' \in \mathbb{N}_0$ *such that* $(m, n) \wr (m', n')$ *and* $(p, q) \wr (p', q')$. *Then*

$$((m, n) + (p, q)) \wr ((m', n') + (p', q')),$$

and
$$((m, n) \cdot (p, q)) \wr ((m', n') \cdot (p', q')).$$

Proof. By assumption, $m \cdot n' = m' \cdot n$ and $p \cdot q' = p' \cdot q$. Then (you explain each step):

$$
\begin{aligned}
(m \cdot q + p \cdot n) \cdot (n' \cdot q') &= (n' \cdot q') \cdot (m \cdot q + p \cdot n) \\
&= (n' \cdot q') \cdot (m \cdot q) + (n' \cdot q') \cdot (p \cdot n) \\
&= (q' \cdot n') \cdot (m \cdot q) + n' \cdot (q' \cdot (p \cdot n)) \\
&= q' \cdot (n' \cdot (m \cdot q)) + n' \cdot ((q' \cdot p) \cdot n) \\
&= q' \cdot ((n' \cdot m) \cdot q) + n' \cdot ((p \cdot q') \cdot n) \\
&= q' \cdot ((m \cdot n') \cdot q) + n' \cdot ((p' \cdot q) \cdot n) \\
&= q' \cdot ((m' \cdot n) \cdot q) + n' \cdot (p' \cdot (q \cdot n))
\end{aligned}
$$

$$= q' \cdot (m' \cdot (n \cdot q)) + n' \cdot (p' \cdot (n \cdot q))$$
$$= (q' \cdot m') \cdot (n \cdot q) + (n' \cdot p') \cdot (n \cdot q)$$
$$= (n \cdot q) \cdot (q' \cdot m') + (n \cdot q) \cdot (n' \cdot p')$$
$$= (n \cdot q) \cdot (m' \cdot q') + (n \cdot q) \cdot (p' \cdot n')$$
$$= (n \cdot q) \cdot (m' \cdot q' + p' \cdot n')$$
$$= (m' \cdot q' + p' \cdot n') \cdot (n \cdot q).$$

This says that

$$(m \cdot q + p \cdot n, n \cdot q), = (m' \cdot q' + p' \cdot n', n' \cdot q'),$$

which in turn says that $((m, n) + (p, q)) \wr ((m', n') + (p', q'))$.

The verification of the easier $((m, n) \cdot (p, q)) \wr ((m', n') \cdot (p', q'))$ is left to the reader. $\qquad \square$

The last theorem proves that the following binary operations $+$ and $\cdot$ on $\mathbb{Q}$ are well-defined:

$$[(m, n)] + [(p, q)] = [(m, n) + (p, q)] = [(m \cdot q + p \cdot n, n \cdot q)],$$
$$[(m, n)] \cdot [(p, q)] = [(m, n) \cdot (p, q)] = [(m \cdot p, n \cdot q)].$$

The $+$ and $\cdot$ on the left are what we are defining here; the $+$ and $\cdot$ in the middle are binary operations on $\mathbb{Z} \times (\mathbb{Z} \setminus \{0\})$, and the $+$ and $\cdot$ on the right are binary operations on $\mathbb{Z}$.

We now move to arithmetic laws on $\mathbb{Q}$.

Theorem 3.6.4. $\mathbb{Q}$ *is a field (see Definition 2.6.1), or explicitly, the following identities hold for all* $m, n, p \in \mathbb{Q}$:

(1) **(Additive identity)** $m + [(0, 1)] = m = [(0, 1)] + m.$

(2) $m \cdot [(0, 1)] = [(0, 1)] = [(0, 1)] \cdot m.$

(3) **(Associativity of addition)** $m + (n + p) = (m + n) + p.$

(4) **(Commutativity of addition)** $m + n = n + m.$

(5) **(Multiplicative identity)** $m \cdot [(1, 1)] = m = [(1, 1)] \cdot m.$

(6) **(Distributivity)** $m \cdot (n + p) = (m \cdot n) + (m \cdot p).$

(7) **(Associativity of multiplication)** $m \cdot (n \cdot p) = (m \cdot n) \cdot p.$

(8) **(Commutativity of multiplication)** $m \cdot n = n \cdot m.$

(9) $[(1, 1)] \neq [(0, 1)].$

(10) **(Existence of additive inverses)** *There exists* $r \in \mathbb{Q}$ *such that* $m + r = r + m = [(0,1)]$.

(11) **(Existence of multiplicative inverses)** *If* $m \neq [(0,1)]$, *there exists* $r \in \mathbb{Q}$ *such that* $m \cdot r = r \cdot m = [(1,1)]$.

Proof. Let $a, b, c, d, e, f \in \mathbb{Z}$ such that $m = [(a,b)]$, $n = [(c,d)]$ and $p = [(e,f)]$. With this, the proofs follow easily from the definitions of $+$ and $\cdot$. For example,

$$m + [(0,1)] = [(a,b)] + [(0,1)] = [(a + 0, b \cdot 1)] = [(a,b)] = m,$$

and similarly $[(0,1)] + m = m$. This proves (1).

Properties (2)–(5) are equally easy to prove.

We check (6), and you explain each step:

$$
\begin{aligned}
m \cdot (n + p) &= [(a,b)] \cdot ([(c,d)] + [(e,f)]) \\
&= [(a,b)] \cdot [(c \cdot f + e \cdot d, d \cdot f)] \\
&= [(a \cdot (c \cdot f + e \cdot d), b \cdot (d \cdot f))] \\
&= [(a \cdot (c \cdot f) + a \cdot (e \cdot d), b \cdot (d \cdot f))] \\
&= [(a,b)] \cdot [(c,d)] + [(a,b)] \cdot [(e,f)] \\
&= m \cdot n + m \cdot p.
\end{aligned}
$$

You verify (7) and (8).

(9) By the properties of $\mathbb{Z}$, $0 \cdot 1 = 0 \neq 0^+ = 1 = 1 \cdot 1$, so that $(0,1) \not\sim (1,1)$, so that $[(0,1)] \neq [(1,1)]$ in $\mathbb{Q}$.

(10) By definition of m and $\mathbb{Q}$, $[(-a,b)] \in \mathbb{Q}$, and $m + [(-a,b)] = [(a \cdot b + (-a) \cdot b, b \cdot b)] = [(b \cdot a + b \cdot (-a), b \cdot b)] = [(b \cdot (a + (-a)), b \cdot b)] = [(b \cdot 0, b \cdot b)] = [(0, b \cdot b)] = [(0,1)]$, and similarly $[(-a,b)] + m = [(0,1)]$. Thus additive inverses exist in $\mathbb{Q}$.

(11) By assumption $[(a,b)] \neq [(0,1)]$, This means that $a \cdot 1 \neq 0 \cdot b$, so that $a \neq 0$. Then $[(b,a)] \in \mathbb{Q}$, and $[(a,b)] \cdot [(b,a)] = [(a \cdot b, b \cdot a)] = [(a \cdot b, a \cdot b)] = [(1,1)]$. Similarly $[(b,a)] \cdot [(a,b)] = [(1,1)]$. $\square$

Notation 3.6.5. The additive inverse of the equivalence class $[(a,b)]$ of (a,b) is $[(-a,b)]$, and the multiplicative inverse of non-zero $[(a,b)]$ is $[(b,a)]$.

For an element m, its additive inverse is always denoted $-m$, and the multiplicative inverse, when it exists, is written m^{-1}. Thus for any $n \in \mathbb{Z}$, by identifying n with $[(n, 1)]$ in $\mathbb{Q}$, the additive inverse is $-n = -[(n, 1)] = [(-n, 1)]$, and if $n \neq 0$, the multiplicative inverse is $n^{-1} = [(n, 1)]^{-1} = [(1, n)]$. Thus, under our identifications, any element $[(m, n)]$ in $\mathbb{Q}$ is equal to

$$[(m, n)] = [(m, 1)] \cdot [(1, n)] = m \cdot n^{-1},$$

and it is standard shorthand to write this as m/n. Thus in the future we write elements $[(m, n)]$ in $\mathbb{Q}$ as m/n, and whenever $m, n \in \mathbb{Z}$ and $n \neq 0$, then $[(m, n)] = m/n \in \mathbb{Q}$. By Remark 3.6.1 we may even assume that $n > 0$.

Theorem 3.6.6. (Cancellation laws in $\mathbb{Q}$) *Let $m, n, p \in \mathbb{Q}$.*
(1) If $m + p = n + p$, then $m = n$.
(2) If $m \cdot p = n \cdot p$ and if $p \neq 0$, then $m = n$.

Proof. Since every $p \in \mathbb{Q}$ has an additive inverse and since every non-zero $p \in \mathbb{Q}$ has a multiplicative inverse, both parts are simply special cases of Theorem 2.5.16.   □

Theorem 3.6.7. $\mathbb{Q}$ *is an ordered field (see Definition 2.7.10) with the order $<$ given by:*

$$[(a, b)] < [(c, d)] \text{ if } \begin{cases} ad < bc; & \text{if } bd > 0; \\ ad > bc; & \text{if } bd < 0. \end{cases}$$

Proof. Theorem 3.6.4 proves that $\mathbb{Q}$ is a field. Now we prove that $\mathbb{Q}$ is ordered.

First we need to prove that $<$ is well-defined. Let $m, n \in \mathbb{Q}$. Write $m = [(a, b)] = [(a', b')]$, $n = [(c, d)] = [(c', d')]$. By definition of $\mathbb{Q}$, $a'b = ab'$ and $c'd = cd'$. By associativity and commutativity of multiplication in $\mathbb{Z}$, $(bd)(b'd')(b'c' - a'd') = bb'dd'(b'c' - a'd') = b(b')^2 d'(dc') - b'd(d')^2(ba') = b(b')^2 d'(cd') - b'd(d')^2(ab') = (b')^2(d')^2(bc - ad)$. Thus the sign of $(bd)(b'd')(b'c' - a'd')$ is the same as the sign of $bc - ad$. In particular, if $bd > 0$, $b'd' > 0$, and $ad < bc$, then necessarily $a'd' < b'c'$; if $bd > 0$, $b'd' < 0$, and $ad < bc$, then necessarily $a'd' > b'c'$; if $bd < 0$, $b'd' > 0$, and $ad < bc$, then necessarily $a'd' > b'c'$; if $bd < 0$, $b'd' < 0$, and $ad < bc$, then necessarily $a'd' < b'c'$. All these cases confirm that the order is independent of the representative of the equivalence class.

Trichotomy follows from trichotomy on $\mathbb{Z}$ (Theorem 3.5.11). The proof of transitivity of $<$, and of compatibility of $<$ with addition and with multiplication by positive rational numbers is left as an exercise. $\square$

Thus the following terms and facts, developed in Sections 2.7 and 2.10, apply to elements of $\mathbb{Q}$: **positive, negative, non-positive, non-negative, intervals, absolute value, triangle inequality, reverse triangle inequality.**

Theorem 3.6.8. (Archimedean property) *Let* $m, n \in \mathbb{Q}$. *If* $m > 0$, *there exists* $p \in \mathbb{N}^+$ *such that* $n < pm$.

Proof. If $n \leq 0$, then take $p = 1$, and $n \leq 0 < m = pm$. So we may assume that $n > 0$. Write $m = a/b$ and $n = c/d$ for some positive $a, b, c, d \in \mathbb{Z}$. Set $p = bc + 1$. Then p is a positive integer, as is ad. By compatibility of $<$ with addition on $\mathbb{Z}$ and multiplication by positive integers, we have that $pad = bcad + ad > bcad \geq bc$, and since $b, d > 0$, it follows that $pm = p(a/b) > c/d = n$. $\square$

Exercises for Section 3.6

3.6.1. Compute $(5/6) \cdot (-3/4)$, $(5/6) + (-3/4)$.

3.6.2. Prove that for all $(m, n) \in \mathbb{Z} \times (\mathbb{Z} \setminus \{0\})$ and all non-zero $p \in \mathbb{Z}$, $(m, n) \wr (m \cdot p, n \cdot p)$. In other words, $m/n = (m \cdot p)/(n \cdot p)$.

3.6.3. Prove (2), (3), (4), (5), (7), (8) in Theorem 3.6.4.

3.6.4. Finish the proof of Theorem 3.6.7.

3.6.5. Let $m, n, p, q \in \mathbb{Z}$ with n, q non-zero. Prove that $m/n + p/q$ and $m/n \cdot p/q$ are rational numbers.

3.6.6. Let $m \in \mathbb{Q}$. Prove that there exists a positive integer N such that $m < 2^N$. (Hint: If $m \leq 0$, this is easy. Now suppose that $m > 0$. Write $m = a/b$ for some positive integers a, b. Prove that $a/b \leq a < 2^a$; perhaps apply or reprove Exercise 3.3.7.)

3.6.7. Find a non-empty subset S of $\mathbb{Q}$ that is bounded above such that S does not contain its least upper bound. (Contrast with Theorem 3.5.12.)

3.7   Construction of the field $\mathbb{R}$ of real numbers

In this section we construct real numbers from $\mathbb{Q}$ via Dedekind cuts: a Dedekind cut stands for a real number and vice versa. We establish the familiar arithmetic on $\mathbb{R}$, and in the subsequent section we show that $\mathbb{R}$ is an ordered field with further good properties. To go through these two sections carefully would take many hours. I typically spend 1.5 class hours going through them, asking the students not to take notes but instead to nod vigorously in agreement and awe at all the constructions and claims. I expect the students to get the gist of the construction, a conviction that any arithmetic and order property of real numbers can be verified if necessary, but I do not expect the students to have gone through all the details. Furthermore, I do not use the standard Dedekind cuts, as arithmetic with them is messier (see Exercise 3.7.12).

Definition 3.7.1. *A* **Dedekind cut** *is a pair* (L, R) *with the following properties:*

(1) L, R *are non-empty subsets of* $\mathbb{Q}$.

(2) *For all* $l \in L$ *and all* $r \in R$, $l < r$.

(3) *For all* $l \in L$, *there exists* $l' \in L$ *such that* $l < l'$.

(4) *For all* $r \in R$, *there exists* $r' \in R$ *such that* $r' < r$.

(5) $L \cup R$ *is either* $\mathbb{Q}$ *or* $\mathbb{Q} \setminus \{m\}$ *for some* $m \in \mathbb{Q}$. *In case* $L \cup R = \mathbb{Q} \setminus \{m\}$, *then for all* $l \in L$ *and all* $r \in R$, $l < m < r$.

By condition (2) and the trichotomy property of $<$ on $\mathbb{Q}$, $L \cap R = \emptyset$.

We can visualize a Dedekind cut (L, R) as the separation of the **rational number line** into the left and right parts, and the separation comes either at a rational number m, or possibly at something else. The letters "L" and "R" are mnemonics for "left" and "right".

An easy consequence of the definition is that if $l \in L$ and $x \in \mathbb{Q}$ with $x < l$, then $x \in L$, and similarly that if $r \in R$ and $x \in \mathbb{Q}$ with $x > l$, then $x \in R$. (See Exercise 3.7.1.)

Examples and non-examples 3.7.2.

(1) $(\mathbb{Q} \setminus \{0\}, \{0\})$ is not a Dedekind cut: it violates conditions (2) and (4).

(2) $(\{x \in \mathbb{Q} : x^2 < 2\}, \{x \in \mathbb{Q} : x^2 > 2\})$ is not a Dedekind cut: it violates condition (2).

(3) $(\{x \in \mathbb{Q} : x < 0\}, \{x \in \mathbb{Q} : x > 0\})$ is a Dedekind cut.

(4) $(\{x \in \mathbb{Q} : x \le 0\}, \{x \in \mathbb{Q} : x > 0\})$ is not a Dedekind cut: it violates condition (3).

Theorem 3.7.3. *Let (L, R) be a Dedekind cut. Let $x \in \mathbb{Q}$.*

(1) Suppose that $x < l$ for some $l \in L$. Then $x \in L$.

(2) Suppose that $x > r$ for some $r \in R$. Then $x \in R$.

Proof. We only prove the first part. By condition (2), x is not in R, and by condition (5), $x \in L$. $\qquad\square$

Definition 3.7.4. *We define $\mathbb{R}$ to be the set of all Dedekind cuts. Elements of $\mathbb{R}$ are therefore Dedekind cuts, but more normally, the elements are called* **real numbers***.*

The following theorem identifies a large collection of Dedekind cuts.

Theorem 3.7.5. *For any $m \in \mathbb{Q}$, $D_m = (\{l \in \mathbb{Q} : l < m\}, \{r \in \mathbb{Q} : r > m\})$ is a Dedekind cut.*

The union of the left and the right parts of D_m equals $\mathbb{Q} \setminus \{m\}$.

Proof. Let $L = \{l \in \mathbb{Q} : l < m\}$, $R = \{r \in \mathbb{Q} : r > m\}$. By Theorem 2.7.11, $1 > 0$, so that by compatibility of $<$ and $+$ on $\mathbb{Q}$, $m+1 > m$ and $m-1 < m$. Thus $m - 1 \in L$ and $m + 1 \in R$. Thus L and R are not empty.

By trichotomy on $\mathbb{Q}$, for every $r \in \mathbb{Q}$, exactly one of the following holds: $r < m$, $r = m$, $r > m$. So $L \cup R = \mathbb{Q} \setminus \{m\}$. By definition, for any $l \in L$ and any $r \in R$, $l < m < r$, and by transitivity, we conclude that $l < r$.

If $l \in L$, then $l < m$, and by Exercise 2.7.15, $l' = (l + m)/2$ is a rational number strictly larger than l and strictly smaller than m. Thus, $l < l' \in L$, and condition (3) of Dedekind cuts holds. Similarly, for any $r \in R$, $(r + m)/2$ is in R and strictly smaller than r, so that condition (4) of Dedekind cuts is satisfied. $\qquad\square$

Theorem 3.7.6. *The function $\mathbb{Q} \to \mathbb{R}$ defined as taking $m \in \mathbb{Q}$ to D_m is injective. Thus we think of $\mathbb{Q}$ as a subset of $\mathbb{R}$.*

Proof. If m, n are distinct in $\mathbb{Q}$, say $m < n$, then by Exercise 2.7.15, $(m+n)/2$ is in the right side of D_m and in the left side of D_n, so that D_m and D_n are also distinct. This proves that the function is injective. □

Many differences between $\mathbb{Q}$ and $\mathbb{R}$ are in Section 3.8. Here is the first one:

Theorem 3.7.7. $\mathbb{R}$ *contains Dedekind cuts other than those in* $\mathbb{Q}$.

Proof. We prove that $(\{x \in \mathbb{Q} : x^2 < 2 \text{ or } x < 0\}, \{x \in \mathbb{Q} : x^2 \geq 2 \text{ and } x \geq 0\})$ is a Dedekind cut that is not of the form D_m for any $m \in \mathbb{Q}$.

Set $L = \{x \in \mathbb{Q} : x^2 < 2 \text{ or } x < 0\}$ and $R = \{x \in \mathbb{Q} : x^2 \geq 2 \text{ and } x \geq 0\}$. Then $0 \in L$ and $2 \in R$, so that L and R are not empty.

Let $l_1, l_2 \in L$ and $r \in R$ with $l_1 < 0$ and $l_2^2 < 2$.

By transitivity of $<$ since $r \geq 0$ we have that $l_1 < r$. By definition, $l_2^2 < 2 \leq r^2$, so that by Theorem 2.9.2, $l_2 < r$. In particular, $0 < r$.

By Exercise 2.7.15, $l_1 < l_1/2 < 0$, and $l_1/2 \in L$. By assumption, $2 - l_2^2 > 0$. Thus by Theorem 3.6.8 there exists $p \in \mathbb{N}^+$ such that $2l_2 < (2 - l_2^2)p$. Hence $2l_2/p - 1/p^2 < 2l_2/p < 2 - l_2^2$, so that $(l_2 + 1/p)^2 = l_2^2 + 2l_2/p - 1/p^2 < 2$. Hence $l_2 + 1/p \in L$ and $l_2 + 1/p > l_2$.

By assumption, $r^2 - 2 > 0$, so that by Theorem 3.6.8 there exist $p_1, p_2 \in \mathbb{N}^+$ such that $2r < p_1(r^2 - 2)$ and $p_2 r > 1$. Set $p = \max\{p_1, p_2\}$. Then $2r < p(r^2 - 2)$ and $pr > 1$. It follows that $2r/p - 1/p^2 < 2r/p < r^2 - 2$. Hence $2 < r^2 - 2r/p + 1/p^2 = (r - 1/p)^2$. By the condition $pr > 1$, it follows that $r - 1/p \in R$ and $r - 1/p < r$.

Let $x \in \mathbb{Q}$. If $x \leq 0$, then $x \in L$. Now suppose that $x > 0$. By trichotomy, one of the following holds: $x^2 = 2$, $x^2 < 0$, $x^2 > 0$. By page 11, 2 is not the square of a rational number, so the first option is not possible, which proves that $x \in L \cup R$.

This proves that (L, R) is a Dedekind cut. Since $L \cup R = \mathbb{Q}$, it follows that (L, R) cannot equal D_m for any $m \in \mathbb{Q}$. □

Theorem 3.7.8. *Let* L *be a non-empty proper subset of* $\mathbb{Q}$ *such that for all* $l \in L$ *and all* $x \in \mathbb{Q}$, *if* $x < l$, *then* $x \in L$. *Then there exists a unique subset* R *of* $\mathbb{Q}$ *such that* (L, R) *is a Dedekind cut.*

Let R be a non-empty proper subset of $\mathbb{Q}$ such that for all $r \in R$ and all $x \in \mathbb{Q}$, if $x > r$, then $x \in R$. Then there exists a unique subset L of $\mathbb{Q}$ such that (L, R) is a Dedekind cut.

Proof. We only prove the first part. Set $R' = \mathbb{Q} \setminus L$.

If condition (4) of Dedekind cuts holds for (L, R'), let $R = R'$. Otherwise, there exists $m \in R'$ such that for all $r' \in R'$, $r' \geq m$. In this case set $R = R' \setminus \{m\}$. Now let $r \in R$. Then $r > m$. By Exercise 2.7.15, $(r + m)/2$ is a rational number strictly larger than m and strictly smaller than r, so that by assumption $(r + m)/2 \in R$. Thus condition (4) of Dedekind cuts holds for (L, R).

Since L is a proper non-empty subset of $\mathbb{Q}$, so is R', and even the modified R in the previous paragraph. If $l \in L$ and $r \in R$, by construction $l \neq r$ and even $l < r$. Thus conditions (1) and (2) hold, and condition (3) holds by assumption. Furthermore, condition (5) holds by construction as well. $\square$

Now we turn to the arithmetic on $\mathbb{R}$.

Definition 3.7.9. *Let A and B be subsets of $\mathbb{Q}$. Define*

$$A + B = \{a + b : a \in A \text{ and } b \in B\},$$
$$- A = \{-a : a \in A\},$$
$$A - B = \{a - b : a \in A \text{ and } b \in B\} = A + (-B),$$
$$A \cdot B = \{a \cdot b : a \in A \text{ and } b \in B\}.$$

Theorem 3.7.10. *Let (L_1, R_1) and (L_2, R_2) be Dedekind cuts. Then $(L_1 + L_2, R_1 + R_2)$ is a Dedekind cut.*

Proof. Since L_1, L_2, R_1, R_2 are non-empty subsets of $\mathbb{Q}$, then $L_1 + L_2, R_1 + R_2$ are non-empty subsets of $\mathbb{Q}$ as well. Let $l \in L_1 + L_2$ and $r \in R_1 + R_2$. Then $l = l_1 + l_2$ and $r = r_1 + r_2$ for some $l_i \in L_i$ and $r_i \in R_i$ ($i = 1, 2$). Then $l_1 < r_1$ and $l_2 < r_2$, so that $l = l_1 + l_2 < l_1 + r_2 < r_1 + r_2 = r$, and by transitivity of $<$ on $\mathbb{Q}$, $l < r$. Thus conditions (1) and (2) of Dedekind cuts hold.

Conditions (3) and (4) hold similarly. Furthermore, if $l \in L_1 + L_2$ and $x \in \mathbb{Q}$ with $x < l$, then we claim that $x \in L_1 + L_2$. Namely, write $l = l_1 + l_2$ for some $l_i \in L_i$, $i = 1, 2$. By assumption we have that $x - l + l_1 < l_1$, and since (L_1, R_1) is a Dedekind cut, by Exercise 3.7.1, we conclude that $x - l + l_1 \in L_1$. Thus $x = x - l + l = x - l + l_1 + l_2 \in L_1 + L_2$. Similarly, whenever $r \in R_1 + R_2$ and $x \in \mathbb{Q}$ with $r < x$, then $x \in R_1 + R_2$.

It remains to prove Condition (5) of Dedekind cuts for $(L_1 + L_2, R_1 + R_2)$. Let $x \in \mathbb{Q}$ and suppose that $x \notin (L_1 + L_2) \cup (R_1 + R_2)$. We will prove that $(L_1 + L_2, R_1 + R_2)$ is the Dedekind cut D_x. For this we need to prove that any $y \in \mathbb{Q}$ not in $(L_1 + L_2) \cup (R_1 + R_2)$ equals x. By Remark 3.6.1 and Notation 3.6.5 we may write $x = \frac{m}{n}$, $y = \frac{p}{n}$ for some $m, n, p \in \mathbb{Z}$, with $n > 0$. By the definition of Dedekind cuts the sets $S_i = \{r \in \mathbb{Z} : \frac{r}{n} \in R_i\}$ are bounded below, so that S_i has a least element $a_i \in \mathbb{Z}$ by Theorem 3.5.12. Thus $\frac{c}{n} \in R_i$ for all integers $c \geq a_i$, and $\frac{c}{n} \notin R_i$ for all integers $c < a_i$. In particular, $\frac{c}{n} \in R_1 + R_2$ for all integers $c \geq a_1 + a_2$. If $(a_i - 1)/n \in L_i$ for $i = 1, 2$, then $\frac{c}{n} \in L_1 + L_2$ for all integers $c \leq a_1 + a_2 - 2$, so that necessarily $x = y = \frac{a_1 + a_2 - 1}{n}$. So say that $(a_1 - 1)/n \notin L_1$. Then by the definition of Dedekind cuts we have that $L_1 = D_{(a_1 - 1)/n}$. If also $(a_2 - 1)/n \notin L_2$, then $L_2 = D_{(a_2 - 1)/n}$, $\frac{c}{n} \in L_1 + L_2$ for all integers $c \leq a_1 + a_2 - 4$, $\frac{a_1 + a_2 - 3}{n} = \frac{2a_1 - 3}{2n} + \frac{2a_2 - 3}{2n} \in L_1 + L_2$, $\frac{a_1 + a_2 - 1}{n} = \frac{2a_1 - 1}{2n} + \frac{2a_2 - 1}{2n} \in R_1 + R_2$, so that $x = y = \frac{a_1 + a_2 - 2}{n}$. The remaining case is where $L_1 = D_{(a_1 - 1)/n}$ and $L_2 \neq D_{(a_2 - 1)/n}$. If $L_2 = D_r$ for some rational number r, then by possibly taking a multiple of n we may assume that r is of the form $\frac{q}{n}$ for some integer q, but then we are in the previous case. So we may suppose that (L_2, R_2) is not a rational number. Then $\frac{2a_2 - 1}{n}$ is either in R_2 or in L_2. If it is in R_2, then $\frac{a_1 + a_2 - 1}{n} = \frac{2a_1 - 1}{2n} + \frac{2a_2 - 1}{2n} \in R_1 + R_2$, so that necessarily $x = y = \frac{a_1 + a_2 - 2}{n}$, and if it is in L_2, then $\frac{a_1 + a_2 - 2}{n} = \frac{2a_1 - 3}{2n} + \frac{2a_2 - 1}{2n} \in L_1 + L_2$, so that $x = y = \frac{a_1 + a_2 - 1}{n}$. Thus Condition (5) holds.   □

Thus we define addition on $\mathbb{R}$ by

$$(L_1, R_1) + (L_2, R_2) = (L_1 + L_2, R_1 + R_2).$$

There are two different meanings to "+" above: on the right side, + acts on two sets of rational numbers, and on the left, + stands for the new sum.

The proof of the following is left as an exercise.

Theorem 3.7.11. *Addition on $\mathbb{R}$, as defined above, is commutative and associative. The additive identity is $D_0 = (\mathbb{Q}^-, \mathbb{Q}^+)$, and the additive inverse of any $(L, R) \in \mathbb{R}$ is $(-R, -L)$ (recall Definition 3.7.9).*

Naturally we write the additive identity D_0 of $\mathbb{R}$ as 0, and for any $x \in \mathbb{R}$, we write its additive inverse as $-x$.

Additions on $\mathbb{Q}$ and $\mathbb{R}$ are compatible, in the following sense:

Theorem 3.7.12. *For any* $m, n \in \mathbb{Q}$, $D_m + D_n = D_{m+n}$ *(the first* $+$ *is in* $\mathbb{R}$, *the second in* $\mathbb{Q}$*)*.

Proof. We know that $D_m + D_n$ is a Dedekind cut, and so we can write it as (L, R).

We first prove that $R = \{r \in \mathbb{Q} : r > m + n\}$. Let $r \in R$. By definition of addition on $\mathbb{R}$, $r = r_m + r_n$ for some $r_m > m$ and $r_n > n$. But then by compatibility of addition in $\mathbb{Q}$ with $>$,

$$r = r_m + r_n > r_m + n > m + n,$$

and by transitivity, $r > m + n$. This proves that $R \subseteq \{r \in \mathbb{Q} : r > m + n\}$. Now suppose that $r \in \mathbb{Q}$ with $r > m + n$. Set $r_m = \frac{r-m-n}{2} + m$ and $r_n = \frac{r-m-n}{2} + n$. By compatibility of $<$ with addition on $\mathbb{Q}$, we know that $r - m - n > 0$. By Theorem 2.7.11 we know that $1 > 0$, so that again by compatibility, $2 = 1 + 1 > 0$. Then by Theorem 2.10.4, $\frac{r-m-n}{2} > 0$, so that $r_m > m$ and $r_n > n$. Finally, $r_m + r_n = r$, which proves that $\{r \in \mathbb{Q} : r > m + n\} \subseteq R$. Thus $R = \{r \in \mathbb{Q} : r > m + n\}$.

Then by uniqueness of Dedekind cuts, $D_m + D_n = D_{m+n}$. $\qquad\square$

Multiplication on $\mathbb{R}$ is more complicated, but if one keeps in mind the "expected" facts about real numbers, the following definition of multiplication on Dedekind cuts is natural:

$$(L, R) \cdot (L', R') = \begin{cases} (\_, R \cdot R'), & \text{if } R \cap \mathbb{Q}^+ = R \text{ and } R' \cap \mathbb{Q}^+ = R'; \\ (R \cdot L', \_), & \text{if } R \cap \mathbb{Q}^+ = R \text{ and } -L' \cap \mathbb{Q}^+ = -L'; \\ (L \cdot R', \_), & \text{if } R' \cap \mathbb{Q}^+ = R' \text{ and } -L \cap \mathbb{Q}^+ = -L; \\ (\_, L \cdot L'), & \text{if } R \cap \mathbb{Q}^+ \neq R \text{ and } R' \cap \mathbb{Q}^+ \neq R', \end{cases}$$

where in each case $\_$ is the unique subset of $\mathbb{Q}$ that makes the result a Dedekind cut. It is only possible to find such a $\_$ if the non-blank part on the right side satisfies the conditions in Theorem 3.7.8. It is left to Exercise 3.7.9 to verify that. (See Exercise 3.7.11 for an alternate formulation of multiplication.)

Clearly $\cdot$ is a commutative operation, and it is easy to show that $D_1 = (\{x \in \mathbb{Q} : x < 1\}, \{x \in \mathbb{Q} : x > 1\})$ is a multiplicative identity (and therefore the unique multiplicative identity by Theorem 2.5.5). The following are multiplicative inverses for $(L, R) \neq D_0$:

$$(L, R)^{-1} = \begin{cases} (\_, \{1/r : r \in R\}), & \text{if } R \cap \mathbb{Q}^+ = R; \\ (\{1/l : l \in L\}, \_), & \text{otherwise.} \end{cases}$$

If all elements of R are positive, we are certainly not dividing by 0 in the first case. In the second case, by Exercise 3.7.1, L contains only negative elements, and so again we are not dividing by 0. It is now straightforward to verify that the listed Dedekind cut is indeed the multiplicative inverse.

I let the reader verify that for all $m, n \in \mathbb{Q}$, $D_m \cdot D_n = D_{mn}$.

Proofs of associativity of multiplication and of distributivity of multiplication over addition are more involved. With the definition above, we would have to verify eight cases for associativity, and many cases for distributivity, depending on whether the left parts of the various Dedekind cuts contain positive elements. Yes, this can be done, but it is tedious work. We provide a different proof in this book that is not a brute-force attack but is more conceptual and gives side results as a bonus. We first need an intermediate result.

Theorem 3.7.13. *For any Dedekind cuts* x, y, $(-x) \cdot y = -(x \cdot y) = x \cdot (-y)$.

Proof. Write $x = (L, R)$ and $y = (L', R')$.

If L, L' contain positive elements, then $-R, -R'$ do not contain positive elements, and by definition of multiplication,

$$(x \cdot y) + ((-x) \cdot y) = ((L, R) \cdot (L', R')) + ((-R, -L) \cdot (L', R'))$$
$$= (\_, R \cdot R') + ((-R) \cdot R', \_)$$
$$= (\_, R \cdot R') + (-(R \cdot R'), \_).$$

By Theorem 3.7.8, the left side U of $(\_, R \cdot R')$ is either $\mathbb{Q} \setminus (R \cdot R')$ or $\mathbb{Q} \setminus ((R \cdot R') \cup \{m\})$ for some rational number m. Since $R \cdot R'$ consists of positive numbers strictly greater than numbers in $\mathbb{Q} \setminus (R \cdot R')$ it follows that the left side is contained in the set $\mathbb{Q}^-$ of all negative rational numbers. Furthermore, let r be any negative rational number. We want to find $x \in U$ and $y \in R \cdot R'$ such that $x - y = r$. Let $x \in U$ and $y \in R \cdot R'$. Write $r = r_0/n$, $x = x_0/n$ and $y = y_0/n$ for some integers r_0, x_0, y_0, n, with n positive. By possibly changing x and y but not n, we may assume that x_0 is the largest possible and y_0 the smallest possible. If $(y_0 - 1)/n \in U$, then

$y_0 = x_0 + 1$, and $y - x = y_0/n - x_0/n = 1/n$. If instead $(y_0 - 1)/n \notin U$, then $(y_0 - 1)/n = m$, so that $(2y_0 - 1)/2n \in R \cdot R'$, $(2y_0 - 3)/2n \in U$, and $(2y_0 - 3)/2n - (2y_0 - 1)/2n = 1/2n$. Thus we have found a representation of r as r_0/n for some positive integer n and some negative integer r_0 such that for some integers x_0, y_0, $x_0/n \in U$, $y_0/n \in R \cdot R'$, and $x_0/n - y_0/n = 1/n$. But then $(y_0 + (r - 1))/n \in R \cdot R'$ and $x_0/n - (y_0 + (r_0 - 1))/n = r_0/n = r$. This proves that the left side of the Dedekind cut in the display equals $\mathbb{Q}^-$. Thus $(x \cdot y) + ((-x) \cdot y) = D_0$. By commutativity of addition this also gives $((-x) \cdot y) + (x \cdot y) = D_0$, $((-x) \cdot y) = -(x \cdot y)$.

If L contains positive elements but L' does not, then

$$(x \cdot y) + ((-x) \cdot y) = ((L, R) \cdot (L', R')) + ((-R, -L) \cdot (L', R'))$$
$$= (R \cdot L', \_) + (\_, -(R \cdot L')),$$

and similarly to the previous case, this is D_0, so again $((-x) \cdot y) = -(x \cdot y)$.

The remaining cases of $((-x) \cdot y) = -(x \cdot y)$ are proved similarly. Then by commutativity of multiplication, $x \cdot (-y) = (-y) \cdot x$, and by the established case this equals $-(y \cdot x) = -(x \cdot y)$.   □

Theorem 3.7.14. $\mathbb{R}$ *is a field properly containing* $\mathbb{Q}$.

Proof. By Theorem 3.7.7, $\mathbb{R}$ properly contains $\mathbb{Q}$. For elements of $\mathbb{Q}$, the arithmetic on $\mathbb{R}$ agrees with the arithmetic on $\mathbb{Q}$: for addition this was done in Theorem 3.7.12, and for multiplication the details are left to the reader.

The only parts left to prove are that multiplication is associative and that it distributes over addition. Let $x = (L, R)$, $y = (L', R')$, $z = (L'', R'')$ be in $\mathbb{R}$.

To prove that $x \cdot (y \cdot z) = (x \cdot y) \cdot z$, it suffices to prove that the additive inverses of the two sides are identical. By Theorem 3.7.13, we may then at will in each product replace $x = (L, R)$ by $-x = (-R, -L)$, and similarly for y by $-y$ and z by $-z$. Note that R does not contain only positive numbers if and only if $-L$ contains only positive numbers, so via these replacements we may assume that R, R', R'' all contain only positive elements. But then $x \cdot (y \cdot z) = (\_, L \cdot (L' \cdot L''))$, which is by associativity of multiplication on $\mathbb{Q}$ the same as $(\_, (L \cdot L') \cdot L'') = (x \cdot y) \cdot z$. This proves associativity of multiplication.

To prove that $x \cdot (y + z) = (x \cdot y) + (x \cdot z)$, we may similarly assume that R and $R' + R''$ contain only positive numbers. Then either R' or

R'' contains only positive elements, and by commutativity of addition we may assume that R' contains only positive elements. From assumptions we have $x \cdot (y + z) = (\_, R \cdot (R' + R''))$. If $a \in R$, $b \in R'$ and $c \in R''$, then $a \cdot (b + c) = a \cdot b + a \cdot c$ by distributivity in $\mathbb{Q}$, which proves that $R \cdot (R' + R'') \subseteq R \cdot R' + R \cdot R''$. We want to prove that equality holds in this inclusion.

Suppose that in addition R'' contains only positive numbers. Let $r_1, r_2 \in R$, $r' \in R'$, $r'' \in R''$. Then $r_1 r' + r_2 r'' \in R \cdot R' + R \cdot R''$, and we want to prove that $r_1 r' + r_2 r'' \in R \cdot (R' + R'')$. We claim that $(r_1 \cdot r' + r_2 \cdot r'')/(r' + r'') \geq \min\{r_1, r_2\}$. If $(r_1 \cdot r' + r_2 \cdot r'')/(r' + r'') < r_1$, then by compatibility of multiplication with $<$ on $\mathbb{Q}$, $r_1 \cdot r' + r_2 \cdot r'' < r_1 \cdot r' + r_1 \cdot r''$, whence $0 < (r_1 - r_2) \cdot r''$, which forces $r_1 > r_2$. But then $(r_1 r' + r_2 r'')/(r' + r'') > (r_2 r' + r_2 r'')/(r' + r'') = r_2$, which proves the claim. It follows by Exercise 3.7.1 that $(r_1 \cdot r' + r_2 \cdot r'')/(r' + r'') \in R$, so that the arbitrary element $r_1 \cdot r' + r_2 \cdot r'' = \frac{r_1 \cdot r' + r_2 \cdot r''}{r' + r''}(r' + r'') \in R$, of $R \cdot R' + R \cdot R''$ is in $R \cdot (R' + R'')$. This proves that $R \cdot (R' + R'') = R \cdot R' + R \cdot R''$. Thus in this case $x \cdot (y + z) = (\_, R \cdot R' + R \cdot R'') = (\_, R \cdot R') + (\_, R \cdot R'') = x \cdot y + x \cdot z$.

Now suppose that R'' does not contain only positive numbers. Then $-L''$ contains only positive numbers. Thus by the previous case $x \cdot (y + z) + x \cdot (-z) = x \cdot ((y + z) + (-z))$. By the already established associativity of addition, this equals $x \cdot y$. Then by Theorem 3.7.13 this says that $x \cdot (y + z) = x \cdot y + x \cdot z$. $\square$

The Dedekind cuts as in Theorem 3.7.5 are the familiar rational numbers, and if we establish some good order on $\mathbb{R}$, we should be able to order the other Dedekind cuts among the rational ones as expected, to get that $\mathbb{R}$ consists of the familiar real numbers.

Exercises for Section 3.7

3.7.1. Let (L, R) be a Dedekind cut, $l \in L$, $r \in R$, and $x \in \mathbb{Q}$. Prove the following:

  i) If $x \leq l$, then $x \in L$.

  ii) If $x \geq r$, then $x \in R$.

3.7.2. Prove that there exists no $R \subset \mathbb{Q}$ such that $(\mathbb{Q}^- \cup \{0\}, R)$ is a Dedekind cut. Similarly, there exists no $L \subset \mathbb{Q}$ such that $(L, \mathbb{Q}^+ \cup \{0\})$ is a Dedekind cut.

3.7.3. Let (L, R) be a Dedekind cut.

i) If $m \in \mathbb{Q}$ is the supremum of L, prove that $R = \mathbb{Q} \setminus (L \cup \{m\})$. If L does not have a supremum in $\mathbb{Q}$, prove that $R = \mathbb{Q} \setminus L$.

ii) If $m \in \mathbb{Q}$ is the infimum of R, prove that $L = \mathbb{Q} \setminus (R \cup \{m\})$. If R does not have an infimum in $\mathbb{Q}$, prove that $L = \mathbb{Q} \setminus R$.

3.7.4. Let (L, R) be a Dedekind cut and let $(L', R') = (L, R) \cdot (L, R)$. Prove that $R' \subseteq \mathbb{Q}_{\geq 0}$.

3.7.5. If m and n are rational numbers, prove that $D_m + D_n = D_{m+n}$.

3.7.6. Let $a = (\{x \in \mathbb{Q} : x < 0 \text{ and } x^2 > 2\}, \{x \in \mathbb{Q} : x^2 \leq 2 \text{ or } x > 0\})$, $b = (\{x \in \mathbb{Q} : x < 0 \text{ or } x^2 < 3\}, \{x \in \mathbb{Q} : x^2 \geq 3 \text{ and } x > 0\})$. Compute $a + b$, $a \cdot b$.

3.7.7. Prove Theorem 3.7.11.

3.7.8. Let $(L_1, R_1), \ldots, (L_n, R_n)$ be Dedekind cuts. Prove that

$$(\cup_{k=1}^n L_k, \cap_{k=1}^n R_k)$$

is a Dedekind cut that equals (L_k, R_k) for some $k \in \{1, \ldots, n\}$.

3.7.9. Prove that multiplication on Dedekind cuts is well-defined.

3.7.10. Let (L, R) be a Dedekind cut. Prove that there exists $z \in \mathbb{Q}$ such that $z \in R$ and $-z \in L$. Prove in addition that for any $a \in \mathbb{Q}$, there exists such z with $z \geq |a|$.

3.7.11. For any set $S \subseteq \mathbb{Q}$, define $S_+ = S \cap \mathbb{Q}_+$ and $S_- = S \cap \mathbb{Q}_-$. Prove that for any Dedekind sets (L, R), (L', R'), $(L, R) \cdot (L', R') = (U, V)$, where

$$U = L_- \cdot R'_+ + R_+ \cdot L'_- + L_+ \cdot L'_+ + R_- \cdot R'_-,$$
$$V = L_- \cdot L'_- + R_- \cdot L'_+ + L_+ \cdot R'_- + R_+ \cdot R'_+,$$

and where $A + \emptyset = A$ and $A \cdot \emptyset = \emptyset$.

3.7.12. Let $x, y \in \mathbb{R}$.

i) Prove that $x^2 = (-x)^2$.

ii) Suppose that $x^2 = y^2$. Prove that either $x = y$ or $x = -y$.

*3.7.13. Let $x, y \in \mathbb{R}$ such that $x^3 = y^3$. Prove that $x = y$.

3.7.14. The **standard Dedekind cut** in the literature is defined as the pair (L, R) with the following properties:

(1) L, R are non-empty subsets of $\mathbb{Q}$.
(2) For all $l \in L$ and all $r \in R$, $l < r$.
(3) For all $l \in L$, there exists $l' \in L$ such that $l < l'$.
(4) $L \cup R = \mathbb{Q}$.

What should be the definition of the sum of two standard Dedekind cuts? With the definition above, why does $(L_1, R_1) + (L_2, R_2) = (L_1 + L_2, R_1 + R_2)$ not work? Why does $-(L, R) = (-R, -L)$ not work?

3.8   Order on $\mathbb{R}$, the Least upper bound theorem

Definition 3.8.1. *We impose the following order on $\mathbb{R}$:* $(L, R) \leq (L', R')$ *if $L \subseteq L'$, and $(L, R) < (L', R')$ if $L \subsetneq L'$.*

Theorem 3.8.2. $\mathbb{R}$ *is an ordered field.*

Proof. By Theorem 3.7.14, $\mathbb{R}$ is a field. Let $(L, R), (L', R'), (L'', R'') \in \mathbb{R}$.

If $L \not\subseteq L'$, let $l \in L \setminus L'$. Let $l' \in L'$. By Exercise 3.7.1 applied to the Dedekind cut (L, R), $l > l'$ and $l' \in L$. This proves that $L' \subseteq L$, and since $L' \neq L$, it follows that $L' \subsetneq L$. This proves that at least one of $L \subsetneq L'$, $L = L'$, and $L' \subsetneq L$ holds. But no two of these conditions can hold simultaneously, so exactly one of the three conditions holds. Hence by definition of order on $\mathbb{R}$, exactly one of the following conditions holds: $(L, R) < (L', R')$, $(L, R) = (L', R')$, $(L, R) > (L', R')$. This proves trichotomy of $<$.

Suppose that $(L, R) < (L', R')$ and $(L', R') < (L'', R'')$. Then by definition, $L \subsetneq L' \subsetneq L''$, so that $L \subsetneq L''$, which says that $(L, R) < (L'', R'')$. Thus transitivity holds for $<$.

Suppose that $(L, R) < (L', R')$. This means that $L \subsetneq L'$, so certainly $L + L'' \subseteq L' + L''$. If equality holds, then $(L, R) + (L'', R'') = (L', R') + (L'', R'')$. By adding the additive inverse of (L'', R'') we then conclude that $(L, R) = (L', R')$, which contradicts the assumption $(L, R) < (L', R')$. Thus necessarily $L + L'' \subsetneq L' + L''$, so that $(L, R) + (L'', R'') < (L', R') + (L'', R'')$. This proves that $<$ is compatible with addition.

Finally, suppose that $(L, R) < (L', R')$ and that $D_0 < (L'', R'')$. Then by compatibility of $<$ with addition, $D_0 < (L', R') - (L, R) =$

$(L', R') + (-R, -L) = (L' - R, R' - L)$. Thus $\{x \in \mathbb{Q} : x < 0\} \subsetneq L' - R$, which means that $R' - L$ contains only positive elements of $\mathbb{Q}$. Thus by definition of multiplication, $((L', R') - (L, R)) \cdot (L'', R'') = (\_, (R' - L) \cdot R'')$, and so $D_0 < ((L', R') - (L, R)) \cdot (L'', R'')$. But then by associativity, $D_0 < ((L', R') \cdot (L'', R'')) - ((L, R) \cdot (L'', R''))$, so that by compatibility of $<$ with addition, $(L, R) \cdot (L'', R'') = D_0 + ((L, R) \cdot (L'', R'')) < (L', R') \cdot (L'', R'')$, which proves that $<$ is compatible with multiplication by positive numbers. $\qquad\square$

Thus the following terms and facts, developed in Sections 2.7 and 2.10, apply to elements of $\mathbb{R}$: **positive, negative, non-positive, non-negative, intervals, absolute value, triangle inequality, reverse triangle inequality.**

The following is a rephrasing of Theorem 2.10.4 in this context: Suppose that $a, b \in \mathbb{R}$ have the property that $|a - b| < \epsilon$ for all real numbers $\epsilon > 0$. Then $a = b$.

Theorem 3.8.3. (Archimedean property) *Let* $m, n \in \mathbb{R}$. *If* $m > 0$, *there exists* $p \in \mathbb{N}^+$ *such that* $n < pm$.

Proof. Write $m = (L, R)$, $n = (L', R')$. Since $m > 0$, L contains a positive rational number x. Let $y \in R'$. Then by the Archimedean property for $\mathbb{Q}$, Theorem 3.6.8, there exists $p \in \mathbb{N}^+$ such that $y < px$. Then

$$n = (L', R')$$
$$< D_y \text{ (since } y \in R')$$
$$< D_{px} \text{ (since } y < px)$$
$$= pD_x \text{ (by multiplication in } \mathbb{Q} \text{ and } \mathbb{R})$$
$$\leq p \cdot (L, R) \text{ (since } x \in L \text{ and by compatibility of } \leq \text{ with multiplication}$$
$$\qquad \text{by positive } p)$$
$$= pm. \qquad\qquad\qquad\qquad\qquad\qquad\qquad\qquad\qquad\qquad\qquad\qquad\qquad\square$$

Theorem 3.8.4. *For any two distinct real numbers there is a rational number strictly between them.*

Proof. Let $a, b \in \mathbb{R}$ with $a < b$. Write $a = (L, R)$ and $b = (L', R')$ as Dedekind cuts. Since $a < b$, it means that $L \subsetneq L'$. Let $r' \in L' \setminus L$. By definition of Dedekind cuts, there exists $r \in L'$ such that $r > r'$. Thus

$L \subsetneq \{x \in R : x < r\} \subsetneq L'$, so that $a < D_r < b$, which by identification $r = D_r$ proves the theorem. □

One of the main properties that distinguishes $\mathbb{R}$ from $\mathbb{Q}$ is the following theorem:

Theorem 3.8.5. (Least upper/greatest lower bound theorem) *Let T be a non-empty subset of $\mathbb{R}$ that is bounded above (resp. below). Then $\sup(T)$ (resp. $\inf(T)$) exists in $\mathbb{R}$.*

Proof. First suppose that T is bounded above. Let $U = \cup_{(L,R)\in T} L$. In other words, U is the union of all the left parts of all the Dedekind cuts in T. Since T is not empty, neither is U. Now let $(U_0, V_0) \in \mathbb{R}$ be an upper bound on T. Then by definition $L \subseteq U_0$ for all $(L, R) \in T$. Thus $U \subseteq U_0$. This implies by Theorem 3.7.8 that there exists $V \subseteq \mathbb{Q}$ be such that (U, V) is a Dedekind cut. By definition, (U, V) is an upper bound for T, and this upper bound is less than or equal to an arbitrary upper bound (U_0, V_0). This proves that $\sup(T)$ exists in $\mathbb{R}$.

The proof for the infimum is left to Exercise 3.8.2. □

By Theorem 2.7.6, $\sup(T)$ and $\inf(T)$ are unique.

Remark 3.8.6. In contrast, note that the non-empty bounded set $\{x \in \mathbb{Q} : x^2 < 2\}$ has no infimum nor maximum in $\mathbb{Q}$. But this same set, as a subset of $\mathbb{R}$, has infimum $(\{x \in \mathbb{Q}^- : x^2 > 2\}, \_)$ and supremum $(\_, \{x \in \mathbb{Q}^+ : x^2 > 2\})$ in $\mathbb{R}$.

We use this Least upper bound theorem in the proofs of the Intermediate value theorem Theorem 5.3.1, the Extreme value theorem Theorem 5.2.2, and in many limit tricks for functions and sequences. Here is the first example of its usage.

Theorem 3.8.7. *The ceiling function makes sense on $\mathbb{R}$, i.e., there exists a function $\lceil \, \rceil : \mathbb{R} \to \mathbb{Z}$ such that for all $x \in \mathbb{R}$, $\lceil x \rceil \geq x$ and there exists no integer n with the property $\lceil x \rceil > n \geq x$.*

Similarly, there exists the floor function $\lfloor \, \rfloor : \mathbb{R} \to \mathbb{Z}$ such that for all $x \in \mathbb{R}$, $\lfloor x \rfloor \leq x$ and there exists no integer n with the property $\lfloor x \rfloor < n \leq x$.

Furthermore, $\lfloor x \rfloor = -\lceil -x \rceil$.

Proof. Let T be the set of all integers that are greater than or equal to x. By the Archimedean principle (Theorem 3.8.3) applied to x and $1 > 0$

there exists a positive integer p such that $x < p \cdot 1$. By the Least upper bound theorem (Theorem 3.8.5), the infimum of T is a real number. We call it r. We need to prove that $r \in T$. Since x is a lower bound on T, necessarily $x \le r$. Since r is the infimum of T, then for every positive number ϵ there exists $t_\epsilon \in T$ such that $0 \le t_\epsilon - r < \epsilon$. Suppose that $r \ne t_1$. Then $0 \le t_{t_1 - r} - r < t_1 - r < 1$, so that by compatibility of $<$ with addition, $0 < t_1 - t_{t_1 - r} \le t_1 - r < 1$, which contradicts Theorem 3.1.7. Thus $r = t_1$ is an integer. By the least upper bound theorem, $x \le r$ and r is the least integer with this property. Hence $r = \lceil x \rceil$.

Now let $m = -\lceil -x \rceil$. Then by definition m is an integer with the properties that $-x \le -m$ and that there is no integer n such that $-x \le n < -m$. Hence by compatibility of $<$ with addition, $x \ge m$ and there is no integer n such that $x \ge -n > m$, i.e., there is no integer n such that $x \ge n > m$. This proves that $m = \lfloor x \rfloor$ and hence finishes the proof. $\quad\square$

Theorem 3.8.8. *Between any two distinct real numbers there is a rational number (strictly between them).*

Proof. Let $x, y \in \mathbb{R}$ with $x < y$. Then $y - x > 0$, and by the Archimedean property, there exists a positive integer p such that $2 < p(y - x)$. Let $r = \frac{1}{p}(\lceil px \rceil + 1)$ (recall that $\lceil px \rceil$ is the ceiling function of px). So r is a rational number. By the definition of the ceiling function, $px \le \lceil px \rceil < \lceil px \rceil + 1$. Since p is positive, so is p^{-1}, and by compatibility of $<$ with multiplication by positive numbers, $x < \frac{1}{p}(\lceil px \rceil + 1) = r$. Furthermore, $px + 1 < 2 + px < py$ by the choice of p, so that $\lceil px \rceil + 1 < py$ and finally $r = \frac{1}{p}(\lceil px \rceil + 1) < y$. $\quad\square$

Remark 3.8.9. It is also true that between any two real numbers there is an irrational number. Namely, let $x < y$ be real numbers. It is proved on page 11 that $\sqrt{2}$ is a positive irrational number. By compatibility of $<$ with multiplication by positive numbers then $x\sqrt{2} < y\sqrt{2}$. By the Archimedean property, there is a rational number r such that $x\sqrt{2} < r < y\sqrt{2}$. If $r = 0$, again by the Archimedean property there exists a rational number s such that $x\sqrt{2} < 0 < s < y\sqrt{2}$. So by possibly replacing r by this s we may assume that r is a non-zero rational number. Then again by compatibility and by Theorem 2.7.11, $x < r/\sqrt{2} < y$. But r is non-zero, so that $r/\sqrt{2}$ is an irrational number strictly between the given real numbers x and y. $\quad\square$

I recommend that in the first pass through the next theorem, the reader works with concrete $n = 2$.

Theorem 3.8.10. (Radicals exist in $\mathbb{R}$.) *Let y be a positive real number and $n \in \mathbb{N}^+$. Then there exists a unique positive real number s such that $s^n = y$.*

Proof. Let $T = \{r \in \mathbb{R} : 0 \leq r \text{ and } r^n \leq y\}$. Then T contains 0, so it is non-empty. If for some $r \in T$, $r > \lceil y \rceil$, then $r > y \geq r^n$, so that by compatibility of multiplication/division by positive numbers, $1 \geq r$. But y is positive, so $\lceil y \rceil$ is bigger than or equal to 1, so that $1 \geq r > \lceil y \rceil \geq 1$, which by transitivity of $>$ contradicts the trichotomy. So necessarily for all $r \in T$, $r \leq \lceil y \rceil$. This means that T is bounded above. By the Least Upper Bound theorem of $\mathbb{R}$, there exists a real number s that is the least upper bound on T. Necessarily $s \geq 0$ as $0 \in T$.

We claim that $s^n = y$.

Suppose first that $s^n < y$. [WE WANT TO FIND A SMALL POSITIVE NUMBER q SUCH THAT $(s + q)^n < y$, WHICH WOULD SAY THAT $s + q \in T$ AND THUS GIVE A CONTRADICTION TO s BEING AN UPPER BOUND ON T. THE SMALL NUMBER q MIGHT AS WELL BE OF THE FORM $\frac{1}{p}$ FOR SOME (LARGE) POSITIVE INTEGER p. BY EXERCISE 1.7.7 WE REALLY WANT TO FIND SOME $p \in \mathbb{N}^+$ SUCH THAT $\sum_{k=0}^{n} \binom{n}{k} s^k / p^{n-k} < y$. IN OTHER WORDS, WE NEED $p \in \mathbb{N}^+$ SUCH THAT $\sum_{k=0}^{n-1} \binom{n}{k} s^k / p^{n-k} < y - s^n$. BUT $\sum_{k=0}^{n-1} \binom{n}{k} s^k / p^{n-k} \leq \sum_{k=0}^{n-1} \binom{n}{k} s^k / p$, SO IT SUFFICES TO FIND p SUCH THAT $\sum_{k=0}^{n-1} \binom{n}{k} s^k / p < y - s^n$.] By the Archimedean property there exists a positive integer p such that $\sum_{k=0}^{n-1} \binom{n}{k} s^k < p(y - s^n)$ We know that $\frac{1}{p} > 0$. Since $p \geq 1$, by compatibility of $<$ (and $\leq$) with multiplication by the positive number $\frac{1}{p}$ we have that $1 \geq \frac{1}{p}$. Thus

$$\sum_{k=0}^{n-1} \binom{n}{k} s^k / p^{n-k} \leq \sum_{k=0}^{n-1} \binom{n}{k} s^k / p < y - s^n,$$

and by compatibility of $<$ with addition, $\sum_{k=0}^{n} \binom{n}{k} s^k / p^{n-k} < y$. Hence by Exercise 1.7.7, $(s + 1/p)^n < y$. Since $s + 1/p > s \geq 0$, it follows that $s + 1/p \in T$, which contradicts the fact that $s = \sup T$. This proves that $s^n \geq y$.

Now suppose that $s^n > y$. Then in particular $s > 0$. By the Archimedean property there exist positive integers p_1 and p_2 such that

$1 < p_1 s$ and $\sum_{k=0}^{n-1} \binom{n}{k} s^k < p_2(s^n - y)$. Let $p = \max\{p_1, p_2\}$. Then $1 < ps$ and $\sum_{k=0}^{n-1} \binom{n}{k} s^k < p(s^n - y)$. Thus $s_0 = s - \frac{1}{p}$ is positive, and

$$s^n - \left(s - \frac{1}{p}\right)^n = \left(s_0 + \frac{1}{p}\right)^n - s_0^n$$

$$= \sum_{k=0}^{n-1} \binom{n}{k} s_0^k / p^{n-k}$$

$$\leq \sum_{k=0}^{n-1} \binom{n}{k} s^k / p \text{ (since } 0 < s_0 < s \text{ and } p^{n-k} \geq p)$$

$$< s^n - y.$$

Hence by compatibility of $<$ with addition, $y < (s - \frac{1}{p})^n$. But then for all $r \in T$, $r^n \leq y < (s - \frac{1}{p})^n$. Since $s - \frac{1}{p}$ is positive, by the increasing property of the power functions on $\mathbb{R}^+$ (Theorem 2.9.2) we have that $r \leq s - \frac{1}{p}$ for all $r \in T$. But then $s - \frac{1}{p}$ is an upper bound on T, which contradicts the fact that s is the supremum of T.

Then by trichotomy on $\mathbb{R}$ we conclude that $s^n = y$.

Now suppose that t is another positive real number such that $t^n = y$. By trichotomy, either $t < s$ or $s < t$. Then by the increasing property of the power functions on $\mathbb{R}^+$, $t^n \neq s^n$, which is a contradiction. This proves that s is unique. $\qquad\square$

We call the number s such that $s^n = y$ the **nth root of** y, and we write it as $\sqrt[n]{y}$ or as $y^{1/n}$.

Exercises for Section 3.8

3.8.1. Prove that the following conditions are equivalent on Dedekind cuts (L, R), (L', R'):
  i) $(L, R) \leq (L', R')$.
  ii) For all $l \in L$ there exists $l' \in L'$ such that $l \leq l'$.
  iii) For all $l \in L$ there exists $l' \in L'$ such that $l < l'$. (Hint: condition (3) of Dedekind cuts.)

3.8.2. Let T be a non-empty subset of $\mathbb{R}$ that is bounded below. Let (U, V) be the Dedekind cut with $U = -\cup_{(L,R)\in T} (-R)$. Prove that (U, V) is the infimum of T.

3.8.3. Let $m \in \mathbb{R}$ be a positive number. Prove that there exists a positive integer N such that $1/2^N < m$.

3.8.4. Compute the least upper bounds and greatest lower bounds, if they exist, of the following subsets of $\mathbb{R}$:

  i) $\{1/n : n \text{ a positive integer}\}$.

  ii) $\{1/n : n \text{ a positive real number}\}$.

  iii) $\{1/n : n > 4\}$.

  iv) $\{(-1)^n : n \in \mathbb{Z}\}$.

  v) $\{2^{-n} : n \in \mathbb{N}_0\}$.

  vi) $\{2^{-n}/n : n \text{ a positive integer}\}$.

  vii) $\{n/(n+1) : n \in \mathbb{N}_0\}$.

3.8.5. Let n be a positive odd integer and let y be a real number. Prove that there exists a real number x such that $x^n = y$.

3.8.6. Let $S = \{\sum_{k=1}^{n}(-1)^k \frac{1}{k} : n \in \mathbb{N}^+\}$. Prove that S is bounded above and below.

3.8.7. Find the least upper bound of $\{\sum_{k=0}^{n} \frac{1}{10^{-k}} : n \in \mathbb{N}^+\}$. (Hint: Example 1.6.4.)

3.9 Complex numbers

By Exercise 3.7.4 there is no real number x such that $x^2 = -1$. In this section we build the smallest possible field containing $\mathbb{R}$ with an element whose square is -1. We proceed by defining a new set $\mathbb{C}$ with new operations $+$ and $\cdot$, we verify that the result is a field that contains $\mathbb{R}$ as a subfield and that has two elements whose squares equal -1. It is left to an interested reader to show that there are no fields strictly between $\mathbb{R}$ and $\mathbb{C}$ (that contain a root of -1).

Definition 3.9.1. *Let* $\mathbb{C} = \mathbb{R} \times \mathbb{R}$ *(the Cartesian product). Elements of* $\mathbb{C}$ *are called* **complex numbers***. Define binary operations* $+$ *and* $\cdot$ *on* $\mathbb{C}$:

$$(a,b) + (c,d) = (a+c, b+d),$$
$$(a,b) \cdot (c,d) = (ac - bd, ad + bc).$$

We represent complex numbers in the real plane like so:

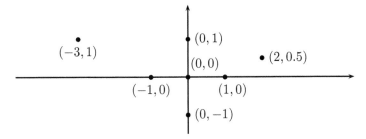

The horizontal axis, on which all complex numbers are of the form $(r, 0)$, is called the **real axis**, and the vertical axis, on which all complex numbers are of the form $(0, r)$, is called the **imaginary axis**.

The figure below shows a geometric interpretation of addition: to add (a, b) and (c, d), draw the parallelogram using these two points and $(0, 0)$ as three of the four vertices. Think through why the sum is the fourth vertex. Because of this figure we loosely say that addition in $\mathbb{C}$ obeys the **parallelogram rule**.

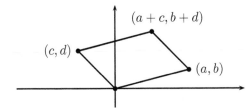

We make algebraic sense of multiplication in Notation 3.9.3 and a geometric interpretation of multiplication is in Theorem 3.12.1.

Theorem 3.9.2. $\mathbb{C}$ *is a field.*

However, $\mathbb{C}$ is not an ordered field in the sense of Definition 2.7.10, and a rigorous proof is left for Exercise 3.9.6.

Rather than giving a formal proof that $\mathbb{C}$ is a field, below is a list of the necessary easy verifications. The reader is encouraged to verify all.

(1) $\cdot, +$ are associative and commutative.

(2) $\cdot$ distributes over $+$.

(3) For all $x \in \mathbb{C}$, $(0, 0) + x = x$. In other words, $\mathbb{C}$ has the additive identity $0 = (0, 0)$. (The additive identity of a field is written as "0" even when it is an ordered pair of real numbers.)

(4) For all $(a, b) \in \mathbb{C}$, $(-a, -b) + (a, b) = (0, 0)$. In other words, every element (a, b) has an additive inverse $-(a, b) = (-a, -b)$. By Theorem 2.5.10, the additive inverse is unique.

(5) For all $x \in \mathbb{C}$, $(1, 0) \cdot x = x$. In other words, $\mathbb{C}$ has the multiplicative identity $1 = (1, 0)$. (The multiplicative identity of a field is written as "1" even when it is an ordered pair of real numbers.)

(6) $(1, 0) \neq (0, 0)$. i.e., $1 \neq 0$.

(7) Every non-zero element has a multiplicative inverse. Namely, for any $(a, b) \neq (0, 0)$, $(\frac{a}{a^2+b^2}, \frac{-b}{a^2+b^2}) \in \mathbb{C}$ and $(\frac{a}{a^2+b^2}, \frac{-b}{a^2+b^2}) \cdot (a, b) = (1, 0)$. By Theorem 2.5.10, the multiplicative inverse is unique so that $(a, b)^{-1} = (\frac{a}{a^2+b^2}, \frac{-b}{a^2+b^2})$.

For example, the multiplicative inverse of $(1, 0)$ is $(1, 0)$, the multiplicative inverse of $(0, 1)$ is $(0, -1)$, and the multiplicative inverse of $(3, 5)$ is $(\frac{3}{34}, -\frac{5}{34})$.

The squares of the complex numbers $(0, 1)$ and $(0, -1)$ are $(-1, 0)$ (check!). Thus we have found two complex roots of the polynomial function $x^2 + (1, 0)$. An identical proof to that of Theorem 2.4.15 shows that these two complex numbers are the only two roots.

Notation 3.9.3. There is another notation for elements of $\mathbb{C}$ that is in some ways better: $(a, b) = a + bi$, with $(a, 0) = a$ and $(0, b) = bi$. This notational convention does not lose any information, but it does save a few writing strokes. Addition is easy: $(a + bi) + (c + di) = (a + b) + (c + d)i$, the additive inverse of $a + bi$ is $-a - bi$, the additive identity is 0. This notation justifies the possibly strange earlier definition of multiplication in $\mathbb{C}$:

$$(a, b) \cdot (c, d) = (a + bi)(c + di) = ac + adi + bic + bidi = (ac - bd, ad + bc).$$

The multiplicative identity is 1 and the multiplicative inverse of a non-zero $a + bi$ is $\frac{a - bi}{a^2 + b^2}$.

Definition 3.9.4. *The* **real part** *of* (a, b) *is* $\mathrm{Re}(a, b) = a$, *and the* **imaginary part** *is* $\mathrm{Im}(a, b) = b$. *In alternate notation,* $\mathrm{Re}(a + bi) = a$ *and* $\mathrm{Im}(a + bi) = b$. *Note that both the real and the imaginary part of a complex number are real numbers.*

We next identify $\mathbb{R}$ as a subset of $\mathbb{C}$: the operative word here is "identify", as $\mathbb{R}$ is **not a subset** of $\mathbb{R} \times \mathbb{R}$, i.e., real numbers are not equal to ordered pairs of real numbers. Nevertheless, with the **natural** identification of any real number r with the complex number $(r, 0) = r + i \cdot 0 = r$, we can **think** of $\mathbb{R}$ as a subset of $\mathbb{C}$.

We thus have the following natural inclusions, all compatible with addition and multiplication:

$$\mathbb{N}_0 \subseteq \mathbb{Z} \subseteq \mathbb{Q} \subseteq \mathbb{R} \subseteq \mathbb{C}.$$

These number systems progressively contain more numbers and more solutions of more equations. For example, the equation $1 + x = 0$ does not have any solutions in $\mathbb{N}_0$ but it does have one in $\mathbb{Z}$; the equation $1 + 2x = 0$ does not have any solutions in $\mathbb{Z}$, but it does have one in $\mathbb{Q}$; the equation $2 - x^2 = 0$ does not have any solutions in $\mathbb{Q}$, but it does have two in $\mathbb{R}$ (namely $\sqrt{2}$ and $-\sqrt{2}$); the equation $2 + x^2 = 0$ does not have any solutions in $\mathbb{R}$, but it does have two in $\mathbb{C}$ (namely $\sqrt{2}i$ and $-\sqrt{2}i$). Furthermore, the standard quadratic formula always yields roots of quadratic equations in $\mathbb{C}$.*

Exercises for Section 3.9

3.9.1. Write the following elements in the form $a + bi$ with $a, b \in \mathbb{R}$:

   i) $\frac{1}{1+i}$.

   ii) $\frac{(4+i)(7-3i)}{(1-i)(3+2i)}$.

   iii) $(2 - 3i)^5$. (Hint: Exercise 1.7.7.)

3.9.2. Let $x = \frac{\sqrt{3}}{2} + \frac{i}{2}$. Draw x, x^2, x^3, x^4, x^5, x^6, x^7. What do you observe?

3.9.3. Draw the following sets in $\mathbb{C}$:

   i) $\{x : \text{the real part of } x \text{ is } 3\}$.

   ii) $\{x : \text{the imaginary part of } x \text{ is } 3\}$.

   iii) $\{x : \text{the product of the real and imaginary parts of } x \text{ is } 3\}$.

   iv) $\{x : \text{the product of the real and imaginary parts of } x \text{ is } 0\}$.

---

   * One of the excellent properties of $\mathbb{C}$ is **the Fundamental Theorem of Algebra**: every polynomial with coefficients in $\mathbb{C}$ or $\mathbb{R}$ has roots in $\mathbb{C}$. The proof of this fact is proved in a junior-level class on complex analysis or in a senior-level class on algebra. The theorem does not say how to find the roots, only that they exist. In fact, there is another theorem in Galois theory that says that in general it is impossible to find roots of a polynomial by using radicals, sums, differences, products, and quotients.

3.9.4. Let $x \in \mathbb{C}$ such that $x^2 = -1$. Prove that either $x = i$ or $x = -i$ by using definitions of squares of complex numbers and without invoking Theorem 2.4.15.

3.9.5. Prove that the only $x \in \mathbb{C}$ with $x^2 = 0$ is $x = 0$.

3.9.6. Prove that $\mathbb{C}$ is not ordered in the sense of Definition 2.7.10. Justify any facts (such as that $0 < 1$).

3.10   Functions related to complex numbers

We have established existence and properties of several functions of complex numbers:

(1) Inclusion of $\mathbb{R}$ into $\mathbb{C}$: $f(r) = r = r + 0 \cdot i = (r, 0)$.

(2) Identity function on $\mathbb{C}$: $\mathrm{id}_{\mathbb{C}}(x) = x$. (Say why this function is different from the function in part (1).)

(3) Real part $\mathrm{Re} : \mathbb{C} \to \mathbb{R}$.

(4) Imaginary part $\mathrm{Im} : \mathbb{C} \to \mathbb{R}$.

(5) Additive inverse $- : \mathbb{C} \to \mathbb{C}$.

(6) Multiplicative inverse $\frac{1}{()} : (\mathbb{C} \setminus \{0\}) \to \mathbb{C}$.

(7) Addition $+ : \mathbb{C} \times \mathbb{C} \to \mathbb{C}$.

(8) Multiplication $\cdot : \mathbb{C} \times \mathbb{C} \to \mathbb{C}$.

(9) Scalar multiplication functions: for any $z \in \mathbb{C}$, multiplication by z is a function with domain and codomain $\mathbb{C}$: $f(x) = zx$.

There are further obvious functions:

(10) A function $f : \mathbb{C} \to \mathbb{C}$ is called **polynomial** if there exist $a_0, a_1, \ldots, a_n \in \mathbb{C}$ such that for all $x \in \mathbb{C}$, $f(x) = a_0 + a_1 x + a_2 x^2 + \cdots + a_n x^n$.

(11) A function is called **rational** if it equals a polynomial function divided by a polynomial function. Examples of rational functions are polynomial functions, as well as such functions as $\frac{1}{x^2+1}$ and $\frac{x^2+i}{x-3i}$.

From now on, when the domain of a function is not given explicitly, we take the domain to be the largest possible subset of $\mathbb{C}$ on which the function makes sense. (Before we took the largest possible subset of $\mathbb{R}$; see Notation 2.4.5.) So, the domain of the last two functions in the previous paragraph are $\mathbb{C} \setminus \{i, -i\}$ and $\mathbb{C} \setminus \{3i\}$.

There is one more very important and powerful function on $\mathbb{C}$, which at first may seem unmotivated:

Definition 3.10.1. *The* **complex conjugate** *of $a + bi$ is $\overline{a + bi} = a - bi$.*

Geometrically, the complex conjugate of a number is the reflection of the number across the real axis.

Theorem 3.10.2. *Let $x, y \in \mathbb{C}$. Then*

(1) $x \neq 0$ if and only if $\overline{x} \neq 0$.

(2) $x = \overline{x}$ if and only if $x \in \mathbb{R}$.

(3) The complex conjugate of the complex conjugate of x equals x. In symbols: $\overline{\overline{x}} = x$.

(4) $\overline{x + y} = \overline{x} + \overline{y}$.

(5) $\overline{x \cdot y} = \overline{x} \cdot \overline{y}$.

(6) If $y \neq 0$, then $\overline{(x/y)} = \overline{x}/\overline{y}$.

Proof. Write $x = a + bi$ and $y = c + di$ for some $a, b, c, d \in \mathbb{R}$. Certainly $x = 0$ if and only if $a = b = 0$, which holds if and only if $\overline{x} = 0$. This proves (1).

Certainly $x = \overline{x}$ if and only if $a + bi = a - bi$, i.e., if and only if $(a, b) = (a, -b)$, and that holds if and only if $b = 0$. Thus $x = \overline{x}$ if and only if $x = a \in \mathbb{R}$. This proves (2).

The following proves (3): $\overline{\overline{x}} = \overline{a - bi} = a + bi = x$. Addition in (4) is straightforward. The following proves (5):

$$
\begin{aligned}
\overline{x \cdot y} &= \overline{(a + bi) \cdot (c + di)} \\
&= \overline{ac - bd + (ad + bc)i} \\
&= ac - bd - (ad + bc)i \\
&= (a - bi) \cdot (c - di) \\
&= \overline{x} \cdot \overline{y}.
\end{aligned}
$$

If $y \neq 0$, then by (1) and (5), $\overline{x} = \overline{(x/y)y} = \overline{x/y} \cdot \overline{y}$, and so (6) follows. $\qquad\square$

Remark 3.10.3. All functions above with domain and codomain equal to $\mathbb{C}$ were given with some sort of algebraic formulation or description.

How else can we represent such a function? We certainly cannot give a tabular function formulation since the domain is infinite. But we cannot draw such a function either: for the domain we would need to draw the two-dimensional real plane, and the same for the codomain, so we would have to draw the four-dimensional picture to see it all, and that is something we cannot do. So we need to be satisfied with the algebraic or verbal descriptions of functions.

Exercises for Section 3.10

3.10.1. Let $x, y \in \mathbb{C}$.

   i) Prove that $x + \overline{x} = 2 \operatorname{Re} x$.

   ii) Prove that $x - \overline{x} = 2i \operatorname{Im} x$.

   iii) Prove that $x \cdot \overline{y} + \overline{x} \cdot y = 2 \operatorname{Re}(\overline{x}y) = 2 \operatorname{Re} x \operatorname{Re} y + 2 \operatorname{Im} x \operatorname{Im} y$.

3.10.2. Let x be a complex number. Prove that $x \cdot \overline{x}$ is real.

3.10.3. Let x and y be non-zero complex numbers such that $x \cdot y$ is real. Prove that there exists a real number r such that $y = r \cdot \overline{x}$.

3.10.4. Let x be non-zero complex number such that x^2 is real. Prove that $\operatorname{Re}(x) \cdot \operatorname{Im}(x) = 0$, i.e., that either $\operatorname{Re}(x)$ or $\operatorname{Im}(x)$ is zero.

3.11   Absolute value in $\mathbb{C}$

We have seen the absolute value function in ordered fields. The Pythagorean theorem in the plane $\mathbb{R} \times \mathbb{R}$ motivates the natural definition of distance in $\mathbb{C}$:

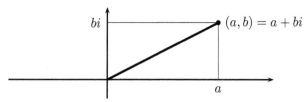

Definition 3.11.1. *The **absolute value** of a complex number $(a, b) = a + bi$ is $|a + bi| = \sqrt{a^2 + b^2} \in \mathbb{R}$. The absolute value is also called the **norm** or the **length**. The **distance** between complex numbers x and y is $|x - y|$.*

Since the absolute value is a real number, this gives a way to partially compare complex numbers, say by their lengths, or by their real components. But recall Exercise 3.9.6: $\mathbb{C}$ is not an ordered field.

The absolute value of $(a, 0) = a$ or $(0, a) = ia$ is $|a|$; the absolute value of $(1, 1) = 1 + i$ is $\sqrt{2}$; the absolute value of $(1, \sqrt{2}) = 1 + i\sqrt{2}$ is $\sqrt{3}$; the absolute value of $(1, \sqrt{3}) = 1 + i\sqrt{3}$ is $\sqrt{4} = 2$, etc.

Theorem 3.11.2. *Let* $x, y \in \mathbb{C}$. *Then*

(1) $|x| = |\overline{x}|$.
(2) $x \cdot \overline{x} = |x|^2$ *is a non-negative real number.*
(3) $x = 0$ *if and only if* $|x| = 0$.
(4) $|\operatorname{Im} x|, |\operatorname{Re} x| \leq |x|$.
(5) *If* $x \neq 0$, *then* $x^{-1} = \overline{x}/|x|^2$.
(6) $|xy| = |x| \, |y|$.
(7) **(Triangle inequality)** $|x \pm y| \leq |x| + |y|$.
(8) **(Reverse triangle inequality)** $|x \pm y| \geq ||x| - |y||$.

Proof. Write $x = a + bi$ for some $a, b \in \mathbb{R}$. Then $|x| = \sqrt{a^2 + b^2} = \sqrt{a^2 + (-b)^2} = |\overline{x}|$, which proves (1). Also, $x \cdot \overline{x} = (a + bi)(a - bi) = a^2 - (bi)^2 = a^2 + b^2 = |x|^2$, and this is the sum of two non-negative real numbers, and is thus non-negative. This proves (2). Furthermore, since $\mathbb{R}$ is an ordered field, by Exercise 2.7.9, $a^2, b^2 \geq 0$, so by Exercise 2.7.10, $a^2 + b^2 = 0$ if and only if $a = b = 0$. This proves (3).

Since $b^2 \geq 0$, it follows that $a^2 \leq a^2 + b^2$, so that by Corollary 2.9.3, $|a| = \sqrt{a^2} \leq \sqrt{a^2 + b^2}$. This proves that $|\operatorname{Re} x| \leq |x|$. Similarly $|\operatorname{Im} x| \leq |x|$. This proves (4).

If $x \neq 0$, then by (3), $|x|$ is a non-zero (real, complex) number, and by (2), $\overline{x}/|x|^2$ is the multiplicative inverse of x. This proves (5).

We could prove (6) with straightforward but laborious algebra by using that $x = a + bi$ and $z = c + di$ for some real numbers a, b, c, d, and expanding the relevant sides, but the following proof is better:

$$|xy|^2 = (xy)\overline{(xy)} \text{ (by (3))}$$
$$= xy\overline{x}\,\overline{y} \text{ (by Theorem 3.10.2)}$$
$$= x\overline{x}y\overline{y} \text{ (by associativity and commutativity of } \cdot \text{ in } \mathbb{C})$$
$$= |x|^2|y|^2 \text{ (by (3))}$$
$$= (|x||y|)^2 \text{ (by associativity and commutativity of } \cdot \text{ in } \mathbb{R}).$$

Now (6) follows by taking square roots of both sides.

To prove the triangle inequality, we also use the squares of the desired quantities to avoid having to write the square root:

$$|x \pm y|^2 = (x \pm y)\overline{(x \pm y)} \text{ (by (2))}$$
$$= (x \pm y)(\overline{x} \pm \overline{y}) \text{ (by Theorem 3.10.2 (2))}$$
$$= x\overline{x} \pm x\overline{y} \pm y\overline{x} + y\overline{y} \text{ (by algebra)}$$
$$= |x|^2 \pm x\overline{y} \pm \overline{x\overline{y}} + |y|^2 \text{ (by Theorem 3.10.2 (2))}$$
$$= |x|^2 \pm 2\operatorname{Re}(x\overline{y}) + |y|^2 \text{ (by Exercise 3.10.1 iii))}$$
$$\leq |x|^2 + 2|\operatorname{Re}(x\overline{y})| + |y|^2 \text{ (comparison of real numbers)}$$
$$\leq |x|^2 + 2|x\overline{y}| + |y|^2 \text{ (by (4))}$$
$$= |x|^2 + 2|x||\overline{y}| + |y|^2 \text{ (by (6))}$$
$$= |x|^2 + 2|x||y| + |y|^2 \text{ (by (1))}$$
$$= (|x| + |y|)^2,$$

and since the squaring function is strictly increasing on the set of non-negative real numbers, it follows that $|x \pm y| \leq |x| + |y|$. This proves (7), and by Theorem 2.10.3 also (8).   □

A consequence of part (6) of this theorem is that if $x \in \mathbb{C}$ has absolute value greater than 1, then positive integer powers of x have increasingly larger absolute values, if $|x| < 1$, then positive integer powers of x have get increasingly smaller than 1, and if $|x| = 1$, then all powers of x have absolute value equal to 1.

The absolute value allows the definition of bounded sets in $\mathbb{C}$ (despite not having an order on $\mathbb{C}$):

Definition 3.11.3. *A subset A of $\mathbb{C}$ is* **bounded** *if there exists a positive real number M such that for all $x \in A$, $|x| \leq M$.*

For example, any set with only finitely many elements is bounded: if $A = \{x_1, \ldots, x_n\}$, set $M = \max\{|x_1|, \ldots, |x_n|\} + 1$, and then certainly for all $x \in A$, $|x| < M$.

The subset $\mathbb{Z}$ of $\mathbb{C}$ is not bounded. The infinite set $\{x \in \mathbb{C} : |x| = 5\}$ is bounded. The set $\{i^n : n \in \mathbb{N}^+\}$ is bounded. The set $\{1/n : n \in \mathbb{N}^+\}$ is bounded. The set of complex numbers at angle $\pi/4$ from the positive real axis is not bounded. (Draw these sets.)

Exercises for Section 3.11

3.11.1. Compute the absolute values of the following complex numbers:

$$1, \ i, \ \pm\sqrt{2}, \ 1+i, \ (1+i)/\sqrt{2}, \ 3+4i.$$

3.11.2. Draw the following sets in $\mathbb{C}$:
  i) $\{x : |x| = 3\}$.
  ii) $\{x : |x - 2 + i| = 3\}$.

3.11.3. Let $a \in \mathbb{C}$, let B be a positive real number, and let $A = \{x \in \mathbb{C} : |x - a| \le B\}$. Draw such a set in the complex plane assuming $a \ne 0$, and prove that A is a bounded set.

3.11.4. Prove that the following are equivalent for a subset A of $\mathbb{C}$.
  i) A is a bounded set.
  ii) There exist $a \in \mathbb{C}$ and a positive real number M such that $A \subseteq B(a, M)$.
  iii) For all $a \in \mathbb{C}$ there exists a positive real number M such that $A \subseteq B(a, M)$.
  iv) For all $a \in \mathbb{R}$ there exists a positive real number M such that $A \subseteq B(a, M)$.
  v) There exist $a \in \mathbb{R}$ and a positive real number M such that $A \subseteq B(a, M)$.

3.11.5. Let F be either $\mathbb{C}$ or an ordered field, so that the absolute value function is defined on F. Let S be a subset of F. Prove that S is bounded if and only if $\{|s| : s \in S\}$ is bounded.

3.11.6. For any subsets A and B of $\mathbb{C}$ and any $c \in \mathbb{C}$, define $A+B = \{a+b : a \in A \text{ and } b \in B\}$, $A \cdot B = \{a \cdot b : a \in A \text{ and } b \in B\}$, $cA = \{c \cdot a : a \in A\}$. Compute $A + B, A \cdot B$, and $c \cdot A$ for the following A, B, c:
  i) $A = \{1, 2, i\}$, $B = \{-1, i\}$, $c = 4$.
  ii) $A = \mathbb{R}^+, B = \mathbb{R}^+, c = -1$.

3.11.7. Let A be a bounded subset of $\mathbb{C}$.
  i) Prove that for any complex number c, $\{ca : a \in A\}$ is bounded.
  ii) Prove that for any complex number c, $\{a + c : a \in A\}$ is bounded.
  iii) Prove that $\{a^2 : a \in A\}$ is bounded.
  iv) Prove that for any positive integer n, $\{a^n : a \in A\}$ is bounded.
  v) Prove that for any polynomial function f, $\{f(a) : a \in A\}$ is bounded.

3.11.8. Let $a, b \in \mathbb{C}$. Suppose that for all real numbers $\epsilon > 0$, $|a - b| < \epsilon$. Prove that $a = b$. (Hint: Theorem 2.10.4.)

*3.11.9. (Keep in mind that a square root function on $\mathbb{C}$ is yet to be discussed carefully; see Exercise 5.4.6.) Discuss correctness/incorrectness issues in the following equalities:

i) $-6 = (\sqrt{3}i)(\sqrt{12}i) = \sqrt{-3}\sqrt{-12} = \sqrt{(-3)(-12)} = \sqrt{36} = 6$.

ii) (R. Bombelli, 1560, when solving the equation $x^3 = 15x + 4$.)

$$4 = \sqrt[3]{2 + \sqrt{-121}} + \sqrt[3]{2 - \sqrt{-121}}.$$

iii) (G. Leibniz, 1675) $\sqrt{1 + \sqrt{-3}} + \sqrt{1 - \sqrt{-3}} = \sqrt{6}$.

3.12   Polar coordinates

So far we have expressed complex numbers with pairs of real numbers either in ordered-pair notation (x, y) or in the form $x + yi$. But a complex number can also be uniquely determined from its absolute value and the angle measured counterclockwise from the positive real axis to the line connecting $(0, 0)$ and (x, y).

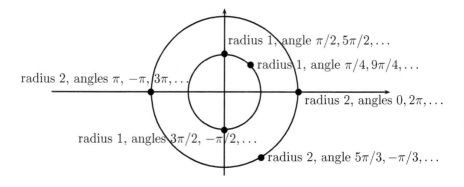

Any choice of θ works for the complex number zero. The angles are measured in radians. (While you may say degrees out loud, get into the habit of writing down radians; later we will see that radians work better.) The angle is not unique; addition of any integer multiple of 2π to it does not change the complex number.

For further examples, $\frac{1+i\sqrt{3}}{2}$ is on the unit circle centered at the origin and is at angle $\pi/3$ counterclockwise from the positive real axis, $\frac{1-i\sqrt{3}}{2}$ is on the same unit circle and at angle $-\pi/3$ counterclockwise from the positive real axis, and $\frac{-1+i\sqrt{3}}{2}$ is on the same circle and at angle $2\pi/3$ counterclockwise from the positive real axis.

We refer to the entries in the ordered pair $(x, y) \in \mathbb{R} \times \mathbb{R} = \mathbb{C}$ as **Cartesian coordinates**. The coordinates (r, θ) consisting of the absolute value r of a complex number and its angle θ (measured counterclockwise from the positive real axis) are referred to as **polar coordinates**.

Numerical conversions between the two coordinate systems use trigonometry. If we know r and θ, then x and y are given by:

$$x = r \cos \theta,$$
$$y = r \sin \theta,$$

and if we know x and y, then r and θ are given by:

$$r = \sqrt{x^2 + y^2},$$

$$\theta = \begin{cases} \text{anything,} & \text{if } x = 0 = y; \\ \frac{\pi}{2}, & \text{if } x = 0 \text{ and } y > 0; \\ \frac{-\pi}{2}, & \text{if } x = 0 \text{ and } y < 0; \\ \arctan(y/x) \in (-\pi/2, \pi/2), & \text{if } x > 0; \\ \arctan(y/x) \in (\pi/2, 3\pi/2), & \text{if } x < 0. \end{cases}$$

Note that the angle is $\pm\pi/2$ precisely when $\operatorname{Re} x = 0$, that the angle is 0 when x is a positive real number, that it is π when x is a negative real number, etc. Furthermore, if the angle is not $\pm\pi/2$, then the tangent of this angle is precisely $\operatorname{Im} x / \operatorname{Re} x$.

We will show in Chapter 9 that the polar coordinates r, θ determine the complex number as $re^{i\theta}$, but at this point we cannot yet make sense out of this exponentiation. Nevertheless, this notation hints at multiplication of complex numbers $re^{i\theta}$ and $se^{i\beta}$ as resulting in $rse^{i(\theta+\beta)}$, confirming that the absolute value of the product is the product of the absolute values, and hinting that the angle of the product is the sum of the two angles of the numbers.

We next prove this beautiful fact of how multiplication works geometrically.

Theorem 3.12.1. (Fun fact) *Let z be a complex number in polar coordinates r and θ. Define functions $M, S, R : \mathbb{C} \to \mathbb{C}$ as follows:*

$M(x) = zx =$ *Multiply x by z,*

$S(x) = rx =$ *Stretch x by a factor of r,*

$R(x) =$ *Rotate x by angle θ counterclockwise around $(0,0)$.*

Then

$$M = S \circ R = R \circ S,$$

or in other words, multiplication by z is the same as stretching by r followed by or preceded by rotating by the angle θ counterclockwise.

Proof. If $z = 0$ or $x = 0$, the conclusion is trivial, so we assume that x and z are non-zero. By the geometry of rotation, rotation and stretching by a positive real number can be done in any order, i.e., $R \circ S(x) = R(rx) = rR(x) = S \circ R(x)$.

So it suffices to prove that $R \circ S(x) = M(x)$ for all x. We first prove this for the special cases $x = 1$ and $x = i$, after which we prove it for general x.

The angle of $M(1) = z$ is θ and its length is r. But $(R \circ S)(1) = R(r)$ also has length r and angle θ, so that $M(1) = (R \circ S)(1)$.

Write $z = (c, d)$ for some $c, d \in \mathbb{R}$. Then $M(i) = (-d, c)$, and we draw a few examples:

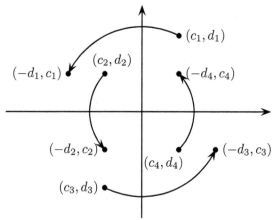

The complex number $M(i) = (-d, c)$ has length equal to $|(c, d)| = r$. The angle between (c, d) and $(-d, c)$ is $90°$, or $\pi/2$ radians, and more precisely, to get from $z = (c, d)$ to $M(i) = (-d, c)$ we have to rotate counterclockwise by $\pi/2$. Thus the angle formed by $M(i)$ counterclockwise from the positive real axis is $\theta + \pi/2$. But $(R \circ S)(i) = R(ri)$ also has the same angle and length as $M(i)$, so that $M(i) = (R \circ S)(i)$.

Now let x be general in $\mathbb{C}$. Write $x = a + bi$ for some $a, b \in \mathbb{R}$. By the geometry of rotation, $R(a + bi) = R(a) + R(bi) = aR(1) + bR(i)$. Then

$$
\begin{aligned}
R \circ S(x) &= S \circ R(x) \text{ (as established from geometry)} \\
&= rR(x) \\
&= rR(a + bi) \\
&= r(aR(1) + bR(i)) \\
&= arR(1) + brR(i) \\
&= aS(R(1)) + bS(R(i)) \\
&= aM(1) + bM(i) \text{ (by previously proved cases)} \\
&= az1 + bzi \\
&= z(a + bi) \\
&= M(x). \qquad \qquad \qquad \square
\end{aligned}
$$

Theorem 3.12.2. *For any non-zero complex number x and any integer n, the angle of x^n counterclockwise away from the positive x-axis is n times the angle of x. Also, $|x^n| = |x|^n$.*

Proof. If $n = 1$, this is trivially true. Now suppose that the theorem is true for some positive integer n. Then the angle of x^{n-1} counterclockwise away from the positive x-axis is $n-1$ times the angle of x, and by Theorem 3.12.1, the angle of $x^n = xx^{n-1}$ is the sum of the angles of x and x^{n-1}, so that it is n times the angle of x. Similarly, by part (6) of Theorem 3.11.2, $|x^n| = |xx^{n-1}| = |x| |x^{n-1}| = |x||x|^{n-1} = |x|^n$.

Thus by induction the theorem is proved for all positive n.

Still keep n positive. Since $1 = x^{-n}x^n$ has angle 0 and x^n has angle n times the angle of x, by Theorem 3.12.1, x^{-n} must have angle $-n$ times the angle of x. Also, by part (6) of Theorem 3.11.2, $1 = |x^{-n}| |x^n| = |x^{-n}| |x|^n$, so that $|x^{-n}| = |x|^{-n}$. Thus the theorem holds for all non-zero n.

Finally, if $n = 0$, then the angle of $x^n = 1$ is 0, which is 0 times the angle of x, and $|x^0| = |1| = 1 = |x|^0$. □

For example, $\frac{-1+\sqrt{3}i}{2}$ is on the unit circle at angle $2\pi/3$ counterclockwise from the positive x-axis (i.e., at angle $120°$ in degrees), and so the second power of $\frac{-1+\sqrt{3}i}{2}$ is on the unit circle at angle $4\pi/3$ counterclockwise from the positive x-axis, and the cube power is on the unit circle at angle 2π, i.e., at angle 0, so that $(\frac{-1+\sqrt{3}i}{2})^3 = 1$.

Corollary 3.12.3. *Let n be a positive integer. Let A be the set of all complex numbers on the unit circle at angles $0, \frac{2\pi}{n}, 2\frac{2\pi}{n}, 3\frac{2\pi}{n}, \ldots, (n-1)\frac{2\pi}{n}$. Then A equals the set of all the complex number solutions to the equation $x^n = 1$.*

Proof. Let $a \in A$. By the previous theorem, a^n has length 1 and angle an integer multiple of 2π, so that $a^n = 1$. If $b \in \mathbb{C}$ satisfies $b^n = 1$, then $|b|^n = |b^n| = 1$, so that the non-negative real number $|b|$ equals 1. Thus b is on the unit circle. If θ is its angle, then the angle of $b^n = 1$ is by the previous theorem equal to $n\theta$, so that $n\theta$ must be an integer multiple of 2π. It follows that θ is an integer multiple of $\frac{2\pi}{n}$, but all those angles appear for the elements of A. Thus every element of A is a root, and every root is an element of A, which proves the corollary. □

Theorem 3.12.4. *Let x be a non-zero complex number and let n be a positive integer. Then there exist exactly n complex numbers whose nth power equals x.*

Proof. By Theorem 3.11.2 we know that $r = |x|$ is positive. By Theorem 3.8.10 there exists a positive real number s such that $s^n = r$. Let α be the angle of x in radians measured counterclockwise from the positive real axis. For any positive integer j let u_j be the complex number on the unit circle whose angle from the positive real axis is $(\alpha + 2\pi j)/n$. By the previous theorem, u_j^n is on the unit circle at angle $\alpha + 2\pi j$ counterclockwise from the positive real axis. But this is the same as the complex number on the unit circle at angle α counterclockwise from the positive real axis. Hence $(su_j)^n = s^n u_j^n = r u_j^n$ is the complex number on the circle of radius r at angle α measured counterclockwise from the positive real axis. This says that $(su_j)^n = x$. By angle considerations, $su_1, su_2, \ldots, su_n$ are distinct. This proves that there exist n complex numbers whose nth power equals x.

Now let y be any complex number whose nth power equals x. By the previous theorem, $|y| = s$. Let β be the angle of y measured counterclockwise from the positive real axis. Since $y^n = x$, by the previous theorem, $n\beta - \alpha = 2\pi k$ for some integer k. Hence $\beta = (\alpha + 2\pi k)/n$. We can write $k = k'n + j$ for some integer $j \in \{1, 2, \ldots, n\}$ and some integer k'. Then

$$\beta = (\alpha + 2\pi k)/n = (\alpha + 2\pi k_1)/n + 2\pi k',$$

so that the angle and the length of y are the same as those of su_j, so that $y = su_j$. Thus there exist exactly n complex numbers whose powers equal x. □

Whereas for non-negative real numbers we choose its non-negative square root as **the** nth root, there is no natural choice for an nth root of a complex number; more on that is in the chapter on continuity in Exercise 5.4.6.

Exercises for Section 3.12

3.12.1. Write each complex number in the form $a + bi$ with $a, b \in \mathbb{R}$:

   i) Complex number of length 1 and at angle $\pi/4$ measured counterclockwise from the positive real axis.

   ii) Complex number of length 1 and at angle $-\pi/4$ measured counterclockwise from the positive real axis.

   iii) The product and the sum of the numbers from the previous two parts.

3.12.2. Draw the following points in the real plane, and think about Theorem 3.12.1:

   i) $3 - 2i, i(3 - 2i)$,

   ii) $-2 - i, i(-2 - i)$.

   iii) $2 - 3i, (1 + i)(2 - 3i)$.

   iv) $1 - i, (1 + i)(1 - i)$.

3.12.3. Draw in $\mathbb{C} = \mathbb{R} \times \mathbb{R}$ the set

$\{x : \text{the angle of } x \text{ counterclockwise from the positive real axis is } \pi/3\}$.

3.12.4. Draw the following sets:

  i) $\{x \in \mathbb{C} : |x| = 3\}$.

  ii) $\{x \in \mathbb{C} : |x| < 3\}$.

  iii) $\{x \in \mathbb{C} : |x - i| = 3\}$.

  iv) $\{x \in \mathbb{C} : |x - i|^2 = 3\}$.

  v) $\{x \in \mathbb{C} : |x^2 - 4x + 4| = 3\}$.

3.12.5. Prove that for any non-zero $z \in \mathbb{C}$ there exist exactly two elements in $\mathbb{C}$ whose square equals z. (Hint: Theorem 3.8.10 and Theorem 3.12.1.)

3.12.6. Let z be non-zero in $\mathbb{C}$ with polar coordinates r and θ and let $n \in \mathbb{N}^+$. For an integer k, let z_k be the complex number whose absolute value equals $r^{1/n}$ and whose angle measured counterclockwise from the positive x axis is $k\theta/n$.

  i) Prove that z_k is uniquely determined.

  ii) Prove that for all k, $z_k^n = z$. (Hint: Theorem 3.12.1.)

  iii) Prove that the set $\{z_k : k \in \mathbb{N}\}$ contains exactly n elements.

3.12.7. "Square" the pentagon drawn below. Namely, estimate the coordinates (real, imaginary or length, angle) of various points on the pentagon, square the point, and draw its image on a different real plane. You need to plot the image not only of the five vertices, but of several representative points from each side to see how the squaring curves the edges. (Hint: You may want to use Theorem 3.12.1.)

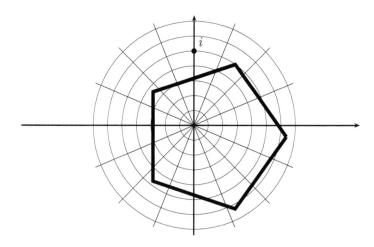

3.13 Topology on the fields of real and complex numbers

When reading this section, absorb the following main points of topology: open ball, open set, limit point, closed set. The main object of this section is to introduce limit points of sets so that we can in subsequent chapters talk about limits of functions, sequences, and series.

By a **topology** on a set we mean that some sets are declared open, subject to the conditions that the empty set and the whole set have to be open, that arbitrary unions of open sets be open, and that finite intersections of open sets be open. In any topology, the complement of an open set is called closed, but a set may be neither open nor closed. A topology can be imposed on any set, not just $\mathbb{R}$ or $\mathbb{C}$, but we focus on these two cases, and in fact we work only with the "standard", or "Euclidean" topology.

Definition 3.13.1. *Let F be $\mathbb{R}$ or $\mathbb{C}$. Let $a \in F$ and let r be a positive real number. An* **open ball with center a and radius r** *is a set of the form*

$$B(a, r) = \{x \in F : |x - a| < r\}.$$

An **open set** *in F is any set that can be written as a union of open balls.*

The following are both $B(0, 1)$, but the left one is a ball in $\mathbb{R}$ and the right one is a ball in $\mathbb{C}$. Note that they are different: by definition the left set is an open subset of $\mathbb{R}$, but if you think of it as a subset of $\mathbb{C}$, it is not open (see Exercise 3.13.1).

 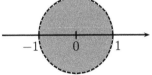

Examples 3.13.2.

(1) $B(a, r)$ is open.

(2) $F = \cup_{a \in F} B(a, 1)$ is an open set.

(3) The empty set is an open set because it is vacuously a union of open sets (see page 49).

(4) For real numbers $a < b$, the interval in $\mathbb{R}$ of the form (a, b) is an open set in $\mathbb{R}$ because it is equal to $B((a + b)/2, (b - a)/2)$. The interval (a, ∞) is open because it equals $\cup_{n=1}^{\infty} B(a + n, 1)$.

(5) The set $A = \{x \in \mathbb{C} : 1 < \operatorname{Re} x < 3 \text{ and } 0.5 < \operatorname{Im} x < 2\}$ is open in $\mathbb{C}$. Namely, this set is the union $\cup_{a \in A} B(a, \min\{\operatorname{Re} a - 1, 3 - \operatorname{Re} a, \operatorname{Im} a - 0.5, 2 - \operatorname{Im} a\}))$.

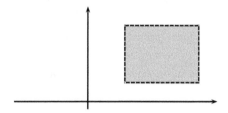

(6) The set $A = \{x \in \mathbb{C} : \operatorname{Re} x < 1 \text{ and } \operatorname{Im} x < 2\}$ is open in $\mathbb{C}$. Namely, this set is the union $\cup_{a \in A} B(a, \min\{1 - \operatorname{Re} a, 2 - \operatorname{Im} a\})$.

Theorem 3.13.3. *$B(a, r)$ has infinitely many points.*

Proof. For each integer $n \geq 2$, $a + r/n \in B(a, r)$. Since $r > 0$, these numbers are all distinct. Since $\mathbb{N}_0$ is infinite, so is the set of all integers that are at least 2. □

Example 3.13.4. Thus if A is an open subset of F, then either A is empty or A has infinitely many points. In particular, $\{a\}$ is not open.

Theorem 3.13.5. *Let A be an open set and let $a \in A$. Then there exists $r > 0$ such that $B(a, r) \subseteq A$.*

Proof. Since A is open, it is a union of open balls. Thus a is an element of one such ball $B(b, s)$, with $B(b, s) \subseteq A$.

Since $a \in B(b, s)$, we have that $|a - b| < s$, so that $r = s - |a - b|$ is a positive real number. (In the illustration, this is the distance between a and the outside of the circle.) We claim that $B(a, r) \subseteq$  $B(b, s)$. To prove this, let $x \in B(a, r)$. Then $|x - b| = |x - a + a - b| \leq |x - a| + |a - b|$ by the triangle inequality, and since $|x - a| < r = s - |a - b|$, it means that $|x - b| < s$, so that $x \in B(b, s)$. This proves the claim, and hence it proves that $B(a, r) \subseteq A$. □

Theorem 3.13.6. (Topology on F)
(1) $\emptyset$ and F are open.

(2) *Arbitrary unions of open sets are open.*

(3) *Finite intersections of open sets are open.*

Proof. The empty set can be written as an empty union of open balls, so it is open vacuously, and $F = \cup_{a \in F} B(a, 1)$, so that F is open. This proves (1).

Every open set is a union of open balls, and so the union of open sets is a union of open balls, hence open. This proves (2).

Now let $A_1, \ldots, A_n$ be open sets. Let $a \in A_1 \cap \cdots \cap A_n$. By Theorem 3.13.5, for each $k = 1, \ldots, n$, there exists $r_k > 0$ such that $B(a, r_k) \subseteq A_k$. Set $r = \min\{r_1, \ldots, r_n\}$. Then $B(a, r) \subseteq \cap_{k=1}^n B(a, r_k) \subseteq \cap_{k=1}^n A_k$. Thus for each $a \in \cap_{k=1}^n A_k$ there exists $r_a > 0$ such that $B(a, r_a) \subseteq \cap_{k=1}^n A_k$. It follows that

$$\cap_{k=1}^n A_k = \cup_{a \in \cap_{k=1}^n A_k} B(a, r_a).$$

This proves that $\cap_{k=1}^n A_k$ is a union of open balls, so it is open. $\quad\square$

An arbitrary intersection of open sets need not be open. For example, $\cap_{n=1}^\infty B(a, 1/n) = \{a\}$ is not open.

Definition 3.13.7. *Let A be an arbitrary subset of F and $a \in F$ (not necessarily in A). We say that a is a **limit point** of A if for all real numbers $s > 0$, $B(a, s)$ contains elements of A different from a.*

Examples 3.13.8.

(1) If $A = \{a\}$, then the set of limit points of A is the empty set.

(2) If $A = \mathbb{Q}$, then the set of limit points of A is $\mathbb{R}$.

(3) The set of limit points of $\emptyset$ is empty, the set of limit points of $\mathbb{R}$ is $\mathbb{R}$, and the set of limit points of $\mathbb{C}$ is $\mathbb{C}$.

(4) The set of limit points of $B(a, r)$ equals $\{x \in F : |x - a| \le r\}$.

Definition 3.13.9. *A set A is a **closed set** if it contains all of its limit points.*

Theorem 3.13.10. *A is open if and only if $F \setminus A$ is closed.*

Proof. Suppose that A is open. Let $x \in A$. By Theorem 3.13.5, there exists $r > 0$ such that $B(x, r) \subseteq A$. Thus $B(x, r) \cap (F \setminus A) = \emptyset$, so that x is not a limit point of $F \setminus A$. Thus no point of A is a limit point of $F \setminus A$, which proves that any limit points of $F \setminus A$ are in $F \setminus A$. Thus $F \setminus A$ is closed.

Now suppose that $F \setminus A$ is closed. Let $x \in A$. Since $F \setminus A$ contains all of its limit points, then x is not a limit point of $F \setminus A$. Thus by the definition of limit points, there exists $r > 0$ such that $B(x, r) \cap (F \setminus A)$ is empty. This means that $B(x, r) \subseteq A$. Thus $A = \cup_{x \in A} B(x, r_x)$ for appropriate $r_x > 0$, so that A is open. □

The following is now almost immediate from previous results:

Theorem 3.13.11. (Topology on F)

(1) $\emptyset$, F are closed sets.

(2) Arbitrary intersections of closed sets are closed.

(3) Finite unions of closed sets are closed.

Proof. Exercise 2.1.8 proves that the union of the complements of two sets equals the complement of the intersection and that the it was proved that the intersection of the complements of two sets equals the complement of the union, and an equally easily proved mathematical truth is the following generalization to possibly many more sets:

$$F \setminus \bigcup_{k \in I} A_k = \bigcap_{k \in I} (F \setminus A_k), \qquad F \setminus \bigcap_{k \in I} A_k = \bigcup_{k \in I} (F \setminus A_k).$$

With this, (2) and (3) follow from the last two theorems, and the proof of (1) is trivial. □

Both $\emptyset$ and F are open and closed, and these turn out to be the only sets that are both open and closed (see Exercise 3.13.2). Some sets are neither open nor closed (see Exercise 3.13.1).

Remark 3.13.12. (This remark puts Theorems 3.13.6 and 3.13.11 in the more general context and is not needed in the first course in analysis.) Any set F (not necessarily a field) is a **topological space** if there exists a collection $\mathcal{T}$ of subsets of F such that the following properties are satisfied:

(1) $\emptyset, F \in \mathcal{T}$,

(2) Arbitrary unions of elements in $\mathcal{T}$ are in $\mathcal{T}$.

(3) Finite intersections of elements in $\mathcal{T}$ are in $\mathcal{T}$.

Elements of $\mathcal{T}$ are called **open**. Subsets of F that are complements of open sets are called **closed**. The proof of Theorem 3.13.11 for closed sets in this topological space are proved in the same way.

Exercises for Section 3.13

3.13.1. Let A be the open ball in $\mathbb{R}$ of radius 1 and centered at 0. Since $\mathbb{R}$ is a subset of $\mathbb{C}$, then A is also a subset of $\mathbb{C}$. Prove that A is neither a closed nor an open subset of $\mathbb{C}$.

3.13.2. Let F be either $\mathbb{R}$ or $\mathbb{C}$, and let A be a subset of F that is both closed and open. and let A be a closed and open subset of F. Prove that $A = \emptyset$ or $A = F$.

3.13.3. Prove that $\mathbb{Q}$ is neither an open nor a closed subset of $\mathbb{R}$ or $\mathbb{C}$.

3.13.4. Sketch the following subsets of $\mathbb{C}$. Determine their sets of limit points, and whether the sets are open, closed, or neither: $\{x \in \mathbb{C} : \operatorname{Im} x = 0 \text{ and } 0 < \operatorname{Re} x < 1\}$, $\{x \in \mathbb{C} : \operatorname{Im} x = 0 \text{ and } 0 \le \operatorname{Re} x \le 1\}$, $\{x \in \mathbb{C} : -2 \le \operatorname{Im} x \le 2 \text{ and } 0 \le \operatorname{Re} x \le 1\}$, $\{x \in \mathbb{C} : 2 \le \operatorname{Im} x \le -2 \text{ and } 0 \le \operatorname{Re} x \le 1\}$, $\{1/n : n \in \mathbb{N}^+\}$.

3.13.5. For each of the following intervals as subsets of $\mathbb{R}$, determine the set of limit points, and whether the set is open, closed, or neither in $\mathbb{R}$:

$$(0, 1), [0, 1], [3, 5), [3, \infty).$$

3.13.6. Let a be a limit point of a set A. Suppose that a set B contains A. Prove that a is a limit point of B.

3.13.7. Give examples of sets $A \subseteq B \subseteq \mathbb{C}$ and $a \in \mathbb{C}$ such that a is a limit point of B but not of A.

3.13.8. Let A be a subset of $\mathbb{C}$ all of whose elements are real numbers. Prove that every limit point of A is a real number.

3.14 The Heine-Borel theorem

Closed and bounded sets in $\mathbb{C}$ and $\mathbb{R}$ have many excellent properties — we will for example see in Section 5.3 that when a good (say continuous) real-valued function has a closed and bounded domain, then that function achieves a maximum and minimum value, etc. The concept of uniform continuity (introduced in Section 5.5) needs the fairly technical Heine-Borel theorems proved in this section.

Construction 3.14.1. (Halving closed and bounded subsets of $\mathbb{R}$ and quartering closed and bounded subsets of $\mathbb{C}$) Let A be a bounded subset of $\mathbb{R}$ or of $\mathbb{C}$, and let P be a property that applies to some subsets of A. Boundedness of A guarantees that A fits inside a closed bounded rectangle R_0 of the form $(a_0, b_0) \times (c_0, d_0)$ in $\mathbb{C}$, with $c_0 = 0 = d_0$ if A is a subset of $\mathbb{R}$. The rectangle can be halved lengthwise and crosswise to get four equal closed subrectangles. In the next iteration we pick, if possible, one of these four closed quarter subrectangles such that its intersection with A has property P. We call this subrectangle R_1. If A is a subset of $\mathbb{R}$, then the length of R_1 is half the length of R_0, and otherwise the area of R_1 is a quarter of the area of R_0. In general, once we have R_n, we similarly pick a subrectangle R_{n+1} such that $R_{n+1} \cap A$ has property P and such that the sides of R_{n+1} are half the lengths of the sides in R_n. Write $R_n = [a_n, b_n] \times [c_n, d_n]$ for some real numbers $a_n \le b_n$ and $c_n \le d_n$. By construction, for all n, $b_n - a_n = (b_0 - a_0)/2^n$, and

$$a_0 \le a_1 \le a_2 \le \cdots \le a_n \le \cdots \le b_n \cdots \le b_2 \le b_1 \le b_0.$$

This means that $\{a_1, a_2, a_3, \ldots\}$ is a non-empty subset of $\mathbb{R}$ that is bounded above, so that by the Least upper bound theorem (Theorem 3.8.5), $a = \sup\{a_1, a_2, a_3, \ldots\}$ is a real number. Similarly, $b = \inf\{b_1, b_2, b_3, \ldots\}$ is a real number. Since $a \le b_1, b_2, b_3, \ldots$, it follows that $a \le b$. Suppose that $a < b$. Then by Exercise 3.8.3, there exists a positive integer N such that $1/2^N < (b-a)/(b_0 - a_0)$. But $a_N \le a \le b \le b_N$, so that $0 \le b - a \le b_N - a_N = (b_0 - a_0)/2^N < b - a$, which contradicts trichotomy. Thus $a = b$, i.e., we just proved that

$$\sup\{a_1, a_2, a_3, \ldots\} = \inf\{b_1, b_2, b_3, \ldots\}.$$

Similarly,

$$c = \sup\{c_1, c_2, c_3, \ldots\} = \inf\{d_1, d_2, d_3, \ldots\}.$$

This means that the intersection of all the R_n equals the set $\{a + ci\}$, consisting of exactly one complex number. By the shrinking property of the subrectangles, for every $\delta > 0$ there exists a positive integer N such that $R_N \cap A \subseteq B(a + ci, \delta)$.

In particular, "quartering" of the closed and bounded region (interval) $A = [a_0, b_0] \subseteq \mathbb{R}$, means halving the rectangle (interval), and the intersection of all the chosen closed half-rectangles is a set with exactly one element. That element is in A, so a real number.

Theorem 3.14.2. (The Heine-Borel theorem (in $\mathbb{R}$, $\mathbb{C}$)) *Let A be a closed and bounded subset of $\mathbb{R}$ or $\mathbb{C}$. For each $c \in A$ let δ_c be a positive number. Then there exists a finite subset S of A such that $A \subseteq \cup_{c \in S} B(c, \delta_c)$.*

Proof. We declare that a subset B of A satisfies (property) P if there exists a finite subset S of B such that $B \subseteq \cup_{c \in S} B(c, \delta_c)$. We want to prove that A has P.

Suppose for contradiction that A does not have P. Since A is closed and bounded, it fits inside a closed rectangle R_0. With Construction 3.14.1, we construct iteratively nested subrectangles $R_0 \supseteq R_1 \supseteq R_2 \supseteq \cdots$. The quarter subrectangles are chosen so that each $R_n \cap A$ does not have P. This is true if $n = 0$ by assumption. Suppose that R_n has been chosen so that $R_n \cap A$ does not have P. If the intersection with A of each of the four quarter subrectangles of R_n has P, i.e., if each of the four subrectangles (as in Construction 3.14.1) intersected with A is contained in the union of finitely many balls $B(c, \delta_c)$, then $R_n \cap A$ is covered by finitely many such balls as well, which contradicts the assumption on R_n. Thus it is possible to choose R_{n+1} so that $R_{n+1} \cap A$ does not have P. By construction, $\cap_{n=1}^{\infty} R_n$ contains exactly one point. Let that point be x. Since each $R_n \cap A$ has infinitely many points, x is a limit point of A, and since A is closed, necessarily $x \in A$.

By the shrinking sizes of the R_n, there exists a positive integer N such that $R_n \subseteq B(x, \delta_x)$. But then $R_n \cap A$ has P, which contradicts the construction. Thus A has P. $\qquad\square$

Remark 3.14.3. Let F be $\mathbb{R}$ or $\mathbb{C}$ and let A be a closed and bounded subset of F. Let T be a collection of open subsets of F such that $A \subset \cup_{U \in T} U$.

This set containment is usually referred to as T being **an open cover** of A. By the definition of set containment, for each $c \in A$ there exists $U_c \in T$ such that $c \in U_c$. Since U_c is open, by Theorem 3.13.5, there exists $\delta_c > 0$ such that $B(c, \delta_c) \subseteq U_c$. The Heine-Borel theorem Theorem 3.14.2 asserts that there exists a finite subset S of A such that $A \subseteq \cup_{c \in S} B(c, \delta_c) \cup_c U_c$. In other words, the Heine-Borel theorem asserts that **every open cover of a closed and bounded subset of $\mathbb{C}$ has a finite subcover.**

Remark 3.14.4. In the more general context of topological spaces as in Remark 3.13.12, there need not be a notion of bounded sets. A set for which every open cover has a finite subcover is called **compact**. So Theorem 3.14.2 proves that every closed and bounded subset of $\mathbb{C}$ is compact.

Theorem 3.14.5. *Let A be a closed and bounded subset of $\mathbb{R}$ or $\mathbb{C}$, and for each $a \in A$ let δ_a be a positive number. Then there exist a finite subset S of A and a positive real number δ such that $A \subseteq \cup_{c \in S} B(c, \delta_c)$ and such that for all $x \in A$ there exists $c \in S$ such that $B(x, \delta) \subseteq B(c, \delta_c)$.*

Proof. By Theorem 3.14.2, there exists a finite subset S of A such that $A \subseteq \cup_{c \in S} B(c, \delta_c/2)$. Let $\delta = \frac{1}{2} \min\{\delta_c : c \in S\}$. Since S is a finite set, δ is a positive real number.

Let $x \in A$. By the choice of S there exists $c \in S$ such that $x \in B(c, \delta_c/2)$. Let $y \in B(x, \delta)$. Then

$$|y - c| = |y - x + x - c| \le |y - x| + |x - c| < \delta + \delta_c/2 \le \delta_c,$$

so that $y \in B(c, \delta_c)$. It follows that $B(x, \delta) \subseteq B(c, \delta_c)$. $\square$

Exercises for Section 3.14

3.14.1. Let $A = \{x \in \mathbb{R} : 2 \le x \le 3\}$. Prove that $A \subseteq \cup_{a \in A} B(a, 1/a)$. Does there exist a finite set S of A such that $A \subseteq \cup_{a \in S} B(a, 1/a)$. If yes, find it, if no, explain why not.

3.14.2. Let $A = \{1/n : n \in \mathbb{N}^+\} \subseteq \mathbb{R}$. Prove that $A \subseteq \cup_{a \in A} B(a, a/3)$. Does there exist a finite set S of A such that $A \subseteq \cup_{a \in S} B(a, a/3)$? Repeat with $A = \mathbb{N}^+$.

3.14.3. (Every open cover of a closed and bounded subset of $\mathbb{C}$ has a finite subcover.) Let A be a closed and bounded subset of $\mathbb{R}$ or $\mathbb{C}$. Let I be a set and for each $k \in I$ let U_k be an open subset of F. Suppose that $A \subseteq \bigcup_{k \in I} U_k$. Prove that there exists a finite subset K of I such that $A \subseteq \bigcup_{k \in K} U_k$. (Hint: Prove that for each $a \in A$ there exists $\delta_a > 0$ such that $B(a, \delta_a)$ is in some U_k. Apply Theorem 3.14.2.)

3.14.4. (Converse of Exercise 3.14.3.) Let A be a subset of $\mathbb{C}$ with the property that whenever $A \subseteq \cup_{k \in I} U_k$ for some open sets U_k as k varies over I, then there exists a finite subset J of I such that $A \subseteq \cup_{k \in J} U_k$. (In other words, every open cover of A has a finite subcover.)

    i) Prove that A is bounded.

    ii) Prove that A is closed.

(Hint: construct appropriate open covers.)

3.14.5. Find subsets $A, U_1, U_2, U_3, \ldots$ of $\mathbb{R}$ such that all U_n are open and $A \subseteq \cup_{n \geq 1} U_n$ but A is not a subset of $\cup_{n=1}^{N} U_n$ for any positive integer N. By Exercise 3.14.3 the set A should not be closed and bounded; try A open and bounded; or try A unbounded.

Chapter 4

Limits of functions

Limits are a foundation of analysis. They are an important mathematical concept and they lend themselves nicely to practicing proofs.

Section 4.1 contains the formal definition together with informal intuitive pictures and explicit examples of limit proofs. Many of the examples are worked out in the longer this-is-how-we-think version as well as in the shorter this-is-how-we-write version. The formal definition is referred to as the "epsilon-delta" definition for obvious reasons. Section 4.2 is a lesson in logical negation on what it means for a number to not be a limit. Section 4.3 looks at the epsilon-delta definition of limits more finely: the order and importance of the quantifiers matter, and small modifications can change the meaning significantly. The lesson of the section is that it is important to remember any statement precisely. The epsilon-delta proofs tend to be time-consuming, so Section 4.4 proves alternative theorems that shortcut that work for many nice functions. But for many limits an epsilon-delta argument is the only possible proof, so it is important to master the method. In the first five sections all is happening in subsets of $\mathbb{C}$, Section 4.5 transitions to limits for real-valued functions being plus or minus infinity, and Section 4.6 for functions whose domain is a subset of $\mathbb{R}$ allows finite or infinite limits to be taken at $\pm\infty$.

4.1 Limit of a function

All calculus classes teach about limits, but the domains there are typically intervals in $\mathbb{R}$. Here we learn a more general definition for (more interesting) domains in $\mathbb{C}$.

Definition 4.1.1. *Let A be a subset of $\mathbb{C}$ and let $f : A \to \mathbb{C}$ be a function. Suppose that a complex number a is a limit point of A (see*

177

Definition 3.13.7). **The limit of f(x) as x approaches a is the complex number L** *if for every real number $\epsilon > 0$ there exists a real number $\delta > 0$ such that for all $x \in A$, if $0 < |x - a| < \delta$, then $|f(x) - L| < \epsilon$.*

When this is the case, we write $\lim_{x \to a} f(x) = L$. Alternatively, in order to not make lines crowded with subscripts, we write $\lim_{x \to a} f(x) = L$.

It is important to note that we are not asking for $f(a)$. For one thing, a may or may not be in the domain of f, we only know that a is a limit point of the domain of f. We are asking for the behavior of the function f at points near a.

We can give a simple geometric picture of this in case the domain and codomain are subsets of $\mathbb{R}$ (refer to Remark 3.10.3 for why we cannot draw functions when domains and codomains are subsets of $\mathbb{C}$). Below are three graphs of real-valued functions defined on a subset of $\mathbb{R}$ and with a being a limit point of the domain. In each, on the graph of $y = f(x)$ we cover the vertical line $x = a$, and with that information, we conclude that $\lim_{x \to a} f(x) = L$.

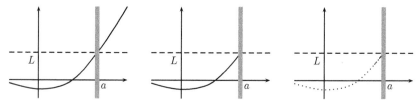

The function f from the first graph above might be any of the following:

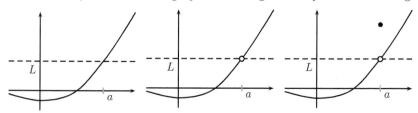

Intuitively we are hoping that $f(x)$ for x near a can predict a trend for the value of f as we get arbitrarily close to a. For example, we may not be able to bring x to 0 Kelvin, but if we can take measurements $f(x)$ for x getting colder and colder, perhaps we can predict what may happen at 0 Kelvin. But how believable is our prediction? Perhaps for our theory to be satisfactory, we need to run experiments at temperatures x that give us $f(x)$ within $\epsilon = 10$ of the predicted value. Or when instruments get better, perhaps ϵ gets smaller, say one thousandth. Or a new material is discovered which allows even smaller ϵ. But no matter what ϵ is determined ahead of time, for the prediction to be believable, we need to determine a fixed range of x, within a δ of a but not equal to a, for which the f-values are within the given ϵ of the prediction.

A graphical way of representing the epsilon-delta definition of limits for real-valued functions with domains in $\mathbb{R}$ is as follows: For every positive ϵ there exists a positive δ such that for all x in the domain with $0 < |x-a| < \delta$ (the $x \neq a$ in the vertical gray band), the value of $f(x)$ is within ϵ of L (in the horizontal grey band):

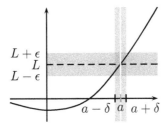

If ϵ gets smaller, δ has to get smaller too; but we may keep the old δ for larger ϵ.

While these pictures can help our intuition, they do not constitute a proof: the definition is an algebraic formulation, and as such it requires algebraic proofs. In the rest of the section we examine many examples algebraically, with the goal of mastering the epsilon-delta proofs. But epsilon-delta proofs are time-consuming, so in the future we will want to replace them with some shortcuts. We will have to prove that those shortcuts are logically correct, and the proofs will require mastering abstract epsilon-delta proofs. Naturally, before we can master abstract epsilon-delta proofs, we need to be comfortable with epsilon-delta proofs on concrete examples. In short, in order to be able to avoid epsilon-delta proofs, we have to master them. (Ha!)

Example 4.1.2. $\lim\limits_{x \to 3}(4x - 5) = 7$.

Proof. The function that takes x to $4x - 5$ is a polynomial function, so it is defined for all complex numbers. Thus the domain of the function is $\mathbb{C}$ and 3 is a limit point of the domain. Let $\epsilon > 0$. [WE ARE PROVING THAT FOR ALL REAL NUMBERS $\epsilon > 0$ SOMETHING-OR-OTHER HOLDS. RECALL THAT

ALL PROOFS OF THIS FORM START WITH "LET ϵ BE AN ARBITRARY POS-
ITIVE REAL NUMBER," OR ABBREVIATED AS WE DID. NOW WE HAVE TO
PROVE THAT THE SOMETHING-OR-OTHER HOLDS. BUT THIS SOMETHING-
OR-OTHER CLAIMS THAT THERE EXISTS A REAL NUMBER $\delta > 0$ WITH A
CERTAIN PROPERTY. THUS WE HAVE TO CONSTRUCT SUCH A δ. IN THIS
FIRST EXAMPLE I SIMPLY PRESENT A δ THAT WORKS, BUT IN SUBSEQUENT
EXAMPLES I SHOW HOW TO FIND A WORKING δ. IN GENERAL, THE PROOF
SHOULD CONTAIN A SPECIFICATION OF δ AND A DEMONSTRATION THAT IT
WORKS, BUT HOW ONE FINDS THAT δ IS IN GENERAL LEFT TO SCRATCH
WORK OR TO INSPIRATION.] Set $\delta = \epsilon/4$. [NEVER MIND HOW THIS MAGIC
$\epsilon/4$ APPEARS HERE; WAIT UNTIL THE NEXT EXAMPLE WHERE I PRESENT
A PROCESS OF FINDING δ.] Then δ is a positive real number. [SURE! —
NOW WE HAVE TO PROVE THAT FOR ALL x, IF $0 < |x - 3| < \delta$, THEN
$|f(x) - L| < \epsilon$. THE PROOF OF "FOR ALL x ..." STARTS WITH:] Let x be an
arbitrary complex number. [FOR THIS x WE NOW HAVE TO PROVE THAT
IF $0 < |x - 3| < \delta$, THEN $|f(x) - L| < \epsilon$. THE PROOF OF "IF P THEN Q"
STARTS WITH "ASSUME P."] Assume that $0 < |x - 3| < \delta$. [NOW WE HAVE
TO PROVE Q, I.E., WE HAVE TO PROVE THAT $|f(x) - L| = |(4x - 5) - 7| < \epsilon$.
WE DO NOT SIMPLY WRITE "$|(4x - 5) - 7| < \epsilon$" BECAUSE WE DO NOT
KNOW THAT YET. WE WRITE THE LEFT SIDE OF THIS INEQUALITY, AND
MANIPULATE IT – ALGEBRAICALLY, OFTEN WITH TRIANGLE INEQUALITIES
AND SEVERAL STEPS, UNTIL WE GET $< \epsilon$.] Then

$$
\begin{aligned}
|(4x - 5) - 7| &= |4x - 12| \\
&= 4|x - 3| \\
&< 4\delta \\
&= 4\frac{\epsilon}{4} \quad [\text{WASN'T } \epsilon/4 \text{ A CLEVER CHOICE OF } \delta?] \\
&= \epsilon. \qquad\qquad \square
\end{aligned}
$$

The commentary in the proof above is describing the thought process
behind the proof but need not and should not be written out in homework
solutions. Below is a homework-style solution:

Proof of Example 4.1.2 without the commentary: The function that takes x
to $4x - 5$ is a polynomial function, so it is defined for all complex numbers.
Thus the domain of the function is $\mathbb{C}$ and 3 is a limit point of the domain.
Let $\epsilon > 0$. Set $\delta = \epsilon/4$. Then δ is a positive real number. Let x be an

arbitrary complex number. Assume that $0 < |x - 3| < \delta$. Then

$$
\begin{aligned}
|(4x - 5) - 7| &= |4x - 12| \\
&= 4|x - 3| \\
&< 4\delta \\
&= 4\frac{\epsilon}{4} \\
&= \epsilon.
\end{aligned}
\qquad \square
$$

In the previous example δ appeared magically as $\epsilon/4$, and it happened to be a positive number depending on ϵ that made the limit proof work. Other numbers would have worked as well, such as $\epsilon/5$, or $\epsilon/(10000+\epsilon)$, and so on. For good style choose simple formulations over complicated ones. If some positive number works as δ, so does any smaller positive number, so there is no smallest possible δ and there are many different correct choices of δ. It is not necessary to find the largest possible δ for a given ϵ, and it is not even necessary to show how you derived or chose your δ. However, it is necessary to show that your pick of δ does satisfy the rest of the defining property of limit.

Think through the comments in the previous paragraph on the example $\lim_{x \to 4} 5 = 5$: any positive number works for δ.

In further examples we show all the necessary work for proofs of limits including how to determine the δ. It is standard to use the proof-writing trick of partial filling-in: when by the definition of limit it is time in the proof to declare what δ is, we typically write "Set $\delta = \underline{\hspace{3cm}}$", and in subsequent reasoning we fill in the underlined blank with whatever restrictions seem necessary. Study the proofs below for how this is done, but let me just say that each restriction puts an upper bound on δ, say bound δ above by 1 and by $\epsilon^2/4$, and this is accomplished by filling in "Set $\delta = \min\{1, \epsilon^2/4 \underline{\hspace{1.5cm}}\}$", with room for possible further restrictions. (In Example 4.1.2 we only had one restriction, namely $\delta \leq \epsilon/4$, so that $\delta = \min\{\epsilon/4\} = \epsilon/4$.)

I find that the following two goals make these proofs more concrete and doable:

- Goal #1: Write $|f(x) - L|$ as less than or equal to (something) times $|x - a|$.

  If f is a rational function, then $|x - a|$ better be a factor of $|f(x) - L|$!

- Goal #2: Find a positive constant upper bound B on the (something) found in Goal #1. Then you make sure that $\delta \leq \epsilon/B$.

You may want to keep these guidelines in mind as you read the proofs below.

Example 4.1.3. $\lim\limits_{x \to -2} (4x^2 - 5x + 2) = 28$.

Proof. The function that takes x to $4x^2 - 5x + 2$ is a polynomial function and it is defined for all complex numbers. Thus -2 is a limit point of the domain. Let $\epsilon > 0$. Set $\delta =$ _____ [δ TO BE DETERMINED STILL; THE FINAL WRITE-UP WILL HAVE THIS FILLED IN, BUT WE DO NOT YET KNOW δ.]. Then δ is a positive real number. [HOPING, ANYWAY.] Let x be any complex number. Suppose that $0 < |x + 2| < \delta$. Then

$$|(4x^2 - 5x + 2) - 28| = |4x^2 - 5x - 26|$$

$$[\text{WANT GOAL \#1: } x - a = x + 2 \text{ BETTER BE A FACTOR.}]$$

$$= |(x + 2)(4x - 13)|$$

$$= |4x - 13| \cdot |x + 2| \quad [\text{GOAL \#1 ACCOMPLISHED}]$$

[GOAL #2: WE WANT TO BOUND ABOVE THE COEFFICIENT $|4x - 13|$ OF $|x-a|$ BY A CONSTANT. BUT OBVIOUSLY $|4x-13|$ IS NOT BOUNDED ABOVE BY A CONSTANT FOR ALL x. SO WE NEED TO MAKE A RESTRICTION ON δ: LET'S MAKE SURE THAT δ IS AT MOST 1 (OR 2, OR 15, IT DOES NOT MATTER WHAT POSITIVE NUMBER YOU PICK IN THIS EXAMPLE). SO ON THE δ LINE WRITE: $\delta = \min\{1,$_____$\}$. THIS GUARANTEES THAT δ WILL BE 1 OR SMALLER, DEPENDING ON WHAT GETS FILLED IN AFTER THE COMMA.]

$$= |4(x + 2 - 2) - 13| \cdot |x + 2| \quad \text{(adding a clever zero)}$$

$$= |4(x + 2) - 21| \cdot |x + 2|$$

$$\leq (|4(x + 2)| + 21) \cdot |x + 2|$$

$$\text{(by the triangle inequality } |a \pm b| \leq |a| + |b|)$$

$$< (4 \cdot 1 + 21) \cdot |x + 2| \text{ (since } |x + 2| < \delta \leq 1)$$

$$= 25 \cdot |x + 2| \qquad [\text{GOAL \#2 ACCOMPLISHED.}]$$

[NOW GO BACK TO SPECIFYING δ AT THE BEGINNING OF THE PROOF BY FILLING IN WITH: $\delta = \min\{1, \epsilon/B\}$. THIS MEANS THAT δ IS THE SMALLER OF 1 AND ϵ/B, AND IN PARTICULAR $\delta \leq 1$ AND $\delta \leq \epsilon/B$.]

$$< 25 \cdot \delta$$
$$\leq 25 \cdot \epsilon/25$$
$$= \epsilon. \qquad \square$$

The final version of the proof of $\lim_{x \to -2}(4x^2 - 5x + 2) = 28$ then looks like this:

The function that takes x to $4x^2 - 5x + 2$ is a polynomial function and it is defined for all complex numbers. Thus -2 is a limit point of the domain. Let $\epsilon > 0$. Set $\delta = \min\{1, \epsilon/25\}$. Then δ is a positive real number. Let x be any complex number. Suppose that $0 < |x + 2| < \delta$. Then

$$
\begin{aligned}
|(4x^2 - 5x + 2) - 28| &= |4x^2 - 5x - 26| \\
&= |(x+2)(4x - 13)| \\
&= |4x - 13| \cdot |x + 2| \\
&= |4(x + 2 - 2) - 13| \cdot |x + 2| \text{ (adding a clever zero)} \\
&= |4(x + 2) - 21| \cdot |x + 2| \\
&\leq (|4(x + 2)| + 21) \cdot |x + 2| \\
&\quad \text{(by the triangle inequality } |a \pm b| \leq |a| + |b|) \\
&< (4 \cdot 1 + 21) \cdot |x + 2| \text{ (since } |x + 2| < \delta \leq 1) \\
&= 25 \cdot |x + 2| \\
&< 25 \cdot \delta \\
&\leq 25 \cdot \epsilon/25 \\
&= \epsilon. \qquad \square
\end{aligned}
$$

The next example has the same type of discovery work with fewer comments.

Example 4.1.4. $\lim_{x \to -1}(4/x^2) = 4$.

Proof. The function that takes x to $4/x^2$ is defined for all non-zero complex numbers, so -1 is a limit point of the domain. Let $\epsilon > 0$. Set $\delta = $ _____. Then δ is a positive real number. Let x be any complex number that satisfies $0 < |x + 1| < \delta$. Then

$$|4/x^2 - 4| = \left| \frac{4 - 4x^2}{x^2} \right|$$

$$= 4 \left| \frac{(1 - x)(1 + x)}{x^2} \right|$$

$$= 4 \left| \frac{1 - x}{x^2} \right| |1 + x|.$$

[GOAL #1 ACCOMPLISHED: SOMETHING TIMES $|x - a|$.]

[NOW WE WANT THE SOMETHING $4\left(\frac{1-x}{x^2}\right)$ TO BE AT MOST SOME CON-
STANT. CERTAINLY IF WE ALLOW x TO GET CLOSE TO 0, THEN $(1-x)/x^2$
IS VERY LARGE, SO IN ORDER TO FIND AN UPPER BOUND, WE NEED TO
MAKE SURE THAT x STAYS AWAY FROM 0. SINCE x IS WITHIN δ OF -1,
IN ORDER TO AVOID 0 WE NEED TO MAKE SURE THAT δ IS STRICTLY
SMALLER THAN 1. FOR EXAMPLE, MAKE SURE THAT $\delta \leq 0.4$. THUS, ON
THE δ LINE WRITE: $\delta = \min\{0.4, \underline{\hspace{2cm}}\}$.]

$$= 4 \left| \frac{2 - (x + 1)}{|x|^2} \right| |1 + x| \text{ (by rewriting } 1 - x = 2 - (x + 1))$$

$$\leq 4 \frac{2 + |x + 1|}{x^2} |1 + x| \text{ (by the triangle inequality)}$$

$$\leq 4 \frac{2.4}{|x|^2} |1 + x| \text{ (since } \delta \leq 0.4)$$

$$= \frac{9.6}{|x + 1 - 1|^2} |1 + x| \text{ (by rewriting } x = x + 1 - 1)$$

$$\leq \frac{9.6}{0.6^2} |1 + x| \text{ (by the reverse triangle inequality because}$$

$$|x + 1 - 1| \geq 1 - |x + 1| > 1 - \delta \geq 1 - 0.4 = 0.6,$$

$$\text{so that } 1/|x + 1 - 1|^2 < 1/0.6^2)$$

[ON THE δ-LINE NOW WRITE: $\delta = \min\{0.4, 0.6^2\epsilon/9.6\}$.]

$$< \frac{9.6}{0.6^2} \delta$$

$$\leq \frac{9.6}{0.6^2} 0.6^2 \epsilon/9.6$$

$$= \epsilon. \qquad \qquad \Box$$

And here is another example with δ already filled in:

Example 4.1.5. $\displaystyle\lim_{x \to 2i} \frac{4x^3+x}{8x-i} = -2.$

Proof. The domain of $\frac{4x^3+x}{8x-i}$ consists of all complex numbers different from $i/8$, so $2i$ is a limit point of the domain. Let $\epsilon > 0$. Set $\delta = \min\{1, \epsilon/9\}$. [IT IS SO OBVIOUS THAT THIS MINIMUM OF TWO POSITIVE NUMBERS IS POSITIVE THAT WE SKIP THE ASSERTION "THUS δ IS A POSITIVE REAL NUMBER." DO NOT OMIT THE ASSERTION OR THE CHECKING OF ITS VERACITY FOR MORE COMPLICATED SPECIFICATIONS OF δ.] Let x be any complex number different from $i/8$ such that $0 < |x - 2i| < \delta$. [HERE WE MERGED: "LET x BE ANY COMPLEX NUMBER DIFFERENT FROM $i/8$. LET x SATISFY $0 < |x - 2i| < \delta$." INTO ONE SHORTER AND LOGICALLY EQUIVALENT STATEMENT "LET x BE ANY COMPLEX NUMBER DIFFERENT FROM $i/8$ SUCH THAT $0 < |x - 2i| < \delta$."] Then

$$\left| \frac{4x^3+x}{8x-i} - (-2) \right| = \left| \frac{4x^3+x}{8x-i} + 2 \right|$$

$$= \left| \frac{4x^3 + x + 16x - 2i}{8x - i} \right|$$

$$= \left| \frac{4x^3 + 17x - 2i}{8x - i} \right|$$

$$= \left| \frac{(4x^2 + 8ix + 1)(x - 2i)}{8x - i} \right|$$

$$= \left| \frac{(4x^2 + 8ix + 1)}{8x - i} \right| |x - 2i|$$

(Goal #1 is accomplished: $x - a$ is a factor.)

$$\leq \frac{|4x^2| + |8ix| + 1}{|8x - i|} |x - 2i| \text{ (by the triangle inequality)}$$

$$\leq \frac{4|(x - 2i + 2i)|^2 + 8|x - 2i + 2i| + 1}{|8(x - 2i) + 15i|} |x - 2i|$$

$$\leq \frac{4(|x - 2i| + 2)^2 + 8(|x - 2i| + 2) + 1}{-8|x - 2i| + 15} |x - 2i|$$

(by the triangle and reverse triangle inequalities)

$$\leq \frac{4(1+2)^2 + 8(1+2) + 1}{-8 + 15} |x - 2i|$$

$$(\text{since } |x - 2i| < \delta \leq 1)$$

$$= \frac{61}{7} |x - 2i|$$

$$< 9\delta$$

$$\leq 9\epsilon/9$$

$$= \epsilon. \qquad \square$$

The next example is of a limit is at a non-specific a.

Example 4.1.6. $\lim\limits_{x \to a} (x^2 - 2x) = a^2 - 2a$.

Proof. Any a is a limit point of the domain of the given polynomial function. Let $\epsilon > 0$. Set $\delta = \min\{1, \epsilon/(1 + |2a - 2|)\}$. Let x satisfy $0 < |x - a| < \delta$. Then

$$\begin{aligned}
|(x^2 - 2x) - (a^2 - 2a)| &= |(x^2 - a^2) - (2x - 2a)| \text{ (by algebra)} \\
&= |(x + a)(x - a) - 2(x - a)| \text{ (by algebra)} \\
&= |(x + a - 2)(x - a)| \text{ (by algebra)} \\
&= |x + a - 2| \, |x - a| \\
&= |(x - a) + 2a - 2| \, |x - a| \text{ (by adding a clever 0)} \\
&\leq (|x - a| + |2a - 2|) \, |x - a| \\
&\qquad \text{(by the triangle inequality)} \\
&\leq (1 + |2a - 2|) \, |x - a| \text{ (since } |x - a| < \delta \leq 1) \\
&< (1 + |2a - 2|)\delta \\
&\leq (1 + |2a - 2|)\epsilon/(1 + |2a - 2|) \\
&= \epsilon. \qquad \square
\end{aligned}$$

Remark 4.1.7. Note that δ depends on ϵ and a, which are constants in the problem; δ is not allowed to depend on x, as the definition goes:

"for all $\epsilon > 0$ there exists $\delta > 0$ such that for all x, etc."

so that x depends on δ, but δ does not depend on x.

By the definition of limits, δ is supposed to be a positive real number, not a function of x. (See also Exercise 4.1.2.)

Remark 4.1.8. In all cases of rational functions, such as in examples above, Goal #1 is to factor $x - a$ from $f(x) - L$: if $x - a$ is not a factor, check your limit or algebra for any mistakes. In the next example, $x - a$ is not a factor, but $\sqrt{x - a}$ is.

Example 4.1.9. $\lim_{x \to 3} \sqrt{2x - 6} = 0$.

Proof. The domain here is all $x \geq 3$. So 3 is a limit point of the domain. Let $\epsilon > 0$. Set $\delta = \epsilon^2/2$. Let $x > 3$ satisfy $0 < |x - 3| < \delta$. Then

$$\sqrt{2x - 6} = \sqrt{2} \cdot \sqrt{x - 3}$$
$$< \sqrt{2} \cdot \sqrt{\delta} \text{ (because } \sqrt{\ } \text{ is an increasing function)}$$
$$= \sqrt{2} \cdot \sqrt{\epsilon^2/2}$$
$$= \epsilon. \qquad \square$$

Often books consider the last example as a case of a one-sided limit (see definition below) since we can only take the x from one side of 3. Our definition handles both-sided and one-sided and all sorts of other limits with one simple notation, but we do have a use for one-sided limits as well, so we define them next.

Definition 4.1.10. *Let $A \subseteq \mathbb{R}$, $a \in \mathbb{R}$, $L \in \mathbb{C}$, and $f : A \to \mathbb{C}$ a function. Suppose that a is a limit point of $\{x \in A : x > a\}$ (resp. of $\{x \in A : x < a\}$). We say that the **right-sided (resp. left-sided) limit of f(x) as x approaches a is L** if for every real number $\epsilon > 0$ there exists a real number $\delta > 0$ such that for all $x \in A$, if $0 < x - a < \delta$ (resp. if $0 < a - x < \delta$) then $|f(x) - L| < \epsilon$. When this is the case, we write $\lim_{x \to a^+} f(x) = L$ (resp. $\lim_{x \to a^-} f(x) = L$).*

With this, Example 4.1.9 can be phrased as $\lim_{x \to 3^+} \sqrt{2x - 6} = 0$, and the proof goes as follows: The domain A consists of all $x \geq 3$, and 3 is a limit point of $A \cap \{x : x > 3\} = A$. Let $\epsilon > 0$. Set $\delta = \epsilon^2/2$. Let x satisfy $0 < x - 3 < \delta$. Then

$$\sqrt{2x - 6} = \sqrt{2} \cdot \sqrt{x - 3}$$
$$< \sqrt{2} \cdot \sqrt{\delta}$$
$$= \sqrt{2} \cdot \sqrt{\epsilon^2/2}$$
$$= \epsilon.$$

Thus, the two proofs are almost identical. Note that $\lim_{x\to 3-} \sqrt{2x-6}$ does not exist because 3 is not the limit point of $A \cap \{x \in \mathbb{R} : x < 3\} = \emptyset$.

One-sided limits can also be used in contexts where $\lim_{x\to a} f(x)$ does not exist. Below is one example.

Example 4.1.11. Let $f : \mathbb{R} \to \mathbb{R}$ be given by $f(x) = \begin{cases} x^2 + 4, & \text{if } x > 1; \\ x - 2, & \text{if } x \leq 1. \end{cases}$
Then $\lim_{x\to 1+} f(x) = 5$, $\lim_{x\to 1-} f(x) = -1$.

Proof. Let $\epsilon > 0$. Set $\delta = \min\{1, \epsilon/3\}$. Let x satisfy $0 < x - 1 < \delta$. Then

$$\begin{aligned}
|f(x) - 5| &= |x^2 + 4 - 5| \text{ (since } x > 1) \\
&= |x^2 - 1| \\
&= |(x+1)(x-1)| \\
&< |x+1|\delta \text{ (since } x > 1, \text{ so } x + 1 \text{ is positive)} \\
&= |x - 1 + 2|\delta \text{ (by adding a clever 0)} \\
&\leq (|x-1| + 2)\delta \text{ (by the triangle inequality)} \\
&< (1 + 2)\delta \text{ (since } 0 < x - 1 < \delta \leq 1) \\
&\leq 3\epsilon/3 \\
&= \epsilon.
\end{aligned}$$

This proves that $\lim_{x\to 1+} f(x) = 5$.

Set $\delta = \epsilon$. Let x satisfy $0 < 1 - x < \delta$. Then

$$\begin{aligned}
|f(x) - (-1)| &= |x - 2 + 1| \text{ (since } x < 1) \\
&= |x - 1| \\
&< \delta \\
&= \epsilon.
\end{aligned}$$

This proves that $\lim_{x\to 1-} f(x) = -3$. $\qquad\square$

Exercises for Section 4.1

4.1.1. Rework Example 4.1.3 with choosing δ to be at most 2 rather than at most 1.

4.1.2. Below is an attempt at a "proof" that $\lim_{x \to 3} x^2 = 9$. Explain how the two starred steps contribute to at least three mistakes total.

Let $\epsilon > 0$. Set $\delta =^* \epsilon/|x+3|$. Then

$$|x^2 - 9| = |x - 3| \cdot |x + 3|$$
$$<^* \delta|x + 3|$$
$$= \epsilon. \qquad \square$$

4.1.3. Fill in the blanks of the following proof that $\lim_{x \to 2}(x^2 - 3x) = -2$. Explain why none of the inequalities can be changed into equalities.

_____ so that 2 is a

_____ . Let $\epsilon > 0$. Set $\delta = $ _____ . Let x satisfy $0 < |x - 2| < \delta$. Then

$$|(x^2 - 3x) - (-2)| = |x^2 - 3x + 2|$$
$$= |x - 1|\,|x - 2| \quad (\text{because } \underline{\hspace{3cm}})$$
$$= |x - 2 + 1|\,|x - 2| \quad (\text{by } \underline{\hspace{2.5cm}})$$
$$\leq (|x - 2| + 1)\,|x - 2| \quad (\text{by } \underline{\hspace{2.5cm}})$$
$$< (3 + 1)\,|x - 2| \quad (\text{because } \underline{\hspace{2.5cm}})$$
$$< 4\delta \quad (\text{because } \underline{\hspace{2.5cm}})$$
$$\leq 4\frac{\epsilon}{4} \quad (\text{because } \underline{\hspace{2.5cm}})$$
$$= \epsilon.$$

i) If the domain of this function is $\mathbb{R}$ as opposed to $\mathbb{C}$, then $|x-2| < 4$ can be shown also with the following proof: Since $|x - 2| < \delta \leq 3$, then $-3 < x - 2 < 3$, so that $-2 < x - 1 < 4$, which means that $|x - 1| < 4$. Say why this argument does not work if the domain is $\mathbb{C}$.

4.1.4. Determine the following limits and prove them with the epsilon-delta proofs.

   i) $\lim_{x \to 1}(x^3 - 4)$.

   ii) $\lim_{x \to 2} \frac{1}{x}$.

   iii) $\lim_{x \to 3} \frac{x-4}{x^2+2}$.

   iv) $\lim_{x \to 4} \sqrt{x + 5}$.

   v) $\lim_{x \to 3} \frac{x^2 - 9}{x - 3}$.

4.1.5. Let $b \in \mathbb{C}$ and $f, g : \mathbb{C} \to \mathbb{C}$ with

$$f(x) = \begin{cases} x^3 - 4x^2, & \text{if } x \neq 5; \\ b, & \text{if } x = 5, \end{cases} \qquad g(x) = \begin{cases} x^3 - 4x^2, & \text{if } x = 5; \\ b, & \text{if } x \neq 5. \end{cases}$$

Prove that the limit of $f(x)$ as x approaches 5 is independent of b, but that the limit of $g(x)$ as x approaches 5 depends on b.

4.1.6. Suppose that a is a limit point of $\{x \in A : x > a\}$ and of $\{x \in A : x < a\}$. Prove that $\lim_{x \to a} f(x) = L$ if and only if $\lim_{x \to a^+} f(x) = L$ and $\lim_{x \to a^-} f(x) = L$.

4.1.7. Suppose that a is a limit point of $\{x \in A : x > a\}$ but not of $\{x \in A : x < a\}$. Prove that $\lim_{x \to a} f(x) = L$ if and only if $\lim_{x \to a^+} f(x) = L$.

4.1.8. Prove that $\lim_{x \to a}(mx + l) = ma + l$, where m and l are constants.

4.1.9. Let $f : \mathbb{R} \to \mathbb{R}$ be given by $f(x) = \frac{x}{|x|}$. Prove that $\lim_{x \to 0^+} f(x) = 1$ and that $\lim_{x \to 0^-} f(x) = -1$.

4.1.10. Find a function $f : \mathbb{R} \to \mathbb{R}$ such that $\lim_{x \to 0^-} f(x) = 2$ and $\lim_{x \to 0^+} \cdot f(x) = -5$.

4.1.11. Find a function $f : \mathbb{R} \to \mathbb{R}$ such that $\lim_{x \to 0^-} f(x) = 2$, $\lim_{x \to 0^+} \cdot f(x) = -5$, $\lim_{x \to 1^-} f(x) = 3$, and $\lim_{x \to 1^+} f(x) = 0$. (Try to define such a function with fewest possible words or symbols, but do use full grammatical sentences.)

4.2   When a number is not a limit

Recall that $\lim_{x \to a} f(x) = L$ means that a is a limit point of the domain of f, and that for all real numbers $\epsilon > 0$ there exists a real number $\delta > 0$ such that for all x in the domain of f, if $0 < |x - a| < \delta$ then $|f(x) - L| < \epsilon$. [THINK OF $\lim_{x \to a} f(x) = L$ AS STATEMENT P, a BEING A LIMIT POINT OF THE DOMAIN AS STATEMENT Q, AND THE EPSILON-DELTA PART AS STATEMENT R. BY DEFINITION, P IS LOGICALLY THE SAME AS THE STATEMENT Q and R.]

Thus if $\lim_{x \to a} f(x) \neq L$, then either a is not a limit point of the domain of f or else it is not true that for all real numbers $\epsilon > 0$ there exists a real number $\delta > 0$ such that for all x in the domain of f, if $0 < |x - a| < \delta$ then $|f(x) - L| < \epsilon$. [THIS SIMPLY SAYS THAT not P IS THE SAME AS (not Q) or (not R).]

In particular,

$$\lim_{x \to a} f(x) \neq L \text{ and } a \text{ is a limit point of the domain of } f$$

means that it is not true that for all real numbers $\epsilon > 0$ there exists a real number $\delta > 0$ such that for all x in the domain of f, if $0 < |x - a| < \delta$ then $|f(x) - L| < \epsilon$. [THIS SAYS THAT (not P) and Q IS THE SAME AS not R. YOU MAY WANT TO WRITE TRUTH TABLES FOR YOURSELF.]

Negations of compound sentences, such as in the previous paragraph, are typically hard to process and to work with in proofs. But by the usual negation rules of compound statements (see chart on page 25), we successively rewrite this last negation into a form that is easier to handle:

not (For all real numbers $\epsilon > 0$ there exists a real number $\delta > 0$ such that for all x in the domain of f, if $0 < |x - a| < \delta$ then $|f(x) - L| < \epsilon$).

[NEGATION OF "FOR ALL z OF SOME KIND, PROPERTY P HOLDS" IS "THERE IS SOME z OF THAT KIND FOR WHICH P IS FALSE." HENCE THE FOLLOWING REPHRASING:]

= There exists a real number $\epsilon > 0$ such that not (there exists a real number $\delta > 0$ such that for all x in the domain of f, if $0 < |x - a| < \delta$ then $|f(x) - L| < \epsilon$).

[NEGATION OF "THERE EXISTS z OF SOME KIND SUCH THAT PROPERTY P HOLDS" IS "FOR ALL z OF THAT KIND, P IS FALSE." HENCE THE FOLLOWING REPHRASING:]

= There exists a real number $\epsilon > 0$ such that for all real numbers $\delta > 0$, not (for all x in the domain of f, if $0 < |x - a| < \delta$ then $|f(x) - L| < \epsilon$).

[NEGATION OF "FOR ALL z OF SOME KIND, PROPERTY P HOLDS" IS "THERE IS SOME z OF THAT KIND FOR WHICH P IS FALSE." HENCE THE FOLLOWING REPHRASING:]

= There exists a real number $\epsilon > 0$ such that for all real numbers $\delta > 0$, there exists x in the

domain of f such that not $\big($if $0 < |x-a| < \delta$
then $|f(x) - L| < \epsilon\big)$.

[NEGATION OF "IF P THEN Q" IS "P AND NOT Q." HENCE THE FOL-
LOWING REPHRASING:]

= There exists a real number $\epsilon > 0$ such that for
all real numbers $\delta > 0$, there exists x in the
domain of f such that $0 < |x - a| < \delta$ and
not $\big(|f(x) - L| < \epsilon\big)$.

= There exists a real number $\epsilon > 0$ such that for
all real numbers $\delta > 0$, there exists x in the
domain of f such that $0 < |x - a| < \delta$ and
$|f(x) - L| \geq \epsilon$.

In summary, we just proved that:

Theorem 4.2.1. *If a is a limit point of the domain of f, then $\lim_{x \to a} f(x) \neq$
L means that there exists a real number $\epsilon > 0$ such that for all real numbers
$\delta > 0$, there exists x in the domain of f such that $0 < |x - a| < \delta$ and
$|f(x) - L| \geq \epsilon$.*

Example 4.2.2. The limit of $\frac{x}{|x|}$ as x approaches 0 does not exist. In other
words, for all complex numbers L, $\lim_{x \to 0} \frac{x}{|x|} \neq L$.

The domain of the function that takes x to $\frac{x}{|x|}$ is the set of all non-
zero complex numbers. For each non-zero x, $\frac{x}{|x|}$ is a complex number of
length 1 and with the same angle as x. Thus the image of this function is
the unit circle in $\mathbb{C}$. Note that it is possible to take two non-zero x very
close to 0 but at different angles so that their images on the unit circle are
far apart. This is a geometric reasoning why the limit cannot exist. Next
we give an epsilon-delta proof.

Proof. The domain of the function that takes x to $\frac{x}{|x|}$ is the set of all non-
zero complex numbers, so that 0 is a limit point of the domain. [THUS IF
THE LIMIT IS NOT L, THEN IT MUST BE THE EPSILON-DELTA CONDITION
THAT FAILS.] Set $\epsilon = 1$. Let $\delta > 0$ be an arbitrary positive number. Let
$x = -\delta/2$ if $\mathrm{Re}(L) \geq 0$, and let $x = \delta/2$ otherwise. Then $0 < |x| = |x-0| <$
δ. If $\mathrm{Re}(L) \geq 0$, then

$$\mathrm{Re}\left(\frac{x}{|x|} - L\right) = \mathrm{Re}\left(\frac{-\delta/2}{|-\delta/2|} - L\right) = -1 - \mathrm{Re}(L) \leq -1,$$

so that $|\frac{x}{|x|} - L| \geq 1 = \epsilon$, and if $\mathrm{Re}(L) < 0$, then

$$\mathrm{Re}\left(\frac{x}{|x|} - L\right) = \mathrm{Re}\left(\frac{\delta/2}{|\delta/2|} - L\right) = 1 - \mathrm{Re}(L) > 1,$$

so that again $|\frac{x}{|x|} - L| > 1 = \epsilon$. This proves the claim of the example. □

Example 4.2.3. For all $L \in \mathbb{C}$, $\lim_{x \to 2} \frac{i}{x-2} \neq L$.

A geometric reason for the non-existence of this limit is that as x gets closer to 2 (but not equal to 2), the size of $\frac{i}{x-2}$ gets larger and larger.

Proof. Set $\epsilon = 1$. Let $\delta > 0$ be an arbitrary positive number. Set $\delta' = \min\{\delta, 1/(|L|+1)\}$. Let $x = 2 + \delta'/2$. Then $0 < |x - 2| < \delta' \leq \delta$, and

$$\left|\frac{i}{x-2} - L\right| = \left|\frac{2i}{\delta'} - L\right|$$

$$\geq \left|\frac{2i}{\delta'}\right| - |L| \text{ (by the reverse triangle inequality)}$$

$$\geq 2(|L|+1) - |L| \text{ (since } \delta' \leq 1/(|L|+1))$$

$$\geq 1$$

$$= \epsilon. \qquad \qquad \square$$

Example 4.2.4. For $f : \mathbb{R} \to \mathbb{R}$ given by the graph below, $\lim_{x \to 2} f(x)$ does not exist because of the jump in the function at 2.

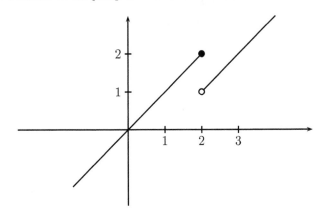

Here is an epsilon-delta proof. Say that the limit exists. Call it L. Set $\epsilon = \frac{1}{4}$. Let δ be an arbitrary positive number. If $L \geq \frac{3}{2}$, set $x = 2 + \min\left\{\frac{1}{4}, \frac{\delta}{2}\right\}$, and if $L < \frac{3}{2}$, set $x = 2 - \min\{\frac{1}{4}, \frac{\delta}{2}\}$. In either case,

$$0 < |x - 2| = \min\left\{\frac{1}{4}, \frac{\delta}{2}\right\} \leq \frac{\delta}{2} < \delta.$$

If $L \geq \frac{3}{2}$, by our choice $x = 2 + \min\{\frac{1}{4}, \frac{\delta}{2}\}$, so that $f(x) = 1 + \min\{\frac{1}{4}, \frac{\delta}{2}\} \leq 1 + \frac{1}{4}$, whence $|f(x) - L| \geq \frac{1}{4} = \epsilon$. Similarly, if $L < \frac{3}{2}$, by our choice $x = 2 - \min\{\frac{1}{4}, \frac{\delta}{2}\}$, so that $f(x) = 2 - \min\{\frac{1}{4}, \frac{\delta}{2}\} \geq 2 - \frac{1}{4} = 1 + \frac{3}{4}$, whence $|f(x) - L| \geq \frac{1}{4} = \epsilon$. Thus no L works, so the limit of $f(x)$ as x approaches 2 does not exist.

Exercises for Section 4.2

4.2.1. Prove that $\lim_{x \to 3}(3x - 4) \neq -3$.

4.2.2. Prove that $\lim_{x \to 1}(x^2 + 4) \neq -5$.

4.2.3. Prove that $\lim_{x \to -3} \frac{x-3}{x^2-9}$ does not exist.

4.2.4. Prove that for all $a \in \mathbb{R}$, $\lim_{x \to a} f(x)$ does not exist, where $f : \mathbb{R} \to \mathbb{R}$ is defined by $f(x) = \begin{cases} 1, & \text{if } x \text{ is rational;} \\ 0, & \text{if } x \text{ is not rational.} \end{cases}$

4.2.5. Prove that $\lim_{x \to 0}(\sqrt{x} - \sqrt{-x})$ does not exist. (Hint: The reason is different from the reasons in other examples in this section.)

4.3 More on the definition of a limit

The purpose of this section is to show that even small changes in the definition of limits affect the meaning significantly. A lesson to be learned is that it is important to remember any formal statement precisely.

Here is a restatement of Definition 4.1.1 for $\lim_{x \to a} f(x) = L$ when a is a limit point of the domain A of f:

$$\forall \epsilon > 0 \quad \exists \delta > 0 \quad \forall x \in A \quad \text{if } 0 < |x - a| < \delta \text{ then } |f(x) - L| < \epsilon.$$
$$(4.3.1)$$

Example 4.3.2. Suppose that in Statement (4.3.1) we switch the order of the first two quantifiers:

$$\exists \delta > 0 \quad \forall \epsilon > 0 \quad \forall x \in A \quad \text{if } 0 < |x - a| < \delta \text{ then } |f(x) - L| < \epsilon.$$

Let $f : \mathbb{C} \to \mathbb{C}$ be the function given by $f(x) = x$. By Exercise 4.1.8, $\lim_{x \to a} f(x) = a$, but this f does not satisfy the modified definition above because no matter what δ is taken, the conditional fails for any $\epsilon < \delta/2$. Thus this modification of the definition of limits changes the meaning.

Example 4.3.3. Suppose that in Statement (4.3.1) we switch the order of the second and third quantifiers:

$$\forall \epsilon > 0 \quad \forall x \in A \quad \exists \delta > 0 \quad \text{if } 0 < |x - a| < \delta \text{ then } |f(x) - L| < \epsilon.$$

Every function $f : A \to \mathbb{C}$ satisfies this statement because after ϵ and x are fixed we may set δ to be $|x - a|/2$ if $x \neq a$ and 1 otherwise. With this, the antecedent $0 < |x - a| < \delta$ is false, which means that the conditional is true, so that f satisfies the statement. In Section 4.2 we saw that limits need not exist, which means that this modification changes the meaning of the definition of limits.

Example 4.3.4. Suppose that in Statement (4.3.1) we replace the first $\forall$ with $\exists$:

$$\exists \epsilon > 0 \quad \exists \delta > 0 \quad \forall x \in A \quad \text{if } 0 < |x - a| < \delta \text{ then } |f(x) - L| < \epsilon.$$

Then whenever $\lim_{x \to a} f(x) = L$, the modified statement is satisfied with any complex number L' in place of L. Namely, set $\epsilon = 1 + |L' - L|$. By the definition of $\lim_{x \to a} f(x) = L$ there exists $\delta > 0$ such that for all $x \in A$, if $0 < |x - a| < \delta$ then $|f(x) - L| < 1$. Then

$$|f(x) - L'| = |f(x) - L + L - L'|$$
$$\leq |f(x) - L| + |L - L'| \text{ (by the triangle inequality)}$$
$$< 1 + |L - L'|$$
$$= \epsilon.$$

This means that with this modification of the definition of limits, limits would not be unique (but they are by Theorem 4.4.1).

Furthermore, the function $\frac{x}{|x|}$, for which we proved in Example 4.2.2 that no limit exists at $a = 0$, satisfies this last modified statement for any L.

Example 4.3.5. Suppose that in Statement (4.3.1) we replace the first $\exists$ with $\forall$:

$$\forall \epsilon > 0 \quad \forall \delta > 0 \quad \forall x \in A \quad \text{if } 0 < |x - a| < \delta \text{ then } |f(x) - L| < \epsilon.$$

We claim that the only functions f that satisfy this statement are those that are constant on $A \setminus \{a\}$. Namely, suppose that $b, c \in A \setminus \{a\}$. Then for any $\epsilon > 0$, whenever $\delta > |b - a| + |c - a|$, we have that $|f(b) - f(c)| = |f(b) - L + L - f(c)| \leq |f(b) - L| + |L - f(c)| < 2\epsilon$. Then by Theorem 2.10.4, $|f(b) - f(c)| = 0$, so that $f(b) - f(c) = 0$. Thus f is constant as claimed. But we know by the first section in this chapter many non-constant functions have limits as well, which means that this modification of the definition of limits changes the meaning.

Example 4.3.6. Suppose that in Statement (4.3.1) we replace the conditional with the conjunction:

$$\forall \epsilon > 0 \quad \exists \delta > 0 \quad \forall x \in A \quad 0 < |x - a| < \delta \text{ and } |f(x) - L| < \epsilon.$$

This modification fails for every function f that is not equal to the constant L on $A \setminus \{a\}$. Namely, for the condition $0 < |x - a| < \delta$ to hold, necessarily δ must be so large so that $A \subseteq B(a, \delta)$; and for $|f(x) - L| < \epsilon$ to hold for all $x \in A \setminus \{a\}$ and all $\epsilon > 0$, by Theorem 2.10.4, $f(x) = L$. Thus again, this modification of the definition of limits changes the meaning.

So far we have examined modifications of Statement (4.3.1) in which we switched the order of quantifiers, we switched a quantifier, or we changed the conditional into a conjunction. In all cases the modification resulted in a different meaning. We can modify Statement (4.3.1) in many other ways and get even further meanings.

Example 4.3.7. Another possible modification of Statement (4.3.1) is to omit a quantifier, such as the quantifier for x:

$$\forall \epsilon > 0 \quad \exists \delta > 0 \quad \text{if } 0 < |x - a| < \delta \text{ then } |f(x) - L| < \epsilon.$$

It is common to treat the occurrence of x in the statement as tacitly assuming "$\forall x \in A$". However, this tacit assumption is not fine! Namely, the negation of this modification is

$$\exists \epsilon > 0 \quad \forall \delta > 0 \quad 0 < |x - a| < \delta \text{ and } |f(x) - L| \geq \epsilon.$$

The lack of the quantifier on x would then again tacitly signal the universal quantifier, whereas the correct negation calls for the existential quantifier. In short, we cannot omit a quantifier.

We finish the section with a modification of Statement (4.3.1) that does not change the meaning: we replace the last two occurrences of "<" in the statement with "≤".

Theorem 4.3.8. *Let $f : A \to \mathbb{C}$ be a function and let $a \in \mathbb{C}$ be a limit point of A. Then $\lim_{x \to a} f(x) = L$ if and only if*

$$\forall \epsilon > 0 \quad \exists \delta > 0 \quad \forall x \in A \quad \text{if } 0 < |x - a| \leq \delta \text{ then } |f(x) - L| \leq \epsilon.$$

Proof. First suppose that $\lim_{x \to a} f(x) = L$. We will prove that the modified statement also holds. Let $\epsilon > 0$. By assumption there exists $\delta' > 0$ such that for all $x \in A$, if $0 < |x - a| < \delta'$ then $|f(x) - L| < \epsilon$. Set $\delta = \delta'/2$. Let $x \in A$ such that $0 < |x - a| \leq \delta$. Then $0 < |x - a| < \delta'$, so that $|f(x) - L| < \epsilon$, and hence $|f(x) - L| \leq \epsilon$. Thus the modified statement holds.

Conversely, suppose that the modified statement holds. We will prove that $\lim_{x \to a} f(x) = L$. Let $\epsilon > 0$. By assumption there exists $\delta > 0$ such that for all $x \in A$, if $0 < |x - a| \leq \delta$ then $|f(x) - L| \leq \epsilon/2$. Let $x \in A$ such that $0 < |x - a| < \delta$. Then $0 < |x - a| \leq \delta$, so that $|f(x) - L| \leq \epsilon/2 < \epsilon$. Thus $\lim_{x \to a} f(x) = L$. □

Exercises for Section 4.3

4.3.1. Prove that the only functions $f : \mathbb{C} \to \mathbb{C}$ that satisfy the statement in Example 4.3.2 at all a in the domain are the constant functions. Find a non-constant function $f : [0, 1] \cup [2, 3] \to \mathbb{C}$ that satisfies that statement at all a in the domain.

4.3.2. Examples 4.3.2 and 4.3.3 modified Statement (4.3.1) in the ordering of the quantifiers. Similarly show that the remaining orderings (all on the list below) also change the meaning.

   i) $\exists \delta > 0 \quad \forall x \in A \quad \forall \epsilon > 0 \quad \text{if } 0 < |x - a| < \delta \text{ then } |f(x) - L| < \epsilon.$

   ii) $\forall x \in A \quad \forall \epsilon > 0 \quad \exists \delta > 0 \quad \text{if } 0 < |x - a| < \delta \text{ then } |f(x) - L| < \epsilon.$

   iii) $\forall x \in A \quad \exists \delta > 0 \quad \forall \epsilon > 0 \quad \text{if } 0 < |x - a| < \delta \text{ then } |f(x) - L| < \epsilon.$

4.3.3. Assume that a is a limit point of the domain A of f.

i) Prove that $\lim_{x \to a} f(x) = L$ if and only if

$$\forall \epsilon > 0 \quad \exists \delta > 0 \quad \forall x \in A \quad \text{if } 0 < |x - a| < \delta \text{ then } |f(x) - L| \le \epsilon.$$

ii) Prove that $\lim_{x \to a} f(x) = L$ if and only if

$$\forall \epsilon > 0 \quad \exists \delta > 0 \quad \forall x \in A \quad \text{if } 0 < |x - a| \le \delta \text{ then } |f(x) - L| < \epsilon.$$

4.4   Limit theorems

While epsilon-delta proofs are a reliable method for proving limits, they do not help in deciding what a limit may be. In this section we prove theorems that will efficiently establish the limits for many functions. The proofs of these theorems require the epsilon-delta machinery — as this is the definition of limits, but subsequent applications of these theorems allow us to omit the time-consuming epsilon-delta proofs.

Theorem 4.4.1. *If a limit exists, it is unique.*

Proof. Suppose that both L_1 and L_2 are limits of $f(x)$ as x approaches a.

First I give a FALSE proof: by assumption, $L_1 = \lim_{x \to a} f(x) = L_2$, so that by transitivity of equality, $L_1 = L_2$.   ⊠

What makes this proof false is that the equal sign in "$\lim_{x \to a} f(x) = L$" signifies, at this point, not numerical equality, but that a is a limit point of the domain of f and that for every $\epsilon > 0$ there exists $\delta > 0$ such that for all x in the domain of f, if $0 < |x - a| < \delta$ then $|f(x) - L| < \epsilon$. Thus, we need a proof that uses this definition.

Real proof of Theorem 4.4.1: Suppose that L_1 and L_2 are both limits. Let ϵ be an arbitrary positive number. Then $\epsilon/2$ is also positive, and by the definition of limits, for each $i = 1, 2$, there exists $\delta_i > 0$ such that for all x in the domain of f, if $0 < |x - a| < \delta_i$, then $|f(x) - L_i| < \epsilon/2$. Set $\delta = \min\{\delta_1, \delta_2\}$. Then δ is a positive real number. Let x in the domain of f satisfy $0 < |x - a| < \delta$. Since a is a limit point of the domain, such x exists. Then

$$|L_1 - L_2| = |L_1 - f(x) + f(x) - L_2| \text{ (by adding a clever 0)}$$
$$\le |L_1 - f(x)| + |f(x) - L_2| \text{ (by the triangle inequality)}$$

$$< 2\epsilon/2 \text{ (since } \delta \leq \delta_1, \delta_2)$$

$$= \epsilon,$$

which says that $|L_1 - L_2| < \epsilon$. Since ϵ was arbitrary, an application of Theorem 2.10.4 gives that $|L_1 - L_2| = 0$, so that $L_1 - L_2 = 0$, and hence that $L_1 = L_2$. □

(If in the definition of $\lim_{x \to a} f(x)$ we did not require that a be a limit point of the domain of f, then x as in the proof above would not exist for small δ, so any complex number L would vacuously satisfy the definition of limits. We would thus not be able to guarantee that limits are unique. Perhaps the definition of limits is technical, but the technicalities are there for a good reason.)

Theorem 4.4.2. *Let a be a limit point of the domain of a function f with $\lim_{x \to a} f(x) = L$. Suppose that $L \neq 0$. Then there exists $\delta > 0$ such that for all x in the domain of f, if $0 < |x - a| < \delta$, then $|f(x)| > |L|/2$.*

In particular, there exists $\delta > 0$ such that for all x in the domain of f, if $0 < |x - a| < \delta$, then $f(x) \neq 0$.

Proof. Since $|L|/2 > 0$, there exists $\delta > 0$ such that for all x in the domain of f, if $0 < |x - a| < \delta$, then $|f(x) - L| < |L|/2$. Hence by the reverse triangle inequality, for the same x, $|L|/2 > |f(x) - L| \geq |L| - |f(x)|$, so that by adding $|f(x)| + |L|/2$ to both sides we get that $|f(x)| > |L|/2$. In particular $f(x) \neq 0$. □

The following theorem is very important, so study it carefully.

Theorem 4.4.3. *Let A be the domain of f and g, and let a be a limit point of A, and let $c \in \mathbb{C}$. Suppose that $\lim_{x \to a} f(x)$ and $\lim_{x \to a} g(x)$ both exist. Then*

(1) (Constant rule) $\lim_{x \to a} c = c$.

(2) (Linear rule) $\lim_{x \to a} x = a$.

(3) (Scalar rule) $\lim_{x \to a} cf(x) = c \lim_{x \to a} f(x)$.

(4) (Sum/difference rule) $\lim_{x \to a} (f(x) \pm g(x)) = \lim_{x \to a} f(x) \pm \lim_{x \to a} g(x)$.

(5) (Product rule) $\lim_{x \to a} (f(x) \cdot g(x)) = \lim_{x \to a} f(x) \cdot \lim_{x \to a} g(x)$.

(6) (Quotient rule) If $\lim_{x \to a} g(x) \neq 0$, *then* $\lim_{x \to a} \dfrac{f(x)}{g(x)} = \dfrac{\lim_{x \to a} f(x)}{\lim_{x \to a} g(x)}$.

Proof. Both (1) and (2) are proved in Exercise 4.1.8.

Set $L = \lim_{x \to a} f(x)$ and $K = \lim_{x \to a} g(x)$.

(3) Let $\epsilon > 0$. Then $\epsilon/(|c| + 1)$ is a positive number. (Note that we did not divide by 0.) Since $L = \lim_{x \to a} f(x)$, there exists $\delta > 0$ such that for all $x \in A$, $0 < |x - a| < \delta$ implies that $|f(x) - L| < \epsilon/(|c| + 1)$. Hence for the same x,

$$|cf(x) - cL| = |c| \cdot |f(x) - L| < |c| \cdot \epsilon/(|c| + 1) < \epsilon,$$

which proves that $\lim_{x \to a} cf(x) = cL = c \lim_{x \to a} f(x)$.

(4) Let $\epsilon > 0$. Since $L = \lim_{x \to a} f(x)$, there exists $\delta_1 > 0$ such that for all $x \in A$, if $0 < |x - a| < \delta_1$, then $|f(x) - L| < \epsilon/2$. Similarly, since $K = \lim_{x \to a} g(x)$, there exists $\delta_2 > 0$ such that for all $x \in A$, if $0 < |x - a| < \delta_2$, then $|g(x) - L| < \epsilon/2$.

Set $\delta = \min\{\delta_1, \delta_2\}$. Then δ is a positive number. Let $x \in A$ satisfy $0 < |x - a| < \delta$. Then (4) follows from:

$$|(f(x) \pm g(x)) - (L \pm K)| = |(f(x) - L) \pm (g(x) - K)| \text{ (by algebra)}$$
$$\leq |f(x) - L| + |g(x) - K| \text{ (by the triangle inequality)}$$
$$< \epsilon/2 + \epsilon/2 \text{ (because } 0 < |x - a| < \delta \leq \delta_1, \delta_2)$$
$$= \epsilon.$$

(5) Let $\epsilon > 0$. Then $\min\{\epsilon/(2|L| + 2), 1\}$, $\min\{\epsilon/(2|K| + 1), 1\}$ are positive numbers. Since $L = \lim_{x \to a} f(x)$, there exists $\delta_1 > 0$ such that for all $x \in A$, if $0 < |x - a| < \delta_1$, then $|f(x) - L| < \min\{\epsilon/(2|K| + 1), 1\}$. Similarly, since $K = \lim_{x \to a} g(x)$, there exists $\delta_2 > 0$ such that for all $x \in A$, if $0 < |x - a| < \delta_2$, then $|g(x) - L| < \epsilon/(2|L| + 2)$.

Set $\delta = \min\{\delta_1, \delta_2\}$. Then δ is a positive number. Let $x \in A$ satisfy $0 < |x - a| < \delta$. Then by the triangle inequality,

$$|f(x)| = |f(x) - L + L| \leq |f(x) - L| + |L| < 1 + |L|,$$

and so

$$|f(x) \cdot g(x) - L \cdot K| = |f(x) \cdot g(x) - f(x)K + f(x)K - L \cdot K|$$
$$\text{(by adding a clever zero)}$$
$$\leq |f(x) \cdot g(x) - f(x)K| + |f(x)K - L \cdot K|$$
$$\text{(by the triangle inequality)}$$

$$= |f(x)| \cdot |g(x) - K| + |f(x) - L| \cdot |K| \text{ (by factoring)}$$

$$< (1 + |L|) \cdot \frac{\epsilon}{2|L| + 2} + \frac{\epsilon}{2|K| + 1}|K| \text{ (since } \delta \le \delta_1, \delta_2)$$

$$< \epsilon/2 + \epsilon/2$$

$$= \epsilon.$$

(6) Let $\epsilon > 0$. Since $K \ne 0$, by Theorem 4.4.2, there exists $\delta_0 > 0$ such that for all $x \in A$, if $0 < |x - a| < \delta_0$, then $|g(x)| > |K|/2$.

The numbers $|K|\epsilon/4$, $|K|^2\epsilon/(4|L| + 1)$ are positive numbers. Since $L = \lim_{x \to a} f(x)$, there exists $\delta_1 > 0$ such that for all $x \in A$, if $0 < |x - a| < \delta_1$, then $|f(x) - L| < |K|\epsilon/4$. Similarly, since $K = \lim_{x \to a} g(x)$, there exists $\delta_2 > 0$ such that for all $x \in A$, if $0 < |x - a| < \delta_2$, then $|g(x) - L| < |K|^2\epsilon/(4|L| + 1)$.

Set $\delta = \min\{\delta_0, \delta_1, \delta_2\}$. Then δ is a positive number. Let $x \in A$ satisfy $0 < |x - a| < \delta$. Then

$$\left| \frac{f(x)}{g(x)} - \frac{L}{K} \right| = \left| \frac{f(x)K - Lg(x)}{Kg(x)} \right| \text{ (by algebra)}$$

$$= \left| \frac{f(x)K - LK + LK - Lg(x)}{Kg(x)} \right| \text{ (by adding a clever zero)}$$

$$\le \left| \frac{f(x)K - LK}{Kg(x)} \right| + \left| \frac{LK - Lg(x)}{Kg(x)} \right|$$

$$\text{(by the triangle inequality)}$$

$$= \left| \frac{f(x) - L}{g(x)} \right| + \frac{|L|}{|K|} \left| \frac{K - g(x)}{g(x)} \right| \text{ (by factoring)}$$

$$< \frac{|K|\epsilon}{4} \cdot \frac{2}{|K|} + \frac{|L|}{|K|} \frac{|K|^2\epsilon}{4|L| + 1} \cdot \frac{2}{|K|} \text{ (since } \delta \le \delta_0, \delta_1, \delta_2)$$

$$< \epsilon/2 + \epsilon/2$$

$$= \epsilon.$$

This proves (6) and thus the theorem. $\square$

Theorem 4.4.4. (Power rule for limits) *Let n be a positive integer. If* $\lim_{x \to a} f(x) = L$, *then* $\lim_{x \to a} f(x)^n = L^n$.

Proof. The case $n = 1$ is the assumption. Suppose that we know the result for $n - 1$. Then

$$\lim_{x \to a} f(x)^n = \lim_{x \to a} \left(f(x)^{n-1} \cdot f(x) \right) \text{ (by algebra)}$$

$$= \lim_{x \to a} \left(f(x)^{n-1} \right) \cdot \lim_{x \to a} \left(f(x) \right) \text{ (by the product rule)}$$

$$= L^{n-1} \cdot L \text{ (by induction assumption)}$$

$$= L^n.$$

So the result holds for n, and we are done by mathematical induction. $\square$

Theorem 4.4.5. (Polynomial function rule for limits) *Let f be a polynomial function. Then for all complex (or real) a, $\lim_{x \to a} f(x) = f(a)$.*

Proof. Because f is a polynomial function, it can be written as

$$f(x) = c_0 + c_1 x + c_2 x^2 + c_3 x^3 + \cdots + c_n x^n$$

for some non-negative integer n and some constants $c_0, c_1, \ldots, c_n$. By the linear rule, $\lim_{x \to a} x = a$. Hence by the power rule, for all $i = 1, \ldots, n$, $\lim_{x \to a} x^i = a^i$. By the constant rule, $\lim_{x \to a} c_i = c_i$, so that by the product rule $\lim_{x \to a} c_i x^i = c_i a^i$. Hence by repeating the sum rule,

$$\lim_{x \to a} f(x) = \lim_{x \to a} \left(c_0 + c_1 x + c_2 x^2 + \cdots + c_n x^n \right)$$

$$= c_0 + c_1 a + c_2 a^2 + \cdots + c_n a^n$$

$$= f(a). \qquad \square$$

Theorem 4.4.6. (Rational function rule for limits) *Let f be a rational function. Then for all complex (or real) a in the domain of f, $\lim_{x \to a} f(x) = f(a)$.*

Proof. Let a be in the domain of f. Write $f(x) = g(x)/h(x)$ for some polynomial functions g, h such that $h(a) \neq 0$. By Theorem 2.4.15, the domain of f is the set of all except finitely many numbers, so that in particular a is a limit point of the domain. By the polynomial function rule for limits, $\lim_{x \to a} g(x) = g(a)$ and $\lim_{x \to a} h(x) = h(a) \neq 0$. Thus by the quotient rule, $\lim_{x \to a} f(x) = g(a)/h(a) = f(a)$. $\square$

Theorem 4.4.7. (Absolute value rule for limits) *For all* $a \in \mathbb{C}$, $\lim_{x \to a} |x| = |a|$.

Proof. This function is defined for all complex numbers, and so every $a \in \mathbb{C}$ is a limit point of the domain. Let $\epsilon > 0$. Set $\delta = \epsilon$. Then for all $x \in \mathbb{C}$, by the reverse triangle inequality,

$$\big| |x| - |a| \big| \leq |x - a| < \delta = \epsilon. \qquad \square$$

Theorem 4.4.8. (Real and imaginary parts of limits) *Let* $f : A \to \mathbb{C}$ *be a function, let* a *be a limit point of* A, *and let* $L \in \mathbb{C}$. *Then* $\lim_{x \to a} f(x) = L$ *if and only if* $\lim_{x \to a} \operatorname{Re} f(x) = \operatorname{Re} L$ *and* $\lim_{x \to a} \operatorname{Im} f(x) = \operatorname{Im} L$.

Proof. First suppose that $\lim_{x \to a} f(x) = L$. Let $\epsilon > 0$. By assumption there exists $\delta > 0$ such that for all $x \in A$, $0 < |x - a| < \delta$ implies that $|f(x) - L| < \epsilon$. Then for the same x,

$$|\operatorname{Re} f(x) - \operatorname{Re} L| = |\operatorname{Re}(f(x) - L)| \leq |f(x) - L| < \epsilon,$$

and similarly $|\operatorname{Im} f(x) - \operatorname{Im} L| < \epsilon$ which proves that $\lim_{x \to a} \operatorname{Re} f(x) = \operatorname{Re} L$ and $\lim_{x \to a} \operatorname{Im} f(x) = \operatorname{Im} L$.

Now suppose that $\lim_{x \to a} \operatorname{Re} f(x) = \operatorname{Re} L$ and $\lim_{x \to a} \operatorname{Im} f(x) = \operatorname{Im} L$. By the scalar and sum rules in Theorem 4.4.3, and by the definition of real and imaginary parts,

$$\begin{aligned}
\lim_{x \to a} f(x) &= \lim_{x \to a} \left(\operatorname{Re} f(x) + i \operatorname{Im} f(x) \right) \\
&= \lim_{x \to a} \operatorname{Re} f(x) + i \lim_{x \to a} \operatorname{Im} f(x) \\
&= \operatorname{Re} L + i \operatorname{Im} L \\
&= L. \qquad \square
\end{aligned}$$

Theorem 4.4.9. (The composite function theorem) *Let* h *be the composition of functions* $h = g \circ f$. *Suppose that* a *is a limit point of the domain of* $g \circ f$, *that* $\lim_{x \to a} f(x) = L$ *and that* $\lim_{x \to L} g(x) = g(L)$. *Then* $\lim_{x \to a} h(x) = \lim_{x \to a} (g \circ f)(x) = g(L)$.

Proof. Let $\epsilon > 0$. Since $\lim_{x \to L} g(x) = g(L)$, there exists $\delta_1 > 0$ such that for all x in the domain of g, if $0 < |x - a| < \delta_1$ then $|g(x) - g(L)| < \epsilon$. Since $\lim_{x \to a} f(x) = L$, there exists $\delta > 0$ such that for all x in the domain of f, if $0 < |x - a| < \delta$ then $|f(x) - L| < \delta_1$. Thus for the same δ, if x is in the domain of h and $0 < |x - a| < \delta$, then $|h(x) - g(L)| = |g(f(x)) - g(L)| < \epsilon$ because $|f(x) - L| < \delta_1$. □

Perhaps the hypotheses on g in the theorem above seem overly restrictive, and you think that the limit of $g(x)$ as x approaches L need not be $g(L)$ but an arbitrary K? Consider the following example which shows that $\lim_{x \to a} g(f(x))$ then need not be $\lim_{x \to L} g(x)$). Let $a \in \mathbb{R}$, $f(x) = 5$ and $g(x) = \begin{cases} 3, & \text{if } x \neq 5; \\ 7, & \text{otherwise.} \end{cases}$ Then

$$\lim_{x \to a} f(x) = 5(= L), \ \lim_{x \to L} g(x) = 3, \ \text{and} \ \lim_{x \to a} g(f(x)) = 7.$$

Theorem 4.4.10. *Suppose that $f, g : A \to \mathbb{R}$, that a is a limit point of A, that $\lim_{x \to a} f(x)$ and $\lim_{x \to a} g(x)$ both exist, and that for all $x \in A$, $f(x) \leq g(x)$. Then*

$$\lim_{x \to a} f(x) \leq \lim_{x \to a} g(x).$$

Proof. Let $L = \lim_{x \to a} f(x)$, $K = \lim_{x \to a} g(x)$. Let $\epsilon > 0$. By assumptions there exists $\delta > 0$ such that for all $x \in A$, if $0 < |x - a| < \delta$, then $|f(x) - L|, |g(x) - K| < \epsilon/2$. Then for the same x,

$$K - L = K - g(x) + g(x) - f(x) + f(x) - L \geq K - g(x) + f(x) - L$$
$$\geq -|K - g(x) + f(x) - L|$$
$$\geq -|K - g(x)| - |f(x) - L| \ \text{(by the triangle inequality)}$$
$$> -\epsilon/2 - \epsilon/2 = -\epsilon.$$

Since this is true for all $\epsilon > 0$, by Theorem 2.10.4, $K \geq L$, as desired. □

Theorem 4.4.11. (The squeeze theorem) *Let A be a set, a a limit point of A, and $f, g, h : A \to \mathbb{R}$ functions such that for all $x \in A \setminus \{a\}$, $f(x) \leq g(x) \leq h(x)$. If $\lim_{x \to a} f(x)$ and $\lim_{x \to a} h(x)$ both exist and are equal, then $\lim_{x \to a} g(x)$ exists as well and*

$$\lim_{x \to a} f(x) = \lim_{x \to a} g(x) = \lim_{x \to a} h(x).$$

Proof. [IF WE KNEW THAT $\lim_{x \to a} g(x)$ EXISTED, THEN BY THE PREVIOUS THEOREM, $\lim_{x \to a} f(x) \leq \lim_{x \to a} g(x) \leq \lim_{x \to a} h(x) = \lim_{x \to a} f(x)$ WOULD GIVE THAT THE THREE LIMITS ARE EQUAL. BUT WE HAVE YET TO PROVE THAT $\lim_{x \to a} g(x)$ EXISTS.]

Let $L = \lim_{x \to a} f(x) = \lim_{x \to a} h(x)$. Let $\epsilon > 0$. By definition of limits there exists $\delta_1 > 0$ such that for all x, if $0 < |x - a| < \delta_1$ then

$|f(x) - L| < \epsilon$. Similarly, since $\lim_{x \to a} h(x) = L$, there exists $\delta_2 > 0$ such that for all x, if $0 < |x - a| < \delta_2$ then $|h(x) - L| < \epsilon$. Now set $\delta = \min\{\delta_1, \delta_2\}$. Let x satisfy $0 < |x - a| < \delta$. Then

$$-\epsilon < f(x) - L \le g(x) - L \le h(x) - L < \epsilon,$$

where the first inequality holds because $\delta \le \delta_1$, and the last inequality holds because $\delta \le \delta_2$. Hence $-\epsilon < g(x) - L < \epsilon$, which says that $|g(x) - L| < \epsilon$, so that $\lim_{x \to a} g(x) = L$. $\qquad\square$

Exercises for Section 4.4

4.4.1. Determine the following limits by invoking appropriate results:

  i) $\lim\limits_{x \to 2} (x^3 - 4x - 27)$, $\lim\limits_{x \to 2} (x^2 + 5)$.

  ii) $\lim\limits_{x \to 2} \dfrac{(x^3 - 4x - 27)^3}{(x^2 + 5)^2}$.

  iii) $\lim\limits_{x \to 2} \dfrac{|x^3 - 4x - 27|}{(x^2 + 5)^3}$.

4.4.2. Assume that a is a limit point of the intersection of the domains of complex-valued functions f and g. Let $\lim\limits_{x \to a} f(x) = L$ and $\lim\limits_{x \to a} g(x) = K$. Prove that $\lim\limits_{x \to a} (3f(x)^2 - 4g(x)) = 3L^2 - 4K$.

4.4.3. Prove that $\lim\limits_{x \to a} x^n = a^n$ for any non-zero complex number a and any integer n.

4.4.4. The following information is known about functions f and g:

a	$f(a)$	$\lim\limits_{x \to a} f(x)$	$g(a)$	$\lim\limits_{x \to a} g(x)$
0	1	2	6	4
1	-1	0	5	5
2	not defined	3	6	-6
3	4	4	not defined	5
4	5	1	4	7

For each part below provide the limit if there is enough information, and justify. (Careful, the answers to the last two parts are different.)

  i) $\lim\limits_{x \to 0} (f(x) - g(x))$.

  ii) $\lim\limits_{x \to 4} (f(x) + g(x))$.

iii) $\lim_{x \to 3} (f(x) \cdot g(x))$.

iv) $\lim_{x \to 2} \frac{f(x)}{g(x)}$.

v) $\lim_{x \to 4} \frac{g(x)}{f(x)}$.

vi) $\lim_{x \to 1} (g \circ f)(x)$.

vii) $\lim_{x \to 4} (g \circ f)(x)$.

4.4.5. By Example 4.2.2 we know that $\lim_{x \to 0} \frac{x}{|x|}$ does not exist.

i) Prove that $\lim_{x \to 0} \left(1 - \frac{x}{|x|}\right)$ does not exist.

ii) Prove that $\lim_{x \to 0} \left(\frac{x}{|x|} + \left(1 - \frac{x}{|x|}\right)\right)$ exists.

iii) Do the previous parts contradict the sum rule for limits? Justify?

4.4.6. Find functions f, g such that $\lim_{x \to a} (f(x)g(x))$ exists but such that $\lim_{x \to a} f(x)$ and $\lim_{x \to a} g(x)$ do not exist. Does this contradict the product rule?

4.4.7. Prove that if $\lim_{x \to a} f(x) = L$, then $\lim_{x \to a} |f(x)| = |L|$. Give an example of a function such that $\lim_{x \to a} |f(x)| = |L|$ and $\lim_{x \to a} f(x)$ does not exist.

4.4.8. Let $A \subseteq B \subseteq \mathbb{C}$, let a be a limit point of A, and let $f : A \to \mathbb{C}$, $g : B \to \mathbb{C}$. Suppose that for all $x \in A \setminus \{x\}$, $g(x) = f(x)$. Prove that if $\lim_{x \to a} g(x) = L$, then $\lim_{x \to a} f(x) = L$. In particular, if $a \in \mathbb{R}$ is a limit point of $A = B \cap \mathbb{R}$ and if $\lim_{x \to a} g(x) = L$, then the restriction of g to A has the same limit point at a.

4.4.9. Let $A \subseteq \mathbb{C}$, let $f, g : A \to \mathbb{C}$, and let a be a limit point of A. Suppose that for all $x \in A \setminus \{a\}$, $f(x) = g(x)$. Prove that $\lim_{x \to a} f(x)$ exists if and only if $\lim_{x \to a} g(x)$ exists, and if they both exist, then the two limits are equal.

4.4.10. Let $f, g : \mathbb{R} \to \mathbb{R}$ be defined by

$$f(x) = \begin{cases} x, & \text{if } x \in \mathbb{Q}; \\ 0, & \text{otherwise,} \end{cases} \qquad g(x) = \begin{cases} 1, & \text{if } x \neq 0; \\ 0, & \text{otherwise.} \end{cases}$$

i) Prove that $\lim_{x \to 0} f(x) = 0$.

ii) Prove that $\lim_{x \to 0} g(x) = 1$.

iii) Prove that $\lim_{x \to 0} (g \circ f)(x)$ does not exist.

iv) Comment on the relevance of Theorem 4.4.9 in this example.

*4.4.11. Let a, L, M be arbitrary real numbers. Construct functions $F, G :$ $\mathbb{R} \to \mathbb{R}$ such that $\lim_{x \to a} F(x) = L$, $\lim_{x \to L} G(x) = M$, but $\lim_{x \to a}(G \circ F)(x) \neq M$. Comment on Theorem 4.4.9. (Hint: modify f and g from the previous exercise.)

4.4.12. (Due to Jonathan Hoseana, "On zero-over-zero form limits of special type", *The College Mathematics Journal* 49 (2018), page 219.) Let $f : [a, b] \to \mathbb{R}$ be a continuous function. Let A be a subset of $\mathbb{C}$, let $a \in \mathbb{C}$ be a limit point of A, and let $f_1, \ldots, f_n, g$ be complex-valued functions defined on A such that for all $k = 1, \ldots, n$, $\lim_{x \to a} f_k(x) = 1$ and $\lim_{x \to a} \frac{f_k(x) - 1}{g(x)}$ exists. Prove that

$$\lim_{x \to a} \frac{f_1(x) f_2(x) \cdots f_k(x) - 1}{g(x)} = \sum_{k=1}^{n} \lim_{x \to a} \frac{f_k(x) - 1}{g(x)}.$$

4.5 Infinite limits (for real-valued functions)

When the codomain of a function is a subset of an ordered field such as $\mathbb{R}$, the values of a function may grow larger and larger with no upper bound, or more and more negative with no lower bound. In that case we may want to declare limit to be ∞ or $-\infty$. Naturally both the definition and how we operate with infinite limits requires different handling.

Definition 4.5.1. *Let $A \subseteq \mathbb{C}$, $f : A \to \mathbb{R}$ a function, and $a \in \mathbb{C}$. Suppose that a is a limit point of A (and not necessarily in A).*

We say that **the limit of f(x) as x approaches a is ∞** *if for every real number $M > 0$ there exists a real number $\delta > 0$ such that for all $x \in A$, if $0 < |x - a| < \delta$ then $f(x) > M$. We write this as $\lim_{x \to a} f(x) = \lim_{x \to a} f(x) = \infty$. Similarly we say that* **the limit of f(x) as x approaches a is $-\infty$** *if for every real number $M < 0$ there exists a real number $\delta > 0$ such that for all $x \in A$, if $0 < |x - a| < \delta$ then $f(x) < M$. We write this as $\lim_{x \to a} f(x) = \lim_{x \to a} f(x) = -\infty$.*

The limit of f(x) as x approaches a from the right is $-\infty$ *if a is a limit point of $\{x \in A : x > a\}$, and if for every real number $M < 0$ there exists a real number $\delta > 0$ such that for all $x \in A$, if $0 < x - a < \delta$ then $f(x) < M$. This is written as $\lim_{x \to a^+} f(x) = \lim_{x \to a^+} f(x) = -\infty$.*

It is left to the reader to spell out the definitions of the following:

$$\lim_{x \to a^+} f(x) = \infty, \ \lim_{x \to a^-} f(x) = \infty, \ \lim_{x \to a^-} f(x) = -\infty.$$

Note that we cannot use epsilon-delta proofs: no real numbers are within ϵ of infinity. So instead we approximate infinity with huge numbers. In fact, infinity stands for that *thing* which is comparable to and larger than any real number. Thus for all M we can find x near a with $f(x) > M$ is simply saying that we can take $f(x)$ arbitrarily large, which is more succinctly expressed as saying that $f(x)$ goes to ∞. (As far as many applications are concerned, a real number larger than the number of atoms in the universe is as close to infinity as realistically possible, but for proofs, the number of atoms in the universe is not large enough.)

Example 4.5.2. $\lim_{x \to 0} \frac{1}{|x|^2} = \infty$.

Proof. 0 is a limit point of the domain (in the field of real or complex numbers) of the function that takes x to $1/|x|^2$. Let M be a positive real number. Set $\delta = 1/\sqrt{M}$. Let x satisfy $0 < |x - 0| < \delta$, i.e., let x satisfy $0 < |x| < \delta$. Then

$$\frac{1}{|x|^2} > \frac{1}{\delta^2} \ (\text{because } 0 < |x| < \delta)$$
$$= M. \qquad \qquad \square$$

Example 4.5.3. $\lim_{x \to 5^+} \frac{x+2}{x^2 - 25} = \infty$.

Proof. Certainly 5 is the limit point of the domain of the given function. Let $M > 0$. Set $\delta = \min\{1, \frac{7}{11M}\}$. Let x satisfy $0 < x - 5 < \delta$. Then

$$\frac{x+2}{x^2 - 25} > \frac{5+2}{x^2 - 25} \ (\text{because } x > 5 \text{ and } x^2 - 25 > 0)$$
$$= \frac{7}{(x+5)} \cdot \frac{1}{x-5} \ (\text{by algebra})$$
$$> \frac{7}{11} \cdot \frac{1}{x-5} \ (\text{because } 0 < x - 5 < \delta \leq 1, \text{ so } 0 < x + 5 < 11)$$
$$> \frac{7}{11} \cdot \frac{1}{\delta} \ (\text{because } 0 < x - 5 < \delta)$$
$$\geq \frac{7}{11} \cdot \frac{1}{\frac{7}{11M}} \ (\text{because } \delta \leq \frac{7}{11M}, \text{ so } 1/\delta \geq 1/(\frac{7}{11M}))$$
$$= M. \qquad \qquad \square$$

Example 4.5.4. $\lim_{x \to 0^+} \frac{1}{x} = \infty$.

Proof. Let $M > 0$. Set $\delta = 1/M$. Then for all x with $0 < x - 0 < \delta$, $\frac{1}{x} > 1/\delta = M$. $\qquad\square$

Example 4.5.5. $\lim_{x \to 0^-} \frac{1}{x} = -\infty$.

Proof. Let $M < 0$. Set $\delta = -1/M$. Then δ is a positive number. Then for all x with $0 < 0 - x < \delta$, and so by compatibility with multiplication by the positive real number $1/(-x\delta)$,

$$\frac{1}{\delta} < -\frac{1}{x},$$

and so by compatibility of order with addition of $\frac{1}{x} - \frac{1}{\delta}$,

$$\frac{1}{x} < -\frac{1}{\delta} = M. \qquad\square$$

Example 4.5.6. We conclude that $\lim_{x \to 0} \frac{1}{x}$ cannot be a real number, and it cannot be either ∞ or $-\infty$. Thus $\lim_{x \to 0} \frac{1}{x}$ does not exist.

The following theorem is straightforward to prove, and it is left to the exercises.

Theorem 4.5.7. *Let* $f, g, h : A \to \mathbb{R}$, *and let* a *be a limit point of* A. *Suppose that* $\lim_{x \to a} f(x) = L \in \mathbb{R}$, $\lim_{x \to a} g(x) = \infty$, $\lim_{x \to a} h(x) = -\infty$. *Then*

(1) (Scalar rule) For any $c \in \mathbb{R}$,

$$\lim_{x \to a} (cg(x)) = \begin{cases} \infty, & \text{if } c > 0; \\ -\infty, & \text{if } c < 0; \\ 0, & \text{if } c = 0; \end{cases} \qquad \lim_{x \to a} (ch(x)) = \begin{cases} -\infty, & \text{if } c > 0; \\ \infty, & \text{if } c < 0; \\ 0, & \text{if } c = 0. \end{cases}$$

(2) (Sum/difference rule)

$$\lim_{x \to a} (f(x) \pm g(x)) = \pm\infty,$$

$$\lim_{x \to a} (f(x) \pm h(x)) = \mp\infty,$$

$$\lim_{x \to a} (g(x) - h(x)) = \infty,$$

$$\lim_{x \to a} (h(x) - g(x)) = -\infty,$$

but we do not have enough information to determine whether the limit of
$g(x) + h(x)$ as x approaches a exists.

(3) *(Product rule)*

$$\lim_{x \to a} (f(x) \cdot g(x)) = \begin{cases} \infty, & \text{if } L > 0; \\ -\infty, & \text{if } L < 0; \end{cases}$$

$$\lim_{x \to a} (f(x) \cdot h(x)) = \begin{cases} -\infty, & \text{if } L > 0; \\ \infty, & \text{if } L < 0; \end{cases}$$

$$\lim_{x \to a} (g(x) \cdot h(x)) = -\infty.$$

We do not have enough information to determine the existence (or value)
of $\lim_{x \to a}(f(x) \cdot g(x))$ and of $\lim_{x \to a}(f(x) \cdot h(x))$ in case $L = 0$.

(4) *(Quotient rule)*

$$\lim_{x \to a} \frac{f(x)}{g(x)} = 0,$$

$$\lim_{x \to a} \frac{f(x)}{h(x)} = 0,$$

$$\lim_{x \to a} \frac{g(x)}{f(x)} = \begin{cases} \infty, & \text{if } L > 0; \\ -\infty, & \text{if } L < 0; \end{cases}$$

$$\lim_{x \to a} \frac{h(x)}{f(x)} = \begin{cases} -\infty, & \text{if } L > 0; \\ \infty, & \text{if } L < 0. \end{cases}$$

We do not have enough information to determine the existence (or value)
of $\frac{\lim_{x \to a} g(x)}{f(x)}$ and of $\frac{\lim_{x \to a} h(x)}{f(x)}$ in case $L = 0$. We also do not have enough
information to determine the existence (or value) of $\frac{\lim_{x \to a} g(x)}{h(x)}$.

Example 4.5.8. Define $g, h_1, h_2, h_3, h_4 : \mathbb{R} \setminus \{0\} \to \mathbb{R}$ by $g(x) = 1/x^2$,
$h_1(x) = -1/x^2$, $h_2(x) = 17 - 1/x^2$, $h_3(x) = -1/x^2 - 1/x^4$, $h_4(x) = -1/x^2 - 1/x^3$. We have seen that $\lim_{x \to 0} g(x) = \infty$, and similarly that
$\lim_{x \to 0} h_1(x) = \lim_{x \to 0} h_2(x) = \lim_{x \to 0} h_3(x) = \lim_{x \to 0} h_4(x) = -\infty$.
However,

$$\lim_{x \to 0} (g(x) + h_1(x)) = \lim_{x \to 0} 0 = 0,$$

$$\lim_{x \to 0} (g(x) + h_2(x)) = \lim_{x \to 0} 17 = 17,$$

$$\lim_{x \to 0} (g(x) + h_3(x)) = \lim_{x \to 0} (-1/x^4) = -\infty,$$

but $\lim_{x \to 0}(g(x) + h_4(x)) = \lim_{x \to 0}(-1/x^3)$ does not exist. This justifies the
"not enough information" line in the sum/difference rule in Theorem 4.5.7.

Other "not enough information" lines are left for the exercises.

Distinguish between the scalar and the product rules: when $\lim_{x \to a} f(x) = 0$, if f is a constant function, then $\lim_{x \to a}(f(x)g(x)) = 0$, but if f is not a constant function, then we do not have enough information for $\lim_{x \to a}(f(x)g(x)) = 0$.

Exercises for Section 4.5

4.5.1. Give definitions for the following limits.

    i) $\lim_{x \to a^+} f(x) = \infty$.

    ii) $\lim_{x \to a^-} f(x) = \infty$.

    iii) $\lim_{x \to a^-} f(x) = -\infty$.

4.5.2. Prove that $\lim_{x \to 0^+} \frac{1}{\sqrt{x}} = \infty$.

4.5.3. Prove that $\lim_{x \to 0^-} \frac{1}{x^3} = -\infty$.

4.5.4. Referring to Theorem 4.5.7, prove the scalar rule, the sum/difference rule, the product rule, and the quotient rule.

4.5.5. Let $g(x) = 1/x^2$, $f_1(x) = x^3$, $f_2(x) = x^2$, $f_3(x) = 17x^2$, $f_4(x) = x$. It is easy to see that $\lim_{x \to 0} g(x) = \infty$, $\lim_{x \to 0} f_1(x) = \lim_{x \to 0} f_2(x) = \lim_{x \to 0} f_3(x) = \lim_{x \to 0} f_4(x) = 0$.

    i) Compute $\lim_{x \to 0}(f_k(x)g(x))$ for $k = 1, 2, 3$.

    ii) Justify the first "not enough information" in the product rule in Theorem 4.5.7.

    iii) Justify the second "not enough information" in the product rule in Theorem 4.5.7.

4.5.6. Justify the three "not enough information" in the quotient rule in Theorem 4.5.7. (Hint: Study the previous exercise.)

4.6   Limits at infinity

Definition 4.6.1. *Let $a \in \mathbb{R}$ and $L \in \mathbb{C}$.*

For $f : (a, \infty) \to \mathbb{C}$, we say that $\lim_{x \to \infty} f(x) = L$ if for every $\epsilon > 0$ there exists a real number $N > 0$ such that for all $x > N$, $|f(x) - L| < \epsilon$.

For $f : (-\infty, a) \to \mathbb{C}$, we say that $\lim_{x \to -\infty} f(x) = L$ if for every $\epsilon > 0$ there exists a real number $N < 0$ such that for all $x < N$, $|f(x) - L| < \epsilon$.

Below we switch to real-valued functions.

For $f : (a, \infty) \to \mathbb{R}$, *we say that* $\lim_{x \to \infty} f(x) = \infty$ *if for every* $M > 0$ *there exists a real number* $N > 0$ *such that for all* $x > N$, $f(x) > M$. *We say that* $\lim_{x \to \infty} f(x) = -\infty$ *if for every* $M < 0$ *there exists a real number* $N > 0$ *such that for all* $x > N$, $f(x) < M$.

For $f : (-\infty, a) \to \mathbb{R}$, *we say that* $\lim_{x \to -\infty} f(x) = \infty$ *if for every* $M > 0$ *there exists a real number* $N < 0$ *such that for all* $x < N$, $f(x) > M$. *We say that* $\lim_{x \to -\infty} f(x) = -\infty$ *if for every* $M < 0$ *there exists a real number* $N < 0$ *such that for all* $x < N$, $f(x) < M$.

Example 4.6.2. $\lim_{x \to \infty} (x^5 - 16x^4) = \infty$.

Proof. Let $M > 0$. Set $N = \max\{17, M^{1/4}\}$. Then for all $x > N$,

$$x^5 - 16x^4 = x^4(x - 16)$$

$$\geq x^4 \ \text{(because } x > N \geq 17)$$

$$> N^4 \ \text{(because } x > N)$$

$$\geq (M^{1/4})^4 \ \text{(because } N \geq M^{1/4})$$

$$= M. \qquad \qquad \square$$

Example 4.6.3. $\lim_{x \to \infty} \frac{x^5 - 16x^4}{x^5 + 4x^2} = 1$.

Proof. Let $\epsilon > 0$. Set $N = \max\{1, 20/\epsilon\}$. Then for all $x > N$,

$$\left| \frac{x^5 - 16x^4}{x^5 + 4x^2} - 1 \right| = \left| \frac{x^5 - 16x^4 - x^5 - 4x^2}{x^5 + 4x^2} \right|$$

$$= \left| \frac{-16x^4 - 4x^2}{x^5 + 4x^2} \right|$$

$$= \left| \frac{16x^4 + 4x^2}{x^5 + 4x^2} \right|$$

$$= \frac{16x^4 + 4x^2}{x^5 + 4x^2} \ \text{(because } x > N \geq 0)$$

$$\leq \frac{16x^4 + 4x^2}{x^5} \ \text{(because } x^5 + 4x^2 \geq x^5 > 0)$$

$$\leq \frac{16x^4 + 4x^4}{x^5} \ \text{(because } x > N \geq 1 \text{ so that } x^2 < x^4)$$

$$= \frac{20x^4}{x^5}$$

$$= \frac{20}{x}$$

$$< \frac{20}{N} \text{ (because } x > N)$$

$$\leq \frac{20}{20/\epsilon} \text{ (because } N \geq 20/\epsilon)$$

$$= \epsilon. \qquad\qquad\qquad \square$$

Exercises for Section 4.6

4.6.1. Prove that $\lim_{x\to\infty}(x^2 - x) = \infty$ and that $\lim_{x\to-\infty}(x^2 - x) = \infty$.

4.6.2. Prove that $\lim_{x\to\infty}(x^3 - x) = \infty$ and that $\lim_{x\to-\infty}(x^3 - x) = -\infty$.

4.6.3. Is there a limit rule of the following form: If $\lim_{x\to\infty} f(x) = \infty$ and $\lim_{x\to\infty} g(x) = \infty$, then $\lim_{x\to\infty}(f(x) - g(x))$ can be determined?

4.6.4. Prove the following limits:

i) $\lim_{x\to\infty} \frac{3x^2-4x+1}{2x^2+2} = 3/2$.

ii) $\lim_{x\to\infty} \frac{3x^3-4x+1}{2x^2+2} = \infty$.

iii) $\lim_{x\to\infty} \frac{3x^2-4x+1}{2x^3+2} = 0$.

iv) $\lim_{x\to\infty} \frac{\sqrt{9x^2+4}}{x+2} = 3$.

v) $\lim_{x\to-\infty} \frac{\sqrt{9x^2+4}}{x+2} = -3$.

4.6.5. For which rational functions f does $\lim_{x\to\infty} f(x) = 0$ hold? Justify.

4.6.6. For which rational functions f does $\lim_{x\to\infty} f(x)$ exist and is not $\pm\infty$? Justify.

4.6.7. Let $f : (a,\infty) \to \mathbb{C}$ and $L \in \mathbb{C}$. Prove that $\lim_{x\to\infty} f(x) = L$ if and only if $\lim_{x\to\infty} \operatorname{Re} f(x) = \operatorname{Re} L$ and $\lim_{x\to\infty} \operatorname{Im} f(x) = \operatorname{Im} L$.

4.6.8. Let $f : (-\infty,b) \to \mathbb{C}$ and $L \in \mathbb{C}$. Prove that $\lim_{x\to-\infty} f(x) = L$ if and only if $\lim_{x\to-\infty} \operatorname{Re} f(x) = \operatorname{Re} L$ and $\lim_{x\to-\infty} \operatorname{Im} f(x) = \operatorname{Im} L$.

Chapter 5

Continuity

Continuous functions from an interval in $\mathbb{R}$ to $\mathbb{R}$ are the ones that we can graph without any holes or jumps, i.e., without lifting the pencil from the paper, so the range of such functions is an interval in $\mathbb{R}$ as well. We make this more formal below, and not just for functions with domains and codomains in $\mathbb{R}$. The formal definition involves limits of functions. All the hard work for that was already done in Chapter 4, so this chapter, after absorbing the definition, is really straightforward. The big new results are the Intermediate value theorem and the Extreme value theorem for real-valued functions (Section 5.3), existence of radical functions (Section 5.4), and the new notion of uniform continuity (Section 5.5).

5.1 Continuous functions

Definition 5.1.1. *A function $f : A \to B$ is **continuous at** $a \in A$ if for all real numbers $\epsilon > 0$ there exists a real number $\delta > 0$ such that for all $x \in A$, $|x - a| < \delta$ implies that $|f(x) - f(a)| < \epsilon$. We say that f is **continuous** if f is continuous at all points in its domain.*

Much of the time for us A is an interval in $\mathbb{R}$, a rectangle in $\mathbb{C}$, or $B(a, r)$ in $\mathbb{R}$ or $\mathbb{C}$, etc. In the more general case, A may contain a point a that is not a limit point of A.

Theorem 5.1.2. *Let $f : A \to B$ be a function and $a \in A$.*
(1) If a is not a limit point of A, then f is continuous at a.
(2) If a is a limit point of A, then f is continuous at a if and only if $\lim_{x \to a} f(x) = f(a)$.

Proof. In the first case, there exists $\delta > 0$ such that $B(a, \delta) \cap A = \{a\}$. Thus by definition, the only $x \in A$ with $|x - a| < \delta$ is $x = a$, whence

215

$|f(x) - f(a)| = 0$ is strictly smaller than an arbitrary positive ϵ. Thus $f : A \to B$ is continuous at a.

Now assume that a is a limit point of A. Suppose that f is continuous at a. We need to prove that $\lim_{x \to a} f(x) = f(a)$. Let $\epsilon > 0$. Since f is continuous at a, there exists $\delta > 0$ such that for all $x \in A$, if $|x - a| < \delta$, then $|f(x) - f(a)| < \epsilon$. Hence for all $0 < |x - a| < \delta$, $|f(x) - f(a)| < \epsilon$, and since a is a limit point of A, this proves that $\lim_{x \to a} f(x) = f(a)$.

Finally suppose that $\lim_{x \to a} f(x) = f(a)$. We need to prove that f is continuous at a. Let $\epsilon > 0$. By assumption there exists $\delta > 0$ such that for all $x \in A$, if $0 < |x - a| < \delta$ then $|f(x) - f(a)| < \epsilon$. If $x = a$, then $|f(x) - f(a)| = 0 < \epsilon$, so that for all $x \in A$, if $|x - a| < \delta$, then $|f(x) - f(a)| < \epsilon$. Thus f is continuous at a.   □

The next theorem is an easy application of Theorems 4.4.3, 4.4.4, 4.4.5, 4.4.6, 4.4.7, 4.4.8, and 4.4.9, at least when the points in question are limit points of the domain. When the points in question are not limit points of the domain, the results below need somewhat different but still easy proofs.

Theorem 5.1.3. *We have:*

    (1) (Constant rule) Constant functions are continuous at all points of the domain.

    (2) (Linear rule) The function $f(x) = x$ is continuous (at all real/ complex numbers).

    (3) (Absolute value rule) The function $f(x) = |x|$ is continuous (at all real/complex numbers).

    (4) (Polynomial function rule) Polynomial functions are continuous (at all real/complex numbers).

    (5) (Rational function rule) Rational functions are continuous (at all points in the domain).

    (6) (Real and imaginary parts) f is continuous at a if and only if $\operatorname{Re} f$ and $\operatorname{Im} f$ are continuous at a.

Suppose that $f : A \to \mathbb{C}$ is continuous at $a \in A$. Then

    (7) (Scalar rule) For any $c \in \mathbb{C}$, cf is continuous at a.

    (8) (Power rule) For any positive integer n, the function that takes x to $(f(x))^n$ is continuous at a.

(9) (The composite function rule) If g is continuous at $f(a)$, then $g \circ f$ is continuous at a.

(10) (The composite power rule) If f is composable with itself i.e., if the range of f is a subset of the domain of f, then $f^n = f \circ f \circ \cdots \circ f$ is continuous at a. (Example of a function that is not composable with itself: $f(x) = \ln x$.)

(11) (Restriction rule) For any subset B of A that contains a and in which a is a limit point, the restriction of f to B is continuous at a. In other words, if $g : B \to \mathbb{C}$ is defined as $g(x) = f(x)$ for all $x \in B$, then g is continuous at a. (Such g is called **the restriction** of f to B. By this rule and by (1), the constant functions with domain a sub-interval in $\mathbb{R}$ are continuous.)

Now suppose that $f, g : A \to \mathbb{C}$, and that f and g are continuous at $a \in A$. Then

(12) (Sum/difference rule) $f + g$ and $f - g$ are continuous at a.

(13) (Product rule) $f \cdot g$ is continuous at a.

(14) (Quotient rule) If $g(a) \neq 0$, then f/g is continuous at a.

The theorem covers many continuous functions, but the following function for example has to be verified differently.

Example 5.1.4. Let $f : \mathbb{C} \to \mathbb{C}$ be defined by

$$f(x) = \begin{cases} x^2 - 4, & \text{if } \operatorname{Re} x > 1; \\ -3x^3, & \text{if } \operatorname{Re} x \leq 1. \end{cases}$$

Then f is continuous at 1, because by the polynomial rules,

$$\lim_{x \to 1, \operatorname{Re} x > 1} f(x) = \lim_{x \to 1, \operatorname{Re} x > 1} (x^2 - 4) = 1^2 - 4 = -3,$$

$$\lim_{x \to 1, \operatorname{Re} x \leq 1} f(x) = \lim_{x \to 1, \operatorname{Re} x \leq 1} -3x^3 = -3 \cdot 1^3 = -3,$$

so that $\lim_{x \to 1} f(x) = -3 = -3 \cdot 1^3 = f(1)$.

Note however that this function is not continuous at $1 + i$:

$$\lim_{x \to 1+i, \operatorname{Re} x > 1} f(x) = \lim_{x \to 1+i, \operatorname{Re} x > 1} (x^2 - 4) = (1 + i)^2 - 4 = -4 + 2i,$$

$$\lim_{x \to 1+i, \operatorname{Re} x \leq 1} f(x) = \lim_{x \to 1+i, \operatorname{Re} x \leq 1} -3x^3 = -3 \cdot (1 + i)^3 = 6 - 6i.$$

so that $\lim_{x \to 1+i} f(x)$ does not exist.

Example 5.1.5. The function $g : \mathbb{R} \to \mathbb{R}$ defined as the restriction of the function in Example 5.1.4 equals

$$g(x) = \begin{cases} x^2 - 4, & \text{if } x > 1; \\ -3x^3, & \text{if } x \leq 1. \end{cases}$$

Since f is continuous at 1, so is g. Explicitly, by the polynomial rules,

$$\lim_{x \to 1^+} g(x) = \lim_{x \to 1^+} (x^2 - 4) = 1^2 - 4 = -3,$$

$$\lim_{x \to 1^-} g(x) = \lim_{x \to 1^-} -3x^3 = -3 \cdot 1^3 = -3,$$

so that (by Exercise 4.1.6), $\lim_{x \to 1} g(x) = -3 = -3 \cdot 1^3 = g(1)$. Thus g is indeed continuous at 1. By the polynomial rules, g is continuous at all other real numbers as well. Thus g is continuous — even if it is a restriction of a non-continuous function. It is worth noting that precisely because of continuity we can write g in the following ways as well:

$$g(x) = \begin{cases} x^2 - 4, & \text{if } x \geq 1; \\ -3x^3, & \text{if } x < 1 \end{cases} = \begin{cases} x^2 - 4, & \text{if } x \geq 1; \\ -3x^3, & \text{if } x \leq 1. \end{cases}$$

Example 5.1.6. Let $f : (-1, 0) \cup (0, 1] \to \mathbb{R}$ be defined by

$$f(x) = \begin{cases} x + 1, & \text{if } x < 0; \\ x - 1, & \text{if } x > 0. \end{cases}$$

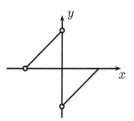

This function is continuous at all $a < 0$ (in the domain) because near such a the function f equals the polynomial/linear function $f(x) = x + 1$. Similarly f is continuous at all $a > 0$. Thus f is continuous at all points in its domain.

Exercises for Section 5.1

5.1.1. Let $c \in \mathbb{C}$.

  i) Prove that the function $f : \mathbb{C} \to \mathbb{C}$ given by $f(x) = c + x$ is continuous.

  ii) Prove that the function $g : \mathbb{C} \to \mathbb{C}$ given by $g(x) = cx$ is continuous.

5.1.2. Give an $\epsilon - \delta$ proof to show that the functions $\mathrm{Re}, \mathrm{Im} : \mathbb{C} \to \mathbb{R}$ are continuous.

5.1.3. Give details of the proofs of Theorem 5.1.3.

5.1.4. The following information is known about functions f and g:

c	$f(c)$	$\lim_{x \to c} f(x)$	$g(c)$	$\lim_{x \to c} g(x)$
0	1	2	3	4
1	-1	0	5	5
2	not defined	$-\infty$	6	-6
3	4	4	not defined	∞
4	2	3	3	2

    i) At which points is f continuous?

    ii) At which points is $|g|$ continuous?

    iii) At which points is fg continuous?

5.1.5. Define a function $f : \mathbb{C} \setminus \{0\} \to \mathbb{R}$ such that for all non-zero $x \in \mathbb{C}$, $f(x)$ is the angle of x counterclockwise from the positive real axis. Argue that f is not continuous.

5.1.6. Let $f : A \to \mathbb{C}$ be a function, and let $a \in A$ be a limit point of A. Let $g : A \to B$ be a continuous invertible function such that its inverse is also continuous. Prove that $\lim_{x \to a} f(x) = L$ if and only if $\lim_{z \to g(a)} f(g^{-1}(z)) = L$.

5.1.7. Let $f : \mathbb{R} \to \mathbb{R}$ be given by $f(x) = \begin{cases} x, & \text{if } x \text{ is rational;} \\ 0, & \text{otherwise.} \end{cases}$

    i) Prove that f is continuous at 0.

    ii) Prove that f is not continuous anywhere else.

*5.1.8. (The Thomae function, also called the popcorn function, the raindrop function, and more) Let $f : \mathbb{R} \to \mathbb{R}$ be defined as

$$f(x) = \begin{cases} \frac{1}{q}, & \text{if } x = \frac{p}{q}, \text{ where } p \in \mathbb{Z} \text{ and } q \in \mathbb{N}^{+} \\ & \qquad \text{and } p \text{ and } q \text{ have no common prime factors;} \\ 0, & \text{if } x \text{ is irrational.} \end{cases}$$

    i) Prove that f is not continuous at any rational number.

    ii) Prove that f is continuous at all irrational numbers.

The exercises below are a further play on Section 4.3: they modify the definition of continuity to get very different types of functions. The moral is that the order of quantifiers and implications is very important!

5.1.9. (This is from page 1177 of the Edward Nelson's article "Internal set theory: a new approach to nonstandard analysis." *Bull. Amer. Math. Soc.* **83** (1977), no. 6, 1165–1198.) A function $f : A \to B$ is **suounitnoc at a** $\in$ **A** if for all real numbers $\epsilon > 0$ there exists a real number $\delta > 0$ such that for all $x \in A$, $|x - a| < \epsilon$ implies that $|f(x) - f(a)| < \delta$.

   i) Prove that if f is suounitnoc at some $a \in A$, it is suounitnoc at every $b \in A$.

   ii) Let $f : \mathbb{R}^+ \to \mathbb{R}$ be given by $f(x) = 1/x$. Prove that at every $a \in \mathbb{R}^+$, f is continuous but not suounitnoc.

   iii) Let $f : \mathbb{R} \to \mathbb{R}$ be given by $f(x) = 1$ if x is irrational and $f(x) = 0$ if x is rational. Prove that at every $a \in \mathbb{R}$, f is suounitnoc but not continuous.

5.1.10. A function $f : A \to B$ is **ticonnuous at a** $\in$ **A** if there exists a real number $\delta > 0$ such that for all real numbers $\epsilon > 0$ and for all $x \in A$, $|x - a| < \delta$ implies that $|f(x) - f(a)| < \epsilon$.

   i) Suppose that f is ticonnuous at some $a \in A$. Prove that there exist a real number $\delta > 0$ and $b \in B$ such that for all $x \in A$, $|x - a| < \delta$ implies that $f(x) = b$.

   ii) Give an example of a continuous function that is not ticonnuous at every a in the domain.

   iii) Prove that every function that is ticonnuous at every point in the domain is continuous.

5.1.11. A function $f : A \to B$ is **connuousti at a** $\in$ **A** if for all real numbers $\epsilon > 0$ there exists a real number $\delta > 0$ such that for all $x \in A$, $|f(x) - f(a)| < \epsilon$ implies that $|x - a| < \delta$.

   i) Let $f : \mathbb{R} \to \mathbb{R}$ be a constant function. Prove that f is not connuousti at any $a \in \mathbb{R}$.

   ii) Let $f : \mathbb{R} \to \mathbb{R}$ be given by $f(x) = 1$ if x is irrational and $f(x) = 0$ if x is rational. Prove that f is not connuousti at any $a \in \mathbb{R}$.

   iii) Let $f : \mathbb{R} \to \mathbb{R}$ be given by $f(x) = x$ if x is irrational and $f(x) = x+1$ if x is rational. Prove that at every $a \in \mathbb{R}$, f is connuousti.

5.2 Topology and the Extreme value theorem

Topology and continuity go hand in hand (see Exercise 5.2.5), but not in the obvious way, as the next two examples show.

(1) If $f : A \to B$ is continuous and A is open, it need not be true that the range of f is open in B, even if A is bounded. For example, let $f : \mathbb{R} \to \mathbb{R}$ or $f : (-1, 1) \to \mathbb{R}$ be the squaring function. Certainly f is continuous and $\mathbb{R}$ is an open set, but the image of f is $[0, \infty)$ or $[0, 1)$, which is not open.

(2) If $f : A \to B$ is continuous and A is closed, it need not be true that the range of f is closed in B. For example, let $f : \mathbb{R} \to \mathbb{R}$ be given by $f(x) = \frac{1}{1+x^2}$. This f is continuous as it is a rational function defined on all of $\mathbb{R}$. The domain $A = \mathbb{R}$ is a closed set (and open), but the image of f is $(0, 1]$, which is not closed (and not open).

However, if f is continuous and its domain A is closed and in addition bounded, then it is true that the image of f is closed. This fact is proved next.

Theorem 5.2.1. *Let A be a closed and bounded subset of $\mathbb{C}$ or $\mathbb{R}$, and let $f : A \to \mathbb{C}$ be continuous. Then the range of f is closed and bounded.*

Proof. We first prove that the range is closed. Let b be a limit point of the range. We want to prove that b is in the range. Since b is a limit point, by definition for every positive real number r, $B(b, r)$ contains an element of the range (even an element of the range that is different from b). In particular, for every positive integer m there exists $x_m \in A$ such that $f(x_m) \in B(b, 1/m)$. If for some m we have $f(x_m) = b$, then we are done, so we may assume that for all m, $f(x_m) \neq b$. Thus there are infinitely many x_m. As in Construction 3.14.1, we can construct nested quarter subrectangles R_n that contain infinitely many x_m. There is a unique complex number c that is contained in all the R_n. By construction, c is the limit point of the set of the x_m, hence of A. As A is closed, $c \in A$. But f is continuous at c, so that for all $\epsilon > 0$ there exists $\delta > 0$ such that for all $x \in A$, if $|x - c| < \delta$ then $|f(x) - f(c)| < \epsilon/2$. In particular, for infinitely many large m, $|x_m - c| < \delta$, so that for these same m, $|f(x_m) - f(c)| < \epsilon/2$. But for all large m we also have $|f(x_m) - b| < \epsilon/2$, so that by the triangle inequality, for $|f(c) - b| < \epsilon$. Since this is true for all positive ϵ, it follows that $f(c) = b$. Thus any limit point of the range is in the range, so that the range is closed.

Next we prove that the range is bounded. If not, then for every positive integer m there exists $x_m \in A$ such that $|f(x_m)| > m$. Again we use Construction 3.14.1, and this time we construct nested quarter subrectangles R_n that contain infinitely many of these x_m. As before, there is a unique complex number c that is contained in all the R_n and in A, and as before (with $\epsilon = 2$), we get that for infinitely many m, $|f(x_m) - f(c)| < 1$. But $|f(x_m)| > m$, so that by the reverse triangle inequality,

$$|f(c)| \geq |f(x_m) - (f(x_m) - f(c))| \geq |f(x_m)| - |f(x_m) - f(c)| > m - 1$$

for infinitely many positive integers m. Since $|f(c)|$ is a fixed real number, it cannot be larger than all positive integers. Thus we get a contradiction to the assumption that the range is not bounded, which means that the range must be bounded. $\qquad\square$

Theorem 5.2.2. (Extreme value theorem) *Let A be a closed and bounded subset of $\mathbb{C}$, and let $f : A \to \mathbb{R}$ be a continuous function. Then there exist $l, u \in A$ such that for all $x \in A$,*

$$f(l) \leq f(x) \leq f(u).$$

In other words, f achieves its maximum value at u and its minimum value at l.

Proof. By Theorem 5.2.1, the range $\{f(x) : x \in A\}$ of f is a closed and bounded subset of $\mathbb{R}$, so that its infimum L and supremum U are real numbers which are by closedness in the range. Thus there exist $u, l \in A$ such that $L = f(l)$ and $U = f(u)$. $\qquad\square$

Example 5.2.3. The function $f : [-2, 2] \to \mathbb{R}$ given by $f(x) = x^2 - 6x + 5$ achieves a minimum and maximum. We can rewrite the function as $f(x) = (x - 3)^2 - 4$, from which it is obvious that the minimum of the function is achieved at 3 – but wait a minute, this function is not defined at 3 and hence cannot achieve a minimum at 3. Here is a correction: the quadratic function $(x-3)^2 - 4$ achieves its minimum at 3 and is decreasing on $(-\infty, 3)$, so that f achieves its minimum on $[-2, 2]$ at -2 and its maximum at 2.

Exercises 5.2.2 and 5.2.3 below give examples of continuous invertible functions whose inverses are not continuous. The next theorem contains instead some positive results of this flavor.

Theorem 5.2.4. *Let $F = \mathbb{R}$ or $F = \mathbb{C}$. Let A, B be subsets of F, let A be closed and bounded, and let $f : A \to B$ be continuous and invertible. Then f^{-1} is continuous.*

Proof. We need to prove that f^{-1} is continuous at every $b \in B$. Let $\epsilon > 0$. Set $a = f^{-1}(b)$. The set $B(a, \epsilon)$ is open, so its complement is closed. Therefore $C = A \setminus B(a, \epsilon) = A \cap (F \setminus B(a, \epsilon))$ is closed. As $C \subseteq A$, then C is also bounded. Thus by Theorem 5.2.1, $\{f(x) : x \in C\}$ is a closed and bounded subset of F. By injectivity of f, this set does not contain $b = f(a)$. The complement of this set contains b and is open, so that there exists $\delta > 0$ such that $B(b, \delta) \subseteq F \setminus \{f(x) : x \in C\}$. Now let $y \in B$ with $|y - b| < \delta$. Since f is invertible, there exists $x \in A$ such that $y = f(x)$. By the choice of δ, $x \notin C$, so that $f^{-1}(y) = x \in B(a, \epsilon) = B(f^{-1}(b), \epsilon)$. In short, for every $\epsilon > 0$ there exists $\delta > 0$ such that for all $y \in B$, if $|y - b| < \delta$, then $|f^{-1}(y) - f^{-1}(a)| < \epsilon$. This proves that f^{-1} is continuous at b. $\qquad\square$

Exercises for Section 5.2

5.2.1. (Compare with Theorem 5.2.1.)

i) Show that $f : (0, 1) \to \mathbb{R}$ defined by $f(x) = \frac{1}{x}$ is a continuous function whose domain is open and bounded but the range is not bounded.

ii) Show that $f : (-1, 1) \to \mathbb{R}$ defined by $f(x) = x^2$ is a continuous function whose domain is open and bounded but the range is not open.

5.2.2. Consider the continuous function f in Example 5.1.6.

i) Prove that f has an inverse function $f^{-1} : (-1, 1) \to (-1, 0) \cup (0, 1]$.

ii) Graph f^{-1} and prove that f^{-1} is not continuous.

iii) Compare with Theorem 5.2.4.

5.2.3. Define $f : \{x \in \mathbb{C} : |x| < 1 \text{ or } |x| \geq 2\} \to \mathbb{C}$ by

$$f(x) = \begin{cases} x, & \text{if } |x| < 1; \\ \frac{1}{2}x, & \text{otherwise.} \end{cases}$$

Prove that f is continuous and invertible but that f^{-1} is not continuous. Compare with Theorem 5.2.4.

5.2.4. Let A be a closed and bounded subset of $\mathbb{C}$, and let $f : A \to \mathbb{C}$ be continuous.

i) Why are we not allowed to talk about f achieving its maximum or minimum?

ii) Can we talk about the maximum and minimum absolute values of f? Justify your answer.

iii) Can we talk about the maximum and minimum of the real or of the imaginary components of f? Justify your answer.

*5.2.5. Let F be $\mathbb{R}$ or $\mathbb{C}$, and let A and B be subsets of F. Prove that $f : A \to B$ is continuous if and only if for every open subset U of F there exists an open subset V of F such that the set $\{x \in A : f(x) \in U\} = V \cap A$.

5.2.6. Let $f : (0,1) \to \mathbb{R}$ be defined by $f(x) = \frac{1}{x}$. Prove that f is continuous and that f has neither a minimum nor a maximum. Explain why this does not contradict the Extreme value theorem (Theorem 5.2.2).

5.2.7. (Outline of another proof of the Extreme value theorem for closed intervals in $\mathbb{R}$, due to Samuel J. Ferguson, "A one-sentence line-of-sight proof of the Extreme value theorem", *American Mathematical Monthly* **121** (2014), page 331.) Let $f : [a,b] \to \mathbb{R}$ be a continuous function. The goal is to prove that f achieves its maximum at a point $u \in [a,b]$. Set $L = \{x \in [a,b] : \text{for all } y \in [a,b], \text{ if } y < x \text{ then } f(y) \leq f(x)\}$.

i) Prove that $a \in L$.

ii) Let $c \in \mathrm{Bd}\,(L)$ (boundary of L). Prove that $c \in [a,b]$.

iii) Suppose that $c \notin L$.

  − Prove that there exists $y \in [a,c)$ such that $f(y) > f(c)$.

  − Prove that there exists $\delta > 0$ such that for all $x \in [a,b] \cap B(c,\delta)$, $|f(x) - f(c)| < (f(y) - f(c))/2$. (Hint: use continuity at c.)

– Prove that there exists $x \in L \cap B(c, \min\{\delta, c - y\})$.

– Prove that $f(y) - f(x) > 0$.

– Prove that $y \geq x$.

– Prove that $c - y > |x - c| \geq c - x \geq c - y$, which is a contradiction.

iv) Conclude that $c \in L$ and that L is a closed set.

v) Let $u = \sup(L)$. Prove that u exists and is an element of L.

vi) Let $k > f(u)$. Prove that the set $S_k = \{x \in [c, b] : f(y) \geq k\}$ is closed, and in particular that S_k, if non-empty, has a minimum.

vii) Suppose that for some $x \in [a, b]$, $f(x) > f(u)$.

– Prove that $x > u$ and $x \notin L$.

– Prove that there exists $x_1 \in (u, x)$ such that $f(x_1) > f(x)$.

– Prove that there exist $x_1 \in (u, x)$, $x_2 \in (u, x_1)$, $x_3 \in (u, x_2)$, $\ldots$, $x_n \in (u, x_{n-1})$, $\ldots$, such that $\cdots > f(x_n) > f(x_{n-1}) > \cdots > f(x_3) > f(x_2) > f(x_1) > f(x) > f(u)$.

– Prove that the set $S_{f(u)}$ does not have a minimum.

viii) Conclude that $f(u)$ is the maximum of f on $[a, b]$.

5.2.8. Let A be a subset of $\mathbb{R}$, and let $f : A \to \mathbb{R}$ be a function with a maximum at $u \in A$. What can you say about the slope of the line between $(a, f(a))$ and $(u, f(u))$ for $a \in A \setminus \{u\}$?

5.3 Intermediate value theorem

In this section, all functions are real-valued. The reason is that we can only make comparisons in an ordered field, and $\mathbb{R}$ is an ordered field (Theorem 3.8.2), and $\mathbb{C}$ is not (Exercise 3.9.6).

Theorem 5.3.1. (Intermediate value theorem) *Let* $a, b \in \mathbb{R}$ *with* $b > a$, *and let* $f : [a, b] \to \mathbb{R}$ *be continuous. Let* k *be a real number strictly between* $f(a)$ *and* $f(b)$. *Then there exists* $c \in (a, b)$ *such that* $f(c) = k$.

Here is a picture that illustrates the Intermediate value theorem: for value k on the y-axis between $f(a)$ and $f(b)$ there happen to be two c for which $f(c) = k$. The Intermediate value theorem guarantees that one such c exists but does not say how many such c exist.

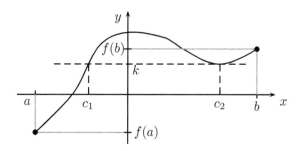

Proof of Theorem 5.3.1: Set $a_0 = a$ and $b_0 = b$. Apply Construction 3.14.1 of halving intervals, with the property P of intervals being that k is strictly between the values of f at the two endpoints. If for any $n > 0$, $f(\frac{a_{n-1}+b_{n-1}}{2}) = k$, then the theorem is proved and we can immediately stop the proof (and the construction).

When k is not equal to $f(\frac{a_{n-1}+b_{n-1}}{2})$, then since k is between $f(a_{n-1})$ and $f(b_{n-1})$, necessarily k is between $f(a_{n-1})$ and $f(\frac{a_{n-1}+b_{n-1}}{2})$ or else between $f(\frac{a_{n-1}+b_{n-1}}{2})$ and $f(b_{n-1})$. Choose that half $[a_n, b_n]$ of $[a_{n-1}, b_{n-1}]$ which says that k is strictly between $f(a_n)$ and $f(b_n)$. Thus for all n, k is strictly between $f(a_n)$ and $f(b_n)$. By construction $c = \sup\{a_1, a_2, a_3, \ldots\} = \inf\{b_1, b_2, b_3, \ldots\}$ is in $[a, b]$. So c is in the domain of f.

We will prove that for all $\epsilon > 0$, $|f(c) - k| < \epsilon$. Let $\epsilon > 0$. Since f is continuous, it is continuous at c, so there exists $\delta > 0$ such that for all $x \in [a, b]$, if $|x - c| < \delta$, then $|f(x) - f(c)| < \epsilon/3$. By Exercise 3.8.3, there exists a positive integer n such that $1/2^n < \delta/(b - a)$. As $a_n \leq c \leq b_n$, we have $|a_n - c| \leq |a_n - b_n| = (b - a)/2^n < \delta$, so that $|f(a_n) - f(c)| < \epsilon/3$. Similarly, $|f(b_n) - f(c)| < \epsilon/3$. Hence by the triangle inequality, $|f(b_n) - f(a_n)| < 2\epsilon/3$. But k is between $f(a_n)$ and $f(b_n)$, and thus both $|f(a_n) - k|$ and $|f(b_n) - k|$ must be less than $2\epsilon/3$. Therefore

$$|f(c) - k| \leq |f(c) - f(a_n)| + |f(a_n) - k| < \epsilon.$$

Since this true for all $\epsilon > 0$, it follows by Theorem 2.10.4 that $f(c) = k$. $\square$

An important application of this theorem is in the next section, introducing the radical functions. (So far we have sporadically used the square roots only, relying on some facts from high school that we have not yet proved.)

Example 5.3.2. There exists a real number c such that $c^5 - 4 = \frac{c^2-2}{c^2+2}$.

Proof. Let $f : \mathbb{R} \to \mathbb{R}$ be defined by $f(x) = x^5 - 4 - \frac{x^2-2}{x^2+2}$. This function is a rational function and defined for all real numbers, so that by Theorem 5.1.3, f is continuous. Note that $f(0) = -3 < 0 < f(2)$, so that by Theorem 5.3.1 there exists c in $(0, 2)$ such that $f(c) = 0$. In other words, $c^5 - 4 = \frac{c^2-2}{c^2+2}$. $\square$

Theorem 5.3.3. *Let I be an interval in $\mathbb{R}$, and let $f : I \to \mathbb{R}$ be continuous. Then the image of f is an interval in $\mathbb{R}$.*

Proof. For any c and d in the image of f, by the Intermediate value theorem (Theorem 5.3.1), any real number between c and d is in the image of f, which proves the theorem. $\square$

However, if $f : A \to B$ is continuous and injective and A is open, then the range of f is open in B. We prove this first for A and B subsets of $\mathbb{R}$, and then for subsets of $\mathbb{C}$.

Compare the next theorem to Example 5.1.6.

Theorem 5.3.4. *Let I be an interval in $\mathbb{R}$, B a subset of $\mathbb{R}$, and let $f : I \to B$ be continuous and invertible. Then f^{-1} is continuous and f, f^{-1} are either both strictly increasing or both strictly decreasing. Furthermore, if I is open, so is B.*

Proof. Let $a < b$ be in I. Since f has an inverse, $f(a) \neq f(b)$, so that by trichotomy, either $f(a) < f(b)$ or $f(a) > f(b)$.

For now we assume that $f(a) < f(b)$. With that we prove that f is an increasing function, i.e., that for any $x, y \in I$, if $x < y$ then $f(x) < f(y)$. First suppose that $x < y < z$ are in I. By invertibility, $f(x), f(y), f(z)$ are distinct. If $f(x)$ is between $f(y)$ and $f(z)$, then an application of the Intermediate value theorem gives $c \in (y, z)$ such that $f(c) = f(x)$. But $x < y < c$, so x and c are distinct, and $f(c) = f(x)$ contradicts invertibility of f. So $f(x)$ is not between $f(y)$ and $f(z)$, and similarly $f(z)$ is not between $f(x)$ and $f(y)$. Thus necessarily $f(x) < f(y) < f(z)$ or $f(x) > f(y) > f(z)$. By setting $x = a$ and $z = b$ we get that f is increasing on $[a, b]$, by setting $y = a$ and $z = b$ we get that f is increasing on $I \cap (-\infty, a]$, and by setting $x = a$ and $y = b$ we get that f is increasing on $I \cap [b, \infty)$. Thus f is increasing on I.

By Theorem 5.3.3 we know that B is an interval. By definition of inverses, f^{-1} is strictly increasing.

If I is in addition open, let $c \in B$ and $a = f^{-1}(c)$. Since I is open and f is increasing, there exist $b_1, b_2 \in A$ such that $b_1 < a < b_2$. Thus $f(b_1) < f(a) = c < f(b_2)$, and by Theorem 5.3.3 we know that $(f(b_1), f(b_2))$ is an open subset of B that contains c. Thus B is open.

Now we prove that f^{-1} is a continuous function at an arbitrary y be in the domain of f^{-1}. Let $\epsilon > 0$. Let $c = f^{-1}(y)$. Let $z \in (c, c + \epsilon) \cap I$. (Of course, if c is the right endpoint of I, there is no such z.) Since f is increasing, $y = f(c) < f(z)$ and we set $\delta_1 = f(z) - y$. Then for any $x \in (y, y + \delta_1) = (f(c), f(z))$, by the Intermediate value theorem x is in the range of f, i.e., x is in the domain of f^{-1}, and since f^{-1} is increasing, $c = f^{-1}(y) < f^{-1}(x) < f^{-1}(f(z)) = z < c + \epsilon$. Thus $|f^{-1}(x) - f^{-1}(y)| < \epsilon$. This proves that for the given $\epsilon > 0$ there exists $\delta_1 > 0$ such that for any $x \in (y, y + \delta_1) \cap I$, $|f^{-1}(x) - f^{-1}(y)| < \epsilon$. Similarly, there exists $\delta_2 > 0$ such that for all $x \in (y - \delta_2, y) \cap I$, $|f^{-1}(y) - f^{-1}(x)| < \epsilon$. Now set $\delta = \min\{\delta_1, \delta_2\}$. By what we just proved, for all x in the domain of f^{-1}, if $|x - y| < \delta$, then $|f^{-1}(x) - f^{-1}(y)| < \epsilon$. Thus f^{-1} is continuous at y, and since y was arbitrary, f^{-1} is continuous.

Finally, suppose that $f(a) > f(b)$ for some $a < b$ in I. Set $g = -f$. Then g is invertible and continuous, $g(a) < g(b)$, and by the work so far we know that g is a strictly increasing function with a continuous strictly increasing inverse g^{-1}. Thus $f = -g$ is a strictly decreasing function, and $f^{-1} = -g^{-1}$ is a continuous strictly decreasing function. In addition, if I is open, so is B. $\square$

Exercises for Section 5.3

5.3.1. Let $f(x) = x^3 + 4$, $g(x) = x^4 + x^2$. Prove that there exists $c \in [-4, 4]$ such that $f(c) = g(c)$.

5.3.2. Let $f : \mathbb{R} \to \mathbb{Q}$ be a continuous function. Prove that f is a constant function.

    i) Prove that any continuous function from $\mathbb{C}$ to $\mathbb{Q}$ is constant.

    ii) Does a continuous function $f : \mathbb{Q} \to \mathbb{R}$ have to be constant?

5.3.3. (Fixed point theorem) Let $f : [a, b] \to [a, b]$ be a continuous function. Prove that there exists $c \in [a, b]$ such that $f(c) = c$.

5.3.4. Find real numbers $a < b$ and a continuous function $f : [a, b] \to [a, b]$ such that $f(c) \neq c$ for all $c \in (a, b)$.

5.3.5. Find real numbers $a < b$ and a continuous function $f : (a, b) \to (a, b)$ such that $f(c) \neq c$ for all $c \in (a, b)$.

5.3.6. The goal of this exercise is to prove that every polynomial of odd degree has a real root. Write $f(X) = a_0 + a_1 X + a_2 X^2 + \cdots + a_n X^n$ for some real numbers $a_0, a_1, a_2, \ldots, a_n$ such that n is odd and $a_n \neq 0$. Set $b = \frac{n}{|a_n|} \max\{|a_0|, |a_1|, \ldots, |a_{n-1}|, |a_n|\}$.

   i) Prove that b is a positive real number. If $f(b) = 0$ or if $f(-b) = 0$, we have found the root. So we may assume that $f(b)$ and $f(-b)$ are not zero.

   ii) Justify all steps below:

$$|a_0 + a_1(\pm b) + a_2(\pm b)^2 + \cdots + a_{n-1}(\pm b)^{n-1}|$$

$$\leq |a_0| + |a_1 b| + |a_2 b^2| + \cdots + |a_{n-1} b^{n-1}|$$

$$\text{(by the triangle inequality)}$$

$$\leq \frac{|a_n| b}{n} + \frac{|a_n| b^2}{n} + \frac{|a_n| b^3}{n} + \cdots + \frac{|a_n| b^n}{n} \quad \text{(by the choice of } b)$$

$$\leq \frac{|a_n| b}{n} \left(1 + b + b^2 + \cdots + b^{n-1}\right)$$

$$\leq \frac{|a_n| b}{n} \left(\underset{0}{b^{n-1}} + \underset{1}{b^{n-1}} + \underset{2}{b^{n-1}} + \cdots + \underset{n-1}{b^{n-1}}\right) \quad \text{(because } b \geq 1) \quad \textit{(place markers)}$$

$$\leq \frac{|a_n| b}{n} n b^{n-1}$$

$$= |a_n| b^n$$

$$= |a_n b^n|.$$

   iii) Justify all steps below:

$$|f(\pm b)| = |a_0 + a_1(\pm b) + a_2(\pm b)^2 + \cdots + a_{n-1}(\pm b)^{n-1} + a_n(\pm b)^n|$$

$$\geq |a_n(\pm b)^n| - |a_0 + a_1(\pm b) + a_2(\pm b)^2 + \cdots + a_{n-1}(\pm b)^{n-1}|$$

$$\geq 0.$$

iv) Prove that $f(b)$ has the same sign (positive or negative) as $a_n b^n$
and that $f(-b)$ has the same sign as $a_n(-b)^n$.

v) Prove that $f(b)$ and $f(-b)$ have opposite signs.

vi) Prove that f has a real root in $(-b, b)$.

5.4   Radical functions

Let n be a positive integer. Define the function $f(x) = x^n$ with
domain $\mathbb{R}$ if n is odd and domain $\mathbb{R}_{\geq 0}$ otherwise. In this section we re-
prove Theorem 3.8.10 that the nth radical function exists, and more.

Theorem 5.4.1. *The range of f is $\mathbb{R}$ if n is odd and is $\mathbb{R}_{\geq 0}$ otherwise.*

Proof. Certainly $0 = 0^n$ is in the range. If $a > 0$, then

$$0^n = 0 < a < a + 1 \leq (a + 1)^n.$$

The last inequality is by Exercise 2.8.1. Since exponentiation by n is a
polynomial function, it is continuous, and so by the Intermediate value
theorem (Theorem 5.3.1), there exists $r \in (0, a+1)$ such that $a = r^n$. Thus
every non-negative real number is in the range of f. If $a < 0$ and n is odd,
then similarly

$$(a - 1)^n \leq a - 1 < a < 0 = 0^n.$$

The Intermediate value theorem guarantees that there exists $r \in (a - 1, 0)$
such that $a = r^n$. So for odd n all real numbers are in the range. If
n is even, then $n/2$ is an integer and for any $x \in \mathbb{R}$, $x^n = (x^2)^{n/2}$ is a
positive-integer power of x^2. By Theorem 2.7.12, x^2 is positive, and so by
Theorem 2.9.2, $x^n \in \mathbb{R}^+ \cup \{0\}$. $\square$

We now re-prove Theorem 3.8.10 that the nth radical function exists when n is a positive integer. Let f be as in the theorem above. By
Theorem 2.9.2 and Exercise 2.9.1, f is strictly increasing. Thus by Theorem 2.9.4, f has an inverse function f^{-1}. By the power rule, f is continuous,
so that by Theorem 5.3.4, f^{-1} is strictly increasing and continuous. We call
this inverse function the **nth radical function**. For any a in its domain
we write $f^{-1}(a) = \sqrt[n]{a}$ or as $f^{-1}(a) = a^{1/n}$, and we call this value the **nth
root of a**.

We just established that for any positive integer n, $\sqrt[n]{\ }$ is defined on $\mathbb{R}_{\geq 0}$, or even on $\mathbb{R}$ if n is odd. We also just established that $\sqrt[n]{\ }$ is a strictly increasing continuous function. By continuity we immediately get the following:

Theorem 5.4.2. (Radical rule for limits) *For any positive integer n and any a in the domain of $\sqrt[n]{\ }$,*

$$\lim_{x \to a} \sqrt[n]{x} = \sqrt[n]{a}. \qquad \square$$

Now let m and n be positive integers. For any $x \in \mathbb{R}^+ \cup \{0\}$, the elements $\sqrt[n]{x^m}$ and $(\sqrt[n]{x})^m$ are well-defined in $\mathbb{R}^+ \cup \{0\}$. Their nth powers are the same value x^m:

$$(\sqrt[n]{x^m})^n = x^m$$

by definition, and

$$((\sqrt[n]{x})^m)^n = (\sqrt[n]{x})^{mn} = ((\sqrt[n]{x})^n)^m = x^m.$$

Thus by uniqueness established in Corollary 2.9.3,

$$\sqrt[n]{x^m} = (\sqrt[n]{x})^m.$$

We record this function as $\sqrt[n/m]{\ } = (\ )^{m/n}$, and call it the **exponentiation** by m/n. This shows that we can handle rational exponents as well.

Theorem 5.4.3. (Exponentiation by rational exponents is continuous.) *Let $r \in \mathbb{Q}$. If $r \geq 0$ let $A = \mathbb{R}_{\geq 0}$ and if $r < 0$ let $A = \mathbb{R}^+$. Let $f : A \to A$ be defined by $f(x) = x^r$. Then f is continuous.*

(1) If $r > 0$, then f is strictly increasing.

(2) If $r < 0$, then f is strictly decreasing.

(3) If $r \neq 0$ then f is invertible and its inverse is exponentiation by $1/r$.

(4) If $r = 0$, then f is a constant function.

Proof. Write $r = m/n$, where n and m are integers and $n \neq 0$. Since $m/n = (-m)/(-n)$, by possibly multiplying by -1 we may assume that $n > 0$. Then f is a composition of exponentiation by m with exponentiation by $1/n$, in either order. Exponentiation by non-negative m is continuous by the constant or power rule, exponentiation by negative m is continuous by the quotient rule, and exponentiation by $1/n$ is continuous by Theorem 5.4.2. Thus f is continuous by the composite rule.

If $r > 0$, then $m, n > 0$, and then f is the composition of two strictly increasing functions, hence strictly increasing. If $r < 0$, then $m < 0$ and $n > 0$, so f is the composition of a strictly increasing and a strictly decreasing function, hence strictly decreasing by Exercise 2.9.7.

For any non-zero integer p, let $g_p, h_p : A \to A$ be defined as $g_p(x) = x^p$ and $h_p(x) = x^{1/p}$. By associativity of function composition (Exercise 2.4.6),

$$
\begin{aligned}
(f(a))^{n/m} &= h_m \circ g_n(f(a)) \\
&= h_m \circ g_n(h_n \circ g_m(a)) \\
&= (h_m \circ (g_n \circ h_n))(g_m(a)) \\
&= h_m \circ g_m(a) \\
&= a,
\end{aligned}
$$

and similarly $f(a^{n/m}) = a$. This proves that exponentiation by $n/m = 1/r$ is the inverse of f. The last part is obvious.   □

Theorem 7.6.4 shows more generally that exponentiation by arbitrary real numbers (not just by rational numbers) is continuous.

Exercises for Section 5.4

5.4.1. Determine the following limits, and justify all steps by invoking the relevant theorems/rules:

   i) $\lim\limits_{x \to 2} \sqrt{x^2 - 3x + 4}$.

   ii) $\lim\limits_{x \to 2} \dfrac{\sqrt{x} - \sqrt{2}}{x^2 + 4}$.

   iii) $\lim\limits_{x \to 2} \dfrac{x - 2}{x^2 - 4}$.

   iv) $\lim\limits_{x \to 2} \dfrac{\sqrt{x} - \sqrt{2}}{x^2 - 4}$.

5.4.2. Let $f(x) = \sqrt{x} - \sqrt{-x}$.

   i) Prove that for all a in the domain of f, $\lim_{x \to a} f(x)$ is inapplicable.

   ii) Prove that f is continuous.

5.4.3. Determine the domain of the function f given by $f(x) = \sqrt{-x^2}$.

5.4.4. Here is an alternate proof of Theorem 5.4.2. Study the proof, and provide any missing commentary. Let A be the domain of $\sqrt[n]{}$.

  i) Prove that an element of A is a limit point of A.

  ii) Suppose that $a = 0$. Set $\delta = \epsilon^n$. Let $x \in A$ satisfy $0 < |x - a| < \delta$. Since the nth root function is increasing, it follows that

$$\left| \sqrt[n]{x} - \sqrt[n]{a} \right| = \left| \sqrt[n]{x} \right| = \sqrt[n]{|x|} < \sqrt[n]{\delta} = \epsilon.$$

  iii) Suppose that $a > 0$. First let $\epsilon' = \min\{\epsilon, \sqrt[n]{a}\}$. So ϵ' is a positive number. Set $\delta = \min\{(\epsilon + \sqrt[n]{a})^n - a, a - (\sqrt[n]{a} - \epsilon')^n\}$. Note that

$$(\epsilon + \sqrt[n]{a})^n - a > (\sqrt[n]{a})^n - a = 0,$$

$$0 \le \sqrt[n]{a} - \epsilon' < \sqrt[n]{a},$$

so that δ is positive. Let $x \in A$ satisfy $0 < |x - a| < \delta$. Then $-\delta < x - a < \delta$ and $a - \delta < x < \delta + a$. Since $\delta = \min\{(\epsilon + \sqrt[n]{a})^n - a, a - (\sqrt[n]{a} - \epsilon')^n\}$, it follows that

$$a - \left(a - (\sqrt[n]{a} - \epsilon')^n\right) \le a - \delta < x < a + \delta \le a + \left((\epsilon + \sqrt[n]{a})^n - a\right),$$

or in other words, $(\sqrt[n]{a} - \epsilon')^n < x < (\epsilon + \sqrt[n]{a})^n$. Since the nth radical function is increasing on $\mathbb{R}^+$, it follows that

$$\sqrt[n]{a} - \epsilon \le \sqrt[n]{a} - \epsilon' < \sqrt[n]{x} < \epsilon + \sqrt[n]{a},$$

whence $-\epsilon < \sqrt[n]{x} - \sqrt[n]{a} < \epsilon$, so that $\left| \sqrt[n]{x} - \sqrt[n]{a} \right| < \epsilon$, as desired.

  iv) Assume that a is negative. Then necessarily n is an odd integer. By what we have proved for positive numbers in the domain, such as for $-a$, there exists $\delta > 0$ such that for all $x \in \mathbb{R}$, if $0 < |(-x) - (-a)| < \delta$, then $\left| \sqrt[n]{-x} - \sqrt[n]{-a} \right| < \epsilon$. But then for $x \in \mathbb{R}$ with $0 < |x - a| < \delta$, since n is odd,

$$\left| \sqrt[n]{x} - \sqrt[n]{a} \right| = \left|(-1)\left(\sqrt[n]{x} - \sqrt[n]{a}\right)\right| = \left| \sqrt[n]{-x} - \sqrt[n]{-a} \right| < \epsilon.$$

5.4.5. Let $n, m \in \mathbb{Q}$, and suppose that a and b are in the domain of exponentiation by n and m. Prove:

  i) $a^n \cdot b^n = (ab)^n$.

  ii) $(a^n)^m = a^{mn}$.

  iii) $a^n \cdot a^m = a^{n+m}$.

  iv) If $a \neq 0$, then $a^{-n} = 1/a^n$.

5.4.6. Recall from Exercise 3.12.5 that for every non-zero complex number a there exist exactly two complex numbers whose squares are a. Let's try to create a continuous square root function $f : \mathbb{C} \to \mathbb{C}$. (We will fail.)

    i) Say that for all a in the first quadrant we choose $f(a)$ in the first quadrant. Where are then $f(a)$ for a in the remaining quadrants?

    ii) Is it possible to extend this square root function to a function f on all of $\mathbb{C}$ (the positive and negative real and imaginary axes) in such a way as to make $\lim_{x \to a} f(x) = f(a)$ for all $a \in \mathbb{C}$?

    iii) Explain away the problematic claims $\sqrt{-4} \cdot \sqrt{-9} = \sqrt{(-4)(-9)} = \sqrt{36} = 6$, $\sqrt{-4} \cdot \sqrt{-9} = 2i \cdot 3i = -6$ from page 11.

    iv) Let D be the set of all complex numbers that are not on the negative real axis. Prove that we can define a continuous square root function $f : D \to \mathbb{C}$.

    v)* Let θ be any real number, and let D be the set of all complex numbers whose counterclockwise angle from the positive real axis is θ. Prove that we can define a continuous square root function $f : D \to \mathbb{C}$.

5.4.7. (The goal of this exercise and the next one is to develop exponential functions without derivatives and integrals. We will see in Section 7.4 that derivatives and integrals give a more elegant approach.) Let $c \in (1, \infty)$ and let $f : \mathbb{Q} \to \mathbb{R}^+$ be the function $f(x) = c^x$.

    i) Why is f a function? (Is it well-defined, i.e., are we allowed to raise positive real numbers to rational exponents?)

    ii) Prove that f is strictly increasing. (Hint: Theorem 5.4.3.)

    iii) Let $\epsilon > 0$. Justify the following:

$$(\epsilon + 1)^{n+1} - 1 = \sum_{i=1}^{n+1} \left((\epsilon + 1)^i - (\epsilon + 1)^{i-1} \right)$$

$$= \sum_{i=1}^{n+1} \epsilon(\epsilon + 1)^{i-1}$$

$$\geq \sum_{i=1}^{n+1} \epsilon$$

$$= (n+1)\epsilon.$$

Use the Archimedean property of $\mathbb{R}$ (Theorem 3.8.3) to prove that the set $\{(\epsilon + 1)^n : n \in \mathbb{N}_0\}$ is not bounded above.

iv) Prove that there exists a positive integer n such that $c^{1/n} < \epsilon + 1$.

v) Prove that there exists $\delta_1 > 0$ such that for all $x \in (0, \delta_1) \cap \mathbb{Q}$, $|c^x - 1| < \epsilon$.

vi) Prove that there exists $\delta_2 > 0$ such that for all $x \in (-\delta_2, 0) \cap \mathbb{Q}$, $|c^x - 1| < \epsilon$.

vii) Prove that $\lim_{x \to 0, x \in \mathbb{Q}} c^x = 1$.

viii) Prove that for any $r \in \mathbb{R}$, $\lim_{x \to r, x \in \mathbb{Q}} c^x$ exists and is a real number.

5.4.8. (Related to the previous exercise.) Let $c \in \mathbb{R}^+$.

i) Prove that for any $r \in \mathbb{R}$, $\lim_{x \to r, x \in \mathbb{Q}} c^x$ exists and is a real number. (Hint: Case $c = 1$ is special; case $c > 1$ done; relate the case $c < 1$ to the case $c > 1$ and the quotient rule for limits.)

ii) We denote the limit in the previous part with c^r. Prove that for all real numbers c, c_1, c_2, r, r_1, r_2, with $c, c_1, c_2 > 0$,

$$c^{-r} = \frac{1}{c^r}, (c_1 c_2)^r = c_1^r c_2^r, c^{r_1 + r_2} = c^{r_1} c^{r_2}, c^{r_1 r_2} = (c^{r_1})^{r_2}.$$

iii) Prove that the function $g : \mathbb{R} \to \mathbb{R}$ given by $g(x) = c^x$ is continuous. (This is easy.)

5.5 Uniform continuity

Definition 5.5.1. *A function f is **uniformly continuous** if for all real numbers $\epsilon > 0$ there exists a real number $\delta > 0$ such that for all x and y in the domain, if $|x - y| < \delta$, then $|f(x) - f(y)| < \epsilon$.*

For example, constant functions are uniformly continuous.

In uniform continuity, given a real number $\epsilon > 0$, there exists $\delta > 0$ that depends only on ϵ that makes some conclusion true, whereas in the definition of continuity at a, given a real number $\epsilon > 0$, there exists $\delta > 0$ that depends on ϵ and on a for the same conclusion to be true (with $y = a$). Thus the following is immediate:

Theorem 5.5.2. *Every uniformly continuous function is continuous.*

The converse is false in general:

Example 5.5.3. Let $f(x) = x^2$, with domain $\mathbb{C}$. Since f is a polynomial function, it is continuous. Suppose that f is also uniformly continuous. Then in particular for $\epsilon = 1$ there exists $\delta > 0$ such that for all $x, y \in A$, if $|x - y| < \delta$ then $|f(x) - f(y)| < 1$. Set $x = \frac{1}{\delta}$ and $y = x + \frac{\delta}{2}$. Then $|x - y| = \delta/2 < \delta$ and

$$|f(x) - f(y)| = |x^2 - y^2| = \left| \delta x + \frac{\delta^2}{4} \right| = 1 + \frac{\delta^2}{4} > 1 = \epsilon.$$

This proves the negation of the definition of uniform continuity.   □

The converse of Theorem 5.5.2 holds with some extra hypotheses.

Theorem 5.5.4. *Let A, B be closed and bounded subsets of $\mathbb{R}$ or $\mathbb{C}$. Let $f : A \to B$ be continuous. Then f is uniformly continuous.*

Proof. Let $\epsilon > 0$. Since f is continuous, for each $a \in A$ there exists $\delta_a > 0$ such that for all $x \in A$, $|x - a| < \delta_a$ implies that $|f(x) - f(a)| < \epsilon/2$. Note that $A \subseteq \cup_{a \in A} B(a, \delta_a)$. By Theorem 3.14.5 there exists $\delta > 0$ such that for all $x \in A$ there exists $a \in A$ such that $B(x, \delta) \subseteq B(a, \delta_a)$. Let $x, y \in A$ with $|x - y| < \delta$. Since $x, y \in B(x, \delta) \subseteq B(a, \delta_a)$, it follows that $|x - a|, |y - a| < \delta_a$. Thus

$$
\begin{aligned}
|f(x) - f(y)| &= |f(x) - f(a) + f(a) - f(y)| \\
&\leq |f(x) - f(a)| + |f(a) - f(y)| \text{ (by the triangle inequality)} \\
&< \epsilon/2 + \epsilon/2 = \epsilon.
\end{aligned}
$$
     □

Example 5.5.5. The continuous function $\sqrt{}$ is uniformly continuous.

Proof. We established in Section 5.4 that $\sqrt{}$ is continuous. Let $\epsilon > 0$. We divide the domain into two regions, one closed and bounded so we can invoke the theorem above, and the other unbounded but on which $\sqrt{}$ does not increase very fast.

The first region is the closed and bounded interval $[0, 2]$. By the previous theorem there exists $\delta_1 > 0$ such that for all $a, x \in [0, 2]$, if $|x - a| < \delta_1$ then $|\sqrt{x} - \sqrt{a}| < \epsilon$.

The second region is the unbounded interval $[1, \infty)$. For $a, x \in [1, \infty)$ with $|x - a| < \epsilon$ we have

$$|\sqrt{x} - \sqrt{a}| = \left| (\sqrt{x} - \sqrt{a}) \frac{\sqrt{x} + \sqrt{a}}{\sqrt{x} + \sqrt{a}} \right|$$

$$= \left| \frac{x - a}{\sqrt{x} + \sqrt{a}} \right|$$

$$\leq \frac{|x - a|}{2} \text{ (because } \sqrt{x}, \sqrt{a} \geq 1)$$

$$< \epsilon.$$

Finally, set $\delta = \min\{\delta_1, \epsilon, 1\}$. Let a and x be in the domain of $\sqrt{}$ such that $|x - a| < \delta$. Since $|x - a| < \delta \leq 1$, necessarily either $x, a \in [0, 2]$ or $x, a \leq [1, \infty)$. We have analyzed both cases, and we conclude that $|\sqrt{x} - \sqrt{a}| < \epsilon$. $\qquad\square$

Theorem 5.5.6. *Let $f : A \to B$ be uniformly continuous and C a subset of A. Let $g : C \to B$ be defined as $g(x) = f(x)$. Then g is uniformly continuous.*

Proof. Let $\epsilon > 0$. Since f is uniformly continuous, there exists $\delta > 0$ such that for all $x, y \in A$, if $|x - y| < \delta$ then $|f(x) - f(y)| < \epsilon$. But then for any $x, y \in C$, if $|x - y| < \delta$, then $|g(x) - g(y)| = |f(x) - f(y)| < \epsilon$. $\qquad\square$

Example 5.5.7. By Example 5.5.3 the squaring function is not uniformly continuous on $\mathbb{C}$ or $\mathbb{R}$, but when the domain is restricted to any bounded subset of $\mathbb{C}$, that domain is a subset of a closed and bounded subset of $\mathbb{C}$, and so since squaring is continuous, it follows by the previous theorem that squaring is uniformly continuous on the closed and bounded set and hence on any subset of that.

Exercises for Section 5.5

5.5.1. Let $f : \mathbb{C} \to \mathbb{C}$ be given by $f(x) = mx + l$ for some constants m, l. Prove that f is uniformly continuous.

5.5.2. Which of the following functions are uniformly continuous? Justify your answers.

    i) $f : B(0, 1) \to \mathbb{C}$, $f(x) = x^2$.

    ii) $f : (0, 1] \to \mathbb{R}$, $f(x) = 1/x$.

    iii) $f : \mathbb{R} \to \mathbb{R}$, $f(x) = \frac{1}{x^2 + 1}$.

    iv) $f : \mathbb{R} \setminus \{0\} \to \mathbb{R}$, $f(x) = \frac{x}{|x|}$.

    v) $f : \mathbb{R} \to \mathbb{R}$, $f(x) = \begin{cases} \frac{x}{|x|}, & \text{if } x \neq 0; \\ 0, & \text{otherwise.} \end{cases}$

vi) $f : \mathbb{R} \to \mathbb{R}$, $f(x) = \begin{cases} 1, & \text{if } x \text{ is rational;} \\ 0, & \text{otherwise.} \end{cases}$

vii) $\mathrm{Re}, \mathrm{Im} : \mathbb{C} \to \mathbb{R}$.

viii) The absolute value function from $\mathbb{C}$ to $\mathbb{R}$.

5.5.3. Suppose that $f, g : A \to \mathbb{C}$ are uniformly continuous and that $c \in \mathbb{C}$.

    i) Prove that cf and $f \pm g$ are uniformly continuous.

    ii) Is $f \cdot g$ uniformly continuous? Prove or give a counterexample.

5.5.4. Let $f : (a, b) \to \mathbb{C}$ be uniformly continuous. Prove that there exists a continuous function $g : [a, b] \to \mathbb{C}$ such that $f(x) = g(x)$ for all $x \in (a, b)$.

5.5.5. Is the composition of two uniformly continuous functions uniformly continuous? Prove or give a counterexample.

5.5.6. Let $f : \mathbb{C} \to \mathbb{C}$ be defined by $f(x) = x^3$.

    i) Prove that f is continuous but not uniformly continuous.

    ii) Find a uniformly continuous function $g : \mathbb{C} \to \mathbb{C}$ such that $g \circ f$ is uniformly continuous.

    iii) Find a uniformly continuous function $g : \mathbb{C} \to \mathbb{C}$ such that $g \circ f$ is not uniformly continuous.

Chapter 6

Differentiation

The geometric motivation for differentiation comes from lines tangent to a graph of a function f at a point $(a, f(a))$. For example, on the graph below are two gray secant lines through $(a, f(a))$:

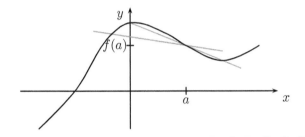

It appears that the line through $(a, f(a))$ and $(x, f(x))$ is closer to the tangent line to the graph of f at $(a, f(a))$ if x is closer to a. Intuitively, the slope of the tangent line is the limit of the slopes of the secant lines.

6.1   Definition of derivatives

Definition 6.1.1. *Let $A \subseteq \mathbb{C}$, and let $a \in A$ be a limit point of A. A function $f : A \to \mathbb{C}$ is* **differentiable** *at a if*

$$\lim_{x \to a} \frac{f(x) - f(a)}{x - a}$$

exists. In this case, we call this limit **the derivative of f at a**, *and we use either Newton's notation $f'(a) = (f(x))'|_{x=a}$ or Leibniz's notation $\frac{df}{dx}(a) = \frac{df(x)}{dx}\big|_{x=a}$.*

A function is **differentiable** *if it is differentiable at all points in its domain.*

An alternative way of computing the derivatives is

$$f'(a) = \frac{df(x)}{dx}\bigg|_{x=a} = \lim_{h\to 0}\frac{f(a+h)-f(a)}{h},$$

as this is simply a matter of writing x as $a+h$, and using that $a+h = x \to a$ if and only if $h \to 0$. (See Exercise 5.1.6.)

Example 6.1.2. Let $f(x) = mx + l$, where m and l are complex numbers. Then for any $a \in \mathbb{C}$,

$$
\begin{aligned}
f'(a) &= \lim_{x\to a}\frac{f(x)-f(a)}{x-a}\\
&= \lim_{x\to a}\frac{(mx+l)-(ma+l)}{x-a}\\
&= \lim_{x\to a}\frac{m(x-a)}{x-a}\\
&= m.
\end{aligned}
$$

Alternatively,

$$
\begin{aligned}
f'(a) &= \lim_{h\to 0}\frac{f(a+h)-f(a)}{h}\\
&= \lim_{h\to 0}\frac{(m(a+h)+l)-(ma+l)}{h}\\
&= \lim_{h\to 0}\frac{mh}{h}\\
&= m.
\end{aligned}
$$

Example 6.1.3. Let $f(x) = x^2$. Then

$$
\begin{aligned}
f'(a) &= \lim_{x\to a}\frac{f(x)-f(a)}{x-a}\\
&= \lim_{x\to a}\frac{x^2-a^2}{x-a}\\
&= \lim_{x\to a}\frac{(x-a)(x+a)}{x-a}\\
&= \lim_{x\to a}(x+a)\\
&= 2a.
\end{aligned}
$$

Alternatively,

$$f'(a) = \lim_{h\to 0}\frac{f(a+h)-f(a)}{h}$$

$$= \lim_{h \to 0} \frac{(a+h)^2 - a^2}{h}$$

$$= \lim_{h \to 0} \frac{a^2 + 2ah + h^2 - a^2}{h}$$

$$= \lim_{h \to 0} \frac{2ah + h^2}{h}$$

$$= \lim_{h \to 0} \frac{(2a+h)h}{h}$$

$$= \lim_{h \to 0} (2a + h)$$

$$= 2a.$$

From now on, we mostly use the alternative notation.

Example 6.1.4. Let $f(x) = 1/x$. Then

$$f'(a) = \lim_{h \to 0} \frac{f(a+h) - f(a)}{h}$$

$$= \lim_{h \to 0} \frac{\frac{1}{a+h} - \frac{1}{a}}{h}$$

$$= \lim_{h \to 0} \frac{\frac{a}{a(a+h)} - \frac{a+h}{a(a+h)}}{h} \quad \text{(common denominator in the fractions)}$$

$$= \lim_{h \to 0} \frac{\frac{a-a-h}{a(a+h)}}{h}$$

$$= \lim_{h \to 0} \frac{-h}{a(a+h)h}$$

$$= \lim_{h \to 0} \frac{-1}{a(a+h)}$$

$$= \frac{-1}{a^2}$$

by the quotient rule for limits.

Example 6.1.5. Let $f(x) = \sqrt{x}$. The domain of f is $\mathbb{R}_{\geq 0}$, and for all $a > 0$,

$$f'(a) = \lim_{h \to 0} \frac{f(a+h) - f(a)}{h}$$

$$= \lim_{h \to 0} \frac{\sqrt{a+h} - \sqrt{a}}{h}$$

$$= \lim_{h \to 0} \frac{\sqrt{a+h} - \sqrt{a}}{h} \cdot \frac{\sqrt{a+h} + \sqrt{a}}{\sqrt{a+h} + \sqrt{a}}$$

$$= \lim_{h \to 0} \frac{(a+h)-a}{h(\sqrt{a+h}+\sqrt{a})} \quad \text{(since } (x-y)(x+y)=x^2-y^2\text{)}$$

$$= \lim_{h \to 0} \frac{h}{h(\sqrt{a+h}+\sqrt{a})}$$

$$= \lim_{h \to 0} \frac{1}{\sqrt{a+h}+\sqrt{a}}$$

$$= \frac{1}{2\sqrt{a}}$$

by the linear, radical, composite, and quotient rules for limits. It is left to Exercise 6.1.4 to show that f is not differentiable at 0.

Example 6.1.6. Let $f(x) = x^{3/2}$. The domain of f is $\mathbb{R}_{\geq 0}$, and for all $a \geq 0$,

$$f'(a) = \lim_{h \to 0} \frac{f(a+h)-f(a)}{h}$$

$$= \lim_{h \to 0} \frac{(a+h)^{3/2}-a^{3/2}}{h}$$

$$= \lim_{h \to 0} \frac{(a+h)^{3/2}-a^{3/2}}{h} \cdot \frac{(a+h)^{3/2}+a^{3/2}}{(a+h)^{3/2}+a^{3/2}}$$

$$= \lim_{h \to 0} \frac{(a+h)^3-a^3}{h((a+h)^{3/2}+a^{3/2})} \quad \text{(since } (x-y)(x+y)=x^2-y^2\text{)}$$

$$= \lim_{h \to 0} \frac{a^3+3a^2h+3ah^2+h^3-a^3}{h((a+h)^{3/2}+a^{3/2})}$$

$$= \lim_{h \to 0} \frac{(3a^2+3ah+h^2)h}{h((a+h)^{3/2}+a^{3/2})}$$

$$= \lim_{h \to 0} \frac{3a^2+3ah+h^2}{(a+h)^{3/2}+a^{3/2}}$$

$$= \begin{cases} \frac{3a^2}{a^{3/2}+a^{3/2}}, & \text{if } a > 0; \\ \lim_{h \to 0} \frac{h^2}{h^{3/2}} = \lim_{h \to 0} h^{1/2} = 0, & \text{if } a = 0; \end{cases}$$

$$= \frac{3}{2}a^{1/2}$$

by the linear, radical, composite, and quotient rules for limits. (Note that this f is differentiable even at 0, whereas the square root function (previous example) is not differentiable at 0.)

Note that in all these examples, f' is a function from some subset of the domain of f to a subset of $\mathbb{C}$, and we can compute f' at a number labeled x rather than a:

$$f'(x) = \frac{df(x)}{dx} = \lim_{h \to 0} \frac{f(x+h) - f(x)}{h} = \lim_{z \to x} \frac{f(z) - f(x)}{z - x}.$$

The h-limit is perhaps preferable to the last limit, where it is z that varies and gets closer and closer to x.

Example 6.1.7. The absolute value function is not differentiable at 0.

Proof. This function is not differentiable whether the domain is $\mathbb{C}$ or $\mathbb{R}$. The reason is that the limit of $\frac{|h| - |0|}{h}$ as h goes to 0 does not exist. Namely, if h varies over positive real numbers, this limit is 1, and if h varies over negative real numbers, the limit is -1, so that the limit indeed does not exist, and hence the derivative does not exist. □

This gives an example of a continuous function that is not differentiable. (Any continuous function with a jagged graph is not differentiable.)

Exercises for Section 6.1

6.1.1. Prove that $f : \mathbb{C} \to \mathbb{C}$ given by $f(x) = x^3$ is differentiable everywhere, and compute the derivative function.

6.1.2. Prove that $f : \mathbb{C} \setminus \{0\} \to \mathbb{C}$ given by $f(x) = 1/x^2$ is differentiable everywhere, and compute the derivative function.

6.1.3. Prove that $f : (0, \infty) \to \mathbb{R}$ given by $f(x) = 1/\sqrt{x}$ is differentiable everywhere, and compute the derivative function.

6.1.4. Prove that the square root function is not differentiable at 0.

6.1.5. Prove that the function $f : [0, \infty) \to \mathbb{R}$ given by $f(x) = x^{3/2}$ is differentiable, including at 0.

6.1.6. Let $f : \mathbb{R} \to \mathbb{R}$ be given by

$$f(x) = \begin{cases} x^2 - 1, & \text{if } x > 1; \\ x^3 - x, & \text{if } x \leq 1. \end{cases}$$

Prove that f is differentiable at 1.

6.1.7. Let $f : \mathbb{R} \to \mathbb{R}$ be given by

$$f(x) = \begin{cases} x^2 - 1, & \text{if } x > 1; \\ x^4 - 4x, & \text{if } x \leq 1. \end{cases}$$

Prove that f is continuous but not differentiable at 1.

6.1.8. Let $f : \mathbb{R} \to \mathbb{R}$ be given by

$$f(x) = \begin{cases} x^2, & \text{if } x > 1; \\ x^3 - x, & \text{if } x \leq 1. \end{cases}$$

Prove that f is not continuous and not differentiable at 1.

6.1.9. Determine if the following function is differentiable at 1: $f(x) = \begin{cases} x^2, & \text{if } x > 1; \\ 2x, & \text{if } x \leq 1. \end{cases}$

6.1.10. Let $f, g : \mathbb{R} \to \mathbb{R}$ be $f(x) = \begin{cases} \frac{1}{2}x^2, & \text{if } x \geq 0; \\ -\frac{1}{2}x^2, & \text{if } x < 0, \end{cases}$ and $g(x) = \begin{cases} \frac{1}{2}x^2 + 3, & \text{if } x \geq 0; \\ -\frac{1}{2}x^2, & \text{if } x < 0. \end{cases}$

  i) Prove that f is differentiable everywhere and that for all $x \in \mathbb{R}$, $f'(x) = |x|$.

  ii) Prove that g is not differentiable at 0.

  iii) Prove that g is differentiable at any non-zero real number x with $g'(x) = |x|$.

6.1.11. Let $f : (0, 1) \to \mathbb{R}$ be the square root function. Verify that f is differentiable, bounded, even uniformly continuous, and that f' is not bounded. (Hint: Example 6.1.5.)

6.2   Basic properties of derivatives

Theorem 6.2.1. *If f is differentiable at a, then f is continuous at a.*

Proof. By definition of differentiability, a is a limit point of the domain of f and a is in the domain of f. Furthermore, since $\lim_{h \to 0} \frac{f(a+h)-f(a)}{h}$ exists, by the product rule for limits also $\lim_{h \to 0}(h \frac{f(a+h)-f(a)}{h})$ exists and equals $0 \cdot f'(a) = 0$. In other words, $\lim_{h \to 0}(f(a + h) - f(a)) = 0$, so that by the sum rule for limits, $\lim_{h \to 0} f(a + h) = \lim_{h \to 0}(f(a + h) - f(a) + f(a)) = 0 + f(a) = f(a)$. Thus by Exercise 5.1.6, $\lim_{x \to a} f(x) = f(a)$, so that f is continuous at a. $\qquad \square$

Theorem 6.2.2. **(Basic properties of derivatives)**
  (1) *(Constant rule) Constant functions are differentiable (at all real/ complex numbers) and the derivative is 0 everywhere.*
  (2) *(Linear rule) The function $f(x) = x$ is differentiable (at all real/complex numbers) and the derivative is 1 everywhere.*

Let A be a subset of $\mathbb{C}$ and let $a \in A$ be a limit point of A. Suppose that $f, g : A \to \mathbb{C}$ are differentiable at a. Then
  (3) *(Scalar rule) For any $c \in \mathbb{C}$, cf is differentiable at a and $(cf)'(a) = cf'(a)$.*
  (4) *(Sum/difference rule) $f \pm g$ is differentiable at a and $(f \pm g)'(a) = f'(a) \pm g'(a)$.*
  (5) *(Product rule) $f \cdot g$ is differentiable at a and $(f \cdot g)'(a) = f'(a)g(a) + f(a)g'(a)$.*
  (6) *(Quotient rule) If $g(a) \neq 0$, then f/g is differentiable at a and $(f/g)'(a) = \frac{f'(a)g(a) - f(a)g'(a)}{(g(a))^2}$.*

Proof. Parts (1) and (2) were already proved in part (1) of Example 6.1.2.
  (3) follows from

$$\lim_{h \to 0} \frac{(cf)(a+h) - (cf)(h)}{h} = \lim_{h \to 0} \frac{cf(a+h) - cf(h)}{h}$$
$$= c \lim_{h \to 0} \frac{f(a+h) - f(h)}{h}$$
$$= cf'(a),$$

and (4) follows from the sum rule for limits and from

$$\lim_{h \to 0} \frac{(f+g)(a+h) - (f+g)(h)}{h}$$
$$= \lim_{h \to 0} \frac{f(a+h) + g(a+h) - f(h) - g(h)}{h}$$
$$= \lim_{h \to 0} \frac{f(a+h) - f(h) + g(a+h) - g(h)}{h}$$
$$= \lim_{h \to 0} \left(\frac{f(a+h) - f(h)}{h} + \frac{g(a+h) - g(h)}{h} \right)$$
$$= f'(a) + g'(a).$$

The following proves the product rule (5):

$$\lim_{h \to 0} \frac{(f \cdot g)(a+h) - (f \cdot g)(a)}{h} = \lim_{h \to 0} \frac{f(a+h) \cdot g(a+h) - f(a) \cdot g(a)}{h}$$

$$= \lim_{h \to 0} \frac{f(a+h)g(a+h) - f(a)g(a+h) + f(a)g(a+h) - f(a)g(a)}{h}$$

(by addition of a clever 0)

$$= \lim_{h \to 0} \left(\frac{(f(a+h) - f(a))g(a+h) + f(a)(g(a+h) - g(a))}{h} \right)$$

(by re-writing)

$$= \lim_{h \to 0} \left(\frac{f(a+h) - f(a)}{h}g(a+h) + f(a)\frac{g(a+h) - g(a)}{h} \right)$$

$$= f'(a)g(a) + f(a)g'(a),$$

where in the last step we used that f and g are differentiable at a and that g is continuous at a (by Theorem 6.2.1).

The proof of the quotient rule (6) is similar. It is written out next without a commentary; the reader is encouraged to supply justification for each step:

$$\lim_{h \to 0} \frac{\frac{f}{g}(a+h) - \frac{f}{g}(a)}{h} = \lim_{h \to 0} \frac{\frac{f(a+h)}{g(a+h)} - \frac{f(a)}{g(a)}}{h}$$

$$= \lim_{h \to 0} \frac{f(a+h)g(a) - f(a)g(a+h)}{hg(a+h)g(a)}$$

$$= \lim_{h \to 0} \frac{f(a+h)g(a) - f(a)g(a) + f(a)g(a) - f(a)g(a+h)}{hg(a+h)g(a)}$$

$$= \lim_{h \to 0} \frac{f(a+h)g(a+h) - f(a)g(a+h) + f(a)g(a) - f(a)g(a+h)}{hg(a+h)g(a)}$$

$$= \lim_{h \to 0} \left(\frac{f(a+h)g(a+h) - f(a)g(a+h)}{hg(a+h)g(a)} - \frac{f(a)g(a+h) - f(a)g(a)}{hg(a+h)g(a)} \right)$$

$$= \lim_{h \to 0} \left(\frac{f(a+h) - f(a)}{h} \cdot \frac{g(a+h)}{g(a+h)g(a)} \right.$$
$$\left. - \frac{f(a)}{g(a+h)g(a)} \cdot \frac{g(a+h) - g(a)}{h} \right)$$

$$= \frac{f'(a)g(a) - f(a)g'(a)}{(g(a))^2}. \qquad \square$$

Theorem 6.2.3. (Power rule) *If n is a positive integer, then $(x^n)' = nx^{n-1}$.*

Proof #1: Part (1) of Example 6.1.2 with $m = 1$ and $l = 0$ proves that $(x^1)' = 1$, so that the theorem is true when $n = 1$. Now suppose that the theorem holds for some positive integer n. Then

$$(x^n)' = (x^{n-1}x)'$$
$$= (x^{n-1})'x + x^{n-1}x' \text{ (by the product rule)}$$
$$= (n-1)x^{n-2}x + x^{n-1} \text{ (by induction assumption)}$$
$$= (n-1)x^{n-1} + x^{n-1}$$
$$= nx^{n-1},$$

so that the theorem holds also for n, and so by induction also for all positive n. □

Proof #2: The second proof uses binomial expansions as in Exercise 1.7.7:

$$(x^n)' = \lim_{h \to 0} \frac{(x+h)^n - x^n}{h}$$
$$= \lim_{h \to 0} \frac{\sum_{k=0}^{n} \binom{n}{k} x^k h^{n-k} - x^n}{h}$$
$$= \lim_{h \to 0} \frac{\sum_{k=0}^{n-1} \binom{n}{k} x^k h^{n-k}}{h}$$
$$= \lim_{h \to 0} \frac{h \sum_{k=0}^{n-1} \binom{n}{k} x^k h^{n-k-1}}{h}$$
$$= \lim_{h \to 0} \sum_{k=0}^{n-1} \binom{n}{k} x^k h^{n-k-1}$$
$$= \sum_{k=0}^{n-1} \binom{n}{k} x^k 0^{n-k-1} \text{ (by the polynomial rule for limits)}$$
$$= \binom{n}{n-1} x^{n-1}$$
$$= nx^{n-1}. \quad \square$$

Theorem 6.2.4. *(Polynomial, rational function rule for derivatives) Polynomial functions are differentiable at all real/complex numbers and rational functions are differentiable at all points in the domain.*

Proof. The proof is an application of the sum, scalar, and power rules from Theorems 6.2.2 and 6.2.3. □

Theorem 6.2.5. (The composite function rule for derivatives, aka the **chain rule**) *Suppose that f is differentiable at a, that g is differentiable at $f(a)$, and that a is a limit point of the domain of $g \circ f$. (If $f : A \to B$, $g : B \to C$, and f is differentiable at a, then automatically a is a limit point of A and hence of the domain of $g \circ f$.) Then $g \circ f$ is differentiable at a, and $(g \circ f)'(a) = g'(f(a)) \cdot f'(a)$.*

Proof. Let $\epsilon > 0$. Since f is differentiable at a, there exists $\delta_1 > 0$ such that for all $a + h$ in the domain of f, if $0 < |h| < \delta_1$ then $|\frac{f(a+h)-f(a)}{h} - f'(a)| < \min\{1, \epsilon/(2|g'(f(a))| + 2)\}$. For all such h, by the triangle inequality, $|\frac{f(a+h)-f(a)}{h}| < |f'(a)| + 1$. By assumption g is differentiable at $f(a)$, so that there exists $\delta_2 > 0$ such that for all x in the domain of g, if $0 < |x - f(a)| < \delta_2$, then $|\frac{g(x)-g(f(a))}{x-f(a)} - g'(f(a))| < \epsilon/(2|f'(a)| + 2)$. Since f is differentiable and hence continuous at a, there exists $\delta_3 > 0$ such that for all x in the domain of f, if $|x - a| < \delta_3$, then $|f(x) - f(a)| < \delta_1$. Set $\delta = \min\{\delta_1, \delta_2, \delta_3\}$. Let $a + h$ be arbitrary in the domain of $g \circ f$ such that $0 < |h| < \delta$. In particular $a + h$ is in the domain of f. If $f(a + h) \neq f(a)$, then

$$\left| \frac{(g \circ f)(a+h) - (g \circ f)(a)}{h} - g'(f(a)) \cdot f'(a) \right|$$

$$= \left| \frac{g(f(a+h)) - g(f(a))}{h} - g'(f(a)) \cdot f'(a) \right|$$

$$= \left| \frac{g(f(a+h)) - g(f(a))}{f(a+h) - f(a)} \cdot \frac{f(a+h) - f(a)}{h} - g'(f(a)) \cdot f'(a) \right|$$

$$= \left| \left(\frac{g(f(a+h)) - g(f(a))}{f(a+h) - f(a)} - g'(f(a)) \right) \cdot \frac{f(a+h) - f(a)}{h} \right.$$
$$\left. + g'(f(a)) \cdot \frac{f(a+h) - f(a)}{h} - g'(f(a)) \cdot f'(a) \right|$$

$$\leq \left| \frac{g(f(a+h)) - g(f(a))}{f(a+h) - f(a)} - g'(f(a)) \right| \cdot \left| \frac{f(a+h) - f(a)}{h} \right|$$
$$+ |g'(f(a))| \left| \frac{f(a+h) - f(a)}{h} - f'(a) \right|$$

$$\leq \frac{\epsilon}{2|f'(a)| + 2} \cdot (|f'(a)| + 1) + |g'(f(a))| \frac{\epsilon}{2|g'(f(a))| + 2}$$

$$< \frac{\epsilon}{2} + \frac{\epsilon}{2}$$

$$= \epsilon.$$

Thus if there exists δ as above but possibly smaller such that $f(a + h) \neq f(a)$ for all $a + h$ in the domain with $0 < |h| < \delta$, the above proves the theorem.

Now suppose that for all $\delta > 0$ there exists h such that $a + h$ is in the domain of f, $0 < |h| < \delta$, and $f(a + h) = f(a)$. Then in particular when h varies over those infinitely many h, $f'(a) = \lim_{h \to 0} \frac{f(a+h) - f(a)}{h} = \lim_{h \to 0} \frac{0}{h} = 0$. Also, for such h, $(g \circ f)(a + h) - (g \circ f)(a) = g(f(a + h)) - g(f(a)) = 0$, so that

$$\left| \frac{(g \circ f)(a + h) - (g \circ f)(a)}{h} - g'(f(a)) \cdot f'(a) \right| = 0 < \epsilon.$$

This analyzes all the cases and finishes the proof of the theorem. $\qquad\square$

Theorem 6.2.6. (Real and imaginary parts) *Let* $A \subseteq \mathbb{R}$, *and let* $a \in A$ *be a limit point of* A. *Let* $f : A \to \mathbb{C}$. *Then* f *is differentiable at* a *if and only if* $\operatorname{Re} f$ *and* $\operatorname{Im} f$ *are differentiable at* a, *and in that case,* $f' = (\operatorname{Re} f)' + i(\operatorname{Im} f)'$.

Proof. Since all h are necessarily real,

$$\frac{f(a + h) - f(a)}{h} = \frac{\operatorname{Re}(f(a + h) - f(a)) + i \operatorname{Im}(f(a + h) - f(a))}{h}$$

$$= \operatorname{Re}\left(\frac{f(a + h) - f(a)}{h} \right) + i \operatorname{Im}\left(\frac{f(a + h) - f(a)}{h} \right),$$

and by the definition of limits of complex functions, the limit of the function on the left exists if and only if the limits of its real and imaginary parts exist. $\qquad\square$

Compare this last theorem with Exercises 6.2.4 and 6.2.5.

Exercises 5.2.2 and 5.2.3 each give an example of a continuous invertible function whose inverse is not continuous. It is also true that the inverse of a differentiable invertible function need not be differentiable. For example, the cubing function on $\mathbb{R}$ ($f(x) = x^3$) is differentiable and invertible, but its inverse, the cube root function, is not differentiable at 0, as the limit of $\frac{\sqrt[3]{0+h} - \sqrt[3]{0}}{h} = h^{-1/3}$ does not exist as h goes to 0. Nevertheless, we can connect differentiability of an invertible function to the derivative of the inverse, as shown in the next theorem.

Theorem 6.2.7. (Derivatives of inverses) *Let $A, B \subseteq \mathbb{C}$, let A be open, and let $f : A \to B$ be an invertible differentiable function whose derivative is never 0. Then for all $b \in B$, f^{-1} is differentiable at b, and*

$$(f^{-1})'(b) = \frac{1}{f'(f^{-1}(b))},$$

or in other words, for all $a \in A$, f^{-1} is differentiable at $f(a)$ and

$$(f^{-1})'(f(a)) = \frac{1}{f'(a)}.$$

Proof. Let $b \in B$. Then $b = f(a)$ for some $a \in A$. Since A is open, there exists $r > 0$ such that $B(a, 2r) \subseteq A$. Let C be any closed and bounded subset of A that contains $B(a, r)$. For example, C could be $\{x \in \mathbb{C} : |x - a| \leq r\}$. Define $g : C \to D = \{f(x) : x \in C\}$ as $g(x) = f(x)$. Then g is invertible and continuous. By Theorem 5.2.4, g^{-1} is continuous. In particular, g^{-1} is continuous at $b = f(a) = g(a)$.

Let $\epsilon > 0$. Since g is differentiable at a, and $g(a) = f(a) = b$ and $g'(a) = f'(a)$, by the limit definition of derivatives, there exists $\delta > 0$ such that for all $x \in C$, if $|x - a| < \delta$ then $\left| \frac{g(x) - b}{x - a} - f'(a) \right| < \epsilon$. By continuity of g^{-1}, there exists $\gamma > 0$ such that for all $y \in D$, if $|y - b| < \gamma$ then $|g^{-1}(y) - a| = |g^{-1}(y) - g^{-1}(b)| < \delta$. In particular, for all such y, $\left| \frac{y - b}{g^{-1}(y) - a} - f'(a) \right| = \left| \frac{g(g^{-1}(y)) - b}{g^{-1}(y) - a} - f'(a) \right| < \epsilon$. This says that for all $h \in \mathbb{C}$, if $|h| < \gamma$ and $b + h \in D$, then $\left| \frac{h}{g^{-1}(b+h) - a} - f'(a) \right| = \left| \frac{g(g^{-1}(b+h)) - b}{g^{-1}(b+h) - a} - f'(a) \right| < \epsilon$. Thus $\lim_{h \to 0} \frac{h}{g^{-1}(b+h) - a} = f'(a)$, and by the quotient rule for limits,

$$\lim_{h \to 0} \frac{g^{-1}(b + h) - g^{-1}(b)}{h} = \lim_{h \to 0} \frac{g^{-1}(b + h) - a}{h} = \frac{1}{f'(a)} = \frac{1}{f'(f^{-1}(b))}.$$

But this holds for every C (and g which is f restricted to C), so that in particular,

$$(f^{-1})'(b) = \lim_{h \to 0} \frac{f^{-1}(b + h) - f^{-1}(b)}{h} = \frac{1}{f'(f^{-1}(b))}. \qquad \square$$

It should be noted that if we know that f^{-1} is differentiable, the proof of the last part of the theorem above goes as follows. For all $x \in B$, $(f \circ f^{-1})(x) = x$, so that $(f \circ f^{-1})'(x) = 1$, and by the chain rule, $f'(f^{-1}(x)) \cdot (f^{-1})'(x) = 1$. Then if f' is never 0, we get that $(f^{-1})'(x) = \frac{1}{f'(f^{-1}(x))}$.

Example 6.2.8. Let $f : [0, \infty) \to [0, \infty)$ be the function $f(x) = x^2$. We know that f is differentiable at all points in the domain and that $f'(x) = 2x$. By Example 6.1.5 and Exercise 6.1.4, the inverse of f, namely the square root function, is differentiable at all positive x, but not at 0. The theorem above applies to positive x (but not to $x = 0$):

$$(\sqrt{x})' = (f^{-1})'(x) = \frac{1}{f'(f^{-1}(x))} = \frac{1}{2f^{-1}(x)} = \frac{1}{2\sqrt{x}}.$$

Theorem 6.2.9. *Let n be a positive integer. Then for all non-zero x in the domain of $\sqrt[n]{\ }$,*

$$(\sqrt[n]{x})' = \frac{1}{n} x^{1/n-1}.$$

Proof. Let $A = \mathbb{R}^+$ if n is even and let $A = \mathbb{R} \setminus \{0\}$ otherwise. Define $f : A \to A$ to be $f(x) = x^n$. We have proved that f is invertible and differentiable. The derivative is $f'(x) = nx^{n-1}$, which is never 0. Thus by the previous theorem, $f^{-1} = \sqrt[n]{\ }$ is differentiable with

$$(\sqrt[n]{x})' = \frac{1}{f'(\sqrt[n]{x})} = \frac{1}{n(\sqrt[n]{x})^{n-1}} = \frac{1}{nx^{(n-1)/n}} = \frac{1}{n} x^{-(n-1)/n} = \frac{1}{n} x^{1/n-1}.$$

$\square$

Theorem 6.2.10. (Generalized power rule) *Let r be an arbitrary rational number and let $f : \mathbb{R}^+ \to \mathbb{R}^+$ be given by $f(x) = x^r$. Then for all x, $f'(x) = rx^{r-1}$.*

Proof. The power rule and quotient rules prove this in case r an integer, and the previous theorem proves it in case r is one over a positive integer. Now suppose that $r = m/n$ for some integers m, n with $n \neq 0$. Since $m/n = (-m)/(-n)$ is also a quotient of two integers, we may write $r = m/n$ so that $m \in \mathbb{Z}$ and n is a positive integer. Thus f is the composition of exponentiation by m and by $1/n$. By the chain rule,

$$f'(x) = m(x^{1/n})^{m-1} \cdot \frac{1}{n} x^{1/n-1}$$
$$= \frac{m}{n} x^{(m-1)/n+1/n-1}$$
$$= rx^{m/n-1/n+1/n-1}$$
$$= rx^{r-1}.$$

This proves the theorem for all rational r. $\square$

The theorem also holds for all real r. But to prove it for all real r one first needs to define exponentiation by non-rational numbers. Such exponentiation was worked through laboriously in Exercises 5.4.7 and 5.4.8, and if we were to continue that kind of laborious treatment, the proof of the form of the derivative of such exponentiation would also be laborious. So we postpone the definition of such exponentiation and the proof of its derivative to Theorem 7.6.4, where with the help of integrals the definition and proofs write themselves elegantly.

Exercises for Section 6.2

6.2.1. Provide the commentary for the proof of the quotient rule in Theorem 6.2.2.

6.2.2. Let f, g, h be differentiable. Prove that $(fgh)' = f'gh + fg'h + fgh'$. More generally, prove a product rule formula for the derivative of the product $f_1 \cdot f_2 \cdots \cdot f_n$.

6.2.3. Prove yet another form of the **general power rule for derivatives**: If f is differentiable at a, then for every positive integer n, f^n is differentiable at a, and $(f^n)'(a) = n(f(a))^{n-1}f'(a)$. (Recall that f^n is the composition of n copies of f, see Remark 2.4.8.)

6.2.4. Prove that the functions $\mathrm{Re}, \mathrm{Im} : \mathbb{C} \to \mathbb{R}$ are **not** differentiable at any a.

6.2.5. (Compare with Theorem 6.2.6.) Prove that the absolute value function on $\mathbb{R}$ is differentiable at all non-zero $a \in \mathbb{R}$. Prove that the absolute value function on $\mathbb{C}$ is not differentiable at any non-zero $a \in \mathbb{C}$. (Hint: Let $h = (r - 1)a$ for r near 1.)

6.2.6. A function f is differentiable on $(-2, 5)$ and $f(3) = 4$, $f'(3) = -1$. Let $g(x) = 3x$. For each of the statements below determine whether it is true, false, or if there is not enough information. Explain your reasoning.

   i) f is constant.

   ii) The slope of the tangent line to the graph of f at 3 is 4.

   iii) f is continuous on $(-2, 5)$.

   iv) The derivative of $(f \circ g)$ at 1 is -3.

   v) $(f + g)'(3) = 2$.

6.2.7. The following information is known:

c	$f(c)$	$f'(c)$	$g(c)$	$g'(c)$
0	1	2	6	4
1	−1	0	5	3
2	2	−3	6	−6
3	4	2	3	5
4	0	8	4	7

For each of the following, either provide the derivative or argue that there is not enough information. In any case, justify every answer.

   i) $(f - g)'(1) =$
   ii) $(f \cdot g)'(2) =$
   iii) $\left(\frac{f}{g}\right)'(3) =$
   iv) $(g \circ f)'(4) =$

6.2.8. Let $f(x) = \frac{x-1}{x-2}$.

   i) Find all a in the domain of f such that the tangent line to the graph of f at a has slope -1.

   ii) Find all a in the domain of f such that the tangent line to the graph of f at a has slope 2. (Solutions need not be real.)

6.3   The Mean value theorem

In this section the domains and codomains of all functions are subsets of $\mathbb{R}$.

Theorem 6.3.1.   *Let $f : [a, b] \to \mathbb{R}$, and let $c \in [a, b]$ such that f achieves an extreme value at c (i.e., either for all $x \in [a, b]$, $f(c) \le f(x)$ or for all $x \in [a, b]$, $f(c) \ge f(x)$). Then at least one of the following holds:*

   (1) $c = a$;
   (2) $c = b$;
   (3) f is not continuous at c;
   (4) f is not differentiable at c;
   (5) f is differentiable at c and $f'(c) = 0$.

Proof. It suffices to prove that if the first four conditions do not hold, then the fifth one has to hold. So we assume that $c \ne a$, $c \ne b$, and that f is differentiable at c.

Suppose that $f'(c) > 0$. By the definition of derivative, $f'(c) = \lim_{x \to c} \frac{f(x)-f(c)}{x-c}$. Thus for all real numbers x very near c but larger than c, $\frac{f(x)-f(c)}{x-c} > 0$, so that $f(x) - f(c) > 0$, so that f does not achieve its maximum at c. Also, for all x very near c but smaller than c, $\frac{f(x)-f(c)}{x-c} > 0$, so that $f(x) - f(c) < 0$, so that f does not achieve its minimum at c. This is a contradiction, so that $f'(c)$ cannot be positive. Similarly, $f'(c)$ cannot be negative. Thus $f'(c) = 0$. □

Thus to find extreme values of a function, one only has to check if extreme values occur at the endpoints of the domain, at points where the function is not continuous or non-differentiable, or where it is differentiable and the derivative is 0. One should be aware that just because any of the five conditions is satisfied, we need not have an extreme value of the function. Here are some examples:

(1) The function $f : [-1,1] \to \mathbb{R}$ given by $f(x) = x^3 - x$ has neither the maximum nor the minimum at the endpoints.

(2) Let $f : [-1,1] \to \mathbb{R}$ be given by $f(x) = \begin{cases} x, & \text{if } x > 0; \\ 1/2, & \text{if } x \le 0. \end{cases}$ Then f is not continuous at 0 but f does not have a minimum or maximum at 0.

(3) Let $f : [-1,1] \to \mathbb{R}$ be given by $f(x) = \begin{cases} x, & \text{if } x > 0; \\ 2x, & \text{if } x \le 0. \end{cases}$ Then f is continuous and not differentiable at 0, yet f does not have a minimum or maximum at 0.

(4) Let $f : [-1,1] \to \mathbb{R}$ be given by $f(x) = x^3$. Then f is differentiable, $f'(0) = 0$, but f does not have a minimum or maximum at 0.

Theorem 6.3.2.   (**Darboux's theorem**) *Let $a < b$ be real numbers, and let $f : [a,b] \to \mathbb{R}$ be differentiable. Then f' has the intermediate value property, i.e., for all k between $f'(a)$ and $f'(b)$ there exists $c \in [a,b]$ such that $f'(c) = k$.*

Proof. If $f'(a) = k$, we set $c = a$, and similarly if $f'(b) = k$, we set $c = b$. So we may assume that k is strictly between $f'(a)$ and $f'(b)$.

The function $g : [a,b] \to \mathbb{R}$ given by $g(x) = f(x) - kx$ is differentiable, hence continuous. Note that $g'(x) = f'(x) - k$, so that 0 is strictly between $g'(a)$ and $g'(b)$. If $g'(a) > 0$, let $c \in [a,b]$ such that g achieves a maximum at c, and if $g'(a) < 0$, let $c \in [a, b]$ such that g achieves a minimum at c. Such c exists by the Extreme value theorem (Theorem 5.2.2). Note that $g'(a) < 0$ if and only if $f'(a) < k$, which holds if and only if $f'(b) > k$, which in turn holds if and only if $g'(b) > 0$. Similarly $g'(a) > 0$ if and only if $g'(b) < 0$. Thus for both choices of c, c cannot be a or b, so that $c \in (a, b)$. By Theorem 6.3.1, $g'(c) = 0$. Hence $f'(c) = k$. $\qquad\square$

Theorem 6.3.3. (**Rolle's theorem**) *Let $a, b \in \mathbb{R}$ with $a < b$, and let $f : [a, b] \to \mathbb{R}$ be a continuous function such that f is differentiable on (a, b). If $f(a) = f(b)$, then there exists $c \in (a, b)$ such that $f'(c) = 0$.*

Proof. By the Extreme value theorem (Theorem 5.2.2) there exist $l, u \in [a, b]$ such that f achieves its minimum at l and its maximum at u. If $f(l) = f(u)$, then the minimum value of f is the same as the maximum value of f, so that f is a constant function, and so $f'(c) = 0$ for all $c \in (a, b)$.

Thus we may assume that $f(l) \neq f(u)$. It may be that f achieves its minimum at the two endpoints, in which case u must be strictly between a and b. Similarly, it may be that f achieves its maximum at the two endpoints, in which case l must be strictly between a and b. In all cases of $f(l) \neq f(u)$, either l or u is not the endpoint.

Say that l is not the endpoint. Then $a < l < b$. For all $x \in [a, b]$, $f(x) \geq f(l)$, so that in particular for all $x \in (a, l)$, $\frac{f(x) - f(l)}{x - l} \leq 0$ and for all $x \in (l, b)$, $\frac{f(x) - f(l)}{x - l} \geq 0$. Since $f'(l) = \lim_{x \to l} \frac{f(x) - f(l)}{x - l}$ exists, it must be both non-negative and non-positive, so necessarily it has to be 0.

If instead u is not the endpoint, then $a < u < b$ and for all $x \in [a, b]$, $f(x) \leq f(u)$. Thus in particular for all $x \in (a, u)$, $\frac{f(x) - f(u)}{x - u} \geq 0$ and for all $x \in (u, b)$, $\frac{f(x) - f(u)}{x - u} \leq 0$. Since $f'(u) = \lim_{x \to u} \frac{f(x) - f(u)}{x - u}$ exists, it must be both non-negative and non-positive, so necessarily it has to be 0. Thus in all cases we found $c \in (a, b)$ such that $f'(c) = 0$. $\qquad\square$

Theorem 6.3.4. (**Mean value theorem**) *Let $a, b \in \mathbb{R}$ with $a < b$, and let $f : [a, b] \to \mathbb{R}$ be a continuous function such that f is differentiable on (a, b). Then there exists $c \in (a, b)$ such that $f'(c) = \frac{f(b) - f(a)}{b - a}$.*

Here is an illustration of this theorem: the slope $\frac{f(b) - f(a)}{b - a}$ of the line from $(a, f(a))$ to $(b, f(b))$ also equals the slope of the tangent line to the graph at some c between a and b:

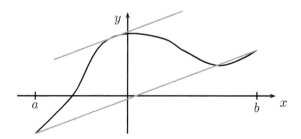

Proof. Let $g : [a, b] \to \mathbb{R}$ be defined by $g(x) = f(x) - \frac{f(b)-f(a)}{b-a}(x - a)$. By the sum and scalar rules for continuity and differentiability, g is continuous on $[a, b]$ and differentiable on (a, b). Also, $g(a) = f(a)$ and $g(b) = f(b) - \frac{f(b)-f(a)}{b-a}(b - a) = f(b) - (f(b) - f(a)) = f(a) = g(a)$. Thus by Rolle's theorem, there exists $c \in (a, b)$ such that $g'(c) = 0$. But

$$g'(x) = f'(x) - \frac{f(b) - f(a)}{b - a},$$

so that $0 = g'(c) = f'(c) - \frac{f(b)-f(a)}{b-a}$, whence $f'(c) = \frac{f(b)-f(a)}{b-a}$.   □

The rest of this section consists of various applications of the Mean value theorem. More concrete examples are left for the exercises.

Theorem 6.3.5. *Let $a, b \in \mathbb{R}$ with $a < b$, and let $f : [a, b] \to \mathbb{R}$ be a continuous function such that f is differentiable on (a, b).*

(1) If $f'(c) \geq 0$ for all $c \in (a, b)$, then f is non-decreasing on $[a, b]$.
(2) If $f'(c) > 0$ for all $c \in (a, b)$, then f is strictly increasing on $[a, b]$.
(3) If $f'(c) \leq 0$ for all $c \in (a, b)$, then f is non-increasing on $[a, b]$.
(4) If $f'(c) < 0$ for all $c \in (a, b)$, then f is strictly decreasing on $[a, b]$.
(5) If $f'(c) = 0$ for all $c \in (a, b)$, then f is a constant function.

Proof of part (2): Let $x, y \in [a, b]$ with $x < y$. By Theorem 6.3.4 there exists $c \in (x, y)$ such that $f'(c) = \frac{f(x)-f(y)}{x-y}$. Since $f'(c) > 0$ and $x < y$, necessarily $f(x) < f(y)$. Since x and y were arbitrary with $x < y$, then f is strictly increasing on $[a, b]$.   □

Example 6.3.6. *For all $x \geq 1$, $4x^3 > 2x + 2$.*

Proof. Let $f(x) = 4x^3 - 2x - 2$. Then f is differentiable and $f'(x) = 12x^2 - 2x = 2x(6x - 1) > 0$ for all $x \geq 1$. By the previous theorem, f is strictly increasing on $[1, \infty)$, so that for all $x > 1$, $4x^3 - 2x - 2 = f(x) > f(1) = 0$ and so $4x^3 > 2x + 2$.   □

Theorem 6.3.7. (Convexity of graphs) *Let f be a function that is continuous on $[a, b]$ and twice-differentiable on (a, b).*

(1) *If $f'' \geq 0$ on (a, b), then the graph of the function lies below the line connecting $(a, f(a))$ and $(b, f(b))$.*

(2) *If $f'' \leq 0$ on (a, b), then the graph of the function lies above the line connecting $(a, f(a))$ and $(b, f(b))$.*

Proof. The equation of the line is $y = \frac{f(b)-f(a)}{b-a}(x - a) + f(a)$.

(i) Let $x \in (a, b)$. We need to prove that $f(x) \leq \frac{f(b)-f(a)}{b-a}(x - a) + f(a)$. By Theorem 6.3.4 there exists $c \in (a, x)$ such that $f'(c) = \frac{f(x)-f(a)}{x-a}$, and there exists $d \in (a, x)$ such that $f'(d) = \frac{f(b)-f(x)}{b-x}$. Since $f'' > 0$ and $c < x < d$, necessarily $\frac{f(x)-f(a)}{x-a} \leq \frac{f(b)-f(x)}{b-x}$. By cross-multiplying by positive $(x - a)(b - x)$ we get that

$$(f(x) - f(a))(b - x) \leq (f(b) - f(x))(x - a),$$

or in other words, that $f(x)(b - a) \leq f(a)(b - x) + f(b)(x - a) = f(a)(b - a) + (f(b) - f(a))(x - a)$, and by dividing by positive $b - a$ yields the desired inequality.

The proof of (ii) is similar.   □

Theorem 6.3.8. (Cauchy's mean value theorem) *Let $a < b$ be real numbers and let $f, g : [a, b] \to \mathbb{R}$ be continuous functions that are differentiable on (a, b). Then there exists $c \in (a, b)$ such that*

$$f'(c)(g(b) - g(a)) = g'(c)(f(b) - f(a)).$$

In particular, if $g'(c) \neq 0$ and $g(b) \neq g(a)$, this says that $\frac{f'(c)}{g'(c)} = \frac{f(b)-f(a)}{g(b)-g(a)}$.

Proof. Define $h : [a, b] \to \mathbb{R}$ by $h(x) = f(x)(g(b) - g(a)) - g(x)(f(b) - f(a))$. Then h is continuous on $[a, b]$ and differentiable on (a, b). Note that $h(a) = f(a)(g(b) - g(a)) - g(a)(f(b) - f(a)) = f(a)g(b) - g(a)f(b) = h(b)$. Then by the Mean value theorem (Theorem 6.3.4) there exists $c \in (a, b)$ such that $h'(c) = 0$, i.e., $0 = f'(c)(g(b) - g(a)) - g'(c)(f(b) - f(a))$.   □

Theorem 6.3.9. (Cauchy's mean value theorem, II) *Let $a < b$ be real numbers and let $f, g : [a, b] \to \mathbb{R}$ be continuous functions that are differentiable on (a, b) and such that g' is non-zero on (a, b). Then*

$g(b) \neq g(a)$, and there exists $c \in (a, b)$ such that

$$\frac{f'(c)}{g'(c)} = \frac{f(b) - f(a)}{g(b) - g(a)}.$$

Proof. By the Mean value theorem (Theorem 6.3.4) there exists $c \in (a, b)$ such that $g'(c) = \frac{g(b) - g(a)}{b - a}$. By assumption, $g'(c) \neq 0$, so that $g(b) \neq g(a)$. The rest follows by Cauchy's mean value theorem (Theorem 6.3.8). $\square$

Exercises for Section 6.3

6.3.1. Let $f : \mathbb{R} \to \mathbb{R}$ be a polynomial function of degree n.

  i) If f has m distinct roots, prove that f' has at least $m - 1$ distinct roots.

 ii) Suppose that f has n distinct roots. Prove that f' has exactly $n-1$ distinct roots. (You may need Theorem 2.4.15.)

iii) Give an example of f with $n = 2$, where f has no roots but f' has 1 root. Why does this not contradict the first part?

6.3.2. Let $f : [0, 1] \to \mathbb{C}$ be given by $f(x) = x^3 + ix^2$. Prove that there exists no c between 0 and 1 such that $f'(c) = \frac{f(1) - f(0)}{1 - 0}$. Does this contradict the Mean value theorem?

6.3.3. Let $f : [a, b] \to \mathbb{C}$ be continuous on $[a, b]$ and differentiable on (a, b). Prove that there exist $c, d \in (a, b)$ such that $\text{Re}(f'(c)) = \text{Re}\left(\frac{f(b) - f(a)}{b - a}\right)$ and $\text{Im}(f'(d)) = \text{Im}\left(\frac{f(b) - f(a)}{b - a}\right)$. Give an example showing that c may not equal to d. Does this contradict the Mean value theorem?

6.3.4. Prove the remaining parts of Theorem 6.3.5.

6.3.5. This water-filled urn is getting drained at a constant rate through a hole at the bottom. Plot the function of depth (d) as a function of time.

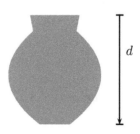

6.3.6. Let A be an open interval in $\mathbb{R}$, and let $f : A \to \mathbb{R}$ be differentiable such that the range of the derivative function is bounded. Prove that f is uniformly continuous.

6.4 L'Hôpital's rule

For L'Hôpital's rule we pass from $\frac{f}{g}$ to $\frac{f'}{g'}$; it is worth reviewing that the quotient rule for derivatives is different:

$$\left(\frac{f}{g}\right)' = \frac{f'g - fg'}{g^2}.$$

Theorem 6.4.1. (L'Hôpital's rule, easiest version) *Let $A \subset \mathbb{C}$, let $a \in A$ be a limit point of A, and $f, g : A \to \mathbb{C}$ such that*

(1) $f(a) = g(a) = 0$.

(2) f and g are differentiable at a.

(3) $g'(a) \neq 0$.

Then $\lim\limits_{x \to a} \dfrac{f(x)}{g(x)} = \dfrac{f'(a)}{g'(a)}$.

Proof. $\lim\limits_{x \to a} \dfrac{f(x)}{g(x)} = \lim\limits_{x \to a} \dfrac{f(x) - f(a)}{g(x) - g(a)}$ (since $f(a) = g(a) = 0$)

$$= \lim_{x \to a} \left(\frac{\frac{f(x) - f(a)}{x - a}}{\frac{g(x) - g(a)}{x - a}} \right)$$

$$= \frac{f'(a)}{g'(a)} \text{ (by the quotient rule for limits as } g'(a) \neq 0). \quad \square$$

The rest of the versions of L'Hôpital's rule in this section only work for real-valued functions on domains that are subsets of $\mathbb{R}$ because the proofs invoke the Cauchy's mean value theorem (Theorem 6.3.8).

Theorem 6.4.2. (L'Hôpital's rule) *Let $a, L \in \mathbb{R}$ and let f, g be real-valued functions defined on an interval containing a such that*

(1) $f(a) = g(a) = 0$.

(2) f and g are differentiable near a but not necessarily at a.

(3) g' is non-zero near a. (This is a condition for x near a but not equal to a.)

(4) $\lim\limits_{x \to a} \dfrac{f'(x)}{g'(x)} = L$.

Then $\lim\limits_{x \to a} \dfrac{f(x)}{g(x)} = L$.

Proof. Let $\epsilon > 0$. Since $\lim_{x \to a} \frac{f'(x)}{g'(x)} = L$, there exists $\delta > 0$ such that for all x in the domain, if $0 < |x - a| < \delta$ then $|\frac{f'(x)}{g'(x)} - L| < \epsilon$. Let x be one such number. By Theorem 6.3.9 there exists a number c strictly between a and x such that $\frac{f'(c)}{g'(c)} = \frac{f(x)-f(a)}{g(x)-g(a)}$. Since $f(a) = g(a) = 0$, this says that $\frac{f'(c)}{g'(c)} = \frac{f(x)}{g(x)}$. Since $0 < |x - a| < \delta$ and c is between a and x, it follows that $0 < |c - a| < \delta$. Hence

$$\left| \frac{f(x)}{g(x)} - L \right| = \left| \frac{f'(c)}{g'(c)} - L \right| < \epsilon. \qquad \square$$

With our definition of derivatives, this last version includes the one-sided cases where the domains for f and g are of the form $[a, b)$, or of the form $(b, a]$.

I should note that some books omit hypothesis (3). A counterexample to the conclusion if we omit this hypothesis but keep all the others is in Exercise 10.4.7.

The versions of L'Hôpital's rule so far deal with limits of the form "zero over zero". There are versions for the form "infinity over infinity"; I write and prove only the right-sided version, but similarly there is a left-sided version and a both-sided version.

Theorem 6.4.3. (L'Hôpital's rule) *Let L and $a < b$ be real numbers. Let $f, g : (a, b) \to \mathbb{R}$ be differentiable with the following properties:*

(1) $\lim_{x \to b^-} f(x) = \lim_{x \to b^-} g(x) = \infty$.

(2) *For all $x \in (a, b)$, $g'(x) \neq 0$.*

(3) $\lim_{x \to b^-} \frac{f'(x)}{g'(x)} = L$.

Then $\lim_{x \to b^-} \frac{f(x)}{g(x)} = L$.

Proof. Let $\epsilon > 0$. By assumption there exists $\delta_1 > 0$ such that for all $x \in (a, b)$, if $0 < b - x < \delta_1$, then $\left| \frac{f'(x)}{g'(x)} - L \right| < \epsilon/4$. By possibly replacing δ_1 by $\min\{\delta_1, (b - a)/2\}$ we may assume that $b - \delta_1 > a$.

Set $a_0 = b - \delta_1$. Fix x such that $0 < b - x < \delta_1$. Then $x \in (a_0, b) \subseteq (a, b)$. Suppose that $g(x) = g(a_0)$. Then by the Mean value theorem (Theorem 6.3.4) there exists $u \in (a_0, x)$ such that $g'(u) = 0$, which contradicts the assumption (2). Thus $g(x) \neq g(a_0)$.

By Theorem 6.3.9 there exists $c \in (a_0, x)$ such that $\frac{f'(c)}{g'(c)} = \frac{f(x)-f(a_0)}{g(x)-g(a_0)}$, and so

$$\left| \frac{f(x) - f(a_0)}{g(x) - g(a_0)} - L \right| = \left| \frac{f'(c)}{g'(c)} - L \right| < \epsilon/4.$$

Since $\lim_{x \to b^-} f(x) = \lim_{x \to b^-} g(x) = \infty$, there exists $\delta_2 > 0$ such that for all x with $0 < b - x < \delta_2$, $f(x)$ and $g(x)$ are non-zero, and so we can define $h : (b - \delta_2, b) \to \mathbb{R}$ as

$$h(x) = \frac{1 - \frac{f(a_0)}{f(x)}}{1 - \frac{g(a_0)}{g(x)}}.$$

By Theorem 4.5.7, $\lim_{x \to b^-} h(x) = \frac{1 - 0}{1 - 0} = 1$. Thus there exists $\delta_3 > 0$ such that for all x, if $0 < b - x < \delta_3$, then $|h(x) - 1| < \min\{\epsilon/4(|L| + 1), \frac{1}{2}\}$. The $(\frac{1}{2})$-restriction in particular means that $h(x) > \frac{1}{2}$ and thus non-zero. Set $\delta = \min\{\delta_1, \delta_2, \delta_3\}$. Then for all x with $0 < b - x < \delta$,

$$\left| \frac{f(x)}{g(x)} - L \right| = \left| \frac{f(x)}{g(x)} h(x) - Lh(x) \right| \frac{1}{|h(x)|}$$

$$= \left| \frac{f(x) - f(a_0)}{g(x) - g(a_0)} - Lh(x) \right| \frac{1}{|h(x)|}$$

$$\leq 2 \left| \frac{f(x) - f(a_0)}{g(x) - g(a_0)} - Lh(x) \right| \quad (\text{since } |h(x)| > 1/2)$$

$$= 2 \left| \frac{f(x) - f(a_0)}{g(x) - g(a_0)} - L + L - Lh(x) \right|$$

(by addition of a clever zero)

$$\leq 2 \left(\left| \frac{f(x) - f(a_0)}{g(x) - g(a_0)} - L \right| + |L - Lh(x)| \right)$$

(by the triangle inequality)

$$= 2 \left(\left| \frac{f'(c)}{g'(c)} - L \right| + |L| \, |1 - h(x)| \right)$$

$$< 2 \left(\frac{\epsilon}{4} + \frac{\epsilon}{4} \right)$$

$$= \epsilon. \qquad \square$$

Example 6.4.4. Compute $\lim_{x \to 1} \frac{x^2-1}{x^3-1}$.

Proof #1: By Example 1.6.4, $\frac{x^2-1}{x^3-1} = \frac{(x-1)(x+1)}{(x-1)(x^2+x+1)}$, so that by Exercise 4.4.8 and Theorem 4.4.6, $\lim_{x\to 1} \frac{x^2-1}{x^3-1} = \lim_{x\to 1} \frac{x+1}{x^2+x+1} = \frac{1+1}{1^2+1+1} = \frac{2}{3}$. $\square$

Proof #2: By Theorem 4.4.5, $\lim_{x\to 1}(x^2-1) = 0 = \lim_{x\to 1}(x^3-1)$, $\lim_{x\to 1}(x^2-1)' = \lim_{x\to 1} 2x = 2$, and $\lim_{x\to 1}(x^3-1)' = \lim_{x\to 1} 3x^2 = 3$. Thus by L'Hôpital's rule (Theorem 6.4.2), $\lim_{x\to 1} \frac{x^2-1}{x^3-1}$ equals $\frac{2}{3}$. $\square$

More forms of L'Hôpital's rule are in the exercises below and also in Sections 7.6 and 10.4 after the exponential and trigonometric functions have been introduced.

Exercises for Section 6.4

6.4.1. Compute and justify $\lim_{x \to -8} \frac{x^2-64}{\sqrt[3]{x}+2}$.

6.4.2. What is wrong with the following: Since $\lim_{x\to 2} \frac{2x}{2} = 2$, by L'Hôpital's rule we conclude that $\lim_{x\to 2} \frac{x^2+3}{2x-3} = 2$.

6.4.3. Let n be a positive integer and $a \in \mathbb{C}$. Use L'Hôpital's rule to prove that $\lim_{x\to a} \frac{x^n-a^n}{x-a} = na^{n-1}$. (We have proved the case of $a = 1$ previously, say how.)

6.4.4. (Here the goal is to prove another version of L'Hôpital's rule.) Let $f, g : (a, \infty) \to \mathbb{R}$ be differentiable. Suppose that $\lim_{x\to\infty} f(x) = \lim_{x\to\infty} g(x) = \infty$, that $g'(x)$ is non-zero for all x, and that $\lim_{x\to\infty} \frac{f'(x)}{g'(x)} = L$ for some $L \in \mathbb{R}$.
   i) Let $\epsilon > 0$. Prove that there exists $N > a$ such that for all $x > N$, $|\frac{f'(x)}{g'(x)} - L| < \epsilon$.
   ii) Prove that there exists $N' > N$ such that for all $x \geq N'$, $f(x)$, $g(x) > 0$.
   iii) Prove that there exists $N'' > N'$ such that for all $x \geq N''$, $f(x) > f(N')$ and $g(x) > g(N')$.
   iv) Prove that for all $x > N''$ there exists $c \in (N', x)$ such that $\frac{f'(c)}{g'(c)} = \frac{f(x)}{g(x)} \frac{1-f(N')/f(x)}{1-g(N')/g(x)}$.
   v) Prove that $\lim_{x\to\infty} \frac{f(x)}{g(x)} = L$.

6.4.5. (Yet another version of L'Hôpital's rule.) Let $f, g : (a, \infty) \to \mathbb{R}$ be differentiable. Suppose that $\lim\limits_{x \to \infty} f(x) = \lim\limits_{x \to \infty} g(x) = 0$, that $g'(x)$ is non-zero for all x, and that $\lim\limits_{x \to \infty} \frac{f'(x)}{g'(x)} = L$. Prove that $\lim\limits_{x \to \infty} \frac{f(x)}{g(x)} = L$.

6.5 Higher-order derivatives, Taylor polynomials

Let f be a function from a subset of $\mathbb{C}$ to $\mathbb{C}$.

If f is continuous at a, then near a, f is approximately the constant function $f(a)$ because $\lim_{x \to a} f(x) = f(a)$. Among all constant functions, the function $y = f(a)$ approximates f at a best.

If f is differentiable near a and the derivative is continuous at a, then for all x near a but not equal to a, $f(x) = f(x) - f(a) + f(a) = \frac{f(x) - f(a)}{x - a}(x - a) + f(a) \cong f'(a)(x - a) + f(a)$, so that for x near a (and possibly equal to a), f is approximately the linear function $f'(a)(x - a) + f(a)$, i.e., f is approximated by its tangent line.

This game keeps going, but for this we need **higher order derivatives**:

Definition 6.5.1. *Let f be differentiable. If f' is differentiable, we write the derivative of f' as f'', or as $f^{(2)}$. If $f^{(n-1)}$ is differentiable, we denote its derivative as $f^{(n)}$. We say that f has the **nth derivative**, or that it has **derivatives up to order n**.*

Using this notation we also write $f^{(1)} = f'$ and $f^{(0)} = f$.

If $f(x) = x^m$, then for all $n \le m$, $f^{(n)}(x) = m(m-1)(m-2) \cdots (m - n + 1)x^{m-n}$.

Definition 6.5.2. *Let f be a function with derivatives of orders $1, 2, \ldots, n$ existing at a point a in the domain. The **Taylor polynomial of f** (centered) at a of order n is*

$$T_{n,f,a}(x) = f(a) + f'(a)(x - a) + \frac{f''(a)}{2!}(x - a)^2 + \cdots + \frac{f^{(n)}(a)}{n!}(x - a)^n$$

$$= \sum_{k=0}^{n} \frac{f^{(k)}(a)}{k!}(x - a)^k.$$

Example 6.5.3. If $f(x) = x^4 - 3x^3 + 4x^2 + 7x - 10$, then with $a = 0$,

$$T_{0,f,0}(x) = -10,$$
$$T_{1,f,0}(x) = -10 + 7x,$$
$$T_{2,f,0}(x) = -10 + 7x + 4x^2,$$
$$T_{3,f,0}(x) = -10 + 7x + 4x^2 - 3x^3,$$
$$T_{n,f,0}(x) = -10 + 7x + 4x^2 - 3x^3 + x^4 \text{ for all } n \geq 4,$$

and for $a = 1$,

$$T_{0,f,1}(x) = -1,$$
$$T_{1,f,1}(x) = -1 + 10(x - 1),$$
$$T_{2,f,1}(x) = -1 + 10(x - 1) + (x - 1)^2,$$
$$T_{3,f,1}(x) = -1 + 10(x - 1) + (x - 1)^2 + (x - 1)^3,$$
$$T_{n,f,1}(x) = -1 + 10(x - 1) + (x - 1)^2 + (x - 1)^3 + (x - 1)^4 \text{ for all } n \geq 4.$$

Note that for all $n \geq 4$, $T_{n,f,0}(x) = T_{n,f,1}(x) = f(x)$.

The following is a generalization of this observation:

Theorem 6.5.4. *If f is a polynomial of degree at most d, then for any $a \in \mathbb{C}$ and any integer $n \geq d$, the nth-order Taylor polynomial of f centered at a equals f.*

Proof. Write $f(x) = c_0 + c_1 x + \cdots + c_d x^d$ for some $c_0, c_1, \ldots, c_d \in \mathbb{C}$. By elementary algebra, it is possible to rewrite f in the form $f(x) = e_0 + e_1(x - a) + e_2(x - a)^2 + \cdots + e_d(x - a)^d$ for some $e_0, e_1, \ldots, e_d \in \mathbb{C}$. Now observe that

$$f(a) = e_0,$$
$$f'(a) = e_1,$$
$$f''(a) = 2e_2,$$
$$f'''(a) = 6e_3 = 3!e_3,$$
$$f^{(4)}(a) = 24e_4 = 4!e_4,$$
$$\vdots$$
$$f^{(k)}(a) = k!e_k \text{ if } k \leq d,$$
$$f^{(k)}(a) = 0 \text{ if } k > d.$$

But then for $n \geq d$,

$$T_{n,f,a}(x) = \sum_{k=0}^{n} \frac{f^{(k)}(a)}{k!}(x-a)^k = \sum_{k=0}^{d} \frac{k!e_k}{k!}(x-a)^k = \sum_{k=0}^{d} e_k(x-a)^k = f(x).$$

$\square$

Theorem 6.5.5. (Taylor's remainder theorem over $\mathbb{R}$) *Let I be an interval in $\mathbb{R}$, and let $(a-r, a+r) = B(a,r) \subseteq I$. Suppose that $f : I \to \mathbb{R}$ has derivatives of orders $1, 2, \ldots, n+1$ on $B(a,r)$. Then for all $x \in B(a,r)$ there exist c, d between a and x such that*

$$f(x) = T_{n,f,a}(x) + \frac{f^{(n+1)}(c)}{n!}(x-a)(x-c)^n,$$

$$f(x) = T_{n,f,a}(x) + \frac{f^{(n+1)}(d)}{(n+1)!}(x-a)^{n+1}.$$

Proof. Let $g : B(a,r) \to \mathbb{R}$ be defined by

$$g(t) = \sum_{k=0}^{n} \frac{f^{(k)}(t)}{k!}(x-t)^k.$$

Then g is differentiable on $B(a,r)$, and

$$g'(t) = \sum_{k=0}^{n} \frac{f^{(k+1)}(t)}{k!}(x-t)^k - \sum_{k=0}^{n} \frac{kf^{(k)}(t)}{k!}(x-t)^{k-1}$$

$$= \frac{f^{(n+1)}(t)}{n!}(x-t)^n + \sum_{k=0}^{n-1} \frac{f^{(k+1)}(t)}{k!}(x-t)^k - \sum_{k=1}^{n} \frac{f^{(k)}(t)}{(k-1)!}(x-t)^{k-1}$$

$$= \frac{f^{(n+1)}(t)}{n!}(x-t)^n.$$

Note that $g(x) = f(x)$, $g(a) = T_{n,f,a}(x)$. By the Mean value theorem (Theorem 6.3.4), there exists c strictly between a and x such that $g'(c)(x-a) = g(x) - g(a)$. In other words, $\frac{f^{(n+1)}(c)}{n!}(x-c)^n(x-a) = f(x) - T_{n,f,a}(x)$, which proves the first formulation.

By Cauchy's mean value theorem (Theorem 6.3.8) applied to functions $g(t)$ and $h(t) = (x-t)^{n+1}$, there exists d between x and a such that $h'(d)(g(x) - g(a)) = g'(d)(h(x) - h(a))$. In other words,

$$-(n+1)(x-d)^n(f(x) - T_{n,f,a}(x)) = \frac{f^{(n+1)}(d)}{n!}(x-d)^n(0 - (x-a)^{n+1}).$$

Since d is not equal to x, we get that $-(n+1)(f(x) - T_{n,f,a}(x)) = \frac{f^{(n+1)}(d)}{n!}(-(x-a)^{n+1})$, which proves the second formulation.   □

Theorem 6.5.6. (**Taylor's remainder theorem over** $\mathbb{C}$) *Let A be a subset of $\mathbb{C}$ or of $\mathbb{R}$, let $a \in A$, and let $r \in \mathbb{R}^+$ such that $B(a,r) \subseteq A$. Let $f : A \to \mathbb{C}$ have higher order derivatives of orders $1, 2, \ldots, n$ on $B(a,r)$ with $n \geq 1$. Then for every $\epsilon > 0$ there exists $\delta > 0$ such that if $x \in B(a,\delta)$, then*

$$|f(x) - T_{n,f,a}(x)| < \epsilon.$$

Proof. Let $\epsilon > 0$. Let $M = 1 + \max\{|f(a)|, |f'(a)|, \ldots, |f^{(n)}(a)|\}$. Since f is differentiable at a, it is continuous at a, so there exists $\delta_1 > 0$ such that for all $x \in B(a,r)$, if $|x - a| < \delta_1$, then $|f(x) - f(a)| < \epsilon/(n+1)$. Set $\delta = \min\{r, \delta_1, \epsilon/(M(n+1)), \sqrt[2]{\frac{\epsilon}{(M(n+1))}}, \ldots, \sqrt[n]{\frac{\epsilon}{(M(n+1))}}\}$. Let x satisfy $|x - a| < \delta$. Then x is in the domain of f and $T_{n,f,a}$, and

$$|f(x) - T_{n,f,a}(x)|$$

$$= \left| f(x) - f(a) - f'(a)(x-a) - \frac{f''(a)}{2!}(x-a)^2 - \cdots - \frac{f^{(n)}(a)}{n!}(x-a)^n \right|$$

$$\leq |f(x) - f(a)| + |f'(a)|\,|x - a| + \left| \frac{f''(a)}{2!} \right| |x-a|^2 + \cdots + \left| \frac{f^{(n)}(a)}{n!} \right| |x-a|^n$$

$$< \frac{\epsilon}{n+1} + M\frac{\epsilon}{M(n+1)} + M\frac{\epsilon}{M(n+1)} + \cdots + M\frac{\epsilon}{M(n+1)}$$

$$= \epsilon.$$

□

More on Taylor polynomials and Taylor series is in the exercises in this section, in Exercise 7.4.9, and in Section 9.3.

Exercises for Section 6.5

6.5.1. (**The second derivative test**) Let $f : [a, b] \to \mathbb{R}$ be differentiable on (a, b) and suppose that f' is differentiable on (a, b) as well. Suppose that $f'(c) = 0$ for some $c \in (a, b)$.

   i) Suppose that $f''(c) > 0$. Prove that f achieves a minimum at c.

   ii) Suppose that $f''(c) < 0$. Prove that f achieves a maximum at c.

   iii) Suppose that $f''(c) = 0$. By examples show that any of the following are possible: 1) f achieves a minimum at c, 2) f achieves a maximum at c, 3) f does not have an extreme point at c.

6.5.2. Prove that if f is a polynomial function, then for every $a \in \mathbb{R}$, $T_{n,f,a} = f$ for all n greater than or equal to the degree of f.

6.5.3. (Generalized product rule.) Suppose that f and g have derivatives of orders $1, \ldots, n$ at a. Prove that

$$(f \cdot g)^{(n)}(a) = \sum_{k=0}^{n} \binom{n}{k} f^{(k)}(a) g^{(n-k)}(a).$$

6.5.4. (Generalized quotient rule.) Prove the following generalization of the product rule: Suppose that f and g have derivatives of orders $1, \ldots, n$ at a and that $g(a) \neq 0$. Find and prove a formula for the nth derivative of the function f/g.

6.5.5. Compute the Taylor polynomial of $f(x) = \sqrt{1+x}$ of degree 5 centered at $a = 0$. Justify your work.

6.5.6. Compute the Taylor polynomial of $f(x) = \sqrt{1-x}$ of degree 10 centered at $a = 0$. Justify your work.

6.5.7. Let $f(x) = \frac{1}{1-x}$.
  i) Compute $f^{(k)}(x)$ for all integers $k \geq 0$.
  ii) Compute the Taylor polynomial of f of arbitrary degree n and centered at $a = 0$.
  iii) Compute $T_{n,f,0}(0.5)$. (Hint: Use Example 1.6.4.)
  iv) Compute $T_{n,f,0}(0.5) - f(0.5)$.
  v) Compute n such that $|T_{n,f,0}(0.5) - f(0.5)| < 0.001$.
  vi) Try to use Theorem 6.5.5 to determine n such that $|T_{n,f,0}(0.5) - f(0.5)| < 0.001$. Note that this usage is not fruitful.
  vii) Use Theorem 6.5.5 to determine n such that $|T_{n,f,0}(0.4) - f(0.4)| < 0.001$.

6.5.8. Let $f : \mathbb{R} \to \mathbb{R}$ be the absolute value function.
  i) Prove that f has derivatives of all orders at all non-zero numbers.
  ii) Compute $T_{2,f,1}$ and $T_{2,f,-1}$.

6.5.9. Let $f : \mathbb{R} \to \mathbb{R}$ be given by $f(x) = |x|^3$.
  i) Prove that f is differentiable, and compute f'.
  ii) Prove that f' is differentiable, and compute f''.
  iii) Prove that f'' is not differentiable.

6.5.10. Find a differentiable function $f : \mathbb{R} \to \mathbb{R}$ such that f' is not differentiable.

6.5.11. Find a function $f : \mathbb{R} \to \mathbb{R}$ that has derivatives of orders $1, 2, 3, 4$, but such that $f^{(4)}$ is not differentiable.

Chapter 7

Integration

The basic motivation for integration is computing areas of regions bounded by graphs of functions. In this chapter we develop the theory of integration for functions whose domains are subsets of $\mathbb{R}$. The first two sections handle only codomains in $\mathbb{R}$, and at the end of Section 7.4 we extend integration to functions with codomains in $\mathbb{C}$. We do not extend to domains being subsets of $\mathbb{C}$ as that would require multi-variable methods and complex analysis, which are not the subject of this course.

7.1 Approximating areas

In this section, domains and codomains of all functions are subsets of $\mathbb{R}$. Thus we can draw the regions and build the geometric intuition together with the formalism.

Let $f : [a, b] \to \mathbb{R}$. The basic aim is to compute the **signed area** of the region bounded by the x-axis, the graph of $y = f(x)$, and the lines $x = a$ and $x = b$. By "signed" area we mean that we add up the areas of the regions above the x-axis and subtract the areas of the regions below the x-axis. Thus a signed area may be positive, negative or zero.

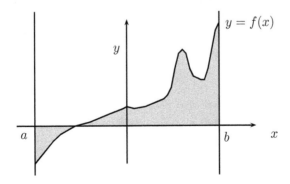

In the plot above, there are many (eight) regions whose boundaries are some of the listed curves, but only the shaded region (comprising two of the eight regions in the count) is bounded as a subset of the plane.

The simplest case of an area is when f is a constant function with constant value c. Then the signed area is $c \cdot (b - a)$, which is positive if $c > 0$ and non-positive if $c \leq 0$.

For a general f, we can try to approximate the area by rectangles, such as in the following approximations with crosshatched rectangles:

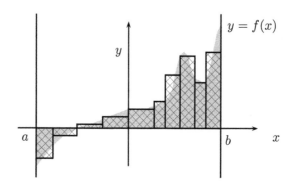

It may be hard to decide how close the approximation is to the true value. But we can approximate the region more systematically, by having heights of the rectangles be either the least possible height or the largest possible height, as below:

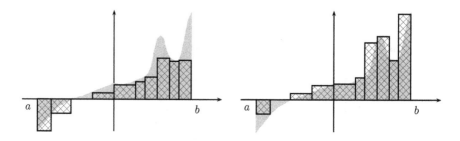

Then clearly the true area is larger than the sum of the areas of the darker rectangles on the left and smaller than the sum of the areas of the darker rectangles on the right.

We establish some notation for all this.

Definition 7.1.1. *A function* $f : A \to \mathbb{C}$ *is* **bounded** *if the range* $\{f(x) : x \in A\}$ *of* f *is a bounded subset of* $\mathbb{C}$.

Remark 7.1.2. If f is a bounded function with codomain $\mathbb{R}$, then by the Least upper bound theorem (Theorem 3.8.5), for any subset B of the domain, $\sup\{f(x) : x \in B\}$ and $\inf\{f(x) : x \in B\}$ are real numbers.

Definition 7.1.3. *A* **partition** *of* $[a, b]$ *is a finite subset of* $[a, b]$ *that contains* a *and* b. *We typically write a partition in the form* $P = \{x_0, x_1, \ldots, x_n\}$, *where* $x_0 = a < x_1 < x_2 < \cdots < x_{n-1} < x_n = b$.

The **sub-intervals of the partition** P *are* $[x_0, x_1]$, $[x_1, x_2], \ldots,$ $[x_{n-1}, x_n]$.

In all illustrated examples above, the partition of $[a, b]$ uses $n = 10$. Note that the sub-interval $[x_1, x_5]$ of the interval $[a, b]$ is **not** a sub-interval of the partition!

Definition 7.1.4. *Let* $P = \{x_0, x_1, \ldots, x_n\}$ *be a partition of* $[a, b]$ *and let* $f : [a, b] \to \mathbb{R}$ *be a bounded function.* **The lower sum of f with respect to P** *is*

$$L(f, P) = \sum_{k=1}^{n} \inf\{f(x) : x \in [x_{k-1}, x_k]\}(x_k - x_{k-1}).$$

The upper sum of f with respect to P *is*

$$U(f, P) = \sum_{k=1}^{n} \sup\{f(x) : x \in [x_{k-1}, x_k]\}(x_k - x_{k-1}).$$

By Remark 7.1.2, $L(f, P)$ and $U(f, P)$ are real numbers.

Clearly if f is a constant function $f(x) = c$ for all x, then all lower and all upper sums are $c(b - a)$, so that every lower and every upper sum equals $c(b - a)$. If instead f is a non-constant function, then for every partition P there exists at least one sub-interval of P on which the supremum of the values of f is strictly bigger than the infimum of such values, so that $L(f, P) < U(f, P)$.

By the geometric set-up for all partitions P of $[a, b]$,

$$L(f, P) \leq \text{ the signed area } \leq U(f, P). \tag{7.1.5}$$

In particular, if $U(f,P) - L(f,P) < \epsilon$, then either $U(f,P)$ or $L(f,P)$ serves as an approximation of the true signed area within ϵ of its true value. For most functions a numerical approximation is the best we can hope for.

Theorem 7.1.6. *If $f(x) = c$ for all $x \in [a,b]$, then for any partition P of $[a,b]$, $L(f,P) = U(f,P) = c(b-a)$.*

Proof. Let $P = \{x_0, \ldots, x_n\}$ be a partition of $[a,b]$. For every $k \in 1, \ldots, n$,

$$\inf\{f(x) : x \in [x_{k-1}, x_k]\} = \sup\{f(x) : x \in [x_{k-1}, x_k]\} = c,$$

so that

$$L(f,P) = U(f,P) = \sum_{k=1}^{n} c(x_k - x_{k-1})$$

$$= c \sum_{k=1}^{n} (x_k - x_{k-1})$$

$$= c(x_n - x_0) = c(b-a). \qquad \square$$

Example 7.1.7. Approximate the area under the curve $y = f(x) = 30x^4 + 2x$ between $x = 1$ and $x = 4$. We first establish a partition $P_n = \{x_0, \ldots, x_n\}$ of $[1,4]$ into n equal sub-intervals. The length of each sub-interval is $(4-1)/n$, and $x_0 = 1$, so that $x_1 = x_0 + 3/n = 1 + 3/n$, $x_2 = x_1 + 3/n = 1 + 2 \cdot 3/n$, and in general, $x_k = 1 + k \cdot 3/n$. Note that $x_n = 1 + n \cdot 3/n = 4$, as needed. Since $f'(x) = 12x^3 + 2$ is positive on $[1,4]$, it follows that f is increasing on $[1,4]$. Thus necessarily for each i, the infimum of all values of f on the ith sub-interval is achieved at the left endpoint, and the supremum at the right endpoint. In symbols, this says that $\inf\{f(x) : x \in [x_{k-1}, x_k]\} = f(x_{k-1})$ and $\sup\{f(x) : x \in [x_{k-1}, x_k]\} = f(x_k)$. For example, with $n = 1$, $L(f, P_1) = f(1) \cdot 3 = 96$ and $U(f, P_1) = f(4) \cdot 3 = 3 \cdot (30 \cdot 4^4 + 2 \cdot 4) = 23064$. Thus the true area is some number between 96 and 23064. Admittedly, this is not much information. The problem is that our partition is too rough. A computer program produced the following better numerical approximations for lower and upper sums with respect to partitions P_n into n equal sub-intervals:

n	$L(f, P_n)$	$U(f, P_n)$
10	5061.2757	7358.0757
100	6038.727	6268.407
1000	6141.5217	6164.48967
10000	6151.8517	6154.148
100000	6152.885	6153.1148
1000000	6152.988	6153.011

Notice how the lower sums get larger and the upper sums get smaller as we take finer partitions. We would like to conclude that the true area is between 6152.988 and 6153.011. This is getting closer but may still be insufficient precision. For more precision, partitions would have to get even finer, but the calculations slow down too.

The observed monotonicity is not a coincidence:

Theorem 7.1.8. *Let P, R be partitions of $[a, b]$ such that $P \subseteq R$. (Then R is called a **refinement** of P, and R is said to be a **finer** partition than P.) Then*
$$L(f, P) \leq L(f, R), \quad U(f, P) \geq U(f, R).$$

Proof. Write $P = \{x_0, x_1, \ldots, x_n\}$. Let $i \in \{1, \ldots, n\}$. Let $R \cap [x_{k-1}, x_k] = \{y_0 = x_{k-1}, y_1, \ldots, y_{m-1}, y_m = x_k\}$. By set inclusion, $\inf\{f(x) : x \in [y_{j-1}, y_j]\} \geq \inf\{f(x) : x \in [x_{k-1}, x_k]\}$, so that

$$L(f, \{y_0, y_1, \ldots, y_{m-1}, y_m\}) = \sum_{j=1}^{m} \inf\{f(x) : x \in [y_{j-1}, y_j]\}(y_j - y_{j-1})$$
$$\geq \sum_{j=1}^{m} \inf\{f(x) : x \in [x_{k-1}, x_k]\}(y_j - y_{j-1})$$
$$= \inf\{f(x) : x \in [x_{k-1}, x_k]\} \sum_{j=1}^{m}(y_j - y_{j-1})$$
$$= \inf\{f(x) : x \in [x_{k-1}, x_k]\}(x_k - x_{k-1}),$$

and hence that

$$L(f, R) = \sum_{k=1}^{n} L(f, R \cap [x_{k-1}, x_k])$$
$$\geq \sum_{k=1}^{n} \inf\{f(x) : x \in [x_{k-1}, x_k]\}(x_k - x_{k-1})$$
$$= L(f, P).$$

In this integral, we call the function f **the integrand**. The proof for upper sums is similar. $\qquad\square$

Example 7.1.9. Let $f(x) = \begin{cases} 1, & \text{if } x \text{ is rational;} \\ 0, & \text{if } x \text{ is irrational.} \end{cases}$ Then for any partition $P = \{x_0, \ldots, x_n\}$ of $[-2, 4]$, for all i, $\inf\{f(x) : x \in [x_{k-1}, x_k]\} = 0$ and $\sup\{f(x) : x \in [x_{k-1}, x_k]\} = 1$, so that $L(f, P) = 0$ and $U(f, P) = \sum_{k=1}^{n} 1 \cdot (x_k - x_{k-1}) = x_n - x_0 = 4 - (-2) = 6$. Thus in this case, changing the partition does not produce better approximations.

Theorem 7.1.10. *For any partitions P and Q of $[a, b]$, and for any bounded function $f : [a, b] \to \mathbb{R}$,*

$$L(f, P) \le U(f, Q).$$

Proof. Let $R = P \cup Q$. (Then R is a refinement of both P and Q.) By Theorem 7.1.8, $L(f, P) \le L(f, R)$ and $U(f, P) \ge U(f, R)$. Since always $L(f, R) \le U(f, R)$, the conclusion follows by transitivity of $\le$. $\square$

It follows that if f is bounded on $[a, b]$, then the set of all lower sums of f as P varies over all the partitions of $[a, b]$ is bounded above, so that the set of all lower sums has a real least upper bound. Similarly, the set of all upper sums is bounded below and has a real greatest lower bound.

Definition 7.1.11. *The **lower integral of f over** $[\mathbf{a}, \mathbf{b}]$ is*

$$L(f) = \sup\{L(f, P) : \text{as } P \text{ varies over partitions of } [a, b]\},$$

*and the **upper integral of f over** $[\mathbf{a}, \mathbf{b}]$ is*

$$U(f) = \inf\{U(f, P) : P \text{ varies over partitions of } [a, b]\}.$$

*We say that f is **integrable over** $[a, b]$ when $L(f) = U(f)$. We call this common value **the integral of f over** $[\mathbf{a}, \mathbf{b}]$ or also **the integral of f from a to b**, and we write it as*

$$\int_a^b f = \int_a^b f(x)\, dx = \int_a^b f(t)\, dt.$$

In this integral, we call the function f **the integrand**.

Theorem 7.1.10 shows that always $L(f) \le U(f)$. Example 7.1.9 shows that sometimes strict inequality holds. Note that we have not yet proved that the function in Example 7.1.7 is integrable, but in Section 7.3 we prove more generally that every continuous function is integrable over a closed bounded interval.

By Exercise 7.1.3, for non-constant functions the upper and lower sums with respect to a partition are never equal. In this case for every partition P, $L(f, P) < U(f, P)$, yet for good, i.e., integrable, functions it can happen that $L(f) = U(f)$.

The simplest examples of integrable functions are constant functions:

Theorem 7.1.12. (Constant rule for integrals.) *If* $f(x) = c$ *for all* $x \in [a, b]$, *then for any partition* P *of* $[a, b]$, $\int_a^b f = L(f) = U(f) = c(b - a)$.

Proof. We apply Theorem 7.1.6 and the definitions.   □

Theorem 7.1.13. *Let* $f : [a, b] \to \mathbb{R}$ *be bounded. Then* f *is integrable over* $[a, b]$ *if and only if for all* $\epsilon > 0$ *there exists a partition* P_ϵ *of* $[a, b]$ *such that* $U(f, P_\epsilon) - L(f, P_\epsilon) < \epsilon$.

Proof. Suppose that f is integrable over $[a, b]$. Then $L(f) = U(f)$. Let $\epsilon > 0$. Since $L(f)$ is the supremum of all lower sums $L(f, P)$ as P varies over partitions of $[a, b]$, there exists a partition P_1 of $[a, b]$ such that $L(f) - L(f, P_1) < \epsilon/2$. Similarly, there exists a partition P_2 of $[a, b]$ such that $U(f, P_2) - U(f) < \epsilon/2$. Let $P = P_1 \cup P_2$. Then P is a partition of $[a, b]$, and by Equation (7.1.5) and by Theorem 7.1.8,

$$L(f, P_1) \leq L(f, P) \leq U(f, P) \leq U(f, P_2).$$

Thus since $L(f) = U(f)$, we have that

$$U(f, P) - L(f, P) \leq U(f, P_2) - L(f, P_1)$$
$$= U(f, P_2) - U(f) + L(f) - L(f, P_1)$$
$$< \epsilon/2 + \epsilon/2 = \epsilon.$$

Now suppose that for every $\epsilon > 0$ there exists a partition P_ϵ of $[a, b]$ such that $U(f, P_\epsilon) - L(f, P_\epsilon) < \epsilon$. By the supremum/infimum definitions of lower and upper integrals,

$$0 \leq U(f) - L(f) \leq U(f, P_\epsilon) - L(f, P_\epsilon) < \epsilon.$$

Since the non-negative constant $U(f) - L(f)$ is strictly smaller than every positive number ϵ, it means, by Theorem 2.10.4, that $U(f) - L(f) = 0$, so that f is integrable.   □

Example 7.1.14. We prove that the following non-continuous function f is integrable over $[0, 1]$ and that $\int_0^1 f = 0$:

$$f(x) = \begin{cases} 1, & \text{if } x = \frac{1}{n} \text{ for some } n \in \mathbb{N}^+; \\ 0, & \text{otherwise.} \end{cases}$$

It is straightforward to verify that this function is not continuous at 0 and at every x of the form $1/n$ for a positive integer n, and that it is continuous everywhere else. For any partition P, $L(f, P) = 0$. So by Theorem 7.1.13 it suffices to prove that for every $\epsilon > 0$ there exists a partition P of $[0, 1]$ such that $U(f, P) < \epsilon$. So let $\epsilon > 0$. We partition $[0, 1]$ so that some sub-intervals are very small and the other sub-intervals have very small (zero) supremum of f. By Theorem 3.8.3, there exists a positive integer p such that $1 < p\epsilon$. Thus $\frac{1}{p} < \epsilon$. Set $d = \frac{1}{2p(p+2)}$. Let $x_0 = 0$, $x_{2p+1} = 1 - d$, $x_{2p+2} = 1$, and for $k = 1, \ldots, p$, set $x_{2k-1} = \frac{1}{p-k+2} - d$ and $x_{2k} = \frac{1}{p-k+2} + d$. Clearly $x_{2k-1} < x_{2k} < 1$, $x_{2p} = \frac{1}{2+d} < 1 - d = x_{2p+1} < x_{2p} = 1$, and $x_{2k} < x_{2k+1}$ because $\frac{1}{p-k+2} - \frac{1}{p-k+1} = \frac{1}{(p-k+2)(p-k+1)} < \frac{1}{(p-1)p} < \frac{1}{p(p+2)} = 2d$. Thus $P = \{x_0, x_1, \ldots, x_{2p+1}\}$ is a partition of $[0, 1]$. For $k = 1, \ldots, p$, the sub-interval $[x_{2k-1}, x_{2k}] = [\frac{1}{p-k+2} - d, \frac{1}{p-k+1} + d]$ is centered on $\frac{1}{p-k+2}$. Thus these p sub-intervals contain the p numbers $1/2, 1/3, \ldots, 1/(p+1)$. The interval $[x_0, x_1] = [0, \frac{1}{p+1} - d] = [0, \frac{1}{p+1} - \frac{1}{2p(p+1)}] = [0, \frac{2p-1}{p+1}]$ contains all $\frac{1}{n}$ for $n \geq p + 2$ (because $\frac{1}{p+2} < \frac{2p-1}{p+1}$), and the subinterval $[x_{2p}, x_{2p+1}]$ contains $1/1$. Consequently, the $p - 1$ sub-intervals $[x_{2k}, x_{2k+1}] = [\frac{1}{p-k+2} + d, \frac{1}{p-k+1} - d]$ for $k = 1, \ldots, p$ contain no elements of the form $1/n$ for a positive integer n. Thus

$$U(f, P) = \frac{1}{p+2} \cdot (x_1 - x_0) + \sum_{k=1}^{p+1} \frac{1}{p-k+2} \cdot (x_{2k} - x_{2k-1})$$

$$+ \sum_{k=1}^{p-1} 0 \cdot (x_{2k+1} - x_{2k}) + (x_{2p+1} - x_{2p})$$

$$\leq d + \left(\sum_{k=1}^{p+1} 2d \right) + d$$

$$= (p+2)(2d)$$

$$= \frac{1}{p}$$

$$< \epsilon.$$

$\square$

Notation 7.1.15. In the definition of integral there is no need to write "dx" when we are simply integrating a function f, as in "$\int_a^b f$": we seek the signed area determined by f over the domain from a to b. For this it does not matter if we like to plug x or t or anything else into f. But when we write "$f(x)$" rather than "f", then we need to write "dx", and the reason is that $f(x)$ is an element of the codomain and is not a function. Why do we have to be pedantic? If x and t are non-dependent variables, by the constant rule (Theorem 7.1.12) we then have

$$\int_a^b f(x)\,dt = f(x)(b - a),$$

and specifically, by geometric reasoning,

$$\int_0^4 x\,dx = 8, \qquad \int_0^4 x\,dt = 4x.$$

Thus writing "dx" versus "dt" is important, and omitting it can lead to confusion: is the answer the constant 8, or is it $4x$ depending on x? Furthermore, if x and t are not independent, we can get further values too. Say if $x = 3t$, then

$$\int_0^4 x\,dt = \int_0^4 3t\,dt = 24.$$

In short, we need to use notation precisely.

The following theorem says that to compute (lower, upper) integrals of integrable functions, we need not use all the possible partitions. It suffices to use, for example, only partitions into equal-length sub-intervals. The second part of the theorem below gives a formulation of integrals that looks very technical but is fundamental for applications (see Section 7.7).

Theorem 7.1.16. *Let $f : [a, b] \to \mathbb{R}$ be bounded and integrable. For each real number $r > 0$ let $P_r = \{x_0^{(r)}, x_1^{(r)}, \ldots, x_{n_r}^{(r)}\}$ be a partition of $[a, b]$ such that each sub-interval $[x_{k-1}^{(r)}, x_k^{(r)}]$ has length at most r. Then*

$$\lim_{r \to 0^+} L(f, P_r) = \lim_{r \to 0^+} U(f, P_r) = \int_a^b f.$$

Furthermore, if for each $r > 0$ and each $k = 1, 2, \ldots, n_r$, $c_k^{(r)}$ is arbitrary in $[x_{k-1}^{(r)}, x_k^{(r)}]$, then

$$\lim_{r \to 0^+} \sum_{k=1}^{n_r} f(c_k^{(r)})(x_k^{(r)} - x_{k-1}^{(r)}) = \int_a^b f.$$

Proof. Let $\epsilon > 0$. By Theorem 7.1.13, there exists a partition P of $[a, b]$ such that $U(f, P) - L(f, P) < \epsilon/2$. Let $P = \{y_0, y_1, \ldots, y_n\}$. Since f is bounded, there exists a positive real number M such that for all $x \in [a, b]$, $|f(x)| < M$. Let r be any positive real number such that

$$r < \frac{1}{2} \min\left\{\frac{\epsilon}{4Mn}, y_1 - y_0, y_2 - y_1, \ldots, y_n - y_{n-1}\right\},$$

and let $P_r = \{x_0, x_1, \ldots, x_{m_r}\}$ (omitting the superscript (r)). By the definition of r, each sub-interval $[x_{k-1}, x_k]$ contains at most one element of P. When it does contain an element of P, we call such k special and we denote its point of P as $y_{j(k)}$. Thus $[x_{k-1}, x_k] = [x_{k-1}, y_{j(k)}] \cup [y_{j(k)}, x_k]$. The elements y_0 and y_n are each in exactly one sub-interval of P_r, whereas the $n - 1$ points $y_1, \ldots, y_{n-1}$ may each be contained in two sub-intervals of P_r (when it is also in P_r), so that there are at most $2 + 2 \cdot (n - 1) = 2n$ special integers. For non-special k there exists a unique $i(k) \in \{1, \ldots, n\}$ such that $[x_{k-1}, x_k] \subseteq [y_{i(k)-1}, y_{i(k)}]$. Then

$$L(f, P_r) = \sum_{k=1}^{m_r} \inf\{f(x) : x \in [x_{k-1}, x_k]\}(x_k - x_{k-1})$$

$$= \sum_{\text{special } k=1}^{m_r} \inf\{f(x) : x \in [x_{k-1}, x_k]\}(x_k - x_{k-1})$$

$$+ \sum_{\text{non-special } k=1}^{m_r} \inf\{f(x) : x \in [x_{k-1}, x_k]\}(x_k - x_{k-1})$$

The last row is greater than or equal to

$$\sum_{\text{non-special } k=1}^{m_r} \inf\{f(x) : x \in [y_{i(k)-1}, y_{i(k)}]\}(x_k - x_{k-1}),$$

which is not quite equal to $L(f, P)$ because for each $i \in \{1, \ldots, n\}$, the union of the sub-intervals $[x_{k-1}, x_k]$ contained in $[y_{i-1}, y_i]$ need not be all of $[y_{i-1}, y_i]$. In fact, $L(f, P)$ equals the summand in the last row plus

$$\sum_{\text{special } k=1}^{m_r} \inf\{f(x) : x \in [y_{j(k)-1}, y_{j(k)}]\}(y_{j(k)} - x_{k-1})$$

$$+ \sum_{\text{special } k=1}^{m_r} \inf\{f(x) : x \in [y_{j(k)}, y_{j(k)+1}]\}(x_k - y_{j(k)}).$$

Thus

$$L(f, P_r) \geq \sum_{\substack{\text{special } k=1}}^{m_r} \inf\{f(x) : x \in [x_{k-1}, x_k]\}(x_k - x_{k-1}) + L(f, P)$$

$$- \sum_{\substack{\text{special } k=1}}^{m_r} \inf\{f(x) : x \in [y_{j(k)-1}, y_{j(k)}]\}(y_{j(k)} - x_{k-1})$$

$$- \sum_{\substack{\text{special } k=1}}^{m_r} \inf\{f(x) : x \in [y_{j(k)}, y_{j(k)+1}]\}(x_k - y_{j(k)}).$$

$$\geq - \sum_{\substack{\text{special } k=1}}^{m_r} M(x_k - x_{k-1}) + L(f, P)$$

$$- \sum_{\substack{\text{special } k=1}}^{m_r} M(y_{j(k)} - x_{k-1}) - \sum_{\substack{\text{special } k=1}}^{m_r} M(x_k - y_{j(k)})$$

$$= L(f, P) - \sum_{\substack{\text{special } k=1}}^{m_r} 2M(x_k - x_{k-1})$$

$$\geq L(f, P) - 2M2nr$$

$$> L(f, P) - \frac{\epsilon}{2}.$$

Since $U(f, P) \geq U(f) = L(f) \geq L(f, P)$ and $U(f, P) - L(f, P) < \frac{\epsilon}{2}$, it follows that $L(f) - L(f, P) < \epsilon/2$. Thus by the triangle inequality and minding which quantities are larger to be able to omit absolute value signs,

$$L(f) - L(f, P_r) < L(f) - L(f, P) + \frac{\epsilon}{2} < \epsilon.$$

This then proves that $\lim_{r \to 0^+} L(f, P_r) = L(f) = \int_a^b f$. Similarly we can prove that $U(f, P_r) - U(f) < \epsilon$ and so that $\lim_{r \to 0^+} U(f, P_r) = U(f) = \int_a^b f$.

Furthermore, by compatibility of order with multiplication by positive numbers and addition, we have that for each r,

$$L(f, P_r) \leq \sum_{k=1}^{n_r} f(c_k^{(r)})(x_k^{(r)} - x_{k-1}^{(r)}) \leq U(f, P_r),$$

so that for all r sufficiently close to 0, $\sum_{k=1}^{n_r} f(c_k^{(r)})(x_k^{(r)} - x_{k-1}^{(r)})$ is within ϵ of $\int_a^b f$. Thus by the definition of limits, $\lim_{r \to 0^+} \sum_{k=1}^{n_r} f(c_k^{(r)})(x_k^{(r)} - x_{k-1}^{(r)}) = \int_a^b f$. $\square$

Exercises for Section 7.1

7.1.1. Prove that if $P = \{x_0, \ldots, x_n\}$ is a partition of $[a, b]$ into n equal parts, then $x_k = a + k\frac{b-a}{n}$.

7.1.2. Let $f(x) = \begin{cases} 1, & \text{if } x < 2; \\ 0, & \text{if } x \geq 2. \end{cases}$ Let P be the partition of $[0, 3]$ into two equal intervals and Q the partition of $[0, 3]$ into three equal intervals.

    i) Compute $L(f, P)$ and $U(f, P)$.

    ii) Compute $L(f, Q)$ and $U(f, Q)$.

    iii) Compare $L(f, P)$ and $L(f, Q)$. Why does this not contradict Theorem 7.1.8?

7.1.3. Let P be a partition of $[a, b]$, and suppose that $L(f, P) = U(f, P)$. Prove that f must be constant on $[a, b]$.

7.1.4. Use geometry to compute the following integrals:

    i) $\displaystyle\int_a^a f =$

    ii) $\displaystyle\int_0^r \sqrt{r^2 - x^2}\, dx =$

    iii) $\displaystyle\int_3^5 (4x - 10)\, dx =$

    iv) $\displaystyle\int_{-1}^5 f$, where $f(x) = \begin{cases} 5, & \text{if } x < -7; \\ 2, & \text{if } -7 \leq x < 1; \\ 3x, & \text{if } 1 \leq x < 3; \\ 9 - x, & \text{if } 3 \leq x. \end{cases}$

7.1.5. Use geometry to compute the following integrals (t and x do not depend on each other). Justify all work.

    i) $\displaystyle\int_0^r \sqrt{r^2 - x^2}\, dt =$

    ii) $\displaystyle\int_3^5 (4x - 10)\, dt =$

    iii) $\displaystyle\int_{-1}^5 f$, where $f(t) = \begin{cases} 5, & \text{if } t < -7; \\ 2, & \text{if } -7 \leq t < 1; \\ 3t, & \text{if } 1 \leq t < 3; \\ 9 - x, & \text{if } 3 \leq t. \end{cases}$

7.1.6. Let P_n be a partition of $[-1, 1]$ into n equal parts. Let $f(x) = 1$ if $x = 0$ and $f(x) = 0$ otherwise.

    i) Graph f and conclude that f is not continuous at 0.

    ii) Compute $L(f, P_n)$ and $U(f, P_n)$.

7.1.7. EasyLanders define Eantegrals as follows: they divide the interval $[a, b]$ into n equal sub-intervals, so each has length $\frac{b-a}{n}$. By Exercise 7.1.1, the kth sub-interval given in this way is $\left[a + (k-1)\frac{b-a}{n}, a + k\frac{b-a}{n}\right]$. Rather than finding the minimum and maximum of f on this sub-interval, they simply Eapproximate f on the sub-interval by plugging in the right endpoint $a + k\frac{b-a}{n}$, so that the signed area over the kth sub-interval is Eapproximately $f\left(a + k\frac{b-a}{n}\right)\frac{b-a}{n}$. Thus the Eapproximate signed area of f over $[a, b]$ via this partition is $\sum_{k=1}^{n} f\left(a + k\frac{b-a}{n}\right)\frac{b-a}{n}$. If this sum has a limit as n goes to infinity, then EasyLanders declare the Eantegral of f over $[a, b]$ to be

$$\int_a^b f = \lim_{n \to \infty} \sum_{k=1}^{n} f\left(a + k\frac{b-a}{n}\right)\frac{b-a}{n}.$$

Suppose that $f : [a, b] \to \mathbb{R}$ is integrable. Prove that the Eantegral $\int_a^b f$ exists and equals the integral $\int_a^b f$.

7.1.8. Use Eantegrability from the previous exercise. Let

$$f(x) = \begin{cases} 1, & \text{if } x \text{ is rational;} \\ 0, & \text{if } x \text{ is irrational.} \end{cases}$$

By Example 7.1.9 we know that f is not integrable over $[0, 2]$.

i) Prove that f is Eantegrable over $[0, 2]$ and find its Eantegral $\int_0^2 f$.

ii) Compute $\int_0^{\sqrt{2}} f$, $\int_{\sqrt{2}}^2 f$, and prove that $\int_0^{\sqrt{2}} f + \int_{\sqrt{2}}^2 f \neq \int_0^2 f$.

7.2 Computing integrals from upper and lower sums

The definition of integrals appears daunting: we seem to need to compute all the possible lower sums to get at the lower integral, all the possible upper sums to get at the upper integral, and only if the lower and upper integrals are the same do we have the precise integral. In Example 7.1.7 in the previous section we have already seen that numerically we can often compute the integral to within desired precision by taking finer and finer partitions. In this section we compute some precise numerical values of integrals, and without computing **all** the possible upper and lower sums. Admittedly, the computations are time-consuming, but the reader is encouraged to read through them to get an idea of what calculations are needed to follow the definition of integrals. In Section 7.4 we will see very efficient shortcuts for computing integrals, but only for easy/good functions.

Example 7.2.1. Let $f(x) = x$ on $[2, 6]$. We know that the area under the curve between $x = 2$ and $x = 6$ is 16. Here we compute that indeed $\int_2^6 f = 16$. For any positive integer n let $P_n = \{x_0, \ldots, x_n\}$ be the partition of $[2, 6]$ into n equal parts. By Exercise 7.1.1, $x_k = 2 + k\frac{4}{n}$. Since f is increasing, on each sub-interval $[x_{k-1}, x_k]$ the minimum is x_{k-1} and the maximum is x_k. Thus

$$U(f, P_n) = \sum_{k=1}^n x_k(x_k - x_{k-1})$$

$$= \sum_{k=1}^n \left(2 + k\frac{4}{n}\right)\frac{4}{n}$$

$$= \sum_{k=1}^n 2\frac{4}{n} + \sum_{k=1}^n k\left(\frac{4}{n}\right)^2$$

$$= 2\frac{4}{n}n + \frac{n(n+1)}{2}\left(\frac{4}{n}\right)^2$$

$$= 8 + \frac{8(n+1)}{n}.$$

It follows that for all n,

$$U(f) = \inf\{U(f, P) : P \text{ varies over partitions of } [2, 6]\}$$
$$\leq U(f, P_n)$$
$$= 8 + \frac{8(n+1)}{n},$$

so that $U(f) \leq 8 + 8 = 16$. Similarly,

$$L(f, P_n) = \sum_{k=1}^n x_{k-1}(x_k - x_{k-1})$$

$$= \sum_{k=1}^n \left(2 + (k-1)\frac{4}{n}\right)\frac{4}{n}$$

$$= \sum_{k=1}^n 2\frac{4}{n} + \sum_{k=1}^n (k-1)\left(\frac{4}{n}\right)^2$$

$$= 2\frac{4}{n}n + \frac{(n-1)n}{2}\left(\frac{4}{n}\right)^2$$

$$= 8 + \frac{8(n-1)}{n},$$

and

$$L(f) = \sup\{L(f, P) : \ P \text{ varies over partitions of } [2, 6]\}$$
$$\geq \sup\{L(f, P_n) : n \in \mathbb{N}^+\}$$
$$= 16.$$

All together this says that

$$16 \leq L(f) \leq U(f) \leq 16,$$

so that $L(f) = U(f) = 16$, and finally that $\int_2^6 f = 16$.

Note that we did not compute all the possible lower and upper sums, but we computed enough of them. We knew that we computed enough of them because as n goes large, $\sup\{L(f, P_n) : n\} = \inf\{U(f, P_n) : n\}$.

Example 7.2.2. We compute the integral for $f(x) = x^2$ over $[1, 7]$. For any positive integer n let $P_n = \{x_0, \ldots, x_n\}$ be the partition of $[1, 7]$ into n equal parts. By Exercise 7.1.1, $x_k = 1 + k\frac{6}{n}$. Since f is increasing on $[1, 7]$, on each sub-interval $[x_{k-1}, x_k]$ the minimum is x_{k-1}^2 and the maximum is x_k^2. Thus, by using Exercise 1.6.1 in one of the steps below:

$$U(f, P_n) = \sum_{k=1}^{n} x_k^2(x_k - x_{k-1})$$

$$= \sum_{k=1}^{n} \left(1 + k\frac{6}{n}\right)^2 \frac{6}{n}$$

$$= \sum_{k=1}^{n} \left(1 + k\frac{12}{n} + k^2\frac{36}{n^2}\right) \frac{6}{n}$$

$$= \sum_{k=1}^{n} \frac{6}{n} + \sum_{k=1}^{n} k\frac{12}{n}\frac{6}{n} + \sum_{k=1}^{n} k^2 \left(\frac{6}{n}\right)^3$$

$$= \frac{6}{n}n + \frac{n(n+1)}{2}\frac{12}{n}\frac{6}{n} + \frac{n(n+1)(2n+1)}{6}\left(\frac{6}{n}\right)^3$$

$$= 6 + \frac{36(n+1)}{n} + \frac{36(n+1)(2n+1)}{n^2},$$

so that $U(f) \leq \inf\{U(f, P_n) : n\} = 6 + 36 + 72 = 114$. Similarly,

$$L(f) \geq L(f, P_n) = 114,$$

whence $114 \leq L(f) \leq U(f) \leq 114$ and $\int_1^7 f = 114$.

Example 7.2.3. The goal of this exercise is to compute $\int_0^2 \sqrt{x}\,dx$. For the first attempt, let $P_n = \{x_0, \ldots, x_n\}$ be the partition of $[0, 2]$ into n equal intervals. By Exercise 7.1.1, $x_k = \frac{2k}{n}$. The square root function is increasing, so that

$$U(f, P_n) = \sum_{k=1}^n \sqrt{x_k}(x_k - x_{k-1})$$

$$= \sum_{k=1}^n \sqrt{\frac{2k}{n}} \frac{2}{n}$$

$$= \left(\frac{2}{n}\right)^{3/2} \sum_{k=1}^n \sqrt{k}.$$

But now we are stuck: we have no simplification of $\sum_{k=1}^n \sqrt{k}$, and we have no other immediate tricks to compute $\inf\{\left(\frac{2}{n}\right)^{3/2} \sum_{k=1}^n \sqrt{k} : n\}$.

But it **is** possible to compute **enough** upper and lower sums for this function to get the integral. Namely, for each positive integer n let $Q_n = \{0, \frac{2}{n^2}, \frac{2\cdot4}{n^2}, \frac{2\cdot9}{n^2}, \ldots, \frac{2\cdot(n-1)^2}{n^2}, \frac{2\cdot n^2}{n^2} = 2\}$. This is a partition of $[0, 2]$ into n (unequal) parts, with $x_k = \frac{2k^2}{n^2}$. Since $\sqrt{}$ is an increasing function, on each sub-interval $[x_{k-1}, x_k]$ the minimum is achieved at x_{k-1} and the maximum at x_k, so that

$$U(f, Q_n) = \sum_{k=1}^n \frac{\sqrt{2}\,k}{n}\left(\frac{2k^2}{n^2} - \frac{2(k-1)^2}{n^2}\right)$$

$$= \frac{2\sqrt{2}}{n^3} \sum_{k=1}^n k\left(k^2 - (k-1)^2\right)$$

$$= \frac{2\sqrt{2}}{n^3} \sum_{k=1}^n k\left(2k - 1\right)$$

$$= \frac{4\sqrt{2}}{n^3} \sum_{k=1}^n k^2 - \frac{2\sqrt{2}}{n^3} \sum_{k=1}^n k$$

$$= \frac{4\sqrt{2}}{n^3} \frac{n(n+1)(2n+1)}{6} - \frac{2\sqrt{2}}{n^3} \frac{n(n+1)}{2} \quad \text{(by Exercise 1.6.1)}$$

$$= \frac{2\sqrt{2}}{6n^2}(n+1)(2(2n+1)-3) \text{ (by factoring)}$$

$$= \frac{\sqrt{2}}{3n^2}(n+1)(4n-1).$$

Thus $U(f) \leq \inf\{U(f, P_n) : n\} = \inf\{\frac{\sqrt{2}}{3n^2}(n+1)(4n-1) : n\} = \frac{4\sqrt{2}}{3}$. Similarly, $L(f) \geq \frac{4\sqrt{2}}{3}$, so that $\int_0^2 \sqrt{x}\, dx = \frac{4\sqrt{2}}{3}$.

Exercises for Section 7.2

7.2.1. Mimic examples in this section to compute $\int_0^2 x^3\, dx$. (You may need Exercise 1.6.2.)

7.2.2. Mimic examples in this section to compute $\int_0^2 \sqrt[3]{x}\, dx$.

7.3 What functions are integrable?

Theorem 7.3.1. *Every continuous real-valued function on $[a, b]$ is integrable over $[a, b]$, where $a, b \in \mathbb{R}$ with $a < b$.*

Proof. Let $f : [a, b] \to \mathbb{R}$ be continuous. We need to prove that $L(f) = U(f)$. By Theorem 2.10.4 it suffices to prove that for all $\epsilon > 0$, $U(f) - L(f) \leq \epsilon$.

So let $\epsilon > 0$. By Theorem 5.5.4, f is uniformly continuous, so there exists $\delta > 0$ such that for all $x, c \in [a, b]$, if $|x - c| < \delta$ then $|f(x) - f(c)| < \epsilon/(b-a)$. Let $P = \{x_0, x_1, \ldots, x_n\}$ be a partition of $[a, b]$ such that for all $k = 1, \ldots, n$, $x_k - x_{k-1} < \delta$. (For example, this can be accomplished as follows: by Theorem 3.8.3, there exists $n \in \mathbb{N}^+$ such that $(b - a) < n\delta$, and then P can be taken to be the partition of $[a, b]$ into n equal parts.) Then

$$U(f, P) - L(f, P)$$

$$= \sum_{k=1}^{n} \left(\sup\{f(x) : x \in [x_{k-1}, x_k]\} - \inf\{f(x) : x \in [x_{k-1}, x_k]\}\right)(x_k - x_{k-1})$$

$$\leq \sum_{k=1}^{n} \frac{\epsilon}{(b-a)}(x_k - x_{k-1})$$

(by uniform continuity since $x_k - x_{k-1} < \delta$)

$$= \frac{\epsilon}{(b-a)} \sum_{k=1}^{n} (x_k - x_{k-1})$$

$$= \epsilon.$$

But $U(f) \leq U(f, P)$ and $L(f) \geq L(f, P)$, so that $0 \leq U(f) - L(f) \leq U(f, P) - L(f, P) \leq \epsilon$. Thus $U(f) = L(f)$ by Theorem 2.10.4, so that f is integrable over $[a, b]$. $\square$

Both the proof above as well as Theorem 7.1.16 prove that for a continuous function f on $[a, b]$,

$$\int_a^b f = \lim_{r \to 0^+} \sum_{k=1}^{n_r} f(c_k^{(r)})(x_k^{(r)} - x_{k-1}^{(r)}),$$

where for each $r > 0$, $\{x_0^{(r)}, x_1^{(r)}, \ldots, x_{n_r}^{(r)}\}$ is a partition of $[a, b]$ into sub-intervals of length at most r, and for each $k = 1, \ldots, n_r$, $c_k \in [x_{k-1}, x_k]$.

Theorem 7.3.2. *Let $f : [a, b] \to \mathbb{R}$ be bounded. Let $S = \{s_0, s_1, \ldots, s_m\}$ be a finite subset of $[a, b]$ with $s_0 = a < s_1 < \cdots < s_m = b$. Suppose that for all $d > 0$ and all k f is integrable over $[s_{k-1} + d, s_k - d]$. (By the previous theorem, Theorem 7.3.1, any f that is continuous at all $x \in [a, b] \backslash S$ satisfies this hypothesis.) Then f is integrable over $[a, b]$, and*

$$\int_a^b f = \sum_{k=1}^{m} \int_{s_{k-1}}^{s_k} f.$$

Proof. Let $\epsilon > 0$. Since f is bounded, there exists a positive real number M such that for all $x \in [a, b]$, $|f(x)| < M$. Let $e = \frac{1}{3} \min\{s_1 - s_0, s_2 - s_1, \ldots, s_m - s_{m-1}\}$ and $d = \min\{e, \epsilon/(4M(m+1)(2m+1))\}$. By assumption, for each $k = 1, \ldots, m - 1$, f is integrable on the interval $[s_{k-1} + d, s_k - d]$. Thus by Theorem 7.1.13, there exists a partition P_k of $[s_{k-1} + d, s_k - d]$ such that $U(f, P_k) - L(f, P_k) < \epsilon/(2m + 1)$. Now let $P = \{a\} \cup P_1 \cup P_2 \cup \cdots \cup P_{m-1} \cup \{b\}$. Then P is a partition of $[a, b]$, and

$U(f, P) - L(f, P)$

$$= \sum_{k=1}^{m} (U(f, P \cap [s_{k-1} + d, s_k - d]) - L(f, P \cap [s_{k-1} + d, s_k - d]))$$

$$+ (U(f, P \cap [a, a + d]) - L(f, P \cap [a, a + d]))$$

$$+ \sum_{k=1}^{m-1} (U(f, P \cap [s_k - d, s_k + d]) - L(f, P \cap [s_k - d, s_k + d]))$$

$$+ (U(f, P \cap [b - d, b]) - L(f, P \cap [b - d, b]))$$

$$\leq \sum_{k=1}^{m} (U(f, P_k) - L(f, P_k)) + 2M \cdot 2d + \sum_{k=1}^{m-1} 2M \cdot 2d + 2M \cdot 2d$$

$$< \sum_{k=1}^{m} \frac{\epsilon}{2m + 1} + 4M(m + 1)d$$

$$< \epsilon.$$

Thus by Theorem 7.3.1, f is integrable. Furthermore, $\int_a^b f - \sum_{k=1}^{m} \int_{s_{k-1}}^{s_k} f$ is bounded above by $U(f, P) - \sum_{k=1}^{m} L(f, P_k)$ and below by $L(f, P) - \sum_{k=1}^{m} U(f, P_k)$, and the upper and lower bounds are within 2ϵ of 0, so that $\int_a^b f - \sum_{k=1}^{m} \int_{s_{k-1}}^{s_k} f$ is within 2ϵ of 0. Since ϵ is arbitrary, it follows by Theorem 2.10.4 that $\int_a^b f = \sum_{k=1}^{m} \int_{s_{k-1}}^{s_k} f$. $\qquad \square$

Notation 7.3.3. What could possibly be the meaning of $\int_b^a f$ if $a < b$? In our definition of integrals, all partitions started from a smaller a to a larger b to get $\int_a^b f$. If we did reverse b and a, then the widths of the sub-intervals in each partition would be negative ($x_k - x_{k-1} = -(x_{k-1} - x_k)$), so that all the partial sums and both the lower and the upper integrals would get the negative value. Thus it seems reasonable to declare

$$\int_b^a f = - \int_a^b f.$$

In fact, this is exactly what makes Theorem 7.3.2 work without any order assumptions on the s_k. For example, if $a < c < b$, by Theorem 7.3.2, $\int_a^b f = \int_a^c f + \int_c^b f$, whence $\int_a^c f = \int_a^b f - \int_c^b f = \int_a^b f + \int_b^c f$.

Theorem 7.3.4. *Suppose that f and g are integrable over $[a, b]$, and that $c \in \mathbb{R}$. Then $f + cg$ is integrable over $[a, b]$ and $\int_a^b (f + cg) = \int_a^b f + c \int_a^b g$.*

Proof. We first prove that $L(cg) = U(cg) = cL(g) = cU(g)$. If $c \geq 0$, then

$$L(cg) = \sup\{L(cg, P) : P \text{ a partition of } [a, b]\}$$
$$= \sup\{cL(g, P) : P \text{ a partition of } [a, b]\}$$
$$= c\sup\{L(g, P) : P \text{ a partition of } [a, b]\}$$
$$= cL(g)$$
$$= cU(g)$$
$$= c\inf\{U(g, P) : P \text{ a partition of } [a, b]\}$$

$$= \inf\{cU(g, P) : P \text{ a partition of } [a, b]\}$$
$$= \inf\{U(cg, P) : P \text{ a partition of } [a, b]\}$$
$$= U(cg),$$

and if $c < 0$, then

$$L(cg) = \sup\{L(cg, P) : P \text{ a partition of } [a, b]\}$$
$$= \sup\{cU(g, P) : P \text{ a partition of } [a, b]\}$$
$$= c\inf\{U(g, P) : P \text{ a partition of } [a, b]\}$$
$$= cU(g)$$
$$= cL(g)$$
$$= c\inf\{L(g, P) : P \text{ a partition of } [a, b]\}$$
$$= \sup\{cL(g, P) : P \text{ a partition of } [a, b]\}$$
$$= \inf\{U(cg, P) : P \text{ a partition of } [a, b]\}$$
$$= U(cg).$$

This proves that cg is integrable with $\int_a^b (cg) = c\int_a^b g$.

Let $\epsilon > 0$. By integrability of f and cg there exist partitions P, Q of $[a, b]$ such that $U(f, P) - L(f, P) < \epsilon/2$ and $U(cg, Q) - L(cg, Q) < \epsilon/2$. Let $R = P \cup Q$. Then R is a partition of $[a, b]$, and by Theorem 7.1.8, $U(f, R) - L(f, R) < \epsilon/2$ and $U(cg, R) - L(cg, R) < \epsilon/2$. By Exercise 2.7.7, for every partition of $[a, b]$, and in particular for the partition R, $L(f + cg, R) \geq L(f, R) + L(cg, R)$, and $U(f + cg, R) \leq U(f, R) + U(cg, R)$. Then

$$0 \leq U(f + cg) - L(f + cg)$$
$$\leq U(f + cg, R) - L(f + cg, R)$$
$$\leq U(f, R) + U(cg, R) - L(f, R) - L(cg, R)$$
$$< \epsilon.$$

Thus $U(f + cg) - L(f + cg) = 0$ by Theorem 2.10.4, and so $f + cg$ is integrable. The inequalities $L(f + cg, R) \geq L(f, R) + L(cg, R)$, and $U(f + cg, R) \leq U(f, R) + U(cg, R)$ furthermore prove that $L(f) + L(cg) \leq L(f + cg) \leq U(f + cg) \leq U(f) + U(cg) = L(f) + L(cg)$, so that finally

$$\int_a^b (f + cg) = \int_a^b f + \int_a^b (cg) = \int_a^b f + c\int_a^b g. \qquad \square$$

Theorem 7.3.5. *Let $a, b \in \mathbb{R}$ with $a < b$. Let $f, g : [a, b] \to \mathbb{R}$ be integrable functions such that $f(x) \leq g(x)$ for all $x \in [a, b]$. Then*

$$\int_a^b f \leq \int_a^b g.$$

Here is a picture that illustrates this theorem: the values of g are at each point in the domain greater than or equal to the values of f, and the area under the graph of g is larger than the area under the graph of f:

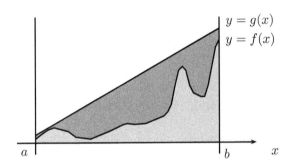

Proof. By assumption on every sub-interval I of $[a, b]$, $\inf\{f(x) : x \in I\} \leq \inf\{g(x) : x \in I\}$. Thus for all partitions P of $[a, b]$, $L(f, P) \leq L(g, P)$. Hence $L(f) \leq L(g)$, and since f and g are integrable, this says that $\int_a^b f \leq \int_a^b g$. $\qquad\square$

Exercises for Section 7.3

7.3.1. The following is known: $\int_0^1 f = 5$, $\int_2^3 f = 6$, $\int_0^3 f = 15$, $\int_1^2 g = -3$. Compute and justify $\int_1^2 (3f - 4g)$, $\int_1^3 5f$, $\int_2^1 3g$.

7.3.2. Let $f(x) = \begin{cases} x, & \text{if } x \text{ is rational;} \\ 0, & \text{if } x \text{ is irrational.} \end{cases}$

    i) Prove that f is not integrable over $[0, 1]$.

    ii) Does this contradict Theorem 7.3.2? Justify.

7.3.3. Let $f : [-a, a] \to \mathbb{R}$ be an integrable odd function. Prove that $\int_{-a}^a f = 0$.

7.3.4. Let $f : [-a, a] \to \mathbb{R}$ be an integrable even function. Prove that $\int_{-a}^a f = 2 \int_0^a f$.

7.3.5. Let $f : [a, b] \to \mathbb{R}$ be piecewise continuous as in Theorem 7.3.2.

i) Prove that $|f|$ is integrable over $[a, b]$.

ii) Prove that $\left| \int_a^b f \right| \le \int_a^b |f|$.

7.3.6. Find a function $f : [0, 1] \to \mathbb{R}$ that is not integrable over $[0, 1]$ but such that $|f|$ is integrable over $[0, 1]$.

7.3.7. So far we have seen that every differentiable function is continuous and that every continuous function is integrable.

i) Give an example of a continuous function that is not differentiable.

ii) Give an example of an integrable function that is not continuous.

*7.3.8. Let $f : [a, b] \to \mathbb{R}$ be monotone. Prove that f is integrable over $[a, b]$.

7.4   The Fundamental theorem of calculus

Despite first appearances, it turns out that integration and differentiation are related. For this we have two versions of the Fundamental theorem of calculus.

Theorem 7.4.1. (The Fundamental theorem of calculus, I) *Let $f, g :$ $[a, b] \to \mathbb{R}$ such that f is integrable over $[a, b]$ and g is differentiable with $g' = f$. Then*

$$\int_a^b f = g(b) - g(a).$$

Proof. Let $P = \{x_0, x_1, \ldots, x_n\}$ be a partition of $[a, b]$. Since g is differentiable on $[a, b]$, it is continuous on each $[x_{k-1}, x_k]$ and differentiable on each (x_{k-1}, x_k). Thus by the Mean value theorem (Theorem 6.3.4), there exists $c_k \in (x_{k-1}, x_k)$ such that $f(c_k) = g'(c_k) = \frac{g(x_k) - g(x_{k-1})}{x_k - x_{k-1}}$. By the definition of lower and upper sums,

$$L(f, P) \le \sum_{k=1}^n f(c_k)(x_k - x_{k-1}) \le U(f, P).$$

But

$$\sum_{k=1}^n f(c_k)(x_k - x_{k-1}) = \sum_{k=1}^n \frac{g(x_k) - g(x_{k-1})}{x_k - x_{k-1}}(x_k - x_{k-1})$$

$$= \sum_{k=1}^n (g(x_k) - g(x_{k-1}))$$

$$= g(x_n) - g(x_0)$$
$$= g(b) - g(a),$$

so that $L(f, P) \leq g(b) - g(a) \leq U(f, P)$, whence

$$L(f) = \sup\{L(f, P) : P \text{ a partition of } [a, b]\}$$
$$\leq g(b) - g(a)$$
$$\leq \inf\{U(f, P) : P \text{ a partition of } [a, b]\}$$
$$= U(f).$$

Since f is integrable over $[a, b]$, by definition $L(f) = U(f)$, and so all inequalities above have to be equalities, so that necessarily $\int_a^b f = g(b) - g(a)$. $\square$

The general notation for applying Theorem 7.4.1 is as follows: if $g' = f$, then

$$\int_a^b f = g(x)\Big|_a^b = g(b) - g(a).$$

For example, since $\frac{1+2x-x^2}{(1+x^2)^2}$ is the derivative of $\frac{x-1}{1+x^2}$, it follows that

$$\int_0^1 \frac{1+2x-x^2}{(1+x^2)^2} dx = \frac{x-1}{1+x^2}\Big|_0^1 = \frac{1-1}{1+1^2} - \frac{0-1}{1+0^2} = 1.$$

If we instead had to compute this integral with upper and lower sums, it would take us a lot longer and a lot more effort to come up with the answer.

In general, upper and lower sums and integrals are time-consuming and we want to avoid them if possible. The fundamental theorem of calculus that we just proved enables us to do that for many functions: to integrate f over $[a, b]$ one needs to find g with $g' = f$. Such g is called an **antiderivative** of f. For example, if r is a rational number different from -1, then by the power rule (Theorem 6.2.10), an antiderivative of x^r is $\frac{x^{r+1}}{r+1}$. By the scalar rule for derivatives, for any constant C, $\frac{x^{r+1}}{r+1} + C$ is also an antiderivative. It does not matter which antiderivative we choose to compute the integral:

$$\int_a^b x^r \, dx = \left(\frac{b^{r+1}}{r+1} + C\right) - \left(\frac{a^{r+1}}{r+1} + C\right) = \frac{b^{r+1}}{r+1} - \frac{a^{r+1}}{r+1},$$

so that the choice of the antiderivative is irrelevant.

Definition 7.4.2. (**Indefinite integral**) *If g is an antiderivative of f, we write also*

$$\int f(x)\,dx = g(x) + C,$$

where C stands for an arbitrary constant.

For example, $\int 3x^2\,dx = x^3 + C$, $\int 3\,dx = 3x + C$, $\int t\,dx = tx + C$, $\int x\,dx = \frac{1}{2}x^2 + C$, and so on. (Study the differences and similarities of the last three.)

So far we have seen 2^x for rational exponents x. Exercise 5.4.7 also allows real exponents, and proves that this function of x is continuous. Thus by Theorem 7.3.1 this function is integrable. We do not yet know $\int 2^x\,dx$, but in Theorem 7.6.5 we will see that $\int 2^x\,dx = \frac{1}{\ln 2}2^x + C$. For $\int 2^{(x^2)}\,dx$ instead, you and I do not know an antiderivative, we will not know one by the end of the course, and there actually is no "**closed-form**" antiderivative. This fact is due to a theory of Joseph Liouville (1809–1882). What is the meaning of "closed-form"? Here is an oblique answer: Exercise 10.1.5 claims that there exists an infinite power series (sum of infinitely many terms) that is an antiderivative of $2^{(x^2)}$. Precisely because of this infinite sum nature, the values of any antiderivative of $2^{(x^2)}$ cannot be computed precisely, only approximately. Furthermore, according to Liouville's theory, that infinite sum cannot be expressed in terms of the more familiar standard functions, and neither can any other expression for an antiderivative. It is in this sense that we say that $2^{(x^2)}$ does not have a "closed-form" antiderivative.

(It is a fact that in the ocean of all functions, those for which there is a "closed-form" antiderivative form only a tiny droplet.)

At this point we know very few methods for computing antiderivatives. We will in time build a bigger stash of functions: see the next section (Section 7.6) and the chapter on power series (Chapter 9).

The simplest method for finding more antiderivatives is to first find a differentiable function and compute its derivative, and voilá, the original function is an antiderivative of its derivative. For example, by the chain and power rules, $(x^2 + 3x)^{100}$ is an antiderivative of $100(x^2 + 3x)^{99}(2x + 3)$.

Theorem 7.4.3. (**The Fundamental theorem of calculus, II**) *Let $f : [a, b] \to \mathbb{R}$ be continuous. Then for all $x \in [a, b]$, f is integrable over*

$[a, x]$, and the function $g : [a, b] \to \mathbb{R}$ given by $g(x) = \int_a^x f$ is differentiable on (a, b) with

$$\frac{d}{dx} \int_a^x f = f(x).$$

Proof. Since f is continuous over $[a, b]$, it is continuous over $[a, x]$, so that by Theorem 7.3.1, f is integrable over $[a, x]$. Thus g is a well-defined function. Let $c \in (a, b)$. We will prove that g is differentiable at c.

Let $\epsilon > 0$. By continuity of f at c, there exists $\delta > 0$ such that for all $x \in [a, b]$, if $|x - c| < \delta$ then $|f(x) - f(c)| < \epsilon$. Thus on $[c - \delta, c + \delta] \cap [a, b]$, $f(c) - \epsilon < f(x) < f(c) + \epsilon$, so that by Theorem 7.3.5,

$$\int_{\min\{x,c\}}^{\max\{x,c\}} (f(c) - \epsilon) \leq \int_{\min\{x,c\}}^{\max\{x,c\}} f \leq \int_{\min\{x,c\}}^{\max\{x,c\}} (f(c) + \epsilon).$$

But by integrals of constant functions,

$$\int_{\min\{x,c\}}^{\max\{x,c\}} (f(c) \pm \epsilon) = (\max\{x, c\} - \min\{x, c\})(f(c) \pm \epsilon) = |x - c|(f(c) \pm \epsilon).$$

Thus

$$|x - c|(f(c) - \epsilon) \leq \int_{\min\{x,c\}}^{\max\{x,c\}} f \leq |x - c|(f(c) + \epsilon).$$

If $x \neq c$, dividing by $|x - c|$ and rewriting the middle term says that

$$f(c) - \epsilon \leq \frac{\int_c^x f}{x - c} \leq f(c) + \epsilon,$$

whence $-\epsilon \leq \frac{\int_c^x f}{x-c} - f(c) \leq \epsilon$, and $\left| \frac{\int_c^x f}{x-c} - f(c) \right| < \epsilon$. Then for all $x \in [a, b]$, if $0 < |x - c| < \delta$, then

$$\left| \frac{g(x) - g(c)}{x - c} - f(c) \right| = \left| \frac{\int_a^x f - \int_a^c f}{x - c} - f(c) \right|$$

$$= \left| \frac{\int_c^x f}{x - c} - f(c) \right|$$

(by Theorem 7.3.2 and Notation 7.3.3)

$$< \epsilon.$$

Thus $\lim_{x \to c} \dfrac{g(x) - g(c)}{x - c}$ exists and equals $f(c)$, i.e., $g'(c) = f(c)$. $\square$

It is probably a good idea to review the notation again. The integral $\int_a^x f$ can also be written as

$$\int_a^x f = \int_a^x f(t)\, dt = \int_a^x f(z)\, dz.$$

This is a function of x because x appears in the bound of the domain of integration. Note similarly that $\int_a^x f(t)\, dz$ are functions that depend on t and on x but not on z. Thus by the Fundamental theorem of calculus, II,

$$\frac{d}{dx}\int_a^x f = f(x), \text{ and } \frac{d}{dx}\int_a^x f(t)\, dz = f(t).$$

Do not write "$\int_a^x f(x)\, dx$": this is trying to say that x varies from a to x, so one occurrence of the letter x is constant and the other occurrence varies from a to that constant, which mixes up the symbols too much.

Corollary 7.4.4. (The Fundamental theorem of calculus, II) *Let $f : [a,b] \to \mathbb{R}$ be continuous and let k, h be differentiable functions with codomain in $[a,b]$. Then*

$$\frac{d}{dx}\int_{k(x)}^{h(x)} f = f(h(x))h'(x) - f(k(x))k'(x).$$

Proof. Let $g(x) = \int_a^x f$. By Theorem 7.4.3, g is differentiable and $g' = f$. Then

$$\frac{d}{dx}\int_{k(x)}^{h(x)} f = \frac{d}{dx}\left(\int_a^{h(x)} f - \int_a^{k(x)} f\right)$$

$$= \frac{d}{dx}\left(g(h(x)) - g(k(x))\right)$$

$$= \frac{d}{dx}\left(g'(h(x))h'(x) - g(k(x))k'(x)\right) \text{ (by the chain rule)}$$

$$= f(h(x))h'(x) - f(k(x))k'(x). \qquad \square$$

Exercises for Section 7.4

7.4.1. Suppose that $f : [a,b] \to \mathbb{R}^+$ is continuous and that $f(c) > 0$ for some $c \in [a,b]$. Prove that $\int_a^b f > 0$.

7.4.2. Compute the integrals below. You may want to use clever guessing and rewriting.

i) $\displaystyle\int_0^1 \sqrt{x}\,dx =$

ii) $\displaystyle\int_0^1 16x(x^2+4)^7\,dx =$

7.4.3. Compute the integrals below, assuming that x is not a function depending on t.

i) $\displaystyle\int_0^1 x^3\,dx =$

ii) $\displaystyle\int_0^1 x^3\,dt =$

iii) $\displaystyle\int_0^x x^3\,dt =$

7.4.4. Below t and x do not depend on each other. Compute the following derivatives, possibly using Theorem 7.4.3.

i) $\displaystyle\frac{d}{dx}\int_0^x t^3\,dt =$

ii) $\displaystyle\frac{d}{dx}\int_0^x x^3\,dt =$

iii) $\displaystyle\frac{d}{dx}\int_0^t x^3\,dx =$

iv) $\displaystyle\frac{d}{dx}\int_0^t t^3\,dx =$

7.4.5. (Integration by substitution) By the chain rule for differentiation, $(f \circ g)'(x) = f'(g(x))g'(x)$.

i) Prove that $\int_a^b f'(g(x))g'(x)\,dx = f(g(b)) - f(g(a))$.

ii) Prove that $\int f'(g(x))g'(x)\,dx = f(g(x)) + C$.

iii) Integrate by substitution by explicitly stating f, g:

$\displaystyle\int_2^3 (2x-4)^{10}\,dx =$

$\displaystyle\int_1^3 \frac{4x+3}{\sqrt{2x^2+3x}^9}\,dx =$

$\displaystyle\int_1^3 (8x+6)\sqrt[3]{2x^2+3x}\,dx =$

7.4.6. (Integration by parts) By the product rule for differentiation, $(f \cdot g)'(x) = f'(x)g(x) + f(x)g'(x)$.

i) Prove that $\int_a^b f'(x)g(x)\,dx = f(b)g(b) - f(a)g(a) - \int_a^b f(x)g'(x)\,dx$.

ii) Prove that $\int f'(x)g(x)\,dx = f(x)g(x) - \int f(x)g'(x)\,dx$.

iii) Compute the following integrals applying this rule explicitly stating f, g:

$$\int_{-1}^{1} (4x+3)(5x+1)^{10}\,dx =$$

$$\int_{-1}^{1} \frac{4x+3}{\sqrt{2x+4}}\,dx =$$

7.4.7. Compute the following derivatives. (Hint: the Fundamental theorem of calculus and the chain rule.)

i) $\dfrac{d}{dx} \displaystyle\int_2^{3x} \sqrt{t^4 + 5\sqrt{t}}\,dt$.

ii) $\dfrac{d}{dx} \displaystyle\int_{-x^2}^{\sqrt{x}} \dfrac{t + 5\sqrt{t}}{t^{100} - 2t^{5}0 + t^7 - 2}\,dt$.

iii) $\dfrac{d}{dx} \displaystyle\int_{h(x)}^{-h(x)} f(t)\,dt$.

7.4.8. (Mean value theorem for integrals) Let $f : [a, b] \to \mathbb{R}$ be continuous. Prove that there exists $c \in (a, b)$ such that

$$f(c) = \frac{1}{b-a} \int_a^b f.$$

7.4.9. Let f have derivatives of order up to $n + 1$ on the interval $[a, b]$.

i) Justify how for any x in $[a, b]$,

$$f(x) = f(a) + \int_a^x f'(t)\,dt.$$

ii) Integrate the integral above by parts, and rewrite, to get that

$$f(x) = f(a) + (x - a)f'(a) + \int_a^x (x - t)f''(t)\,dt.$$

iii) Use induction, integration by parts, and rewritings, to get that $f(x)$ equals

$$f(a) + \frac{f'(a)}{1!}(x-a) + \cdots + \frac{f^{(n)}(a)}{n!}(x-a)^n + \int_a^x \frac{(x-t)^n}{n!} f^{(n+1)}(t)\,dt.$$

iv) Say why you cannot apply the Fundamental theorem of calculus II, to compute $\frac{d}{dx} \int_a^x (x-t)^{n+1} f^{(n+1)}(t)\, dt$.

v) (Taylor' remainder formula in integral form) Consult Section 6.5 for Taylor polynomials to prove that

$$f(x) = T_{n,f,a}(x) + \int_a^x \frac{(x-t)^n}{n!} f^{(n+1)}(t)\, dt.$$

7.4.10. (Improper integral, unbounded domain) Let $f : [a, \infty) \to \mathbb{R}$ be continuous.

i) Discuss how our construction/definition of integrals fails when the domain is not bounded.

ii) Prove that for all $N \in [a, \infty)$, $\int_a^N f$ exists.

iii) If $\lim_{N \to \infty} \int_a^N f$ exists, we call the limit the **(improper) integral** of f over $[a, \infty)$. We denote it $\int_a^\infty f$. Observe that this is a limit of limits. Similarly formulate the definition of $\int_{-\infty}^b g$ for a continuous function $g : (-\infty, b] \to \mathbb{R}$.

iv) Let $f(x) = \begin{cases} 3x + 1, & \text{if } x < 10; \\ 0, & \text{otherwise.} \end{cases}$. Compute $\int_0^\infty f$, and justify.

7.4.11. Use Exercise 7.4.10.

i) Compute $\int_1^\infty \frac{1}{x^2}\, dx$.

ii) For any rational number $p < -1$, compute $\int_1^\infty x^p\, dx$. (The same is true for real $p < -1$, but we have not developed enough properties for such functions.)

7.4.12. (Improper integral, unbounded domain) Let $f : \mathbb{R} \to \mathbb{R}$ be bounded such that for some $c \in \mathbb{R}$, $\int_{-\infty}^c f$ and $\int_c^\infty f$ exist in the sense of Exercise 7.4.10.

i) Prove that for all $e \in \mathbb{R}$, $\int_{-\infty}^e f$ and $\int_e^\infty f$ exist and that

$$\int_{-\infty}^c f + \int_c^\infty f = \int_{-\infty}^e f + \int_e^\infty f.$$

We denote this common value as $\int_{-\infty}^\infty f$.

ii) Prove that $\int_{-\infty}^\infty f = \lim_{N \to \infty} \int_{-N}^N f$.

7.4.13. Let $f : \mathbb{R} \to \mathbb{R}$ over $\mathbb{R}$. This exercise is meant to show that integrability of f over $\mathbb{R}$ cannot be simply defined as the existence of the limit $\lim_{N \to \infty} \int_{-N}^N f$. Namely, let $f(x) = \begin{cases} 1, & \text{if } \lfloor x \rfloor \text{ (the floor of } x) \text{ is even;} \\ -1, & \text{otherwise.} \end{cases}$

    i) Sketch the graph of this function.

    ii) Prove that for every positive real number N, $\int_{-N}^{N} f = 0$.

    iii) Prove that f is not integrable over $[0, \infty)$ or over $(-\infty, 0)$ in the sense of Exercise 7.4.10.

7.4.14. (Improper integral, unbounded function) Let $f : (a, b] \to \mathbb{R}$ be continuous.

    i) Discuss how our construction/definition of integrals fails when the domain does not include the boundaries of the domain.

    ii) Prove that for all $N \in (a, b)$, $\int_{N}^{b} f$ exists.

    iii) If $\lim_{N \to a^+} \int_{N}^{b} f$ exists, we call this limit the **(improper) integral** of f over $[a, b]$, and we denote it $\int_{a}^{b} f$. Similarly formulate $\int_{a}^{b} f$ if the domain of f is $[a, b)$ or (a, b).

7.4.15. Use Exercise 7.4.14.

    i) Compute $\int_{0}^{1} \frac{1}{\sqrt{x}} dx$.

    ii) For any rational number $p > -1$, compute $\int_{0}^{1} x^p \, dx$. (The same is true for real $p > -1$, but we have not developed enough properties for such functions.)

    iii) Let $f : (a, b) \to \mathbb{R}$ be given by $f(x) = 1$. Prove that $\int_{a}^{b} f = b - a$.

7.5   Integration of complex-valued functions

    So far we have defined integrals of real-valued functions. By Theorem 7.4.1, if $g' = f$, then $\int_{a}^{b} f = g(b) - g(a)$, i.e., $\int_{a}^{b} g' = g(b) - g(a)$. If g is complex-valued, we know that $g' = (\operatorname{Re} g)' + i(\operatorname{Im} g)'$, so that it would make sense to define $\int_{a}^{b} g'$ as the integral of $(\operatorname{Re} g)'$ plus i times the integral of $(\operatorname{Im} g)'$. Indeed, this is the definition:

Definition 7.5.1. Let $f : [a, b] \to \mathbb{C}$ be a function such that $\operatorname{Re} f$ and $\operatorname{Im} f$ are integrable over $[a, b]$. **The integral of f over $[a, b]$** is

$$\int_{a}^{b} f = \int_{a}^{b} \operatorname{Re} f + i \int_{a}^{b} \operatorname{Im} f.$$

    The following are then immediate generalizations of the two versions of the fundamental theorem of calculus Theorems 7.4.1 and 7.4.3:

Theorem 7.5.2. (The Fundamental theorem of calculus, I, for complex-valued functions) *Let $f, g : [a, b] \to \mathbb{C}$ such that f is continuous and g is differentiable with $g' = f$. Then*

$$\int_a^b f = g(b) - g(a).$$

Theorem 7.5.3. (The Fundamental theorem of calculus, II, for complex-valued functions) *Let $f : [a, b] \to \mathbb{C}$ be continuous. Then for all $x \in [a, b]$, f is integrable over $[a, x]$, and the function $g : [a, b] \to \mathbb{R}$ given by $g(x) = \int_a^x f$ is differentiable on (a, x) with*

$$\frac{d}{dx} \int_a^x f = f(x).$$

Exercises for Section 7.5

7.5.1. Work out Exercises 7.3.3 and 7.3.4 for complex-valued functions.

7.5.2. This is a generalization of Exercise 7.3.5. Let $f : [a, b] \to \mathbb{C}$ be integrable. Prove that $\left| \int_a^b f \right| \leq \int_a^b |f|$. (Hint: Write $\int_a^b f = s + ti$ for some $s, t \in \mathbb{R}$. Integrate $(s - ti)f$.)

7.6   Natural logarithm and the exponential functions

The function that takes a non-zero x to $1/x$ is continuous everywhere on its domain since it is a rational function. Thus by Theorem 7.3.1 and Notation 7.3.3, for all $x > 0$, $\int_1^x \frac{1}{x} dx$ is well-defined. This function has a familiar name:

Definition 7.6.1. *The **natural logarithm** is the function*

$$\ln x = \int_1^x \frac{1}{t} dt$$

for all $x > 0$.

We prove below all the familiar properties of this familiar function.

Remark 7.6.2.

(1) $\ln 1 = \int_1^1 \frac{1}{t} dt = 0$.

(2) By geometry, for $x > 1$, $\ln x = \int_1^x \frac{1}{t} dt > 0$, and for $x \in (0,1)$, $\ln x = \int_1^x \frac{1}{t} dt = -\int_x^1 \frac{1}{t} dt < 0$.

(3) By the Fundamental theorem of calculus (Theorem 7.4.3), for all $b \in \mathbb{R}^+$, $\ln$ is differentiable on $(0,b)$, so that $\ln$ is differentiable on $\mathbb{R}^+$. Furthermore, $\ln'(x) = \frac{1}{x}$.

(4) $\ln$ is continuous (since it is differentiable) on $\mathbb{R}^+$.

(5) The derivative of $\ln$ is always positive. Thus by Theorem 6.3.5, $\ln$ is everywhere increasing.

(6) Let $c \in \mathbb{R}^+$, and set $g(x) = \ln(cx)$. By the chain rule, g is differentiable, and $g'(x) = \frac{1}{cx}c = \frac{1}{x} = \ln'(x)$. Thus the function $g - \ln$ has constant derivative 0. It follows by Theorem 6.3.5 that $g - \ln$ is a constant function. Hence for all $x \in \mathbb{R}^+$,

$$\ln(cx) - \ln(x) = g(x) - \ln(x) = g(1) - \ln(1) = \ln(c) - 0 = \ln(c).$$

This proves that for all $c, x \in \mathbb{R}^+$,

$$\ln(cx) = \ln(c) + \ln(x).$$

(7) By the previous part, for all $c, x \in \mathbb{R}^+$,

$$\ln\left(\frac{c}{x}\right) = \ln(c) - \ln(x).$$

(8) For all non-negative integers n and all $c \in \mathbb{R}^+$, $\ln(c^n) = n \ln(c)$. We prove this by mathematical induction. If $n = 0$, then $\ln(c^n) = \ln(1) = 0 = 0 \ln c = n \ln c$. Now suppose that equality holds for some $n - 1$. Then $\ln(c^n) = \ln(c^{n-1}c) = \ln(c^{n-1}) + \ln(c)$ by what we have already established, so that by the induction assumption $\ln(c^n) = (n-1)\ln(c) + \ln(c) = n \ln(c)$.   □

(9) For all rational numbers r and all $c \in \mathbb{R}^+$, $\ln(c^r) = r \ln(c)$. Here is a proof. By (8), this holds in case r is a non-negative integer. If r is a negative integer, then $-r$ is a positive integer, so that by (1), (7) and (8), $\ln(c^r) = \ln(1/c^{-r}) = \ln(1) - \ln(c^{-r}) = 0 - (-r)\ln(c) = r \ln(c)$, which proves the claim for all integers. Now write $r = \frac{m}{n}$ for some integers m, n with $n \neq 0$. Then $n \ln(c^r) = n \ln(c^{m/n}) = \ln(c^m) = m \ln(c)$, so that $\ln(c^r) = \frac{m}{n} \ln(c) = r \ln(c)$.   □

(10) The range of ln is $\mathbb{R} = (-\infty, \infty)$. Here is a proof. By geometry, $\ln(0.5) < 0 < \ln(2)$. Let $y \in \mathbb{R}^+$. By Theorem 3.8.3, there exists $n \in \mathbb{N}^+$ such that $y < n\ln(2)$. Hence $\ln 1 = 0 < y < n\ln(2) = \ln(2^n)$, so that since ln is continuous, by the Intermediate value theorem (Theorem 5.3.1), there exists $x \in (1, 2^n)$ such that $\ln(x) = y$. If $y \in \mathbb{R}^-$, then by the just proved we have that $-y = \ln(x)$ for some $x \in \mathbb{R}^+$, so that $y = -\ln(x) = \ln(x^{-1})$. Finally, $0 = \ln(1)$. Thus every real number is in the range of ln.   □

(11) Thus $\ln : \mathbb{R}^+ \to \mathbb{R}$ is a strictly increasing continuous and surjective function. Thus by Theorem 2.9.4, ln has an inverse $\ln^{-1} : \mathbb{R} \to \mathbb{R}^+$. By Theorem 5.3.4, $\ln^{-1}$ is increasing and continuous.

(12) By Theorem 6.2.7, the derivative of $\ln^{-1}$ is

$$(\ln^{-1})'(x) = \frac{1}{\ln'(\ln^{-1}(x))} = \ln^{-1}(x).$$

(13) For all $x, y \in \mathbb{R}$,

$$\frac{\ln^{-1}(x)}{\ln^{-1}(y)} = \ln^{-1}\left(\ln\left(\frac{\ln^{-1}(x)}{\ln^{-1}(y)}\right)\right)$$
$$= \ln^{-1}\left(\ln\left(\ln^{-1}(x)\right) - \ln\left(\ln^{-1}(y)\right)\right)$$
$$= \ln^{-1}(x - y).$$

We have proved that for all $c \in \mathbb{R}^+$ and $r \in \mathbb{Q}$, $\ln^{-1}(r\ln(c)) = \ln^{-1}(\ln(c^r)) = c^r$, and we have proved that for all $r \in \mathbb{R}$, $\ln^{-1}(r\ln(c))$ is well-defined. This allows us to define exponentiation with real (not just rational) exponents:

Definition 7.6.3. *Let $c \in \mathbb{R}^+$ and $r \in \mathbb{R}$. Set*

$$c^r = \ln^{-1}(r\ln(c)).$$

This immediately gives rise to two functions:

(1) The **generalized power function** with **exponent** r when c varies and r is constant;

(2) The **exponential function** with **base** c when r varies and c is constant.

(3) We refer to the exponential function $\ln^{-1}$ with base $c = \ln^{-1}(1)$ as **the exponential function** (so the base is implicit).

Theorem 7.6.4. *Let $r \in \mathbb{R}$. The function $f : \mathbb{R}^+ \to \mathbb{R}^+$ given by $f(x) = x^r$ is differentiable, with $f'(x) = rx^{r-1}$. This function is increasing if $r > 0$ and decreasing if $r < 0$.*

Proof. By definition, $f(x) = \ln^{-1}(r\ln(x))$, which is differentiable by the chain and scalar rules and the fact that $\ln$ and its inverse are differentiable. Furthermore, the derivative is $f'(x) = \ln^{-1}(r\ln(x)) \cdot \frac{r}{x} = r\frac{\ln^{-1}(r\ln(x))}{\ln^{-1}(\ln(x))} = r\ln^{-1}(r\ln(x) - \ln(x)) = r\ln^{-1}((r-1)\ln(x)) = rx^{r-1}$. The monotone properties then follow from Theorem 6.3.5. $\qquad\square$

Theorem 7.6.5. *Let $c \in \mathbb{R}^+$. The function $f : \mathbb{R} \to \mathbb{R}^+$ given by $f(x) = c^x$ is differentiable, and $f'(x) = (\ln(c))c^x$. This function is increasing if $c > 1$ and decreasing if $c \in (0,1)$.*

Proof. By definition, $f(x) = \ln^{-1}(x\ln(c))$, which is differentiable by the chain and scalar rules and the fact that $\ln^{-1}$ is differentiable. Furthermore, the derivative is $f'(x) = \ln^{-1}(x\ln(c)) \cdot \ln(c) = f(x) \cdot \ln(c) = (\ln(c))c^x$. The monotone properties then follow from Theorem 6.3.5. $\qquad\square$

We next give a more concrete form to the exponential function $\ln^{-1}$.

Definition 7.6.6. *Let $e = \ln^{-1}(1)$ (so that $\ln(e) = 1$). The constant e is called **Euler's constant**.*

Since $\ln(e) = 1 > 0 = \ln(1)$, by the increasing property of $\ln$ it follows that $e > 1$.

Now let $f(x) = 1/x$. This function is non-negative on $[1, \infty)$. If P is a partition of $[1, 3]$ into 5 equal parts, then $L(f, P) \cong 0.976934$, if P is a partition of $[1, 3]$ into 6 equal parts, then $L(f, P) \cong 0.995635$, and if P is a partition of $[1, 3]$ into 7 equal parts, then $L(f, P) \cong 1.00937$. This proves that $L(f) > 1$ over the interval $[1, 3]$. On $[1, 3]$, the function f is continuous and thus integrable, so that $\ln 3 = \int_1^3 f > 1 = \ln e$. Since $\ln$ is an increasing function, this means that $e < 3$. By geometry $\ln(2) < U(f, \{1, 2\}) = 1 = \ln e$, so that similarly $e > 2$. We conclude that e is a number strictly between 2 and 3.

Note that $U(f, \{1, 1.25, 1.5, 1.75, 2, 2.25, 2.5\}) = 0.25(1 + \frac{1}{1.25} + \frac{1}{1.5} + \frac{1}{1.75} + \frac{1}{2} + \frac{1}{2.25}) = \frac{2509}{2520} < 1$, so that $\ln 2.5 = \int_1^{2.5} f$ is strictly smaller than this upper sum. It follows that $\ln 2.5 < 1 = \ln e$ and $2.5 < e$. If P is a partition of $[1, 2.71828]$ into a million pieces of equal length, a computer gives that $U(f, P)$ is just barely smaller than 1, so that $2.71828 < e$. If P is a partition of $[1, 2.718285]$ into a million pieces of equal length, then $L(f, P)$ is just barely bigger than 1, so that $e < 2.718285$. Thus $e \cong 2.71828$.

A reader may want to run further computer calculations for greater precision. A different and perhaps easier computation is in Exercise 7.6.14.

Theorem 7.6.7. (The exponential function.) *For all $x \in \mathbb{R}$, $\ln^{-1}(x) = e^x$.*

Proof. By the definitions,

$$e^x = \ln^{-1}(x \cdot \ln(e)) = \ln^{-1}(x \cdot \ln(\ln^{-1}(1))) = \ln^{-1}(x \cdot 1) = \ln^{-1}(x). \quad \square$$

We proved on page 301 that the derivative of $\ln^{-1}$ is $\ln^{-1}$, which immediately yields the following result:

Theorem 7.6.8. *For all $x \in \mathbb{R}$, $(e^x)' = e^x$.*

Exercises for Section 7.6

7.6.1. Prove that for all $x > 0$, $1/(1+x) < \ln(1+x) - \ln x < 1/x$. (Hint: Use the geometry of the definition of ln.)

7.6.2. Let $c \in \mathbb{R}^+$.
  i) Prove that for all $x \in \mathbb{R}$, $c^x = e^{\ln(c^x)} = e^{x \ln(c)}$.
  ii) Prove that if $c \neq 1$, then $\int_a^b c^x \, dx = \frac{1}{\ln(c)}(c^b - c^a)$.

7.6.3. Use integration by substitution (Exercise 7.4.5) to compute $\int \frac{x}{x^2+4} \, dx$.

7.6.4. Let $c \in \mathbb{R}^+$. Use integration by substitution (Exercise 7.4.5) to compute $\int x c^{(x^2)} \, dx$.

7.6.5. Let $c \in \mathbb{R}^+$. Use integration by parts (Exercise 7.4.6) to compute $\int x c^x \, dx$.

7.6.6. Use integration by parts (Exercise 7.4.6) to compute $\int \ln(x) \, dx$. (Hint: $\ln(x) = 1 \cdot \ln(x)$.)

7.6.7. Prove by integration by parts the following improper integral value for a non-negative integer n:

$$\int_0^1 (-x^{2n} \ln x) dx = \frac{1}{(2n+1)^2}.$$

7.6.8. (Logarithmic differentiation) Sometimes it is hard or even impossible to compute the derivative of a function. Try for example $f(x) = x^x$, $f(x) = (x^2 + 2)^{x^3+4}$, or $f(x) = \frac{x^2(x-1)^3(x+4)^3\sqrt{x^2+1}}{\sqrt[3]{x+2}(x+7)^3(x-2)^4}$. There is another way if the range of the function consists of positive real numbers: Apply ln of both sides, take derivatives of both sides, and solve for $f'(x)$. For example, if $f(x) = x^x$, then $\ln(f(x)) = \ln(x^x) = x\ln(x)$, so that

$$\frac{f'(x)}{f(x)} = (\ln(f(x)))' = (x\ln(x))' = \ln(x) + \frac{x}{x} = \ln(x) + 1,$$

so that $f'(x) = x^x(\ln(x) + 1)$.

   i) Compute and justify the derivative of $f(x) = (x^2 + 2)^{x^3+4}$.

   ii) Compute and justify the derivative of $f(x) = \frac{x^2(x-1)^3(x+4)^3\sqrt{x^2+1}}{\sqrt[3]{x+2}(x+7)^3(x-2)^4}$.

7.6.9. This exercise is about applying versions of L'Hôpital's rule.

   i) Prove that $\lim\limits_{x\to\infty} \frac{\ln x}{x} = 0$.

   ii) Prove that $\lim\limits_{x\to 0^+} x\ln x = 0$. (Hint: Use one of $x\ln x = \frac{x}{1/\ln x}$ or $x\ln x = \frac{\ln x}{1/x}$. Perhaps one works and the other does not; we can also learn from attempts that do not lead to a successful completion.)

   iii) Compute and justify $\lim\limits_{x\to 0^+} \ln(x^x)$.

   iv) Compute and justify $\lim\limits_{x\to 0^+} x^x$.

7.6.10. Prove that $\lim\limits_{x\to 0} \frac{1}{x} \ln(1 + x) = 1$. (Hint: L'Hôpital's rule.)

   i) Prove that for any $c \in \mathbb{R}^+$, $\lim\limits_{x\to\infty} x\ln\left(1 + \frac{c}{x}\right) = c$. (Hint: Change variables.)

   ii) Prove that for any $c \in \mathbb{R}^+$, $\lim\limits_{x\to\infty} \left(1 + \frac{c}{x}\right)^x = e^c$.

7.6.11. Prove that $f(x) = (1 + \frac{1}{x})^x$ is a strictly increasing function. (Hint: use logarithmic differentiation and Exercise 7.6.1.)

7.6.12. Prove that for all $x \in \mathbb{R}$, $(1 + \frac{1}{x})^x < e$. (Hint: previous two exercises.)

7.6.13. Apply L'Hôpital's rule (Theorem 6.4.2):

   i) Prove that $\lim_{x\to 0} \frac{e^x-1}{x} = 1$.

   ii) Prove that $\lim_{x\to 0} \frac{e^x-1-x}{x^2} = 1/2$.

   iii) Prove that $\lim_{x\to 0} \frac{e^x-1-x-x^2/2}{x^3} = 1/6$.

7.6.14. Let f be the exponential function with $f(x) = e^x$ for all $x \in \mathbb{R}$.

i) For any positive integer n, compute the Taylor polynomial $T_{n,f,0}$ for e^x of degree n centered at 0 (Definition 6.5.2).

ii) Use Taylor's remainder theorem (Theorem 6.5.5) to show that for any $x \in \mathbb{R}$ and any $\epsilon > 0$ there exists $n \in \mathbb{N}$ such that $|f(x) - T_{n,f,0}(x)| < \epsilon$.

iii) Use the Taylor's remainder theorem (Theorem 6.5.5) to prove that $|e - T_{8,f,0}(1)| = |e^1 - T_{8,f,0}(1)| < 0.00001$.

iv) Compute $T_{8,f,0}(1)$ to 7 significant digits.

v) (Unusual) We computed some digits of e on page 303. On a computer, try to get more digits of e with those methods and separately with methods in this exercise. Which method is faster? Can you streamline either method?

7.6.15. Let $f(x) = e^x$. Use L'Hôpital's rule (Theorem 6.4.2) and induction on n to prove that

$$\lim_{x \to 0} \frac{e^x - T_{n,f,0}(x)}{x^{n+1}} = \frac{1}{(n+1)!}.$$

7.6.16. Let $f : \mathbb{R} \to \mathbb{R}$ be given by $f(x) = \begin{cases} e^{-1/x^2}, & \text{if } x \neq 0; \\ 0, & \text{if } x = 0. \end{cases}$

i) Prove by induction on $n \geq 0$ that for each n there exists a polynomial function $h_n : \mathbb{R} \to \mathbb{R}$ such that

$$f^{(n)}(x) = \begin{cases} h_n(\frac{1}{x}) \cdot e^{-1/x^2}, & \text{if } x \neq 0; \\ 0, & \text{if } x = 0. \end{cases}$$

(At non-zero x you can use the chain rule, the derivative of the exponential function, and the power rule for derivatives. However, $f^{(n+1)}(0)$ is (but of course) computed as $\lim_{h \to 0} \frac{f^{(n)}(h) - f^{(n)}(0)}{h}$, and then you have to use L'Hôpital's rule. You do not have to be explicit about the polynomial functions h_n.)

ii) Compute the nth Taylor polynomial for f centered at 0.

7.6.17. (A friendly competition between e and π) Use calculus and not a calculator to determine which number is bigger, e^π or π^e. You may assume that $1 < e < \pi$. (Hint: compute the derivative of some function.)

7.7   Applications of integration

The Fundamental theorems of calculus relate integration with differentiation. In particular, to compute $\int_a^b f$, if we know an antiderivative g of f, then the integral is easy. However, as already mentioned after the Fundamental theorem of calculus I, many functions do not have a "closed-form" antiderivative. One can still compute definite integrals up to a desired precision, however: we take finer and finer partitions of $[a, b]$, and when $U(f, P)$ and $L(f, P)$ are within a specified distance from each other, we know that the true integral is somewhere in between, and hence up to the specified precision either $L(f, P)$ or $U(f, P)$ stands for $\int_a^b f$. In applications, such as in science and engineering, many integrals have to be and are computed in this way because of the lack of closed-form antiderivatives.

In this section we look at many applications that exploit the original definition of integrals via sums over finer and finer partitions. For many concrete examples we can then solve the integral via antiderivatives, but for many others, we have to make do with numerical approximation.

7.7.1   Length of a curve

Let $f : [a, b] \to \mathbb{R}$ be a continuous function. If the graph of f is a line, then by the Pythagorean theorem the length of the curve is $\sqrt{(b - a)^2 + (f(b) - f(a))^2}$. For a general curve it is harder to determine its length from $(a, f(a))$ to $(b, f(b))$. But we can do the standard calculus trick: let $P = \{x_0, x_1, \ldots, x_n\}$ be a partition of $[a, b]$; on each sub-interval $[x_{k-1}, x_k]$ we "approximate" the curve with the line $(x_{k-1}, f(x_{k-1}))$ to $(x_k, f(x_k))$, compute the length of that line as $\sqrt{(x_k - x_{k-1})^2 + (f(x_k) - f(x_{k-1}))^2}$, and sum up all the lengths:

$$\sum_{k=1}^n \sqrt{(x_k - x_{k-1})^2 + (f(x_k) - f(x_{k-1}))^2}.$$

Whether this is an approximation of the true length depends on the partition, but geometrically it makes sense that the true length of the curve equals

$$\lim \sum_{k=1}^n \sqrt{(x_k - x_{k-1})^2 + (f(x_k) - f(x_{k-1}))^2},$$

as the partitions $\{x_0, x_1, \ldots, x_n\}$ get finer and finer. But this is not yet in form of Theorem 7.1.16. For that we need to furthermore assume that f is **differentiable** on (a, b). Then by the Mean value theorem (Theorem 6.3.4) for each $k = 1, \ldots, n$ there exists $c_k \in (x_{k-1}, x_k)$ such that $f(x_k) - f(x_{k-1}) = f'(c_k)(x_k - x_{k-1})$. If in addition we assume that f' is continuous, then it is integrable by Theorem 7.3.1, and by Theorem 7.1.16 the true length of the curve equals

$$\lim_{P=\{x_0, x_1, \ldots, x_n\} \text{finer}, c_k \in [x_{k-1}, x_k]} \sum_{k=1}^{n} \sqrt{(x_k - x_{k-1})^2 + (f'(c_k)(x_k - x_{k-1}))^2}$$

$$= \lim_{P=\{x_0, x_1, \ldots, x_n\} \text{finer}, c_k \in [x_{k-1}, x_k]} \sum_{k=1}^{n} \sqrt{1 + (f'(c_k))^2}(x_k - x_{k-1})$$

$$= \int_a^b \sqrt{1 + (f'(x))^2} \, dx.$$

We just proved:

Theorem 7.7.1. *If* $f : [a, b] \to \mathbb{R}$ *is continuous and differentiable such that* f' *is continuous on* $[a, b]$, *then the length of the curve from* $(a, f(a))$ *to* $(b, f(b))$ *is*

$$\int_a^b \sqrt{1 + (f'(x))^2} \, dx.$$

7.7.2 Volume of a surface area of revolution, disk method

Let $f : [a, b] \to \mathbb{R}$ be continuous. We rotate the region between $x = a$ and $x = b$ and bounded by the x-axis and the graph of f around the x-axis. If the graph is a horizontal line, then the rotated region is a disk of height $b - a$ and radius $f(a)$, so that its volume is $\pi(f(a))^2(b - a)$. For a general f we let $P = \{x_0, x_1, \ldots, x_n\}$ be a partition of $[a, b]$; on each sub-interval $[x_{k-1}, x_k]$ we "approximate" the curve with the horizontal line $y = f(c_k)$ for some $c_k \in [x_{k-1}, x_k]$, we compute the volume of the solid of revolution obtained by rotating that approximated line over the interval $[x_{k-1}, x_k]$ around the x-axis, and sum up all the volumes:

$$\sum_{k=1}^{n} \pi(f(c_k))^2(x_k - x_{k-1}).$$

Geometrically it makes sense that the true volume equals

$$\lim_{P=\{x_0,x_1,\ldots,x_n\}\text{finer},c_k\in[x_{k-1},x_k]} \sum_{k=1}^{n} \pi(f(c_k))^2(x_k - x_{k-1}),$$

and by Theorems 7.1.16 and 7.3.1, this equals $\pi \int_a^b (f(x))^2\,dx$. This proves:

Theorem 7.7.2. *If* $f : [a,b] \to \mathbb{R}$ *is continuous, then the volume of the solid of revolution obtained by rotating around the* x-*axis the region between* $x = a$ *and* $x = b$ *and bounded by the* x-*axis and the graph of* f *is*

$$\pi \int_a^b (f(x))^2\,dx.$$

7.7.3   Volume of a surface area of revolution, shell method

Let $0 \le a \le b$, and let $f, g : [a,b] \to \mathbb{R}$ be continuous such that for all $x \in [a,b]$ $f(x) \le g(x)$. We rotate the region between $y = a$ and $y = b$ and bounded by the graphs of $x = f(y)$ and $x = g(y)$ around the x-axis. If $f(x) = c$ and $g(x) = d$ are constant functions, then the solid of revolution is a hollowed disk, with the outer border of height $d - c$ and radius b and the hole inside it has the same height but radius a. Thus the volume is $\pi(d-c)(b^2 - a^2)$. For general f and g we let $P = \{y_0, y_1, \ldots, y_n\}$ be a partition of $[a,b]$; on each sub-interval $[y_{k-1}, y_k]$ we "approximate" the curve f, g with the horizontal line $f(c_k), g(c_k)$ for some $c_k \in [y_{k-1}, y_k]$, we compute the volume of the solid of revolution obtained by rotating that approximated region over the interval $[y_{k-1}, y_k]$ around the x-axis, and sum up all the volumes:

$$\sum_{k=1}^{n} \pi(g(c_k) - f(c_k))(y_k^2 - y_{k-1}^2).$$

Geometrically it makes sense that the true volume equals

$$\lim_{P=\{y_0,y_1,\ldots,y_n\}\text{finer},c_k\in[y_{k-1},y_k]} \sum_{k=1}^{n} \pi(g(c_k) - f(c_k))(y_k^2 - y_{k-1}^2)$$

$$= \lim_{P=\{y_0,y_1,\ldots,y_n\}\text{finer},c_k\in[y_{k-1},y_k]} \sum_{k=1}^{n} \pi(g(c_k) - f(c_k))(y_k + y_{k-1})(y_k - y_{k-1}),$$

and by Theorems 7.1.16 and 7.3.1, this equals $\pi \int_a^b (g(y) - f(y)) 2y \, dy$. This proves:

Theorem 7.7.3. *If $0 \le a \le b$ and $f, g : [a, b] \to \mathbb{R}$ are continuous, then the volume of the solid of revolution obtained by rotating around the x-axis the region between $y = a$ and $y = b$ and bounded by the graphs of $x = f(y)$ and $x = g(y)$, is*

$$2\pi \int_a^b y(g(y) - f(y)) \, dy.$$

Example 7.7.4. The volume of the sphere of radius r is $\frac{4}{3}\pi r^3$.

Proof. We rotate the upper-half circle of radius r centered at the origin around the x-axis. The circle of radius r centered at the origin consists of all points (x, y) such that $x^2 + y^2 = r^2$, so we have $g(y) = \sqrt{r^2 - y^2}$ and $f(y) = -\sqrt{r^2 - y^2}$. Thus the volume is

$$4\pi \int_0^r y\sqrt{r^2 - y^2} \, dy = \left(-\frac{4}{3}\pi(r^2 - y^2)^{3/2} \right) \Big|_0^r = \frac{4}{3}\pi(r^2)^{3/2} = \frac{4}{3}\pi r^3. \qquad \square$$

7.7.4   Surface area of the surface area of revolution

It takes quite a few steps to analyze the simple case.

We first rotate the line segment $y = mx$ from $x = 0$ to $x = b > 0$ around the x-axis. We assume for now that $m \ne 0$.

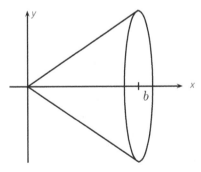

In this way we obtain a right circular cone of height b and base radius $|m|b$. The perimeter of that base circle is of course $2\pi|m|b$. If we cut the cone in a straight line from a side to the vertex, we cut along an edge of length $\sqrt{b^2 + (mb)^2}$, and we get the wedge as follows:

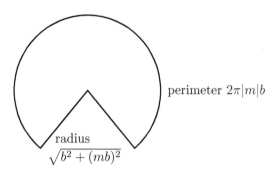

perimeter $2\pi|m|b$

radius
$\sqrt{b^2 + (mb)^2}$

Without the clip in the disc, the perimeter would be $2\pi\sqrt{b^2 + (mb)^2}$, but our perimeter is only $2\pi|m|b$. Thus the angle subtended by the wedge is by proportionality equal to $\frac{2\pi|m|b}{2\pi\sqrt{b^2+(mb)^2}}2\pi = \frac{2\pi|m|}{\sqrt{1+m^2}}$. The area of the full circle is radius squared times one half of the full angle, and so proportionally the area of our wedge is radius squared times one half of our angle, i.e., the surface area is $(\sqrt{b^2 + (mb)^2})^2\frac{2\pi|m|}{2\sqrt{1+m^2}} = \pi|m|b^2\sqrt{1 + m^2}$. Note that even if $b < 0$, the surface area is the absolute value of $\pi mb^2\sqrt{1 + m^2}$.

Thus if m is not zero and $0 \le a < b$ or $a < b \le 0$, then the surface area of revolution obtained by rotating the line $y = mx$ from $x = a$ to $x = b$ equals the absolute value of $\pi m(b^2-a^2)\sqrt{1 + m^2}$. Note the geometric requirement that at a and b the line is on the same side of the x-axis.

Now suppose that we rotate the line $y = mx + l$ around the x-axis, with $m \ne 0$ and $a < b$ and both on the same side of the intersection of the line with the x-axis. This intersection is at $x = -l/m$. By shifting the graph by l/m to the right, this is the same as rotating the line $y = mx$ from $x = a+l/m$ to $x = b+l/m$, and by the previous case the surface area of this is the absolute value of

$$\pi m((b + l/m)^2 - (a + l/m)^2)\sqrt{1 + m^2}$$
$$= \pi m \left(b^2 + 2bl/m + l^2/m^2 - (a^2 + 2al/m + l^2/m^2)\right)\sqrt{1 + m^2}$$
$$= \pi m \left(b^2 - a^2 + 2(b - a)l/m\right)\sqrt{1 + m^2}$$
$$= \pi m(b - a)(b + a + 2l/m)\sqrt{1 + m^2}$$
$$= \pi(b - a)(m(b + a) + 2l)\sqrt{1 + m^2}.$$

If instead we rotate the line $y = l$ (with $m = 0$) around the x-axis, we get a ring whose surface area is $(b - a)2\pi|l|$, which is the absolute value of $\pi(b - a)(m(b + a) + 2l)\sqrt{1 + m^2}$. Thus for all m, the surface area of the surface of revolution obtained by rotating the line $y = mx + l$ from $x = a$ to $x = b$ around the x-axis is the absolute value of

$$\pi(b - a)(m(b + a) + 2l)\sqrt{1 + m^2},$$

with the further restriction in case $m \neq 0$ that a and b are both on the same side of the x-intercept.

Now let $f : [a, b] \to \mathbb{R}_{\geq 0}$ be a differentiable (and not necessarily linear) function. Let $P = \{x_0, x_1, \ldots, x_n\}$ be a partition of $[a, b]$; on each sub-interval $[x_{k-1}, x_k]$ we approximate the curve with the line from $(x_{k-1}, f(x_{k-1}))$ to $(x_k, f(x_k))$. By the assumption that f takes on only non-negative values we have that both x_{k-1}, x_k are both on the same side of the x-intercept of that line. The equation of the line is $y = \frac{f(x_k) - f(x_{k-1})}{x_k - x_{k-1}}(x - x_k) + f(x_k)$, so that $m = \frac{f(x_k) - f(x_{k-1})}{x_k - x_{k-1}}$ and $l = -\frac{f(x_k) - f(x_{k-1})}{x_k - x_{k-1}}x_k + f(x_k)$. Since f is differentiable, by the Mean value theorem (Theorem 6.3.4) there exists $c_k \in (x_{k-1}, x_k)$ such that $f'(c_k) = \frac{f(x_k) - f(x_{k-1})}{x_k - x_{k-1}}$. Thus $m = f'(c_k)$ and $l = f(x_k) - f'(c_k)x_k$.

We rotate that line segment around the x-axis, compute its volume of that solid of revolution, and add up the volumes for all the subparts:

$$\sum_{k=1}^{n} \left| \pi(x_k - x_{k-1})\left(f'(c_k)(x_k + x_{k-1}) + 2\left(f(x_k) - f'(c_k)x_k\right)\right)\sqrt{1 + (f'(c_k))^2} \right|$$

$$= \sum_{k=1}^{n} \pi \left|(f'(c_k)(-x_k + x_{k-1}) + 2f(x_k))\right| \sqrt{1 + (f'(c_k))^2}(x_k - x_{k-1}).$$

By Theorems 7.1.16 and 7.3.1 we get the following:

Theorem 7.7.5. *If* $f : [a, b] \to \mathbb{R}_{\geq 0}$ *is differentiable with continuous derivative, then the surface area of the solid of revolution obtained by rotating around the x-axis the curve* $y = f(x)$ *between* $x = a$ *and* $x = b$ *is*

$$\pi \int_a^b |f'(x)(-x + x) + 2f(x)| \sqrt{1 + (f'(x))^2}\, dx = 2\pi \int_a^b f(x)\sqrt{1 + (f'(x))^2}\, dx.$$

Exercises for Section 7.7

7.7.1. Use the methods from this section to compute the following:

    i) The perimeter of the circle.

    ii) The volume of the sphere of radius r.

    iii) The volume of the ellipsoid whose boundary satisfies $\frac{x^2}{a^2} + \frac{y^2}{b^2} + \frac{z^2}{b^2} = 1$. (We need multi-variable calculus to be able to compute the volume of the ellipsoid whose boundary satisfies $\frac{x^2}{a^2} + \frac{y^2}{b^2} + \frac{z^2}{c^2} = 1$.)

    iv) The volume of a conical pyramid with base radius r and height h.

    v) The volume of a doughnut (you specify its dimensions).

    vi) The surface area of the sphere of radius r.

7.7.2. The moment of inertia of a tiny particle of mass m rotating around a circle of radius r is $I = mr^2$. Likewise, the moment of inertia of a (circular) hoop of radius r and of mass m rotating around its center is $I = mr^2$. The goal of this exercise is to derive the moment of inertia of a thin circular plate of radius $5a$ meters and mass 4 kilograms rotating about its diameter. (Let b be the thickness of the plate and ρ the mass divided by the volume.)

    i) Center the plate at the origin. Let the axis of rotation be the y-axis.

    ii) Let P be a partition of $[0,5]$. Prove that the moment of inertia of the sliver of the plate between x_{k-1} and x_k is approximately $\frac{4}{5^2}\sqrt{x_k^2 - x_{k-1}^2}c_k^2$, where $c_k \in [x_{k-1}, x_k]$.

    iii) Prove that the moment of inertia of the rotating circular plate is 50 kilogram meters squared.

7.7.3. If a constant force F moves an object by d units, the work done is $W = Fd$. Suppose that F depends on the position as $F(x) = kx$ for some constant k. Find the total work done between $x = a$ and $x = b$.

7.7.4. In hydrostatics, (constant) force equals (constant) pressure times (constant) area, and (constant) pressure equals the weight density of the water w times (constant) depth h below the surface. But most of the time we do not have tiny particles but large objects where depth, pressure, and surface areas vary. For example, an object is completely submerged under water from depth a to depth b. At depth h, the cross section area of the object is $A(h)$. Compute the total force exerted on the object by the water.

Chapter 8

Sequences

In this chapter, Sections 8.5 and 8.4 contain identical results in identical order, but the proofs are different. You may want to learn both perspectives, or you may choose to omit one of the two sections.

8.1 Introduction to sequences

Definition 8.1.1. *An* **infinite sequence** *is a function with domain* $\mathbb{N}^+$ *and codomain* $\mathbb{C}$. *If* s *is such a function, instead of writing* $s(n)$, *it is common to write* s_n, *and the sequence* s *is commonly expressed also in all of the following notations (and obviously many more):*

$$s = \{s_1, s_2, s_3, \ldots\} = \{s_n\}_{n=1}^{\infty} = \{s_n\}_{n \geq 1} = \{s_n\}_{n \in \mathbb{N}^+}$$
$$= \{s_n\}_n = \{s_{n+3}\}_{n=-2}^{\infty} = \{s_{n-4}\}_{n>4} = \{s_n\}.$$

The n*th element* $s_n = s(n)$ *in the ordered list is called the* **nth term of the sequence.**

The notation $\{s_1, s_2, s_3, \ldots\}$ usually stands for the **set** consisting of the elements $s_1, s_2, s_3, \ldots$, and the order of a listing of elements in a set is irrelevant. Here, however, $\{s_1, s_2, s_3, \ldots\}$ stands for the **sequence**, and the order matters. When the usage is not clear from the context, we add the word "sequence" or "set" as appropriate.

The first term of the sequence $\{2n - 1\}_{n \geq 4}$ is 7, the second term is 9, etc. The point is that even though the notating of a sequence can start with an arbitrary integer, the counting of the terms always starts with 1.

Note that s_n is the nth term of the sequence s, whereas $\{s_n\} = \{s_n\}_{n \geq 1}$ is the sequence in which n plays a dummy variable. Thus

$$s = \{s_n\} \neq s_n.$$

313

Examples and notation 8.1.2.

(1) The terms of a sequence need not be distinct. For any complex number c, $\{c\} = \{c, c, c, \ldots\}$ is called a **constant sequence**.

(2) For any complex numbers c and d, the sequence $\{c, d, c, d, c, d, \ldots\}$ can be written more concisely as $\{\frac{d-c}{2}(-1)^n + \frac{c+d}{2}\}_n$.

(3) The sequence $\{(-1)^n\} = \{-1, 1, -1, 1, \ldots\}$ has an infinite number of terms, and its **range** is the finite set $\{-1, 1\}$. The range of the sequence $\{i^n\}$ is the set $\{i, -1, -i, 1\}$.

(4) The sequence s with $s(n) = n + 4$ for all $n \geq 1$ can be written as $\{n + 4\} = \{n + 4\}_{n \geq 1} = \{5, 6, 7, \ldots\} = \{n\}_{n \geq 5}$. This sequence is different from the sequence $\{n\} = \{1, 2, 3, \ldots\}$.

(5) The sequence $\{2n\}$ is the sequence $\{2, 4, 6, 8, \ldots\}$. It is not the same as the sequence $\{4, 2, 8, 6, 12, 10, \ldots\}$, namely, if we scramble the order of the terms, we change the sequence.

(6) The sequence of all odd positive integers (in the natural order) can be written as $\{2n - 1\} = \{2n - 1\}_{n \geq 1} = \{2n + 1\}_{n \geq 0}$ (and in many other ways).

(7) Terms of a sequence can be defined **recursively**. For example, let $s_1 = 1$, and for each $n \geq 1$ let $s_{n+1} = 2s_n$. Then the sequence of these s_n has a non-recursive form $s_n = 2^{n-1}$. If however, $s_1 = 1$ and for each $n \geq 1$, $s_{n+1} = 2s_n^3 + \sqrt{s_n + 2} + \ln(s_n) + 1$, then we can certainly compute any one s_n by invoking s_k for $k < n$, but we do not get a closed-form for s_n as in the previous example.

(8) **(Fibonacci numbers)** Let $s_1 = 1$, $s_2 = 1$, and for all $n \geq 2$, let $s_{n+1} = s_n + s_{n-1}$. This sequence starts with 1, 1, 2, 3, 5, 8, 13, 21, 34, ..., and obviously keeps growing. (For more on these numbers, see Exercise 1.6.30, where in particular it is proved that for all integers n, $s_n = \frac{1}{\sqrt{5}}\left(\frac{1+\sqrt{5}}{2}\right)^n - \frac{1}{\sqrt{5}}\left(\frac{1-\sqrt{5}}{2}\right)^n$.)

(9) Some sequences, just like functions, do not have an algebraic expression for terms. For example, let s be the sequence whose nth term is the nth prime number. This sequence starts with 2, 3, 5, 7, 11, 13, 17, 19, 23, and we could write many more terms out explicitly, but we do not have a formula for them. (This s is indeed an infinite sequence since there are infinitely many primes, as proved on page 25.)

(10) Note that $\{n\}_{n\in\mathbb{Z}}$ is NOT a sequence because the list has no first term.

(11) We can scramble the set $\mathbb{Z}$ of all integers into a sequence, for example as follows: $\{0, 1, -1, 2, -2, 3, -3, 4, -4, \ldots\}$, which algebraically equals

$$s_n = \begin{cases} n/2, & \text{if } n \text{ is even;} \\ -(n-1)/2, & \text{otherwise.} \end{cases}$$

("Otherwise" applies to odd (positive) integers.)

(12) One can scramble the set $\mathbb{Q}^+$ of all positive rational numbers into a sequence via a diagonal construction as follows. First of all, each positive rational number can be written in the form a/b for some positive integers a, b. Rather than plotting the fraction a/b, we plot the point (a, b) in the plane. Refer to Plot 8.1.4: the bold points are elements of $\mathbb{N}^+ \times \mathbb{N}^+$, and each such (a, b) is identified with the fraction a/b. Every positive rational number appears in this way somewhere as a bold point, and all appear multiple times because $\frac{a}{b} = \frac{ac}{bc}$. Now we want to systematically enumerate these bold points/rational numbers. If we first enumerate all of them in the first row, and then proceed to the second row, well, actually, we never get to the second row as we never finish the first row. So we need a cleverer way of counting, and that is done as follows. We start counting at $(1, 1)$, which stands for $1/1 = 1$. We then proceed through all the other integer points in the positive quadrant of the plane via diagonals as in Plot 8.1.4. The given instructions would enumerate positive rational numbers as $1/1, 2/1, 1/2, 1/3, 2/2$. Ah, but $2/2$ has already been counted as $1/1$, so we do not count $2/2$. Thus, the proper counting of positive rational numbers in this scheme starts with:

$$1/1, 2/1, 1/2, 1/3, \cancel{2/2}, 3/1, 4/1, 3/2, 2/3, 1/4, 1/5,$$
$$\cancel{2/4}, \cancel{3/3}, \cancel{4/2}, 5/1, 6/1, 5/2, 4/3, 3/4, 2/5, 1/6,$$

etc., where the crossed out numbers are not part of the sequence because they had been counted earlier. Thus in this count the fifth term is 3.

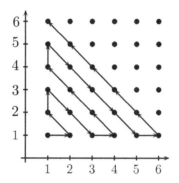

Plot 8.1.3 Counting of the positive rational numbers.

It is important to note that every positive rational number appears on this list, and because we are skipping any repetitions, it follows that every positive rational number appears on this list exactly once. Thus this gives an enumeration of positive rational numbers.*

A different enumeration of $\mathbb{Q}^+$ is given with an algebraic formulation in Exercise 2.4.28.

(13) If $\{s_n\}$ is a sequence of positive rational numbers in which every positive rational number appears exactly (or at least) once, we can construct from it a sequence in which **every** rational number appears exactly (or at least) once as follows:

$$q_n = \begin{cases} 0, & \text{if } n = 1; \\ s_{n/2}, & \text{if } n \text{ is even}; \\ s_{(n-1)/2}, & \text{if } n \geq 3 \text{ is odd.} \end{cases}$$

This sequence starts with $0, s_1, -s_1, s_2, -s_2, s_3, -s_3, \ldots$. Since every positive rational number is one of the s_n (exactly once/at least once), so every rational number is on this new list (exactly once/at least once).

---

* Here is a fun exercise: look at the ordered list of positive rational numbers above, including the crossed-out fractions. Verify for a few of them that n/m is in position $\frac{(n+m-2)^2}{2} + \frac{3n+m-4}{2} + 1$ on the list. Namely, it is a fact that $f(x,y) = \frac{(x+y-2)^2}{2} + \frac{3x+y-4}{2} + 1$ gives a bijection of $(\mathbb{N}^+)^2$ with $\mathbb{N}^+$. This was first proved by Rudolf Fueter and George Pólya, but the proof is surprisingly hard, using transcendence of e^r for algebraic numbers r, so do not attempt to prove this without more number theory background.

Incidentally, it is impossible to scramble $\mathbb{R}$ or $\mathbb{C}$ into a sequence. This can be proved with a so-called **Cantor's diagonal argument**, which we are not presenting here, but an interested reader can consult other sources.

(14) Sequences are functions, and if all terms of the sequence are real numbers, we can plot sequences in the usual manner for plotting functions. The following is part of a plot of the sequence $\{1/n\}$.

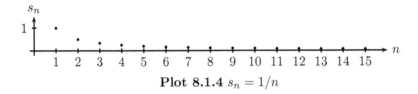

Plot 8.1.4 $s_n = 1/n$

(15) Another way to plot a sequence is to simply plot and label each s_n in the complex plane or on the real line. We plot three examples below.

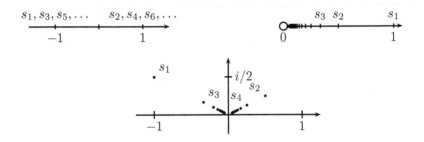

Plot 8.1.5 Image sets of $\{(-1)^n\}$, $\{1/n\}$, $\{(-1)^n/n + 0.5i/n\}$.

There is an obvious arithmetic on sequences (just like there is on functions):

$$\{s_n\} \pm \{t_n\} = \{s_n \pm t_n\},$$

$$\{s_n\} \cdot \{t_n\} = \{s_n \cdot t_n\},$$

$$c\{s_n\} = \{cs_n\},$$

$$\{s_n\}/\{t_n\} = \{s_n/t_n\} \text{ (if } t_n \neq 0 \text{ for all } n\text{)}.$$

One has to make sure to add/multiply/divide equally numbered terms of the two sequences, such as in the following:

$$\{n\}_{n\geq 3} + \{n\}_{n\geq 2} = \{n+1\}_{n\geq 2} + \{n\}_{n\geq 2} = \{2n+1\}_{n\geq 2}.$$

Here are a few further examples of arithmetic on sequences, with $+$ and $\cdot$ binary operations on the set of sequences:

$$\{2^n\} + \{-2^n\} = \{0\},$$
$$\{2^n\} + \{(-2)^n\} = \{0, 8, 0, 32, 0, 128, 0, 512, 0, \ldots\},$$
$$2\{2^n\} = \{2^{n+1}\},$$
$$\{2^n\} \cdot \{2^{-n}\} = \{1\},$$
$$\{2n-1\} + \{1\} = \{2n\},$$
$$\{(-1)^n n\} + \{2/n\} = \{(-1)^n n + 2/n\},$$
$$\{i^n\}/\{(-i)^n\} = \{(-1)^n\}.$$

Exercises for Section 8.1

8.1.1. Express algebraically the ordered sequence of all positive integer multiples of 3.

8.1.2. Think of a sequence whose terms are all between 2 and 3.

8.1.3. Plot the sequences $\{n^2\}$ and $\{1/n^2\}$. Compare the plots.

8.1.4. Plot the sequence $\{(-1)^n\}$.

8.1.5. Plot the image set of the sequence $s = \{i^n\}$: draw the real and imaginary axes and the unit circle centered at the origin; on this circle, plot $i^1, i^2, i^3, i^4, i^5, i^6$, and label each correspondingly with "s_1", "s_2", "s_3", "s_5", "s_6". Label also $s_{20}, s_{100}, s_{101}, s_{345}$.

8.1.6. Prove that the difference of the ordered sequence of all positive odd integers and the constant sequence $\{1\}$ is the ordered sequence of all non-negative even integers.

8.1.7. Let S be the set of infinite sequences.
  i) Prove that $\{0\}$ is the identity for $+$ and $\{1\}$ the identity for $\cdot$.
  ii) Prove that every infinite sequence has an inverse for $+$.
  iii) Show that not every infinite sequence has an inverse for $\cdot$. What infinite sequences have an inverse for this operation?

8.1.8. Sequences can also be **finite**. Two examples of finite sequences are: (i) last exam scores in a class arranged in alphabetic order by the student; and (ii) last exam scores in a class arranged in ascending order by score. Give two more examples of finite sequences.

8.2   Convergence of infinite sequences

Definition 8.2.1. *A sequence $s = \{s_n\}$ converges to $\mathbf{L} \in \mathbb{C}$ if for every real number $\epsilon > 0$ there exists a positive real number N such that for all integers $n > N$, $|s_n - L| < \epsilon$.*

If s converges to L, we also say that L is the limit of $\{s_n\}$. We use the following notations for this:

$$s_n \to L, \; \{s_n\} \to L,$$

$$\lim s = L, \; \lim s_n = L, \; \lim\{s_n\} = L, \; \lim(s_n) = L,$$

$$\lim_{n\to\infty} s_n = L, \; \lim_{n\to\infty}\{s_n\} = L, \; \lim_n\{s_n\} = L, \; \lim_{n\to\infty}(s_n) = L,$$

and, to save vertical space, just like for limits of functions, we also use a variation on the last three: $\lim_{n\to\infty} s_n = L$, $\lim_{n\to\infty}\{s_n\} = L$, $\lim_{n\to\infty}(s_n) = L$.

We say that a sequence is **convergent** *if it has a limit.*

For example, the constant sequence $s = \{c\}$ converges to $L = c$ because for all n, $|s_n - L| = |c - c| = 0$ is strictly smaller than any positive real number ϵ.

The sequence $s = \{300, -5, \pi, 4, 0.5, 10^6, 2, 2, 2, 2, 2, \ldots\}$ converges to $L = 2$ because for all $n \geq 7$, $|s_n - L| = |2 - 2| = 0$ is strictly smaller than any positive real number ϵ.

In conceptual terms, a sequence $\{s_n\}$ converges if the tail end of the sequence gets closer and closer to L; you can make all s_n with $n > N$ get arbitrarily close to L by simply increasing N a sufficient amount.

We work out examples of epsilon-N proofs; they are similar to the epsilon-delta proofs, and we go through them slowly at first. Depending on the point of view of your class, the reader may wish to skip the rest of this section for an alternative treatment in Section 8.5 in terms of limits of functions. More epsilon-N proofs are in Section 8.4. Be aware that this section is more concrete; the next section assumes greater ease with abstraction.

Example 8.2.2. Consider the sequence $s = \{1/n\}$. Plot 8.1.4 gives a hunch that $\lim s_n = 0$, and now we prove it. [(Recall that any text between square brackets in this font is what should approximately be going through your thoughts, but it is not something to write down in a final solution.) By the definition of convergence, we have to show that for all $\epsilon > 0$ some property holds. All proofs of this form start with:] Let ϵ be an arbitrary positive number. [Now we have to show that there exists an N for which some other property holds. Thus we have to construct an N. Usually this is done in retrospect, one cannot simply guess an N, but in the final write-up, readers see simply that educated guess — more about how to guess educatedly later:] Set $N = \frac{1}{\epsilon}$. Then N is a positive real number. [Now we have to show that for all integers $n > N$, $|s_n - 0| < \epsilon$. All proofs of statements of the form "for all integers $n > N$" start with:] Let n be an (arbitrary) integer with $n > N$. [Finally, we have to prove the inequality $|s_n - 0| < \epsilon$. We do that by algebraically manipulating the left side until we get the desired final $< \epsilon$:]

$$
\begin{aligned}
|s_n - 0| &= |1/n - 0| \\
&= |1/n| \\
&= 1/n \quad \text{(because n is positive)} \\
&< 1/N \quad \text{(because $n > N > 0$)} \\
&= \frac{1}{(1/\epsilon)} \quad \text{(because $N = 1/\epsilon$) [That was a clever guess!]} \\
&= \epsilon.
\end{aligned}
$$

So we conclude that $|s_n - 0| < \epsilon$, which proves that $\lim s_n = 0$.

Just as in the epsilon-delta proofs where one has to find a δ, similarly how does one divine an N? In the following two examples we indicate this step-by-step, not as a book or your final homework solution would have it recorded.

Example 8.2.3. Let $s_n = \{\frac{1}{n}((-1)^n + i(-1)^{n+1})\}$. If we write out the first few terms, we find that $\{s_n\} = \{-1 + i, 1/2 - i/2, -1/3 + i/3, \ldots\}$, and we may speculate that $\lim s_n = 0$. Here is plot of the image set of this sequence in the complex plane:

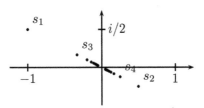

We prove that $\lim s_n = 0$. Let $\epsilon > 0$. Set $N = $ _____. [We will eventually fill in what the positive real number N should be, but at this point of the proof simply leave some blank space. Assuming that N is in place, we next need to prove that for every integer $n > N$, the relevant property as in the $\epsilon - N$ definition of limits holds. The proof of "for every integer $n > N$" always starts with:] Let n be an integer strictly bigger than N. [We want to make sure eventually that n is positive, which is guaranteed if N is positive, but with blank N, we will simply assume in the algebra below that N is positive.] Then

$$|s_n - 0| = \left| \frac{1}{n}((-1)^n + i(-1)^{n+1}) - 0 \right|$$

$$= \left| \frac{1}{n} \right| \cdot |(-1)^n + i(-1)^{n+1}| \text{ (because } |ab| = |a||b|)$$

$$= \frac{1}{n} \cdot |(-1)^n + i(-1)^{n+1}| \text{ (because } n \text{ is positive)}$$

$$= \frac{1}{n} \cdot \sqrt{((-1)^n)^2 + ((-1)^{n+1})^2}$$

$$= \frac{1}{n} \cdot \sqrt{2}$$

$$< \frac{1}{N} \cdot \sqrt{2} \text{ (because } n > N)$$

[Aside: we want/need $\sqrt{2}/N \leq \epsilon$, and $\sqrt{2}/N = \epsilon$ is a possibility, so set $N = \sqrt{2}/\epsilon$. Now go ahead, write that missing information

ON N IN LINE 1 OF THIS PROOF!]

$$= \frac{\sqrt{2}}{\sqrt{2}/\epsilon} \quad (\text{because } N = \sqrt{2}/\epsilon)$$

$$= \epsilon,$$

which proves that for all $n > 1/\epsilon$, $|s_n - 0| < \epsilon$. Since ϵ is arbitrary, this proves that $\lim s_n = 0$. □

Thus a polished version of the example above looks like this:

We prove that $\lim \left\{ \frac{1}{n}((-1)^n + i(-1)^{n+1}) - 0 \right\} = 0$. Let $\epsilon > 0$. Set $N = \sqrt{2}/\epsilon$. Then N is a positive real number. Let n be an integer strictly bigger than N. Then

$$|s_n - 0| = \left| \frac{1}{n}((-1)^n + i(-1)^{n+1}) - 0 \right|$$

$$= \left| \frac{1}{n} \right| \cdot |(-1)^n + i(-1)^{n+1}| \quad (\text{because } |ab| = |a||b|)$$

$$= \frac{1}{n} \cdot |(-1)^n + i(-1)^{n+1}| \quad (\text{because } n \text{ is positive})$$

$$= \frac{1}{n} \cdot \sqrt{((-1)^n)^2 + ((-1)^{n+1})^2}$$

$$= \frac{1}{n} \cdot \sqrt{2}$$

$$< \frac{1}{N} \cdot \sqrt{2} \quad (\text{because } n > N)$$

$$= \frac{\sqrt{2}}{\sqrt{2}/\epsilon} \quad (\text{because } N = \sqrt{2}/\epsilon)$$

$$= \epsilon,$$

which proves that for all $n > \sqrt{2}/\epsilon$, $|s_n - 0| < \epsilon$. Since ϵ is arbitrary, this proves that $\lim s_n = 0$. □

Example 8.2.4. Claim: $\lim \frac{2n+3n^2}{3+4n+n^2} = 3$. Proof: Let $\epsilon > 0$. Set $N = $ _____. Let n be an integer strictly bigger than N. Then

$$\left| \frac{2n + 3n^2}{3 + 4n + n^2} - 3 \right| = \left| \frac{2n + 3n^2}{3 + 4n + n^2} - \frac{3(3 + 4n + n^2)}{3 + 4n + n^2} \right|$$

$$= \left| \frac{-9 - 10n}{3 + 4n + n^2} \right|$$

$$= \frac{9 + 10n}{3 + 4n + n^2} \quad \text{(because } n > 0\text{)}$$

[ASSUMING THAT $N > 0$.]

$$\leq \frac{n + 10n}{3 + 4n + n^2} \quad \text{(because } n \geq 9\text{)}$$

[ASSUMING THAT $N \geq 8$.]

$$= \frac{11n}{3 + 4n + n^2}$$

$$\leq \frac{11n}{n^2} \quad \text{(because } 3 + 4n + n^2 > n^2,$$

$$\text{so } 1/(3 + 4n + n^2) < 1/n^2\text{)}$$

$$= \frac{11}{n}$$

$$< \frac{11}{N} \quad \text{(because } n > N\text{)}$$

$$\leq \frac{11}{11/\epsilon} \quad \text{(because } N \geq 11/\epsilon \text{ so } 1/N \leq 1/(11/\epsilon)\text{)}$$

[ASSUMING THIS.]

$$= \epsilon,$$

which was desired. Now (on scratch paper) we gather all the information we used about N: $N > 0$, $N \geq 8$, $N \geq 11/\epsilon$, and that is it. Thus on the first line we fill in the blank part: Set $N = \max\{8, 11/\epsilon\}$, which says that N is either 8 or $11/\epsilon$, whichever is greater, so that $N \geq 8$ and $N \geq 11/\epsilon$.

The polished version of this proof would go as follows:

Example 8.2.5. Claim: $\lim \frac{2n + 3n^2}{3 + 4n + n^2} = 3$. Proof: Let $\epsilon > 0$. Set $N = \max\{8, 11/\epsilon\}$. Then N is a positive real number. Let n be an integer strictly bigger than N. Then

$$\left| \frac{2n + 3n^2}{3 + 4n + n^2} - 3 \right| = \left| \frac{2n + 3n^2}{3 + 4n + n^2} - \frac{3(3 + 4n + n^2)}{3 + 4n + n^2} \right|$$

$$= \left| \frac{-9 - 10n}{3 + 4n + n^2} \right|$$

$$= \frac{9 + 10n}{3 + 4n + n^2} \quad \text{(because } n > 0\text{)}$$

$$\leq \frac{n + 10n}{3 + 4n + n^2} \quad \text{(because } n \geq 9\text{)}$$

$$= \frac{11n}{3 + 4n + n^2}$$

$$\leq \frac{11n}{n^2} \quad (\text{because } 3 + 4n + n^2 > n^2,$$

$$\text{so } 1/(3 + 4n + n^2) < 1/n^2)$$

$$= \frac{11}{n}$$

$$< \frac{11}{N} \quad (\text{because } n > N)$$

$$\leq \frac{11}{11/\epsilon} \quad (\text{because } N \geq 11/\epsilon \text{ so } 1/N \leq 1/(11/\epsilon))$$

$$= \epsilon,$$

which proves that for all $n > N$, $|s_n - 3| < \epsilon$. Since ϵ is arbitrary, this proves that the limit of this sequence is 3. □

Below is a polished proof of a very similar problem.

Example 8.2.6. Claim: $\lim \frac{-2n+3n^2}{9-4n+n^2} = 3$. Proof: Let $\epsilon > 0$. Set $N = \max\{2, 20/\epsilon\}$. Then N is a positive real number. Let n be an integer strictly bigger than N. Then

$$\left| \frac{-2n + 3n^2}{9 - 4n + n^2} - 3 \right| = \left| \frac{-2n + 3n^2}{9 - 4n + n^2} - \frac{3(9 - 4n + n^2)}{9 - 4n + n^2} \right|$$

$$= \left| \frac{-27 + 10n}{9 - 4n + n^2} \right|$$

$$= \frac{-27 + 10n}{9 - 4n + n^2} \quad (\text{because } n > N \geq 2, \text{ so } n \geq 3,$$

$$\text{so } 10n - 27 > 0 \text{ and } n^2 - 4n + 9 = (n-3)^2 + 2n > 0)$$

$$< \frac{10n}{9 - 4n + n^2} \quad (\text{because } 10n - 27 < 10n)$$

$$= \frac{10n}{9 - 4n + (1/\sqrt{2})n^2 + (1 - 1/\sqrt{2})n^2}$$

$$< \frac{10n}{(1 - 1/\sqrt{2})n^2}$$

$$(\text{because } 0 < (n - 4)^2 + 2 = 2(9 - 4n + (1/\sqrt{2})n^2))$$

$$< \frac{10}{0.5n} \quad (\text{because } 1 - 1/\sqrt{2} > 0.5, \text{ so } \frac{1}{1 - 1/\sqrt{2}} < \frac{1}{0.5})$$

$$= \frac{20}{n}$$

$$< \frac{20}{N} \text{ (because } n > N)$$

$$\leq \frac{20}{20/\epsilon} \text{ (because } N \geq 20/\epsilon \text{ so } 1/N \leq 1/(20/\epsilon))$$

$$= \epsilon,$$

which proves that for all $n > N$, $|s_n - 3| < \epsilon$. Since ϵ is arbitrary, this proves that the limit of this sequence is 3. □

Example 8.2.7. Let $r \in \mathbb{C}$ such that $|r| < 1$. Then $\lim_{n \to \infty} r^n = 0$.

Proof. If $r = 0$, the sequence $\{r^n\}$ is the constant zero sequence so certainly the limit is 0.

So we may assume without loss of generality that $r \neq 0$. Let $\epsilon > 0$. Set $N = \frac{\ln(\min\{\epsilon, 0.5\})}{\ln|r|}$. Since ln of numbers between 0 and 1 is negative, we have that N is a positive number. Let n be an integer with $n > N$. Then

$$|r^n - 0| = |r|^n$$

$$= |r^{n-N}| \, |r|^N$$

$$< 1^{n-N} \, |r|^N \text{ (by Theorem 7.6.5)}$$

$$= \min\{\epsilon, 0.5\}$$

$$\leq \epsilon.$$

Example 8.2.8. $\lim \left\{ \left(1 + \frac{1}{n}\right)^n \right\} = e$.

Proof. All the hard work for this has been done already in Exercise 7.6.10. Let $\epsilon > 0$. By Exercise 7.6.10, $\lim_{x \to \infty} \left\{ \left(1 + \frac{1}{x}\right)^x \right\} = e$. Thus there exists $N > 0$ such that for all $x > N$, $\left| \left\{ \left(1 + \frac{1}{x}\right)^x \right\} - e \right| < \epsilon$. In particular, for any integer $n > N$, $\left| \left\{ \left(1 + \frac{1}{n}\right)^n \right\} - e \right| < \epsilon$. □

Euler's constant e was defined in Definition 7.6.6 via integrals, and lower and upper sums of the relevant integrals allowed us to compute some numerical approximations of e in paragraphs after Definition 7.6.6. The example above allows us to approximate e with these sequences: $(1 + 1/1)^1 = 2$, $(1 + 1/2)^2 = 2.25$, $(1 + 1/3)^3 \cong 2.37037$, $(1 + 1/4)^4 = 2.44140625$, $(1 + 1/5)^5 = 2.48832$, ..., $(1 + 1/100)^{100} \cong 2.716924$, By Exercise 7.6.11, this sequence is increasing, so $e > 2.716924$.

Example 8.2.9. $\lim n^{1/n} = 1$.

Proof. For all $n \geq 2$, by Exercise 1.7.7,

$$n = (n^{1/n})^n = (n^{1/n} - 1 + 1)^n = \sum_{k=0}^{n} \binom{n}{k} (n^{1/n} - 1)^k.$$

Each of the summands is non-negative, and if we only use the summands with $k = 0$ and $k = 2$, we then get that

$$n \geq 1 + \frac{1}{2} n(n-1)(n^{1/n} - 1)^2.$$

By subtracting 1 we get that $n - 1 \geq \frac{1}{2} n(n-1)(n^{1/n} - 1)^2$, so that for $n \geq 2$, $\frac{2}{n} \geq (n^{1/n} - 1)^2$, and hence that $\frac{\sqrt{2}}{\sqrt{n}} \geq n^{1/n} - 1$. Certainly $n^{1/n} - 1 \geq 0$ for all $n \geq 1$. It follows that for all $n \geq 2$, and even for all $n \geq 1$, $0 \leq n^{1/n} - 1 \leq \frac{\sqrt{2}}{\sqrt{n}}$. Now let $\epsilon > 0$. Set $N = \max\{2, 2/\epsilon^2\}$. Then N is a positive real number. Let $n > N$ be an integer. Then $0 \leq n^{1/n} - 1 \leq \frac{\sqrt{2}}{\sqrt{n}} < \frac{\sqrt{2}}{\sqrt{N}} = \epsilon$, which proves that $|n^{1/n} - 1| < \epsilon$, and hence proves this limit. □

Example 8.2.10. Let M be a positive real number. Then $\lim M^{1/n} = 1$.

Proof. First suppose that $M \geq 1$. Then certainly for all integers $n \geq M$, we have that $1 \leq M^{1/n} \leq n^{1/n}$. Let $\epsilon > 0$. By the previous example, there exists $N > 0$ such that for all integers $n > N$, $0 < n^{1/n} - 1 < \epsilon$. Then for all integers $n > \max\{M, N\}$, $0 \leq M^{1/n} - 1 \leq n^{1/n} - 1 < \epsilon$, which implies that $|M^{1/n} - 1| < \epsilon$. This proves the example in case $M \geq 1$.

Now suppose that $M < 1$. By assumption, $1/M > 1$, so by the previous case (and by using that $(1/M)^{1/n} = 1/M^{1/n}$), there exists $N > 0$ such that for all integers $n > N$, $0 \leq 1/M^{1/n} - 1 < \epsilon$. By adding 1 to all three parts in this inequality we get that $1 \leq 1/M^{1/n} < \epsilon + 1$, so that by compatibility of $<$ with multiplication by positive numbers, $\frac{1}{\epsilon+1} < M^{1/n} \leq 1$. Hence by compatibility of $<$ with addition,

$$0 \leq 1 - M^{1/n} < 1 - \frac{1}{\epsilon+1} = \frac{\epsilon}{\epsilon+1} < \epsilon,$$

since $\epsilon + 1 > 1$. Thus $|1 - M^{1/n}| < \epsilon$, which proves that $\lim M^{1/n} = 1$. □

This proves that $\lim_{M \to 0^+} (\lim_{n \to \infty} M^{1/n}) = 1$, and it is easy to see that $\lim_{n \to \infty} (\lim_{M \to 0^+} M^{1/n}) = 0$. This gives an important lesson that the order of limits cannot be switched arbitrarily!

Exercises for Section 8.2

8.2.1. Let $s_n = 1/n^2$. Fill in the blanks in the following proof that $\lim\{s_n\} = 0$.

Let $\epsilon > 0$. Set $N = $ _____ . Then if $n > N$,

$$|s_n - 0| = \left|\frac{1}{n^2} - 0\right|$$

$$= \frac{1}{n^2} \quad (\text{because} \text{_____})$$

$$< \frac{1}{N^2} \quad (\text{because} \text{_____})$$

$$= \text{_____} \quad (\text{because} \text{_____})$$

$$= \epsilon.$$

8.2.2. Prove that for any real number L, $\lim\{\frac{1}{n} + L\} = L$.

8.2.3. Prove that for any positive real number k, $\lim\{1/n^k\} = 0$.

8.2.4. Let $f : \mathbb{N} \to \mathbb{C}$ be a bounded function. Prove that $\lim \frac{f(n)}{n} = 0$.

8.2.5. Prove that $\lim \frac{2n+1}{n^2-4} = 0$.

8.2.6. Prove that $\lim \frac{3n+4}{2n} = \frac{3}{2}$.

8.2.7. Prove that $\lim \{\sqrt{n+1} - \sqrt{n}\} = 0$.

8.2.8. Prove that for every positive integer k, $\lim \{\sqrt{n+k} - \sqrt{n}\} = 0$.

8.2.9. Suppose that the terms of a sequence are given by $s_n = \sum_{k=1}^{n} \frac{1}{k(k+1)}$.

i) Using induction on n, prove that $\sum_{k=1}^{n} \frac{1}{k(k+1)} = 1 - \frac{1}{n+1}$.

ii) Use this to find and prove the limit of $\{s_n\}$.

8.2.10. Prove the following limits:

i) $\lim \frac{\ln(n)}{\ln(n+1)} = 1$. (Hint: epsilon-N proof, continuity of ln.)

ii) $\lim \frac{\ln(\ln(n))}{\ln(\ln(n+1))} = 1$.

8.2.11. Let $r \in \mathbb{R}$. By Theorem 3.8.8, for every positive integer n there exists a rational number $s_n \in (r - \frac{1}{n}, r + \frac{1}{n})$. Prove that $\{s_n\}$ converges to r (regardless of the choice of the s_n).

8.2.12. Let r be a real number with a known decimal expansion. Let s_n be the rational number whose digits $n+1, n+2, n+3$, etc., beyond the decimal point are all 0, and all other digits agree with the digits of r. (For example, if $r = \pi$, then $s_1 = 3.1$, $s_2 = 3.14$, $s_7 = 3.1415926$, etc.) Prove that $\lim\{s_n\} = r$. (Repeat with binary expansions if you know what a binary expansion is.)

8.2.13. Prove that $\lim_n\{ \sqrt[n!]{n!} \} = 1$.

8.2.14. What is wrong with the following "proof" that $\lim_{n\to\infty} \frac{n}{2n+1} = \frac{1}{2}$.

"Proof." Let $\epsilon > 0$. Set $N = \frac{\frac{1}{2\epsilon}-1}{2}$. Let $n > N$. Then

$$\left| \frac{n}{2n+1} - \frac{1}{2} \right| = \left| \frac{2n - (2n+1)}{2(2n+1)} \right|$$

$$= \left| \frac{-1}{2(2n+1)} \right|$$

$$= \frac{1}{2(2n+1)}$$

$$< \frac{1}{2(2N+1)} \quad \text{(because all terms are positive)}$$

$$= \epsilon. \qquad \square$$

8.3 Divergence of infinite sequences and infinite limits

The sequence $\{(-1)^n\}$ alternates in value between -1 and 1, and does not seem to converge to a single number. The following definition addresses this situation.

Definition 8.3.1. *A sequence* **diverges** *if it does not converge. In other words, $\{s_n\}$ diverges if for all complex numbers L, $\lim\{s_n\} \neq L$.*

By the usual negation of statements (see chart on page 25), $\lim\{s_n\} \neq L$ means:

> not (For all real numbers $\epsilon > 0$ there exists a positive real number N such that for all integers $n > N$, $|s_n - L| < \epsilon$.)

> $=$ There exists a real number $\epsilon > 0$ such that
>   not (there exists a positive real number N

such that for all integers $n > N$, $|s_n - L| < \epsilon$.)

$=$ There exists a real number $\epsilon > 0$ such that for all positive real numbers N, not (for all integers $n > N$, $|s_n - L| < \epsilon$.)

$=$ There exists a real number $\epsilon > 0$ such that for all positive real numbers N there exists an integer $n > N$ such that not $(|s_n - L| < \epsilon$.)

$=$ There exists a real number $\epsilon > 0$ such that for all positive real numbers N there exists an integer $n > N$ such that $|s_n - L| \geq \epsilon$.

Example 8.3.2. $\{(-1)^n\}$ is divergent. Namely, for all complex numbers L, $\lim s_n \neq L$.

Proof. Set $\epsilon = 1$ (half the distance between the two values of the sequence). Let N be an arbitrary positive number. If $\mathrm{Re}(L) > 0$, let n be an odd integer greater than N, and if $\mathrm{Re}(L) \leq 0$, let n be an even integer greater than N. In either case, $|s_n - L| \geq |\mathrm{Re}(s_n) - \mathrm{Re}(L)| \geq 1 = \epsilon$. □

The sequence in the previous example has no limit, whereas the sequence in the next example has no finite limit:

Example 8.3.3. For all complex numbers L, $\lim\{n\} \neq L$.

Proof. Set $\epsilon = 53$ (any positive number works). Let N be a positive real number. Let n be any integer that is strictly bigger than N and strictly bigger than $|L| + 53$ (say strictly bigger than $N + |L| + 53$). Such an integer exists. Then by the reverse triangle inequality, $|n - L| \geq |n| - |L| \geq 53 = \epsilon$. □

The last two examples are different: the first one has no limit at all since the terms oscillate wildly, but for the second example we have a sense that its limit is infinity. We formalize this:

Definition 8.3.4. *A real-valued sequence* $\{s_n\}$ **diverges to** ∞ *if for every positive real number* M *there exists a positive number* N *such that for all integers* $n > N$, $s_n > M$. *We write this as* $\lim s_n = \infty$.

A real-valued sequence $\{s_n\}$ **diverges to** $-\infty$ *if for every negative real number* M *there exists a positive real number* N *such that for all integers* $n > N$, $s_n < M$. *We write this as* $\lim s_n = -\infty$.

Example 8.3.5. $\lim n = \infty$.

Proof. Let $M > 0$. Set $N = M$. (As in epsilon-delta or epsilon-N proofs, we must figure out what to set N to. In this case, $N = M$ works). Let $n \in \mathbb{N}^+$ with $n > N$. Since $N = M$, we conclude that $n > M$, and the proof is complete. $\qquad\square$

Example 8.3.6. $\lim \sqrt[n]{n!} = \infty$.

Proof. Let $M > 0$. Let M_0 be an integer that is strictly greater than M. By Example 8.2.10 there exists $N_1 > 0$ such that for all integers $n > N_1$,

$$\left| \sqrt[n]{\frac{M_0!}{(M_0 + 1)^{M_0}}} - 1 \right| < \frac{M_0 + 1 - M}{M_0 + 1}.$$

Since $M_0! = M_0(M_0 - 1)(M_0 - 2) \cdots 3 \cdot 2 \cdot 1 < (M_0 + 1)^{M_0}$, this means that

$$0 \leq 1 - \sqrt[n]{\frac{M_0!}{(M_0 + 1)^{M_0}}} < \frac{M_0 + 1 - M}{M_0 + 1} = 1 - \frac{M}{M_0 + 1},$$

so that $\sqrt[n]{\frac{M_0!}{(M_0+1)^{M_0}}} > \frac{M}{M_0+1}$. Set $N = \max\{N_1, M_0\}$. Then for all integers $n > N$,

$$n! = n(n-1) \cdots (M_0+1) \cdot M_0! \geq (M_0+1)^{n-M_0} \cdot M_0! = (M_0+1)^n \frac{M_0!}{(M_0 + 1)^{M_0}},$$

so that

$$\sqrt[n]{n!} \geq (M_0 + 1) \sqrt[n]{\frac{M_0!}{(M_0 + 1)^{M_0}}} > (M_0 + 1) \frac{M}{M_0 + 1} = M. \qquad\square$$

Theorem 8.3.7. (Comparison theorem (for sequences with infinite limits)) *Let* $\{s_n\}, \{t_n\}$ *be real-valued sequences such that for all sufficiently large n (say for $n \geq N$ for some fixed N), $s_n \leq t_n$.*
   (1) If $\lim s_n = \infty$, *then* $\lim t_n = \infty$.
   (2) If $\lim t_n = -\infty$, *then* $\lim s_n = -\infty$.

Proof. (1) By assumption $\lim s_n = \infty$ for every positive M there exists a positive N' such that for all integers $n > N'$, $s_n > M$. Hence by assumption $t_n \geq s_n$ for all $n \geq N$ we get that for every positive M and for all integers $n > \max\{N, N'\}$, $t_n > M$. Thus by definition $\lim t_n = \infty$.
   Part (2) has an analogous proof.   □

Example 8.3.8. $\lim \frac{n^2+1}{n} = \infty$.

Proof. Note that for all $n \in \mathbb{N}^+$, $\frac{n^2+1}{n} = n + \frac{1}{n} \geq n$, and since we already know that $\lim n = \infty$, it follows by the comparison theorem above that $\lim \frac{n^2+1}{n} = \infty$.   □

Example 8.3.9. $\lim \frac{n^2-1}{n} = \infty$.

Proof. Note that for all integers $n > 2$, $\frac{n^2-1}{n} = n - \frac{1}{n} \geq \frac{n}{2}$. We already know that $\lim n = \infty$, and it is straightforward to prove that $\lim \frac{n}{2} = \infty$. Hence by the comparison theorem, $\lim \frac{n^2-1}{n} = \infty$.
   Or, we can give an $M - N$ proof. Let $M > 0$. Set $N = \max\{2, 2M\}$. Let n be an integer strictly bigger than N. Then

$$\frac{n^2 - 1}{n} = n - \frac{1}{n} \geq \frac{n}{2} > \frac{N}{2} \geq M.$$ □

Theorem 8.3.10. *Let* $\{s_n\}$ *be a sequence of positive numbers. Then* $\lim s_n = \infty$ *if and only if* $\lim \frac{1}{s_n} = 0$.

Proof. Suppose that $\lim s_n = \infty$. Let $\epsilon > 0$. By the definition of infinite limits, there exists a positive number N such that for all integers $n > N$, $s_n > 1/\epsilon$. Then for the same n, $0 < \frac{1}{s_n} < \epsilon$, so that $|\frac{1}{s_n}| < \epsilon$. This proves that $\lim \frac{1}{s_n} = 0$.
   Now suppose that $\lim \frac{1}{s_n} = 0$. Let M be a positive number. By assumption $\lim \frac{1}{s_n} = 0$ there exists a positive number N such that for all integers $n > N$, $|\frac{1}{s_n} - 0| < 1/M$. Since each s_n is positive, it follows that for the same n, $\frac{1}{s_n} < 1/M$, so that $s_n > M$. This proves that $\lim s_n = \infty$.

□

Exercises for Section 8.3

8.3.1. Prove that the following sequences diverge:

   i) $\{\sqrt{n}\}$.

   ii) $\{2^{\sqrt{n}}\}$.

   iii) $\{(n+1)^3 - n^3\}$.

   iv) $\{(-1)^n + 3/n\}$.

8.3.2. Suppose that $\{s_n\}$ diverges and $\{t_n\}$ converges. Prove that $\{s_n \pm t_n\}$ diverges.

8.3.3. Suppose that $\{s_n\}$ diverges and $\{t_n\}$ converges to a non-zero number. Prove that $\{s_n \cdot t_n\}$ diverges.

8.3.4. Give an example of two divergent sequences $\{s_n\}$, $\{t_n\}$ such that $\{s_n + t_n\}$ converges.

8.3.5. Let $\{s_n\}$ be a sequence of negative numbers. Prove that $\lim s_n = -\infty$ if and only if $\lim \frac{1}{s_n} = 0$.

8.3.6. Given the following sequences, find and prove the limits, finite or infinite, if they exist. Otherwise, prove divergence:

   i) $\left\{\frac{n+5}{n^3-5}\right\}$.

   ii) $\left\{\frac{2n^2-n}{3n^2-5}\right\}$.

   iii) $\left\{\frac{(-1)^n}{n-3}\right\}$.

   iv) $\left\{\frac{(-1)^n n}{n+1}\right\}$.

   v) $\left\{\frac{n^3-8n}{n^2+8n}\right\}$.

   vi) $\left\{\frac{1-n^2}{n}\right\}$.

   vii) $\left\{\frac{2^n}{n!}\right\}$.

   viii) $\left\{\frac{n!}{(n+1)!}\right\}$.

8.3.7. Prove or give a counterexample:

   i) If $\{s_n\}$ and $\{t_n\}$ both diverge, then $\{s_n + t_n\}$ diverges.

   ii) If $\{s_n\}$ converges and $\{t_n\}$ diverges, then $\{s_n + t_n\}$ diverges.

   iii) If $\{s_n\}$ and $\{t_n\}$ both diverge, then $\{s_n \cdot t_n\}$ diverges.

   iv) If $\{s_n\}$ converges and $\{t_n\}$ diverges, then $\{s_n \cdot t_n\}$ diverges.

8.3.8. Find examples of the following:

   i) A sequence $\{s_n\}$ of non-zero terms such that $\lim\{\frac{s_n}{s_{n+1}}\} = 0$.

   ii) A sequence $\{s_n\}$ of non-zero terms such that $\lim\{\frac{s_{n+1}}{s_n}\} = 1$.

   iii) A sequence $\{s_n\}$ of non-zero terms such that $\lim\{\frac{s_{n+1}}{s_n}\} = \infty$.

   iv) A sequence $\{s_n\}$ such that $\lim\{s_{n+1} - s_n\} = 0$.

   v) A sequence $\{s_n\}$ such that $\lim\{s_{n+1} - s_n\} = \infty$.

8.3.9. Suppose that the sequence $\{s_n\}_n$ diverges to ∞ (or to $-\infty$). Prove that $\{s_n\}_n$ diverges. (The point of this exercise is to parse the definitions correctly.)

8.4 Convergence theorems via epsilon-N proofs

All of the theorems in this section are also proved in Section 8.5 with a different method; here we use the epsilon-N formulation for proofs without explicitly resorting to functions whose domains have a limit point. I recommend reading this section and omitting Section 8.5.

Theorem 8.4.1. *If a sequence converges, then its limit is unique.*

Proof. Let $\{s_n\}$ be a convergent sequence. Suppose that $\{s_n\}$ converges to both L and L'. Then for any $\epsilon > 0$, there exists an N such that $|s_n - L| < \epsilon/2$ for all $n > N$. Likewise, for any $\epsilon > 0$, there exists an N' such that $|s_n - L'| < \epsilon/2$ for all $n > N'$. Then by the triangle inequality, $|L - L'| = |L - s_n + s_n - L'| \leq |L - s_n| + |s_n - L'| < \epsilon/2 + \epsilon/2 = \epsilon$. Since ϵ is arbitrary, by Theorem 2.10.4 it must be the case that $|L - L'| = 0$, i.e., that $L = L'$. $\square$

Theorem 8.4.2. *Suppose that $\lim\{s_n\} = L$ and that $L \neq 0$. Then there exists a positive number N such that for all integers $n > N$, $|s_n| > |L|/2$. In particular, there exists a positive number N such that for all integers $n > N$, $s_n \neq 0$.*

Proof. Note that $p = |L|/2$ is a positive real number. Since $\lim\{s_n\} = L$, it follows that there exists a real number N such that for all integers $n > N$, $|s_n - L| < |L|/2$. Then by the reverse triangle inequality (proved in Theorem 2.10.3),

$$|s_n| = |s_n - L + L| = |(s_n - L) + L| \geq |L| - |s_n - L| > |L| - |L|/2 = |L|/2. \quad \square$$

Theorem 8.4.3. *Suppose that* $\lim s_n = L$ *and* $\lim t_n = K$. *Then*

(1) *(Constant rule) For any complex number* c, $\lim\{c\} = c$.

(2) *(Linear rule)* $\lim\{1/n\} = 0$.

(3) *(Sum/difference rule)* $\lim\{s_n \pm t_n\} = L \pm K$.

(4) *(Scalar rule) For any complex number* c, $\lim\{cs_n\} = cL$.

(5) *(Product rule)* $\lim\{s_n t_n\} = LK$.

(6) *(Quotient rule) If* $t_n \neq 0$ *for all* n *and* $K \neq 0$, *then* $\lim\{s_n/t_n\} = L/K$.

(7) *(Power rule) For all positive integers* m, $\lim\{s_n^m\} = L^m$.

Proof. Part (1) was proved immediately after Definition 8.2.1. Part (2) was Example 8.2.2.

Part (3): Let $\epsilon > 0$. Since $\lim s_n = L$, there exists a positive real number N_1 such that for all integers $n > N_1$, $|s_n - L| < \epsilon/2$. Since $\lim t_n = K$, there exists a positive real number N_2 such that for all integers $n > N_2$, $|t_n - K| < \epsilon/2$. Let $N = \max\{N_1, N_2\}$. Then for all integers $n > N$,

$$|(s_n \pm t_n) - (L \pm K)| = |(s_n - L) \pm (t_n - K)|$$
$$\leq |s_n - L| + |t_n - K| \text{ (by the triangle inequality)}$$
$$< \epsilon/2 + \epsilon/2 \text{ (since } n > N \geq N_1, N_2)$$
$$= \epsilon.$$

This proves (3).

Part (4): Let $\epsilon > 0$. Note that $\epsilon/(|c| + 1)$ is a positive number. Since $\lim s_n = L$, there exists N such that for all integers $n > N$, $|s_n - L| < \epsilon/(|c| + 1)$. Then for the same n, $|cs_n - cL| = |c| \cdot |s_n - L| \leq |c|\epsilon/(|c| + 1) = \frac{|c|}{|c|+1}\epsilon < \epsilon$. Since ϵ is any positive number, we conclude that cs_n converges to cL.

Part (5): Let $\epsilon > 0$. Since $\lim\{s_n\} = L$, there exists N_1 such that for all integers $n > N_1$, $|s_n - L| < 1$. Thus for all such n, $|s_n| = |s_n - L + L| \leq |s_n - L| + |L| < 1 + |L|$. There also exists N_2 such that for all integers $n > N_2$, $|s_n - L| < \epsilon/(2|K| + 1)$. By assumption $\lim\{t_n\} = K$ there exists N_3 such that for all integers $n > N_3$, $|t_n - K| < \epsilon/(2|L| + 2)$. Let $N = \max\{N_1, N_2, N_3\}$. Then for all integers $n > N$,

$$|s_n t_n - LK| = |s_n t_n - s_n K + s_n K - LK| \text{ (by adding a clever 0)}$$
$$= |(s_n t_n - s_n K) + (s_n K - LK)|$$

$$\leq |s_n t_n - s_n K| + |s_n K - LK| \quad \text{(by the triangle inequality)}$$
$$= |s_n(t_n - K)| + |(s_n - L)K|$$
$$= |s_n| \cdot |t_n - K| + |s_n - L| \cdot |K|$$
$$< (|L| + 1) \cdot \frac{\epsilon}{2|L| + 2} + \frac{\epsilon}{2|K| + 1} \cdot |K|$$
$$< \frac{\epsilon}{2} + \frac{\epsilon}{2}$$
$$= \epsilon,$$

which proves (5).

Part (6): Let $\epsilon > 0$. Since $\lim\{s_n\} = L$, there exists N_1 such that for all integers $n > N_1$, $|s_n - L| < 1$. Thus for all such n, $|s_n| = |s_n - L + L| \leq |s_n - L| + |L| < 1 + |L|$. There also exists N_2 such that for all integers $n > N_2$, $|s_n - L| < \frac{|K|\epsilon}{2}$. By assumption $\lim\{t_n\} = L$ there exists N_3 such that for all integers $n > N_3$, $|t_n - K| < \frac{|K|^2\epsilon}{4(1+|L|)}$. Since $K \neq 0$, by Theorem 8.4.2 there exists a positive number N_4 such that for all integers $n > N_4$, $|t_n| > |K|/2$. Let $N = \max\{N_1, N_2, N_3, N_4\}$. Then for all integers $n > N$,

$$\left|\frac{s_n}{t_n} - \frac{L}{K}\right| = \left|\frac{K \cdot s_n - L \cdot t_n}{K \cdot t_n}\right|$$
$$= \left|\frac{K \cdot s_n - t_n s_n + t_n s_n - L \cdot t_n}{K \cdot t_n}\right| \quad \text{(by adding a clever 0)}$$
$$\leq \left|\frac{K \cdot s_n - t_n s_n}{K \cdot t_n}\right| + \left|\frac{t_n s_n - L \cdot t_n}{K \cdot t_n}\right| \quad \text{(by the triangle inequality)}$$
$$= |K - t_n||s_n|\frac{1}{|K|}\frac{1}{|t_n|} + |L - s_n|\frac{1}{K}$$
$$< \frac{|K|^2\epsilon}{4(1+|L|)}(1+|L|)\frac{1}{|K|}\frac{2}{|K|} + \frac{|K|\epsilon}{2}\frac{1}{|K|}$$
$$= \frac{\epsilon}{2} + \frac{\epsilon}{2}$$
$$= \epsilon,$$

which proves (6).

Part (7): The proof is by induction on m, the base case being the assumption. If $\lim_{n\to\infty}\{1/n^{m-1}\} = 0$, then by the product rule, $\lim_{n\to\infty}\{1/n^m\} = \lim_{n\to\infty}\{1/n^{m-1} \cdot 1/n\} = \lim_{n\to\infty}\{1/n^{m-1}\} \cdot \lim_{n\to\infty}\{1/n\} = 0$. This proves (7). $\qquad\square$

Example 8.4.4. Suppose $s_n = \frac{5n-2}{3n+4}$. To prove that $\lim s_n = 5/3$, we note that $s_n = \frac{5n-2}{3n+4}\frac{1/n}{1/n} = \frac{5-2/n}{3+4/n}$. By the linear rule, $\lim(1/n) = 0$, so by the scalar rule, $\lim(2/n) = \lim(4/n) = 0$. Thus by the constant, sum, and difference rules, $\lim\{5 - 2/n\} = 5$ and $\lim\{3 + 4/n\} = 3$, so that by the quotient rule, $\lim s_n = 5/3$.

Example 8.4.5. Let $s_n = \frac{3n+2}{n^2-3}$. Note that $s_n = \frac{3n+2}{n^2-3} \cdot \frac{1/n^2}{1/n^2} = \frac{3/n+2/n^2}{1-3/n^2}$. By the linear rule, $\lim 1/n = 0$, so that by the scalar rule, $\lim 3/n = 0$, and by the product and scalar rules, $\lim 2/n^2 = \lim 3/n^2 = 0$. Thus by the sum and difference rules, $\lim\{3/n + 2/n^2\} = 0$ and $\lim\{1 - 3/n^2\} = 1$. Finally, by the quotient rule, $\lim s_n = 0/1 = 0$.

Theorem 8.4.6. (Power, polynomial, rational rules for sequences) *For any positive integer m, $\lim_{n\to\infty}\{1/n^m\} = 0$. If f is a polynomial function, then $\lim\{f(1/n)\} = f(0)$. If f is a rational function that is defined at 0, then $\lim\{f(1/n)\} = f(0)$.*

Proof. By the linear rule (in Theorem 8.4.3), $\lim_{n\to\infty}\{1/n\} = 0$, and by the power rule, for all positive integers m, $\lim_{n\to\infty}\{1/n^m\} = 0$.

Now write $f(x) = a_0 + a_1 x + \cdots + a_k x^k$ for some non-negative integer k and some complex numbers $a_0, a_1, \ldots, a_k$. By the constant, power and repeated sum rules,

$$\lim_{n\to\infty} f(1/n) = \lim_{n\to\infty}\left\{a_0 + a_1(1/n) + a_2(1/n)^2 + \cdots + a_k(1/n)^k\right\}$$
$$= a_0 + \lim_{n\to\infty} a_1(1/n) + \lim_{n\to\infty} a_2(1/n)^2 + \cdots + \lim_{n\to\infty} a_k(1/n)^k$$
$$= a_0 + a_1 \cdot 0 + a_2 \cdot 0^2 + \cdots + a_k \cdot 0^k$$
$$= a_0 = f(0).$$

This proves the polynomial rule.

Finally, let f be a rational function. Write $f(x) = g(x)/h(x)$, where g, h are polynomial functions with $h(0) \neq 0$. By the just-proved polynomial rule, $\lim_{n\to\infty} g(1/n) = g(0)$, $\lim_{n\to\infty} h(1/n) = h(0) \neq 0$, so that by the quotient rule, $\lim_{n\to\infty} f(1/n) = g(0)/h(0) = f(0)$. $\qquad\square$

Theorem 8.4.7. (The composite rule for sequences) *Suppose that* $\lim s_n = L$. *Let g be a function whose domain contains L and all terms s_n. Suppose that g is continuous at L. Then $\lim g(s_n) = g(L)$.*

Proof. Let $\epsilon > 0$. Since g is continuous at L, there exists a positive number $\delta > 0$ such that for all x in the domain of g, if $|x - L| < \delta$ then $|g(x) - g(L)| < \epsilon$. Since $\lim s_n = L$, there exists a positive number N such that for all integers $n > N$, $|s_n - L| < \delta$. Hence for the same n, $|g(s_n) - g(L)| < \epsilon$. $\qquad\square$

In particular, since the absolute value function, the real part, and the imaginary part functions are continuous everywhere, we immediately conclude the following:

Theorem 8.4.8. *Suppose that $\lim s_n = L$. Then*

(1) $\lim |s_n| = |L|$.

(2) $\lim \operatorname{Re} s_n = \operatorname{Re} L$.

(3) $\lim \operatorname{Im} s_n = \operatorname{Im} L$. $\qquad\square$

Furthermore, since the real and imaginary parts determine a complex number, we moreover get:

Theorem 8.4.9. *A sequence $\{s_n\}$ of complex numbers converges if and only if the sequences $\{\operatorname{Re} s_n\}$ and $\{\operatorname{Im} s_n\}$ of real numbers converge.*

Proof. By Theorem 8.4.8 it suffices to prove that if $\lim\{\operatorname{Re} s_n\} = a$ and $\lim\{\operatorname{Im} s_n\} = b$, then $\lim\{s_n\} = a + bi$. Let $\epsilon > 0$. By assumptions there exist positive real numbers N_1, N_2 such that for all integers $n > N_1$ $|\operatorname{Re} s_n - a| < \epsilon/2$ and such that for all integers $n > N_2$ $|\operatorname{Im} s_n - b| < \epsilon/2$. Set $N = \max\{N_1, N_2\}$. Then for all integers $n > N$,

$$
\begin{aligned}
|s_n - (a + bi)| &= |\operatorname{Re} s_n + i \operatorname{Im} s_n - a - bi| \\
&\leq |\operatorname{Re} s_n - a||i||\operatorname{Im} s_n - b| \\
&< \epsilon/2 + \epsilon/2 \\
&= \epsilon.
\end{aligned}
$$

$\qquad\square$

Theorem 8.4.10. (Comparison of sequences) *Let s and t be convergent sequences of complex numbers. Suppose that $|s_n| \leq |t_n|$ for all except finitely many n. Then $|\lim s_n| \leq |\lim t_n|$.*

If in addition for all except finitely many n, s_n, t_n are real numbers with $s_n \leq t_n$, then $\lim s_n \leq \lim t_n$.

Proof. Let $L = \lim s_n$, $K = \lim t_n$. By Theorem 8.4.8, $\lim |s_n| = |L|$, $\lim |t_n| = |K|$.

Suppose that $|L| > |K|$. Set $\epsilon = (|L| - |K|)/2|$. By the definition of convergence, there exist $N_1, N_2 > 0$ such that if n is an integer, $n > N_1$ implies that $-\epsilon < |s_n| - |L| < \epsilon$, and $n > N_2$ implies that $-\epsilon < |t_n| - |K| < \epsilon$. Let N_3 be a positive number such that for all integers $n > N_3$, $|s_n| \leq |t_n|$. If we let $N = max\{N_1, N_2, N_3\}$, then for integers $n > N$ we have that $|t_n| < |K| + \epsilon = (|L| + |K|)/2 = |L| - \epsilon < |s_n|$. This contradicts the assumption $|s_n| \leq |t_n|$, so that necessarily $|L| \leq |K|$.

The proof of the second part is similar and left to the exercises. $\square$

Theorem 8.4.11. (The squeeze theorem for sequences) *Suppose that s, t, u are sequences of real numbers and that for all $n \in \mathbb{N}^+$, $s_n \leq t_n \leq u_n$. If $\lim s$ and $\lim u$ both exist and are equal, then $\lim t$ exists as well and*

$$\lim s = \lim t = \lim u.$$

Proof. Set $L = \lim s = \lim u$. Let $\epsilon > 0$. Since $\lim s = L$, there exists a positive N_1 such that for all integers $n > N_1$, $|s_n - L| < \epsilon$. Since $\lim u = L$, there exists a positive N_2 such that for all integers $n > N_2$, $|u_n - L| < \epsilon$. Set $N - max\{N_1, N_2\}$. Let n be an integer strictly greater than N. Then $-\epsilon < s_n - L \leq t_n - L \leq u_n - L < \epsilon$, so that $|t_n - L| < \epsilon$. Since ϵ was arbitrary, this proves that $\lim t = L$. $\square$

Example 8.4.12. $\lim(n + 1)^{1/n} = 1$.

Proof. For $n \geq 2$, $1 \leq \left(\frac{n+1}{n}\right)^{1/n} \leq \left(\frac{3}{2}\right)^{1/n} \leq n^{1/n}$. Thus by the previous theorem and by Example 8.2.9, $\lim\{\left(\frac{n+1}{n}\right)^{1/n}\}_n = 1$. Hence by Theorem 8.4.3 and by Example 8.2.9, $\lim\{(n + 1)^{1/n}\} = \lim\left\{\left(\frac{n+1}{n}\right)^{1/n} n^{1/n}\right\} = \lim\left\{\left(\frac{n+1}{n}\right)^{1/n}\right\} \lim\{n^{1/n}\} = 1$. $\square$

Exercises for Sections 8.4 and 8.5

8.4.1. Compute the following limits, and justify:

  i) $\lim \frac{1}{n^2-2}$.

  ii) $\lim \frac{3n^2+20}{2n^3-20}$.

  iii) $\lim \frac{2n^{-2}}{-5n^{-2}-n^{-1}}$.

  iv) $\lim \frac{n^2+4n-2}{4n^2+2}$.

  v) $\lim \frac{-3n^3+2}{2n^3+n}$.

  vi) $\lim \frac{2n^{-4}+2}{3n^{-2}+n^{-1}+2}$.

8.4.2. Give an example of a sequence $\{s_n\}$ and a number L such that $\lim |s_n| = |L|$ but $\{s_n\}$ does not converge. Justify your example.

8.4.3. Prove that $\lim \left\{ \frac{(-1)^n}{n} \right\} = 0$.

†8.4.4. Let $\{s_n\}$ be a convergent sequence of positive real numbers. Prove that $\lim s_n \geq 0$.

8.4.5. Prove that for all integers m, $\lim_{n \to \infty} \left(\frac{n+1}{n} \right)^m = 1$.

8.4.6. Prove that $\lim_{n \to \infty} \{3 - \frac{2}{n}\} = 3$.

8.4.7. Prove that $\lim \{ \sqrt[3]{n+1} - \sqrt[3]{n} \} = 0$.

8.4.8. By Example 8.2.8, $\lim \left\{ \left(1 + \frac{1}{n}\right)^n \right\} = e$. Determine the following limits:

  i) $\left\{ \left(\frac{n+1}{n} \right)^n \right\}$.

  ii) $\left\{ \left(1 + \frac{1}{2n} \right)^{2n} \right\}$.

  iii) $\left\{ \left(1 + \frac{1}{n} \right)^{2n} \right\}$.

  iv) $\left\{ \left(1 + \frac{1}{2n} \right)^{n} \right\}$.

  v) $\left\{ \left(1 + \frac{1}{n} \right)^{n+1} \right\}$.

8.4.9. Let s_1 be a positive real number. For each $n \geq 1$, let $s_{n+1} = \sqrt{s_n}$. Prove that $\lim s_n = 1$. (Hint: Prove that $\lim s_n = \lim s_n^2$.)

8.5   Convergence theorems via functions

The results in this section are the same as those in Section 8.4, but here they are proved with theorems about limits of functions that were proved in Chapter 4. So, a connection is made between limits of functions and limits of sequences. The reader may omit this section (or the previous one). This section is more abstract; one has to keep in mind connections with functions as well as theorems about limits of functions to get at theorems about limits of sequences. Exercises for this section appear at the end of Section 8.4.

For any sequence s we can define a function $f : \{1/n : n \in \mathbb{N}^+\} \to \mathbb{C}$ with $f(1/n) = s_n$. Conversely, for every function $f : \{1/n : n \in \mathbb{N}^+\} \to \mathbb{C}$ we can define a sequence s with $s_n = f(1/n)$.

The domain of f has exactly one limit point, namely 0. With this we have the usual notion of $\lim_{x \to 0} f(x)$ with standard theorems from Section 4.4.

Theorem 8.5.1. *Let s, f be as above. Then $\lim s_n = L$ if and only if $\lim_{x \to 0} f(x) = L$.*

Proof. ($\Rightarrow$) Suppose that $\lim s_n = L$. We have to prove that $\lim_{x \to 0} f(x) = L$. Let $\epsilon > 0$. By assumption $\lim s_n = L$, there exists a positive real number N such that for all integers $n > N$, $|s_n - L| < \epsilon$. Let $\delta = 1/N$. Then δ is a positive real number. Let x be in the domain of f such that $0 < |x - 0| < \delta$. Necessarily $x = 1/n$ for some positive integer n. Thus $0 < |x - 0| < \delta$ simply says that $1/n < \delta = 1/N$, so that $N < n$. But then by assumption $|f(1/n) - L| = |s_n - L| < \epsilon$, which proves that $\lim_{x \to 0} f(x) = L$.

($\Leftarrow$) Now suppose that $\lim_{x \to 0} f(x) = L$. We have to prove that $\lim s_n = L$. Let $\epsilon > 0$. By assumption $\lim_{x \to 0} f(x) = L$ there exists a positive real number δ such that for all x in the domain of f, if $0 < |x - 0| < \delta$ then $|f(x) - L| < \epsilon$. Set $N = 1/\delta$. Then N is a positive real number. Let n be an integer greater than N. Then $0 < 1/n < 1/N = \delta$, so that by assumption $|f(1/n) - L| < \epsilon$. Hence $|s_n - L| = |f(1/n) - L| < \epsilon$, which proves that $\lim s_n = L$.   □

Example 8.5.2. (Compare the reasoning in this example with the epsilon-N proofs of Section 8.2.) Let $s_n = \frac{5n-2}{3n+4}$. We note that $s_n = \frac{5n-2}{3n+4} \cdot \frac{1/n}{1/n} = \frac{5-2/n}{3+4/n}$. The corresponding function $f : \{1/n : n \in \mathbb{N}^+\}$ is $f(x) = \frac{5-2x}{3+4x}$, and by the scalar, sum, difference, and quotient rules for limits of functions, $\lim_{x\to 0} f(x) = \frac{5-0}{3+0} = 5/3$, so that by Theorem 8.5.1, $\lim s_n = 5/3$.

Example 8.5.3. Suppose $s_n = \frac{3n+2}{n^2-3}$. Note that $s_n = \frac{3n+2}{n^2-3} \cdot \frac{1/n^2}{1/n^2} = \frac{3/n+2/n^2}{1-3/n^2}$. The corresponding function $f : \{1/n : n \in \mathbb{N}^+\}$ is $f(x) = \frac{3x+2x^2}{1-3x^2}$, and by the scalar, sum, difference, product, and quotient rules for limits of functions, $\lim s_n = \lim_{x\to 0} f(x) = \frac{0+0}{1-0} = 0$.

Theorem 8.5.4. *The limit of a converging sequence is unique.*

Proof. Let $\{s_n\}$ be a convergent sequence. Suppose that $\{s_n\}$ converges to both L and L'. Let $f : \{1/n : n \in \mathbb{N}^+\}$ be the function corresponding to s. By Theorem 8.5.1, $\lim_{x\to 0} f(x) = L$ and $\lim_{x\to 0} f(x) = L'$. By Theorem 4.4.1, $L = L'$. This proves uniqueness of limits for sequences.   □

Theorem 8.5.5. *Suppose that $\lim\{s_n\} = L$ and that $L \neq 0$. Then there exists a positive number N such that for all integers $n > N$, $|s_n| > |L|/2$. In particular, there exists a positive number N such that for all integers $n > N$, $s_n \neq 0$.*

Proof. Let f be the function corresponding to s. Then $\lim_{x\to 0} f(x) = L$, by Theorem 4.4.2 there exists $\delta > 0$ such that for all x in the domain of f, if $x < \delta$ then $|f(x)| > |L|/2$. Set $N = 1/\delta$. Let n be an integer strictly greater than N. Then $1/n < 1/N = \delta$, so $|s_n| = |f(1/n)| > |L|/2$.   □

Theorem 8.5.6. *Suppose that $\lim s_n = L$ and $\lim t_n = K$. Then*
  (1) *(Constant rule) For any complex number c, $\lim\{c\} = c$.*
  (2) *(Linear rule) $\lim\{1/n\} = 0$.*
  (3) *(Sum/difference rule) $\lim\{s_n \pm t_n\} = L \pm K$.*
  (4) *(Scalar rule) For any complex number c, $\lim\{cs_n\} = cL$.*
  (5) *(Product rule) $\lim\{s_n t_n\} = LK$.*
  (6) *(Quotient rule) If $t_n \neq 0$ for all n and $K \neq 0$, then $\lim\{s_n/t_n\} = L/K$.*
  (7) *(Power rule) For all positive integers m, $\lim\{s_n^m\} = L^m$.*

Proof. Let $f, g : \{1/n : n \in \mathbb{N}^+\} \to \mathbb{C}$ be given by $f(1/n) = s_n$, $g(1/n) = t_n$. By Theorem 8.5.1, $\lim_{x \to 0} f(x) = L$ and $\lim_{x \to 0} g(x) = K$. Theorem 4.4.3 proves parts (3), (4), (5), (6) for f, g, hence via Theorem 8.5.1 also for s, t. Theorem 4.4.4 finishes the proof of (7).

For part (1), declare $f(1/n) = c$ and for part (2), declare $f(1/n) = 1/n$. Again Theorems 4.4.3 and 8.5.1 easily finish the proofs of (1), (2). $\square$

The following theorem for sequences follows immediately from the corresponding power, polynomial, and rational rules for functions:

Theorem 8.5.7. (Power, polynomial, rational rules for sequences) *Let f be a polynomial function. Then $\lim\{f(1/n)\} = f(0)$. In particular, for any positive integer m, $\lim_{n \to \infty}\{1/n^m\} = 0$. If f is a rational function that is defined at 0, then $\lim\{f(1/n)\} = f(0)$.*

Theorem 8.5.8. (The composite rule for sequences) *Suppose that $\lim s_n = L$. Let g be a function whose domain contains L and all terms s_n. Suppose that g is continuous at L. Then $\lim g(s_n) = g(L)$.*

Proof. Let $f(1/n) = s_n$. By Theorem 8.5.1, $\lim_{x \to 0} f(x) = L$, and by assumption $\lim_{x \to L} g(x) = g(L)$. Thus by the composite function theorem (Theorem 4.4.9), $\lim_{x \to 0} g(f(x)) = g(L)$, so that by Theorem 8.5.1, $\lim g(s_n) = g(L)$. $\square$

In particular, since the absolute value function, the real part, and the imaginary part functions are continuous everywhere, we immediately conclude the following:

Theorem 8.5.9. *Suppose that $\lim s_n = L$. Then*
  (1) $\lim |s_n| = |L|$.
  (2) $\lim \operatorname{Re} s_n = \operatorname{Re} L$.
  (3) $\lim \operatorname{Im} s_n = \operatorname{Im} L$.

Since the real and imaginary parts determine a complex number, we also have:

Theorem 8.5.10. *A sequence $\{s_n\}$ of complex numbers converges if and only if the sequences $\{\operatorname{Re} s_n\}$ and $\{\operatorname{Im} s_n\}$ of real numbers converge.*

Proof. Let $f(1/n) = s_n$. By Theorem 4.4.8, $\lim_{x \to 0} f(x) = L$ if and only if $\lim_{x \to 0} \operatorname{Re} f(x) = \operatorname{Re} L$ and $\lim_{x \to 0} \operatorname{Im} f(x) = \operatorname{Im} L$, which is simply a restatement of the theorem. □

Theorem 8.5.11. (Comparison of sequences) *Let s and t be convergent sequences of complex numbers. Suppose that $|s_n| \le |t_n|$ for all except finitely many n. Then $|\lim s_n| \le |\lim t_n|$.*

If in addition for all except finitely many n, s_n, t_n are real numbers with $s_n \le t_n$, then $\lim s_n \le \lim t_n$.

Proof. Let A be the set of those $1/n$ for which $|s_n| \le |t_n|$ in the first case and for which $s_n \le t_n$ in the second case. Let $f, g : A \to \mathbb{R}$ be the functions $f(1/n) = |s_n|$, $g(1/n) = |t_n|$ in the first case, and $f(1/n) = s_n$, $g(1/n) = t_n$ in the second case. By assumption for all x in the domain, $f(x) \le g(x)$. Since 0 is a limit point of the domain (despite omitting finitely many $1/n$) and since by Theorem 8.5.1, $\lim_{x \to 0} f(x)$ and $\lim_{x \to 0} g(x)$ both exist, by Theorem 4.4.10, $\lim_{x \to 0} f(x) < \lim_{x \to 0} g(x)$. In the first case, this translates to $|\lim s_n| \le |\lim t_n|$, and in the second case it translates to $\lim s_n \le \lim t_n$. □

Theorem 8.5.12. (The squeeze theorem for sequences) *Suppose that s, t, u are sequences of real numbers and that for all $n \in \mathbb{N}^+$, $s_n \le t_n \le u_n$. If $\lim s$ and $\lim u$ both exist and are equal, then $\lim t$ exists as well and*

$$\lim s = \lim t = \lim u.$$

Proof. Let $f, g, h : \{1/n : n \in \mathbb{N}^+\} \to \mathbb{C}$ be functions defined by $f(1/n) = s_n$, $g(1/n) = t_n$, $h(1/n) = u_n$. The assumption is that for all x in the domain of f, g, h, $f(x) \le g(x) \le h(x)$, and by Theorem 8.5.1 that $\lim_{x \to 0} f(x) = \lim_{x \to 0} h(x)$. Then by the squeeze theorem for functions (Theorem 4.4.11), $\lim_{x \to 0} f(x) = \lim_{x \to 0} g(x) = \lim_{x \to 0} h(x)$. Hence by Theorem 8.5.1, $\lim s = \lim t = \lim u$. □

Example 8.5.13. $\lim(n+1)^{1/n} = 1$.

Proof. For $n \geq 2$, $1 \leq \left(\frac{n+1}{n}\right)^{1/n} \leq \left(\frac{3}{2}\right)^{1/n} \leq n^{1/n}$. Thus by the previous theorem and by Example 8.2.9, $\lim\left\{\left(\frac{n+1}{n}\right)^{1/n}\right\} = 1$. Hence by Theorem 8.5.6 and by Example 8.2.9, $\lim\{(n+1)^{1/n}\} = \lim\left\{\left(\frac{n+1}{n}\right)^{1/n} n^{1/n}\right\} = \lim\left\{\left(\frac{n+1}{n}\right)^{1/n}\right\} \lim\{n^{1/n}\} = 1$. $\qquad\square$

Exercises for this section appear at the end of Section 8.4.

8.6   Bounded and monotone sequences, ratio test

Definition 8.6.1. *A sequence* $\{s_n\}$ *is* **bounded** *if there exists a positive real number* B *such that for all integers* n, $|s_n| \leq B$.

If all s_n *are real numbers, we say that* $\{s_n\}$ *is* **bounded above** *(resp.* **below***) if there exists a real number* M *such that for all positive integers* n, $s_n \leq M$ *(resp.* $s_n \geq M$*). In other words,* $\{s_n\}$ *is bounded if the set* $\{s_n : n \in \mathbb{N}^+\}$ *is a subset of* $B(0, M)$ *for some real number* M.

Sequences $\{4 + 5i\}$, $\{1/n\}$, $\left\{\left(\frac{4+i}{5+2i}\right)^n\right\}$, $\{(-1)^n\}$, $\{3 - 4i^n\}$, $\{\frac{4n}{3n^2-4}\}$, $\{\sqrt[n]{n}\}$ are bounded (the latter by Example 1.6.7), but $\{n\}$, $\{(-1)^n n\}$, $\{\frac{n^2+1}{n}\}$, $\{\sqrt[n]{n!}\}$, $\{2^{\sqrt{n}}\}$ are not bounded. The next theorem provides many examples of bounded sequences.

Theorem 8.6.2. *Every convergent sequence is bounded.*

Proof. Let $\{s_n\}$ be a convergent sequence with limit L. Thus there exists a positive integer N such that for all integers $n > N$, $|s_n - L| < 1$. Set $B = \max\{|s_1|, |s_2|, |s_3|, \ldots, |s_{N+1}|, |L| + 1\}$. Then for all positive integers $n \leq N$, $|s_n| \leq B$ by definition of B, and for $n > N$, $|s_n| = |s_n - L + L| \leq |s_n - L| + |L| < 1 + |L| \leq B$. $\qquad\square$

Another proof of the theorem above is given in the next section via Cauchy sequences.

Definition 8.6.3. *A sequence $\{s_n\}$ of real numbers is called* **non-decreasing** *(resp.* **non-increasing, strictly increasing, strictly decreasing***) if for all n, $s_n \leq s_{n+1}$ (resp. $s_n \geq s_{n+1}$, $s_n < s_{n+1}$, $s_n > s_{n+1}$). Any such sequence is called* **monotone***.*

Sequences $\{1/n\}, \{-n\}$ are strictly decreasing, $\{\frac{n^2+1}{n}\}$ is strictly increasing, $\{(-1)^n n\}$ is neither increasing nor decreasing, $\{\frac{n^2+5}{n}\}_{n \geq 1}$ is neither increasing nor decreasing, but $\{\frac{n^2+5}{n}\}_{n \geq 2}$ is strictly increasing.

Theorem 8.6.4. (Bounded monotone sequences) *Let $\{s_n\}$ be a bounded sequence of real numbers such that for some integer N, $\{s_n\}_{n \geq N}$ is non-decreasing (resp. non-increasing). Then $\lim s_n$ exists, and equals the least upper bound (resp. greatest lower bound) of the set $\{s_N, s_{N+1}, s_{N+2}, \ldots\}$.*

Proof. Suppose that for all $n \geq N$, $s_n \leq s_{n+1}$. By the Least upper bound theorem (Theorem 3.8.5), the least upper bound of the set $\{s_N, s_{N+1}, s_{N+2}, \ldots\}$ exists. Call it L.

Let $\epsilon > 0$. Since L is the least upper bound, there exists a positive integer $N' \geq N$ such that $0 \leq L - s_{N'} < \epsilon$. Hence for all integers $n > N'$, $s_{N'} \leq s_n$, so that

$$0 \leq L - s_n \leq L - s_{N'} < \epsilon,$$

which proves that for all $n > N'$, $|s_n - L| < \epsilon$. Thus $\lim s_n = L$.

The proof of the case of $s_n \geq s_{n+1}$ for all $n \geq N$ is similar. □

The theorem below was already proved in Example 8.2.7.

Theorem 8.6.5. (Ratio test for sequences) *Let $r \in \mathbb{C}$ with $|r| < 1$. Then $\lim r^n = 0$.*

Proof. If $r = 0$, the sequence is the constant zero sequence, so of course its limit is 0. Now suppose that $r \neq 0$. By Exercise 2.8.2, for all positive integers n, $0 < |r|^{n+1} < |r|^n$. Thus the sequence $\{|r|^n\}$ is a non-increasing sequence that is bounded below by 0 and above by 1. By Theorem 8.6.4, $L = \lim |r|^n = \inf\{|r|^n : n \in \mathbb{N}^+\}$. Since 0 is a lower bound and L is the greatest of lower bounds of $\{|r|^n : n \in \mathbb{N}^+\}$, necessarily $0 \leq L$.

Suppose that $L > 0$. Then $L(1 - |r|)/(2|r|)$ is a positive number. Since L is the infimum of the set $\{|r|, |r|^2, |r|^3, \ldots\}$, there exists a positive integer $N > N_1$ such that $0 \le |r|^N - L < L\frac{1-|r|}{2|r|}$. By multiplying by $|r|$ we get that

$$|r|^{N+1} < |r|L + L\frac{1-|r|}{2} = L\left(|r| + \frac{L(1-|r|)}{2}\right)$$

$$= L\frac{(1+|r|)}{2} < L\frac{(1+1)}{2} = L,$$

and since L is the infimum of all powers of $|r|$, we get that $L \le |r|^{N+1} < L$, which is a contradiction. So necessarily $L = 0$. Hence for every $\epsilon > 0$ there exists $N \in \mathbb{R}^+$ such that for all integers $n > N$, $||r^n| - 0| < \epsilon$. But this says that $|r^n - 0| < \epsilon$, so that $\lim r^n = 0$ as well. $\qquad\square$

Theorem 8.6.6. (Ratio test for sequences) *Let $\{s_n\}$ be a sequence of non-zero complex numbers and let L be a real number in the interval $[0, 1)$. Assume that $\lim \left|\frac{s_{n+1}}{s_n}\right| = L$ or that there exists a positive integer K such that for any integer $n \ge K$, $\left|\frac{s_{n+1}}{s_n}\right| \le L$.*
Then $\lim s_n = 0$.

Proof. Let r be a real number strictly between L and 1. Then r and $r - L$ are positive numbers.

Under the first (limit) condition, there exists a positive number K such that for all integers $n > K$, $\left|\left|\frac{s_{n+1}}{s_n}\right| - L\right| < r - L$. Then for all $n > K$,

$$\left|\frac{s_{n+1}}{s_n}\right| = \left|\frac{s_{n+1}}{s_n}\right| - L + L \le \left|\left|\frac{s_{n+1}}{s_n}\right| - L\right| + L \le r - L + L = r.$$

Thus both conditions say that there exists a positive K such that for all integers $n > K$, $\left|\frac{s_{n+1}}{s_n}\right| \le r$. We may replace this K by any larger number, and the conclusion still holds. So from now on we assume that K is a positive integer.

Let $\epsilon > 0$. By Theorem 8.6.5, $\lim r^n = 0$. Thus there exists $M \ge 0$ such that for all integers $n > M$, $|r^n - 0| < \frac{\epsilon r^K}{|s_K|}$. Set $N = \max\{M, K\}$. Let n be an integer strictly greater than N. Then

$$|s_n - 0| = \left|\frac{s_n}{s_{n-1}} \cdot \frac{s_{n-1}}{s_{n-2}} \cdot \frac{s_{n-2}}{s_{n-3}} \cdot \ \cdots \ \cdot \frac{s_{K+1}}{s_K} \cdot s_K\right|$$

$$= \left|\frac{s_n}{s_{n-1}}\right| \cdot \left|\frac{s_{n-1}}{s_{n-2}}\right| \cdot \left|\frac{s_{n-2}}{s_{n-3}}\right| \cdot \ \cdots \ \cdot \left|\frac{s_{K+1}}{s_K}\right| \cdot |s_K|$$

$$\leq r^{n-K}|s_K|$$
$$= r^n|s_K|r^{-K}$$
$$< \epsilon. \qquad \square$$

Exercises for Section 8.6

8.6.1. Compute and justify the limits. (Use the comparison test, ratio test, decreasing property, or other tricks.)

i) $\lim_{n\to\infty} \frac{1}{2^n}$.

ii) $\lim_{n\to\infty} \frac{(-1)^n}{3^n}$.

iii) $\lim_{n\to\infty} \frac{(-4)^n}{7^n}$.

iv) $\lim_{n\to\infty} \frac{k^n}{n!}$ for all $k \in \mathbb{C}$.

v) $\lim_{n\to\infty} \frac{n}{k^n}$ for all non-zero $k \in \mathbb{C}$ with $|k| > 1$.

vi) $\lim_{n\to\infty} \frac{n^m}{k^n}$ for all non-zero $k \in \mathbb{C}$ with $|k| > 1$ and all integers m.

8.6.2. Give an example of a sequence $\{s_n\}$ of non-zero complex numbers such that $\lim\{s_{n+1}/s_n\}$ exists and has absolute value strictly smaller than 1.

8.6.3. Give an example of a sequence $\{s_n\}$ of non-zero complex numbers such that $\lim\{s_{n+1}/s_n\} = i/2$.

8.6.4. Let s_n be the nth Fibonacci number. Prove that $\lim \frac{s_{n+1}}{s_n} = \frac{1+\sqrt{5}}{2}$. (Hint: Exercise 1.6.30.)

8.6.5. Let $r \in \mathbb{C}$ satisfy $|r| > 1$. Prove that the sequence $\{r^n\}$ is not bounded.

8.6.6. (Ratio test again, and compare to Theorem 8.6.6) Let $\{s_n\}$ be a sequence of non-zero complex numbers such that $\lim \frac{s_{n+1}}{s_n} = L$. Suppose that $|L| < 1$. Prove that $\lim s_n = 0$.

8.6.7. Consider sequences $\{(-1)^n\}$ and $\{1\}$. For each, determine whether it converges, and the limit of ratios $\left|\frac{s_{n+1}}{s_n}\right|$. Comment why the Ratio Test does not apply.

8.6.8. Let $\{s_n\}$ be a sequence of real numbers such that for all $n \geq 1$, $s_n \leq s_{n+1}$ (resp. $s_n \geq s_{n+1}$). Prove that $\{s_n\}$ is convergent if and only if it is bounded.

8.6.9. Prove that the sequence $\left\{\left(1+\frac{1}{n}\right)^n\right\}_n$ is strictly increasing. (Hint: Exercise 7.6.11.)

8.6.10. Prove that the sequence $\{\sqrt[n]{n!}\}$ is strictly increasing.

8.6.11. Let $\{s_n\}$ be a non-decreasing sequence. Prove that the set $\{s_1, s_2, s_3, \ldots\}$ is bounded below.

8.6.12. (**Monotone sequences**) Let $\{s_n\}$ be a monotone sequence of real numbers.

  i) Suppose that $\{s_n\}$ is not bounded above. Prove that $\{s_n\}$ is non-decreasing and that $\lim s_n = \infty$.

  ii) Suppose that $\{s_n\}$ is not bounded below. Prove that $\{s_n\}$ is non-increasing and that $\lim s_n = -\infty$.

  iii) Prove that $\lim s_n$ is a real number if and only if $\{s_n\}$ is bounded.

8.7   Cauchy sequences, completeness of $\mathbb{R}$, $\mathbb{C}$

So far, to determine convergence of a sequence required knowing the limit. In this section we prove an alternate machinery for deciding that a sequence converges without knowing what its limit is. This is used for subsequential limits and in comparison theorems for convergence of series in the next chapter.

Definition 8.7.1. *A sequence $\{s_n\}$ is **Cauchy** if for all $\epsilon > 0$ there exists a positive real number N such that for all integers $m, n > N$, $|s_n - s_m| < \epsilon$.*

Theorem 8.7.2. *Every Cauchy sequence is bounded.*

Proof. Let $\{s_n\}$ be a Cauchy sequence. Thus for $\epsilon = 1$ there exists a positive integer N such that for all integers $m, n > N$, $|s_n - s_m| < 1$. Then the set $\{|s_1|, |s_2|, \ldots, |s_N|, |s_{N+1}|\}$ is a finite and hence a bounded subset of $\mathbb{R}$. Let M' an upper bound of this set, and let $M = M' + 1$. It follows that for all $n = 1, \ldots, N$, $|s_n| < M$, and for $n > N$, $|s_n| = |s_n - s_{N+1} + s_{N+1}| \le |s_n - s_{N+1}| + |s_{N+1}| < 1 + M' = M$. Thus $\{s_n\}$ is bounded by M. $\square$

Theorem 8.7.3. *Every convergent sequence is Cauchy.*

Proof. Let $\{s_n\}$ be a convergent sequence. Let L be the limit. Let $\epsilon > 0$. Since $\lim s_n = L$, there exists a positive real number N such that for all $n > N$, $|s_n - L| < \epsilon/2$. Thus for all integers $m, n > N$,

$$|s_n - s_m| = |s_n - L + L - s_m| \leq |s_n - L| + |L - s_m| < \epsilon/2 + \epsilon/2 = \epsilon. \quad \square$$

Remark 8.7.4. It follows that every convergent sequence is bounded (this was already proved in Theorem 8.6.2).

The converse of Theorem 8.7.3 is not true if the field in which we are working is $\mathbb{Q}$. For example, let s_n be the decimal approximation of $\sqrt{2}$ to n digits after the decimal point. Then $\{s_n\}$ is a Cauchy sequence of rational numbers: for every $\epsilon > 0$, let N be a positive integer such that $1/10^N < \epsilon$. Then for all integers $n, m > N$, s_n and s_m differ at most in digits $N + 1, N + 2, \ldots$ beyond the decimal point, so that $|s_n - s_m| \leq 1/10^N < \epsilon$. But $\{s_n\}$ does not have a limit in $\mathbb{Q}$, so that $\{s_n\}$ is a Cauchy but not a convergent sequence. Another way of writing this is:

$$\lim \left\{ \frac{\lfloor 10^n \sqrt{2} \rfloor}{10^n} \right\} = \sqrt{2}.$$

Over $\mathbb{R}$ and $\mathbb{C}$, all Cauchy sequences are convergent, as we prove next.

Theorem 8.7.5. (Completeness of $\mathbb{R}$, $\mathbb{C}$) *Every Cauchy sequence in $\mathbb{R}$ or $\mathbb{C}$ is convergent.*

Proof. First let $\{s_n\}$ be a Cauchy sequence in $\mathbb{R}$ (as opposed to in $\mathbb{C}$). By Theorem 8.7.2, $\{s_n\}$ is bounded. It follows that all subsets $\{s_1, s_2, s_3, \ldots\}$ are bounded too. In particular, by the Least upper bound theorem (Theorem 3.8.5), $u_n = \sup\{s_n, s_{n+1}, s_{n+2}, \ldots\}$ is a real number. For all n, $u_n \geq u_{n+1}$ because u_n is the supremum of a larger set. Any lower bound on $\{s_1, s_2, s_3, \ldots\}$ is also a lower bound on $\{u_1, u_2, u_3, \ldots\}$. Thus by Theorem 8.6.4, the monotone sequence $\{u_n\}$ has a limit $L = \inf\{u_1, u_2, u_3, \ldots\}$.

We claim that $L = \lim\{s_n\}$. Let $\epsilon > 0$. Since $\{s_n\}$ is Cauchy, there exists $N_1 > 0$ such that for all integers $m \geq n > N_1$, $|s_n - s_m| < \epsilon/2$. Thus if we fix $n > N_1$, then for all $m \geq n$, we have that $s_m < s_n + \epsilon/2$. But u_n is the least upper bound on all s_m for $m \geq n$, so that $s_m \leq u_n < s_n + \epsilon/2$, and in particular, $s_n \leq u_n < s_n + \epsilon/2$. It follows that $|s_n - u_n| < \epsilon/2$ for all integers $n > N_1$. Since $L = \inf\{u_1, u_2, u_3, \ldots\}$, there exists an integer N_2 such that $0 \leq u_{N_2} - L < \epsilon/2$. Set $N = \max\{N_1, N_2\}$. Let $n > N$ be an integer. By the definition of the u_n, $L \leq u_n \leq u_N \leq u_{N_2}$, so that $0 \leq u_n - L \leq u_{N_2} - L < \epsilon/2$. Hence

$$|s_n - L| = |s_n - u_n + u_n - L| \leq |s_n - u_n| + |u_n - L| < \epsilon/2 + \epsilon/2 = \epsilon.$$

This proves that every real Cauchy sequence converges.

Now let $\{s_n\}$ be a Cauchy sequence in $\mathbb{C}$. We leave it to Exercise 8.7.1 that then $\{\operatorname{Re} s_n\}$ and $\{\operatorname{Im} s_n\}$ are Cauchy. By what we have proved for real sequences, there exist $a, b \in \mathbb{R}$ such that $\lim \operatorname{Re} s_n = a$ and $\lim \operatorname{Im} s_n = b$. Then by Theorem 8.5.10 or Theorem 8.4.9, $\{s_n\}$ converges to $a + bi$. $\square$

Example 8.7.6. (Harmonic sequence) The sequence $\{1 + \frac{1}{2} + \frac{1}{3} + \cdots + \frac{1}{n}\}_n$ is not Cauchy, does not converge, is monotone, and is not bounded.

Proof. Let s_n be the nth term of the sequence. If $m > n$, then

$$s_m - s_n = \frac{1}{m} + \frac{1}{m-1} + \frac{1}{m-2} + \cdots + \frac{1}{n+1} \geq \frac{m-n}{m} = 1 - \frac{n}{m}.$$

Thus if $m = 1000n$, then $s_m - s_n > 1 - 1/1000 = 0.999$. In particular, if $\epsilon = 1/2$, then $|s_m - s_n| \not< \epsilon$. Thus the sequence is not Cauchy. By Theorem 8.7.3 it is not convergent, and by Theorem 8.7.2 it is not bounded.

The terms of $\{1 + \frac{1}{2} + \frac{1}{3} + \cdots + \frac{1}{n}\}$ get larger and larger beyond bound. But this sequence gets large notoriously slowly — the 1000th term of the sequence is smaller than 8, the 10000th term is smaller than 10, the 100000th term is barely larger than 12. (One could lose patience in trying to **see** how this sequence grows without bound.)

Here is a related thought experiment: we can stack finitely many books to form a bridge to the Moon and beyond! All books are identical and stacked so that each one protrudes out from the heap below as much as possible and with keeping the center of mass stable. The topmost book, by the uniform assumption, protrudes out $\frac{1}{2}$ of its length. Actually, for the book to be stable this should actually be $\frac{1}{2}$ minus a tad, but that tad can be taken to be say $\frac{1}{\text{googolplex}}$, say, or much much less. We will ignore this tiny tad, but an interested reader may wish to work through the stacking below with (varied and tinier!) corrections being incorporated in all the centers of mass and all the protrusions.

This diagram shows the top two books.

The center of mass of the top two books is clearly at $\frac{3}{4}$ from the right-hand edge of the bottom book, so that the third book from the top down should protrude from underneath the second one $\frac{1}{4}$.

The center of mass of this system, measured from the rightmost edge, is $\frac{2\cdot1+1\cdot\frac{1}{2}}{3} = \frac{5}{6}$, so that the fourth book has to protrude out $\frac{1}{6}$.

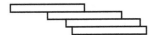

The center of mass of this system, measured from the rightmost edge, is $\frac{3\cdot1+1\cdot\frac{1}{2}}{4} = \frac{7}{8}$, so that the fifth book has to protrude out $\frac{1}{8}$ units.

In general, the center of mass of the top n books is at $\frac{(n-1)\cdot1+1\cdot\frac{1}{2}}{n} = \frac{2n-1}{2n}$ measured from the rightmost edge, so that the $(n+1)$st book should protrude out by $\frac{1}{2n}$ units.

Thus the total protrusion of the top n books equals

$$\frac{1}{2}\left(1 + \frac{1}{2} + \frac{1}{3} + \cdots + \frac{1}{n-1}\right)$$

units. This sum is not bounded by Example 8.7.6, and so in particular we can reach the Moon with enough books (and a platform to stand on).

Exercises for Section 8.7

†**8.7.1.** (Invoked in Theorems 8.8.4 and 8.7.5.) Prove that $\{s_n\}$ is Cauchy if and only if $\{\operatorname{Re} s_n\}$ and $\{\operatorname{Im} s_n\}$ are Cauchy.

8.7.2. Let $\{s_n\}$ and $\{t_n\}$ be Cauchy sequences.

　i) Prove that $\{s_n \pm t_n\}$ is a Cauchy sequence.

　ii) Prove that $\{s_n \cdot t_n\}$ is a Cauchy sequence.

　iii) Suppose that for some positive number B, $|t_n| \geq B$ for all $n \in \mathbb{N}^+$. Prove that $\{s_n/t_n\}$ is Cauchy.

　iv) Prove that for all $c \in \mathbb{C}$, $\{cs_n\}$ is a Cauchy sequence.

8.7.3. Give examples of non-Cauchy sequences $\{s_n\}$, $\{t_n\}$ such that $\{s_n + t_n\}$ is a Cauchy sequence. Why does this not contradict the previous exercise? Repeat for $\{s_n \cdot t_n\}$.

8.7.4. Let $r \in \mathbb{C}$ satisfy $|r| > 1$. Prove that $\lim r^n$ diverges. (Hint: Check the Cauchy property.)

8.7.5. Review the proof in Example 8.3.2 that $\{(-1)^n\}$ is divergent. Give another proof using the contrapositive of Theorem 8.7.3.

8.7.6. Give examples of sequences with the listed properties.

   i) A Cauchy sequence in $\mathbb{Q}$ that is not convergent.

   ii) A bounded sequence that is not convergent.

   iii) A bounded sequence that is not Cauchy.

   iv) A real increasing sequence that is not Cauchy.

   v) A real Cauchy sequence that is increasing.

8.7.7. Say why the sequences below do not exist.

   i) A convergent sequence that is not Cauchy.

   ii) A Cauchy sequence in $\mathbb{R}$ that is not convergent.

   iii) A bounded real increasing sequence that is not Cauchy.

8.8   Subsequences

Definition 8.8.1. *A* **subsequence** *of an infinite sequence* $\{s_n\}$ *is an infinite sequence* $\{s_{k_1}, s_{k_2}, s_{k_3}, \ldots\}$ *where* $1 \leq k_1 < k_2 < k_3 < \cdots$ *are integers. Notations for such a subsequence are:*

$$\{s_{k_n}\}, \quad \{s_{k_n}\}_n, \quad \{s_{k_n}\}_{n \geq 1}^{\infty}, \quad \{s_{k_n}\}_{n \geq 1}, \quad \{s_{k_n}\}_{n \in \mathbb{N}^+}.$$

Note that for all n, $k_n \geq n$.

Examples 8.8.2. Every sequence is a subsequence of itself. Sequences $\{1/2n\}$, $\{1/3n\}$, $\{1/(2n+1)\}$ are subsequences of $\{1/n\}$, and $\{1\}$, $\{-1\}$, $\{(-1)^{n+1}\}$ are subsequences of $\{(-1)^n\}$. The constant sequence $\{1\}$ is not a subsequence of $\{1/n\}$, because the latter sequence does not have infinitely many terms equal to 1. If $\{s_n\}$ is the sequential enumeration of $\mathbb{Q}^+$ on page 315, then $\{1/n\}$, $\{n\}$, $\{\frac{n}{n+1}\}$, $\{\frac{1}{n^2}\}$ are subsequences, but $\{\frac{n}{n+2}\} = \{1/3, 1/2, \ldots\}$ is not.

Theorem 8.8.3. *A subsequence of a convergent sequence is convergent, with the same limit. A subsequence of a Cauchy sequence is Cauchy.*

Proof. Let $\{s_n\}$ be a convergent sequence, with limit L, and let $\{s_{k_n}\}$ be a subsequence. Let $\epsilon > 0$. By assumption there exists a positive number N such that for all integers $n > N$, $|s_n - L| < \epsilon$. Since $n \leq k_n$, it follows that $|s_{k_n} - L| < \epsilon$. Thus $\{s_{k_n}\}$ converges. The proof of the second part is similar. □

Theorem 8.8.4. *Every bounded sequence has a Cauchy subsequence.*

Proof. This proof uses the halving construction already encountered in Construction 3.14.1, and here the property P is that the subset contains infinitely many elements of the sequence.

Let $\{s_n\}$ be a bounded sequence (of real or complex numbers). Let M be a positive real number such that for all n, $|s_n| \leq M$. Let $a_0 = c_0 = -M$ and $b_0 = d_0 = M$. The sequence $\{s_n\}$ has infinitely many (all) terms in the rectangle $R_0 = [a_0, b_0] \times [c_0, d_0]$. Set $l_0 = 0$. (If all s_n are real, we may take $c_0 = d_0 = 0$, or perhaps better, ignore the second coordinates.)

We prove below that for all $m \in \mathbb{N}^+$ there exists a subsequence $\{s_{k_n}\}$ all of whose terms are in the rectangle $R_m = [a_m, b_m] \times [c_m, d_m]$, where $b_m - a_m = 2^{-m}(b_0 - a_0)$, $[a_m, b_m] \subseteq [a_{m-1}, b_{m-1}]$, $d_m - c_m = 2^{-m}(d_0 - c_0)$, $[c_m, d_m] \subseteq [c_{m-1}, d_{m-1}]$. Furthermore, we prove that there exists $l_m > l_{m-1}$ such that $s_{l_m} \in R_m$.

Namely, given the $(m-1)$st rectangle R_{m-1}, integer l_{m-1} such that $s_{l_{m-1}} \in R_{m-1}$, and a subsequence $\{s_{k_n}\}$ all of whose terms are in R_{m-1}, divide R_{m-1} into four equal-sized subrectangles. Necessarily at least one of these four subrectangles contains infinitely many elements of $\{s_{k_n}\}$, so pick one such subrectangle, and call it R_m. Therefore there exists a subsequence of $\{s_{k_n}\}$ that is contained in R_m, and that subsequence of $\{s_{k_n}\}$ is also a subsequence of $\{s_n\}$. We call it $\{s_{k'_n}\}$. Since we have infinitely many k'_n, in particular there exists $k'_n > l_{m-1}$, and we set $l_m = k'_n$. Thus $s_{l_m} \in R_m$.

By construction, $\{s_{l_n}\}_n$ is a subsequence of $\{s_n\}_n$. We next prove that $\{s_{l_n}\}_n$ is a Cauchy sequence. Let $\epsilon > 0$. Since the either side length of the mth subrectangle R_m equals the corresponding side length of R_0 divided by 2^m, by Exercise 3.8.3 there exists a positive integer N such that any side length of R_N is strictly smaller than the constant $\epsilon/2$. Let $m, n > N$ be integers. Then s_{l_m}, s_{l_n} are in R_N, so that

$$|s_{l_m} - s_{l_n}| \leq \sqrt{(\text{one side length of } R_n)^2 + (\text{other side length of } R_n)^2} < \epsilon.$$

$\square$

The following is now an immediate consequence of Theorem 8.7.5:

Theorem 8.8.5. *Every bounded sequence in* $\mathbb{C}$ *has a convergent subsequence.*

Example 8.8.6. We work out the construction of a subsequence as in the proof on the bounded sequence $\{(-1)^n - 1\}$. For example, all terms lie on the interval $[a_0, b_0] = [-4, 4]$. Infinitely many terms lie on $[a_1, b_1] = [-4, 0]$, and on this sub-interval I arbitrarily choose the second term, which equals 0. Infinitely many terms lie on $[a_2, b_2] = [-4, -2]$, in particular, I choose the third term -2. After this all terms of the sequence in $[a_2, b_2]$ are -2, so that we have built the Cauchy subsequence $\{0, -2, -2, -2, \ldots\}$ (and subsequent $[a_n, b_n]$ all have $b_n = -2$). We could have built the Cauchy subsequence $\{-2, -2, \ldots\}$, or, if we started with the interval $[-8, 8]$, we could have built the Cauchy subsequences $\{0, 0, -2, -2, \ldots\}$ or $\{-2, 0, 0, -2, -2, \ldots\}$, and so on.

Definition 8.8.7. *A* **subsequential limit** *of a sequence* $\{s_n\}$ *is a limit of any subsequence of* $\{s_n\}$. *Thus a subsequential limit can be a complex number, or as in Definition 8.3.4, it can be* $\pm\infty$.

Here are a few examples in tabular form:

sequence $\{s_n\}$	the set of subsequential limits of $\{s_n\}$
convergent $\{s_n\}$	$\{\lim s_n\}$
$\{(-1)^n\}$	$\{-1, 1\}$
$\{(-1)^n - 1\}$	$\{0, -2\}$
$\{(-1)^n + 1/n\}$	$\{-1, 1\}$
$\{(-1)^n n + n + 3\}$	$\{3, \infty\}$
$\{i^n\}$	$\{i, -1, -i, 1\}$
$\{n\}$	$\{\infty\}$
$\{(-1)^n n\}$	$\{-\infty, \infty\}$

Theorem 8.8.8. *Every unbounded sequence of real numbers has a subsequence that has limit $-\infty$ or ∞.*

Proof. If $\{s_n\}$ is not bounded, choose $k_1 \in \mathbb{N}^+$ such that $|s_{k_1}| \geq 1$, and once k_{n-1} has been chosen, choose an integer $k_n > k_{n-1}$ such that $|s_{k_n}| \geq n$. Now $\{s_{k_n}\}_n$ is a subsequence of $\{s_n\}$. Either infinitely many among the s_{k_n} are positive or else infinitely many among the s_{k_n} are negative. Choose a subsequence $\{s_{l_n}\}_n$ of $\{s_{k_n}\}_n$ such that all terms in $\{s_{l_n}\}$ have the same sign. If they are all positive, then since $s_{l_n} \geq n$ for all n, it follows that $\lim_{n\to\infty} s_{l_n} = \infty$, and if they are all negative, then since $s_{l_n} \leq -n$ for all n, it follows that $\lim_{n\to\infty} s_{l_n} = -\infty$. $\qquad\square$

Exercises for Section 8.8

8.8.1. Determine the subsequential limit sets of the following sequences, and justify:

  i) $\{(1/2 + i\sqrt{3}/2)^n\}$, $\{(1/2 - i\sqrt{3}/2)^n\}$.
  ii) $\{1/2^n\}$.
  iii) $\{(1/2 + i\sqrt{3}/2)^n + 1/2^n\}$.
  iv) $\{(1/2 + i\sqrt{3}/2)^n \cdot 1/2^n\}$.
  v) $\{(1/2 + i\sqrt{3}/2)^n + (1/2 - i\sqrt{3}/2)^n\}$.
  vi) $\{(1/2 + i\sqrt{3}/2)^n \cdot (1/2 - i\sqrt{3}/2)^n\}$.

8.8.2. Consider the sequence enumerating $\mathbb{Q}^+$ as on page 315. Prove that the set of its subsequential limits equals $\mathbb{R}_{\geq 0}$.

8.8.3. Prove that the real-valued sequence $\left\{ \frac{n^2 + (-1)^n n(n+1)}{n+1} \right\}$ is unbounded. Find a subsequence that diverges to ∞ and a subsequence that converges to -1.

8.8.4. Let $m \in \mathbb{N}^+$ and let c be the complex number of absolute value 1 at angle $2\pi/m$ counterclockwise from the positive real axis. Find the set of subsequential limits of $\{c^n\}_n$.

8.8.5. Give examples of sequences with the listed properties, if they exist. If they do not exist, justify.

  i) A sequence with no convergent subsequences.
  ii) A sequence whose set of subsequential limits equals $\{1\}$.
  iii) A sequence whose set of subsequential limits equals $\{1, 3\}$.
  iv) A sequence whose set of subsequential limits equals $\{1, 2, 12\}$.

8.8.6. Suppose that a Cauchy sequence has a convergent subsequence. Prove that the original sequence is convergent as well.

8.8.7. Prove that a sequence of real numbers contains a monotone subsequence.

8.9   Liminf, limsup for real-valued sequences

Recall that by Theorem 3.8.5 every non-empty subset T of $\mathbb{R}$ bounded above has a least upper bound $\sup T = \operatorname{lub} T$ in $\mathbb{R}$, and that every non-empty subset T bounded below has a greatest lower bound $\inf T = \inf T$ in $\mathbb{R}$. We extend this definition by declaring

$$\sup T = \infty \text{ if } T \text{ is not bounded above,}$$
$$\inf T = -\infty \text{ if } T \text{ is not bounded below.}$$

The same definitions apply to sequences thought of as sets: $\sup\{1/n\}_n = 1$, $\inf\{1/n\}_n = 0$, $\sup\{n\}_n = \infty$, $\inf\{n\}_n = 1$, $\sup\{(-1)^n\}_n = 1$, $\inf\{(-1)^n\}_n = -1$, etc.

Much analysis of sequences has to do with their long-term behavior rather than with their first three, first hundred, or first million terms — think convergence or the Cauchy property of sequences. Further usage of such tail-end analysis is in the next chapter (for convergence criteria for series). Partly with this goal in mind, we apply infima and suprema to sequences of tail ends of sequences:

Definition 8.9.1. *Let* $\{s_n\}$ *be a real-valued sequence. The* **limit superior** *limsup and* **limit inferior** *liminf of* $\{s_n\}$ *are:*

$$\limsup s_n = \inf\{\sup\{s_n : n \geq m\} : m \geq 1\},$$
$$\liminf s_n = \sup\{\inf\{s_n : n \geq m\} : m \geq 1\}.$$

For any positive integers $m_1 < m_2$,

$$\inf\{s_n : n \geq m_2\} \leq \sup\{s_n : n \geq m_2\} \leq \sup\{s_n : n \geq m_1\},$$
$$\inf\{s_n : n \geq m_1\} \leq \inf\{s_n : n \geq m_2\} \leq \sup\{s_n : n \geq m_2\}.$$

Hence for any positive integers m_1, m_2, $\inf\{s_n : n \geq m_2\} \leq \sup\{s_n : n \geq m_1\}$, so that $\liminf s_n = \sup\{\inf\{s_n : n \geq m\} : m \geq 1\} \leq \inf\{\sup\{s_n : n \geq m\} : m \geq 1\} = \limsup s_n$.

If $\{s_n\}$ is bounded by A below and B above, then also $A \leq$ liminf $s_n \leq$ limsup $s_n \leq B$, so that in particular by the Least upper bound theorem (Theorem 3.8.5), liminf s_n and limsup s_n are real numbers.

In other words, limsup s_n is the infimum of the set of all the suprema of all the tail-end subsequences of $\{s_n\}$, and analogously, liminf s_n is the supremum of the set of all the infima of all the tail-end subsequences of $\{s_n\}$. In the plot below, the sequence $\{s_n\}$, drawn with thick dots, oscillates between positive and negative values with peaks and valleys getting smaller and smaller. The connected non-increasing dashed top line denotes the sequence $\{\sup\{s_m : m \geq n\}\}_n$ and the connected bottom non-decreasing line represents the sequence $\{\inf\{s_m : m \geq n\}\}_n$.

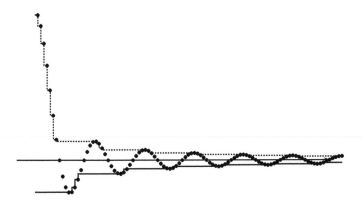

For liminf s_n to be a real number it is not enough for the sequence to be bounded below: for example, $\{n\}$ is bounded below but liminf$\{n\} =$ sup$\{\inf\{n : n \geq m\} : m \geq 1\} =$ sup$\{m : m \geq 1\} = \infty$. Similarly, for limsup s_n to be a real number it is not enough for the sequence to be bounded above.

Theorem 8.9.2. *If* $\{s_n\}$ *converges to* L, *then* liminf $s_n =$ limsup $s_n = L$.

Proof. Let $\epsilon > 0$. Since $\lim s_n = L$, there exists $N > 0$ such that for all integers $n > N$, $|s_n - L| < \epsilon$. Thus for all integers $n \geq N$, so that $L - \epsilon \leq s_n \leq L + \epsilon$, and so $L - \epsilon \leq$ limsup $s_n \leq L + \epsilon$. Since this is true for all $\epsilon > 0$, it follows by Theorem 2.10.4 that $L =$ limsup s_n. The proof for the liminf part is similar. $\qquad\square$

Remark 8.9.3. (Ratio test for sequences) With the new language, Theorem 8.6.6 can be rephrased as follows: If $\{s_n\}$ is a sequence of non-zero complex numbers such that $\limsup\{|s_{n+1}/s_n|\} < 1$, then $\lim s_n = 0$. The proof there already accomplishes this. On the other hand, the ratio test for divergence in Exercise 8.6.6 is not phrased in the most general form. One generalization is that if $\limsup\{|s_{n+1}/s_n|\} > 1$, then $\{s_n\}$ diverges. The proof is simple. Let $r \in (1, \liminf\{|s_{n+1}/s_n|\})$. By definition of liminf as supremum of some infima, this means that there exists an integer m such that $\inf\{|s_{n+1}/s_n| : n \geq m\} > r$. Thus by an easy induction, for all $n > m$, $|s_n| > r^{n-m}|s_m|$, and then by the Comparison test (Theorem 8.3.7), $\{|s_n|\}$ diverges to infinity, hence $\{s_n\}$ does not converge to a complex number. $\square$

It turns out that there is an important connection between limsup, liminf, and subsequential limits:

Theorem 8.9.4. *Let $\{s_n\}$ be a bounded sequence of real numbers. Then the supremum of the set of all subsequential limits equals* $\limsup s_n$, *and the infimum of the set of all subsequential limits equals* $\liminf s_n$.

Proof of of the limsup part only: Let $A = \limsup\{s_n\}$, let S be the set of all subsequential limits of $\{s_n\}$, and let $U = \sup(S)$. Since the sequence is bounded, A and U are real numbers.

Let $\epsilon > 0$. Since $A = \inf\{\sup\{s_n : n \geq m\} : m \geq 1\}$, there exists $m_0 \geq 1$ such that $\sup\{s_n : n \geq m_0\} - A < \epsilon$. Thus for all $n \geq m_0$, $s_n - A < \epsilon$. Then any subsequential limit of $\{s_n\}$ is a subsequential limit of $\{s_n\}_{n \geq m_0}$, so that this limit must be at most $A + \epsilon$. Thus $A + \epsilon$ is an upper bound on all subsequential limits of $\{s_n\}$, so that $U \leq A + \epsilon$. Since ϵ is an arbitrary positive number, by Theorem 2.10.4 this means that $U \leq A$.

By definition of U, there exists a convergent subsequence $\{s_{k_n}\}$ such that $U - \lim\{s_{k_n}\}_n < \epsilon/2$. Let $L = \lim\{s_{k_n}\}_n$. So $U - L < \epsilon/2$, and there exists a positive real number N such that for all integers $n > N$, $|s_{k_n} - L| < \epsilon/2$. Thus for all $n > N$,

$$s_{k_n} > L - \epsilon/2 > (U - \epsilon/2) - \epsilon/2 = U - \epsilon.$$

Thus for any integer $m \geq N$, the supremum of $\{s_n : n > m\}$ must be at least $U - \epsilon$, so that A, the limsup of $\{s_n\}$ must be at least $U - \epsilon$. Hence by Theorem 2.10.4 this means that $A \leq U$.

It follows that $A = U$. $\square$

Theorem 8.9.5. *Let* $\{s_n\}$, $\{t_n\}$ *be bounded sequences in* $\mathbb{R}$. *Then* $\limsup s_n + \limsup t_n \geq \limsup(s_n + t_n)$ *and* $\liminf s_n + \liminf t_n \leq \liminf(s_n + t_n)$.

Proof. Let $a = \limsup s_n$, $b = \limsup t_n$, $c = \limsup(s_n + t_n)$. By boundedness, $a, b, c \in \mathbb{R}$. Let $\epsilon > 0$. Recall that $a = \inf\{\sup\{s_n : n \geq m\} : m \geq 1\}$. Then there exists a positive integer m such that

$$\sup\{s_n : n \geq m\} - \epsilon < a \leq \sup\{s_n : n \geq m\}.$$

In particular for all $n > m$, $s_n - \epsilon < a$. By possibly increasing m, we similarly get that for all $n > m$ in addition $t_n - \epsilon < b$. Thus for all $n > m$, $s_n + t_n - 2\epsilon < a + b$. Thus $c = \inf\{\sup\{s_n + t_n : n \geq m\} : m \geq 1\} \leq a + b + 2\epsilon$. Since ϵ is arbitrary, it follows that $c \leq a + b$, which proves the first part. The rest is left as an exercise. $\square$

Theorem 8.9.6. *Let* $\{s_n\}$ *and* $\{t_n\}$ *be sequences of non-negative real numbers such that* $\lim s_n$ *is a positive real number* L. *Then* $\limsup(s_n t_n) = L \limsup t_n$ *and* $\liminf(s_n t_n) = L \liminf t_n$.

Proof. Let $\epsilon > 0$. Set $\epsilon' = \min\{L/2, \epsilon\}$. By assumption there exists $N > 0$ such that for all integers $n > N$, $|s_n - L| < \epsilon'$. Then $L - \epsilon' < s_n < L + \epsilon'$. Thus each s_n is positive, and in fact $|s_n| > L/2$. It follows that $(L - \epsilon')t_n \leq s_n t_n \leq (L + \epsilon')t_n$. But then since $L \pm \epsilon' \geq 0$,

$$
\begin{aligned}
(L - \epsilon')\limsup t_n &= \limsup(L - \epsilon')t_n \\
&\leq \limsup s_n t_n \\
&\leq \limsup(L + \epsilon')t_n \\
&= (L + \epsilon')\limsup t_n.
\end{aligned}
$$

The proof of the liminf part is similar. $\square$

Exercises for Section 8.9

8.9.1. Let $\{s_n\}$ and $\{t_n\}$ be sequences of real numbers such that $\lim s_n$ is a positive real number L. Prove that $\limsup(s_n + t_n) = L + \limsup t_n$ and $\liminf(s_n + t_n) = L + \liminf t_n$.

8.9.2. Compute liminf and limsup for the following sequences. Justify your work.

i) $\{\frac{f(n)}{n}\}$ where f is a bounded function.

ii) $\{(-1)^n n!\}$.

iii) $\{2^{-n}\}$.

iv) $\{1, 2, 3, 1, 2, 3, 1, 2, 3, 1, 2, 3, 1, 2, 3, 1, 2, 3, \ldots\}$.

v) $\{(4 + \frac{1}{n})(-1)^n\}$.

vi) The sequence of all positive prime numbers.

vii) The sequence of all multiplicative inverses of positive prime numbers.

8.9.3. Prove the following for any sequence $\{s_n\}$:

i) $\inf(-s_n) = -\sup(s_n)$.

ii) $\liminf(-s_n) = -\limsup(s_n)$.

8.9.4. Suppose that $\{s_n\}$ converges to L. Finish the proof of Theorem 8.9.2, namely prove that $\liminf s_n = L$.

8.9.5. For every positive integer n, let $s_{2n-1} = 1$ and $s_{2n} = 1/2^n$.

i) Compute $\liminf s_n$, $\liminf s_n$.

ii) Prove that $\limsup \frac{s_{n+1}}{s_n} > 1$.

iii) Does this contradict the ratio test as in Remark 8.9.3.

8.9.6. Find bounded real-valued sequences $\{s_n\}, \{t_n\}$ such that $\limsup s_n + \limsup t_n > \limsup(s_n + t_n)$. (Compare with Theorem 8.9.5.)

8.9.7. Let $\{s_n\}, \{t_n\}$ be bounded sequences in $\mathbb{R}$.

i) Finish the proof of Theorem 8.9.5, namely prove that $\liminf s_n + \liminf t_n \le \liminf(s_n + t_n)$.

ii) Find such $\{s_n\}, \{t_n\}$ so that $\liminf s_n + \liminf t_n < \liminf(s_n + t_n)$.

8.9.8. Suppose that $\lim s_n = \infty$. Prove that the set of subsequential limits of $\{s_n\}$ is empty.

8.9.9. Finish the proof of Theorem 8.9.4, namely prove that the infimum of the set of all subsequential limits of a bounded sequence equals the liminf of the sequence.

8.9.10. Let $\{s_n\}$ be a sequence of positive real numbers. Prove that $\limsup \frac{1}{s_n} = \frac{1}{\liminf s_n}$.

Chapter 9

Infinite series and power series

In this section we introduce infinite sums, what it means for them to make sense, and we introduce functions that are infinite sums of higher and higher powers of a variable x. The work horse of infinite sums are the geometric series, and they are almost the only type of infinite sums that we can compute numerically. The more technical sections on differentiability of power series then allow us to compute many more infinite sums.

Warning: Finite sums are possible by the field axioms, but infinite sums need not make any sense at all. For example,

$$1 + (-1) + 1 + (-1) + 1 + (-1) + 1 + (-1) + 1 + (-1) + 1 + (-1) + 1 + (-1) \cdots$$

may be taken to be 0 or 1 depending on which consecutive pairs are grouped together in a sum, or it could even be summed to exactly 3 by taking the first three positive 1s, and then matching each successive -1 in the sum with the next not-yet-used $+1$. In this way each ± 1 in the expression is used exactly once, so that the sum can indeed be taken to be 3. Similarly, we can make the limit be 4, -17, etc.

This should convince you that in infinite sums the order of addition matters. For more on the order of addition, see Theorem 9.2.8 and Exercise 9.2.16.

Infinite sums require special handling. Limits of sequences prepared the ground.

9.1 Infinite series

Definition 9.1.1. *For an infinite sequence $\{a_n\}$ of complex numbers, define the corresponding* **sequence of partial sums**

$$\{a_1, a_1 + a_2, a_1 + a_2 + a_3, a_1 + a_2 + a_3 + a_4, \ldots\}.$$

We denote the nth term of this sequence $s_n = \sum_{k=1}^{n} a_k$. **The (infinite) series** corresponding to the sequence $\{a_n\}$ is $\sum_{k=1}^{\infty} a_k$ (whether this "infinite sum" makes sense or not).

When the range of indices is clear, we write simply $\sum_k a_k$ or $\sum a_k$.

Example 9.1.2. For the sequence $\{1\}$, the sequence of partial sums is $\{n\}$. If $a \neq 1$, by Example 1.6.4 the sequence of partial sums of $\{a^n\}$ is $\{\sum_{k=1}^{n} a^k\}_n = \{\frac{a^{n+1}-a}{a-1}\}_n$. In particular, the sequence of partial sums of $\{(-1)^n\}$ is $\{\frac{(-1)^n-1}{2}\}_n = \{-1, 0, -1, 0, -1, 0, \ldots\}$.

We have encountered shifted sequences, such as $\{a_n\}_{n \geq m}$, and similarly there are shifted series: $\sum_{k=m}^{\infty} a_k$ stands for the limit of the sequence of partials sums, but in this case, the nth **partial sum** is $s_n = a_m + a_{m+1} + a_{m+2} + \cdots + a_{m+n-1}$.

Definition 9.1.3. (Most of the time and by default we take $m = 1$.) The series $\sum_{k=m}^{\infty} a_k$ **converges** to $L \in \mathbb{C}$ if the sequence $\{\sum_{k=m}^{n-m+1} a_k\}_n$ converges to L. We say then that L is the **sum** of the series and we write $\sum_{k=m}^{\infty} a_k = L$.

If the series does not converge, it **diverges**.

Just like for sequences, when a series diverges, it may diverge to ∞ or to $-\infty$, or it may simply have no limit.

Since a sequence $\{s_n\}$ converges if and only if $\{s_n + c\}$ converges (where c is any constant), it follows that $\sum_{k=1}^{\infty} a_k$ converges if and only if $\sum_{k=m}^{\infty} a_k$ converges, and then

$$\sum_{k=1}^{\infty} a_k = a_1 + a_2 + \cdots + a_{m-1} + \sum_{k=m}^{\infty} a_k.$$

The following follows immediately from the corresponding results for sequences:

Theorem 9.1.4. Let $A = \sum_{k=1}^{\infty} a_k$, $B = \sum_{k=1}^{\infty} b_k$, and $c \in \mathbb{C}$.

(1) If $A, B \in \mathbb{C}$, then

$$\sum_{k=1}^{\infty} (a_k + cb_k) = A + cB.$$

(2) If all a_k, b_k are real numbers and $A, B \in \mathbb{R} \cup \{\infty, -\infty\}$, then

$$\sum_{k=1}^{\infty}(a_k + cb_k) = \begin{cases} \infty, & \text{if } A = \infty, B \in \mathbb{R} \cup \{\infty\} \text{ and } c \geq 0; \\ \infty, & \text{if } A = \infty, B \in \mathbb{R} \cup \{-\infty\} \text{ and } c \leq 0; \\ -\infty, & \text{if } A = -\infty, B \in \mathbb{R} \cup \{-\infty\} \text{ and } c \geq 0; \\ -\infty, & \text{if } A = -\infty, B \in \mathbb{R} \cup \{\infty\} \text{ and } c \leq 0; \\ \infty, & \text{if } A \in \mathbb{R}, B = \infty \text{ and } c > 0; \\ -\infty, & \text{if } A \in \mathbb{R}, B = \infty \text{ and } c < 0; \\ \infty, & \text{if } A \in \mathbb{R}, B = -\infty \text{ and } c < 0; \\ -\infty, & \text{if } A \in \mathbb{R}, B = -\infty \text{ and } c > 0; \\ A, & \text{if } c = 0. \end{cases}$$

Proof. (1) By assumption, the sequences $\{\sum_{k=1}^{n} a_k\}$ and $\{\sum_{k=1}^{n} b_k\}$ converge to A and B in $\mathbb{C}$, respectively. By the theorem on the convergence of sums of sequences (Theorem 8.4.3) then $\sum_{k=1}^{n}(a_k + cb_k) = \sum_{k=1}^{n} a_k + c\sum_{k=1}^{n} b_k$ converges to $A + cB$.

Other parts are proved similarly.   □

Remark 9.1.5. This theorem justifies the binary operation of addition on the set of convergent infinite series:

$$\left(\sum_{k=1}^{\infty} a_k\right) + \left(\sum_{k=1}^{\infty} b_k\right) = \sum_{k=1}^{\infty}(a_k + b_k).$$

It is hard to immediately present examples of this because we know so few limits of infinite series. There are examples in the exercises.

Theorem 9.1.6. *If $r \in \mathbb{C}$ satisfies $|r| < 1$, then the* **geometric series** $\sum_{k=1}^{\infty} r^{k-1}$ *converges to* $\frac{1}{1-r}$*, so* $\sum_{k=1}^{\infty} r^k = \frac{r}{1-r}$.

Proof. By Example 1.6.4, $\sum_{k=1}^{n} r^{k-1} = \frac{1-r^n}{1-r}$. By Theorem 8.6.5, $\lim r^n = 0$. Thus by the scalar and sum rules for limits of sequences (Theorem 8.5.6 or Theorem 8.4.3),

$$\lim_{n \to \infty} \sum_{k=1}^{n} r^{k-1} = \lim_{n \to \infty} \frac{1-r^n}{1-r} = \lim_{n \to \infty} \left(\frac{1}{1-r} - \frac{r^n}{1-r}\right) = \frac{1}{1-r}.$$

Thus $\lim_{n \to \infty} \sum_{k=1}^{n} r^k = r \lim_{n \to \infty} \sum_{k=1}^{n} r^{k-1} = \frac{r}{1-r}$.   □

In particular, the familiar decimal expansion $0.33333\cdots$ of $1/3$ can be thought of as the infinite sum $\sum_{k=1}^{\infty} \frac{3}{10^k}$. The sequence of its partial sums is $\{0.3, 0.33, 0.333, 0.3333, \ldots\}$, and by the theorem above,

$$\sum_{k=1}^{\infty} \frac{3}{10^k} = \frac{3}{10} \sum_{k=1}^{\infty} \left(\frac{1}{10}\right)^{k-1} = \frac{3}{10} \cdot \frac{1}{1 - \frac{1}{10}} = \frac{3}{10} \cdot \frac{10}{9} = \frac{1}{3}.$$

A little more work is expressing $5.\overline{523} = 5.523523523523\cdots$ (repeating 523) as a fraction:

$$5.\overline{523} = 5 + \frac{523}{1000} + \frac{523}{1000^2} + \frac{523}{1000^3} + \cdots = 5 + 523 \sum_{k=1}^{\infty} \left(\frac{1}{1000^k}\right)$$

$$= 5 + 523 \frac{\frac{1}{1000}}{1 - \frac{1}{1000}} = 5 + 523 \frac{1}{1000 - 1} = 5 + \frac{523}{999} = \frac{4995 + 523}{999}$$

$$= \frac{5518}{999}.$$

Example 9.1.7. The **harmonic series** $\sum_{k=1}^{\infty} \frac{1}{k}$ diverges to ∞ by Example 8.7.6.

Example 9.1.8. $\sum_{k=1}^{\infty} \frac{1}{k \cdot (k+1)} = 1$.

Proof. By Exercise 8.2.9, $\sum_{k=1}^{n} \frac{1}{k(k+1)} = 1 - \frac{1}{n+1}$. Thus

$$\sum_{k=1}^{\infty} \frac{1}{k(k+1)} = \lim_{n \to \infty} \sum_{k=1}^{n} \frac{1}{k(k+1)} = \lim_{n \to \infty} \left(1 - \frac{1}{n+1}\right) = 1. \qquad \square$$

Example 9.1.9. The series $\sum_{k=1}^{\infty} \frac{1}{k^2}$ converges.

Proof #1: By Exercise 1.6.11,

$$0 \leq \sum_{k=1}^{n} \frac{1}{k^2} \leq 2 - \frac{1}{n},$$

so that $\{\sum_{k=1}^{n} \frac{1}{k^2}\}_n$ is a bounded increasing sequence of real numbers. By Theorem 8.6.4, the sequence has a limit that is at most 2, so that $\sum_{k=1}^{\infty} \frac{1}{k^2}$ converges. $\qquad \square$

Proof #2: Since $k^2 > k(k-1)$, we have that for $k > 2$, $\frac{1}{k^2} < \frac{1}{k(k-1)}$. Thus $\sum_{k=1}^{n} \frac{1}{k^2} = 1 + \sum_{k=2}^{n} \frac{1}{k^2} < 1 + \sum_{k=2}^{n} \frac{1}{k(k-1)}$. By Example 9.1.8, the series $\sum_{k=2}^{\infty} \frac{1}{k(k-1)} = \sum_{k=1}^{\infty} \frac{1}{k(k+1)}$ converges, so that the increasing sequence $\{\sum_{k=1}^{n} \frac{1}{k^2}\}$ of partial sums is bounded above, so that by Theorem 8.6.4, $\sum_{k=1}^{\infty} \frac{1}{k^2}$ converges. $\qquad \square$

It turns out that $\sum_{k=1}^{\infty} \frac{1}{k^2} = \frac{\pi^2}{6}$, but this is harder to prove. (Three different proofs can be found in Exercises 10.5.1, 10.5.4 and 10.5.3.)

Theorem 9.1.10. *If $\sum_{k=1}^{\infty} a_k$ converges, then $\lim a_n = 0$, and the sequence $\{a_n\}$ is bounded.*

Proof. By assumption, $\sum_{k=1}^{\infty} a_k$ converges, so that by definition, the sequence $\{s_n\}$ of partial sums converges, and is thus Cauchy. In particular, for every $\epsilon > 0$ there exists $N > 0$ such that for all integers $n > N + 1$, $|a_n| = |s_n - s_{n-1}| < \epsilon$. Thus $\lim a_n = 0$, and by Theorem 8.6.2, $\{a_n\}$ is bounded. $\qquad\square$

The converse of this theorem is of course false; see Example 9.1.7.

Theorem 9.1.11. *Let $\{a_n\}$ be a sequence of non-negative real numbers. If the sequence $\{a_1 + a_2 + \cdots + a_n\}$ of partial sums is bounded above, then $\sum a_n$ converges.*

Proof. The sequence $\{a_1 + a_2 + \cdots + a_n\}$ of partial sums is monotone and bounded above, so it converges by Theorem 8.6.4. $\qquad\square$

Theorem 9.1.12. *Let $\{a_n\}$ be a sequence of complex numbers, and let m be a positive integer. Then $\sum_{k=1}^{\infty} a_n$ converges if and only if $\sum_{k=m}^{\infty} a_n$ converges. Furthermore in this case, $\sum_{k=1}^{\infty} a_n = (a_1 + a_2 + \cdots + a_{m-1}) + \sum_{k=m}^{\infty} a_n$.*

Proof. Let $s_n = a_1 + a_2 + \cdots + a_n$, and $t_n = a_m + a_{m+1} + \cdots + a_n$. By the constant and sum rules for sequences (Theorem 8.5.6 or Theorem 8.4.3), the sequence $\{s_n\}_n = \{a_1 + a_2 + \cdots + a_{m-1}\}_n + \{t_n\}_n$ converges if the sequence $\{t_n\}$ converges, and similarly, $\{t_n\}_n = \{s_n\}_n - \{a_1 + a_2 + \cdots + a_{m-1}\}_n$ converges if $\{s_n\}$ converges. $\qquad\square$

Exercises for Section 9.1

9.1.1. Let $r \in \mathbb{C}$ satisfy $|r| \geq 1$. Prove that $\sum_{k=1}^{\infty} r^k$ diverges.

9.1.2. Compute and justify the following sums: $\sum_{k=1}^{\infty} \frac{1}{2^k}$, $\sum_{k=6}^{\infty} \frac{1}{3^k}$, $\sum_{k=8}^{\infty} \frac{2}{5^k}$.

9.1.3. Prove that $\sum_{k=1}^{\infty} \frac{2k+1}{k^2 \cdot (k+1)^2}$ converges, and find the sum. (Hint: Do some initial experimentation with partial sums, find a pattern for partial sums, and prove the pattern with mathematical induction.)

9.1.4. Let $a_n = (-1)^n$. Prove that the sequence of partial sums $\{a_1 + a_2 + \cdots + a_n\}$ is bounded but does not converge. How does this not contradict Theorem 9.1.11?

9.1.5. For each $k \in \mathbb{N}^+$ let x_k be an integer between 0 and 9.

i) Prove that $\sum_{k=1}^{\infty} \frac{x_k}{10^k}$ converges.

ii) What does this say about decimals?

iii) Find the sum if $x_k = 4$ for all k. Express the sum with its decimal expansion and also as a ratio of two positive integers.

iv) Find the sum if $\{x_n\} = \{1, 2, 3, 1, 2, 3, 1, 2, 3, 1, 2, 3, \ldots\}$. Express the sum with its decimal expansion and also as a ratio of two positive integers.

v)* Prove that whenever the sequence $\{x_n\}$ is eventually periodic, then $\sum_{k=1}^{\infty} \frac{x_k}{10^k}$ is a rational number.

9.1.6. Prove that $\sum_{k=1}^{\infty} a_k$ converges if and only if $\sum_{k=1}^{\infty} \operatorname{Re} a_k$ and $\sum_{k=1}^{\infty} \operatorname{Im} a_k$ converge. Furthermore, $\sum_{k=1}^{\infty} a_k = \sum_{k=1}^{\infty} \operatorname{Re} a_k + i \sum_{k=1}^{\infty} \operatorname{Im} a_k$.

9.1.7. Suppose that $\lim a_n \neq 0$. Prove that $\sum_{k=1}^{\infty} a_k$ diverges.

9.1.8. Determine with proof which series converge.

i) $\sum_{k=1}^{\infty} \frac{1}{k^k}$.

ii) $\sum_{k=1}^{\infty} \left(\frac{1}{k^3} + ik \right)$.

iii) $\sum_{k=1}^{\infty} \frac{1}{k!}$.

9.1.9. Let $\{a_n\}$ and $\{b_n\}$ be complex sequences, and let $m \in \mathbb{N}^+$ such that for all $n \geq 1$, $a_n = b_{n+m}$. Prove that $\sum_{k=1}^{\infty} a_k$ converges if and only if $\sum_{k=1}^{\infty} b_k$ converges.

9.2   Convergence and divergence theorems for series

Theorem 9.2.1. (Cauchy's criterion for series) *The infinite series $\sum a_k$ converges if and only if for all real numbers $\epsilon > 0$ there exists a real number $N > 0$ such that for all integers $n \geq m > N$,*

$$|a_{m+1} + a_{m+2} + \cdots + a_n| < \epsilon.$$

Proof. Suppose that $\sum a_k$ converges. This means that the sequence $\{s_n\}$ of partial sums converges, and by Theorem 8.7.3 this means that $\{s_n\}$ is a Cauchy sequence. Thus for all $\epsilon > 0$ there exists $N > 0$ such that for all integers $m, n > N$, $|s_n - s_m| < \epsilon$. In particular for $n \geq m > N$, $|a_{m+1} + a_{m+2} + \cdots + a_n| = |s_n - s_m| < \epsilon$.

Now suppose that for all real numbers $\epsilon > 0$ there exists a real number $N > 0$ such that for all integers $n \geq m > N$, $|a_{m+1} + a_{m+2} + \cdots + a_n| < \epsilon$. This means that the sequence $\{s_n\}$ of partial sums is Cauchy. By Theorem 8.7.5, $\{s_n\}$ is convergent. Then by the definition of series, $\sum_k a_k$ converges. $\square$

Theorem 9.2.2. (Comparison test (for series)) *Let $\{a_n\}$ be a real and $\{b_n\}$ a complex sequence. such that for all n, $a_n \geq |b_n|$.*

(1) If $\sum a_k$ converges then $\sum b_k$ converges.

(2) If $\sum b_k$ diverges then $\sum a_k$ diverges.

Proof. Note that all a_n are non-negative, and for all integers $n \geq m$,

$$|b_{m+1} + b_{m+2} + \cdots + b_n| \leq |b_{m+1}| + |b_{m+2}| + \cdots + |b_n|$$
$$\leq a_{m+1} + a_{m+2} + \cdots + a_n$$
$$= |a_{m+1} + a_{m+2} + \cdots + a_n|.$$

If $\sum a_k$ converges, by Theorem 9.2.1, for every $\epsilon > 0$ there exists $N > 0$ such that for all integers $n > m > N$, $|a_{m+1} + a_{m+2} + \cdots + a_n| < \epsilon$, and hence $|b_{m+1} + b_{m+2} + \cdots + b_n| < \epsilon$. Thus again by Theorem 9.2.1, $\sum b_k$ converges. This proves the first part.

The second part is the contrapositive of the first. $\square$

Theorem 9.2.3. (Ratio test) *(Compare to Remark 8.9.3.) Let $\{a_n\}$ be a sequence of non-zero complex numbers.*

(1) If $\limsup \left| \frac{a_{n+1}}{a_n} \right| < 1$, then $\sum |a_k|$ and $\sum a_k$ converge.

(2) If $\liminf \left| \frac{a_{n+1}}{a_n} \right| > 1$, then $\sum |a_k|$ and $\sum a_k$ diverge.

Proof. Let $L = \limsup |\frac{a_{n+1}}{a_n}|$. Suppose that $L < 1$. Let r be a real number in the open interval $(L, 1)$. Since $L = \inf\{\sup\{\frac{|a_{n+1}|}{|a_n|} : n \geq m\} : m \geq 1\}$ and $r > L$, it follows that there exists $m \geq 1$ such that $r > \sup\{\frac{|a_{n+1}|}{|a_n|} : n \geq m\}$. Thus for all $n \geq m$, $|a_{n+1}| < r|a_n|$. Thus by Exercise 1.6.22, $|a_{m+n}| < r^n |a_m|$. The geometric series $\sum_k r^k$ converges by Theorem 9.1.6, and by Theorem 9.1.4, $\sum_k a_m r^k$ converges. Thus by Theorem 9.2.2, $\sum_{k=1}^{\infty} |a_{m+k}|$ and $\sum_{k=1}^{\infty} a_{m+k}$ converge. Hence by Theorem 9.1.12, $\sum |a_k|$ and $\sum a_k$ converge. This proves (1).

Now let $L = \liminf |\frac{a_{n+1}}{a_n}|$, and suppose that $L > 1$. Let r be a real number in the open interval $(1, L)$. Since $L = \sup\{\inf\{\frac{|a_{n+1}|}{|a_n|} : n \geq m\} : m \geq 1\}$ and $r < L$, it follows that there exists $m \geq 1$ such that $r < \inf\{\frac{|a_{n+1}|}{|a_n|} : n \geq m\}$. Thus for all $n \geq m$, $|a_{n+1}| > r|a_n|$. Thus by a straightforward modification of Exercise 1.6.22, $|a_{m+n}| > r^n|a_m|$. The geometric series $\sum_k r^k$ diverges by Exercise 9.1.1, and by Theorem 9.1.4, $\sum_k a_m r^k$ diverges. Thus by Theorem 9.2.2, $\sum_{k=1}^{\infty} |a_{m+k}|$ and $\sum_{k=1}^{\infty} a_{m+k}$ diverge. Hence by Theorem 9.1.12, $\sum |a_k|$ and $\sum a_k$ diverge, proving (2).

$\square$

This ratio test for convergence of series does not apply when $\limsup |\frac{a_{n+1}}{a_n}| = 1$ or $\liminf |\frac{a_{n+1}}{a_n}| = 1$. The reason is that under these assumptions the series $\sum_k |a_k|$ and $\sum_k a_k$ sometimes converge and sometimes diverge. For example, if $a_n = 1/n$ for all n, $\limsup |\frac{a_{n+1}}{a_n}| = \liminf |\frac{a_{n+1}}{a_n}| = 1$, and $\sum_{k=1}^{n} \frac{1}{k}$ diverges; whereas if $a_n = 1/n^2$ for all n, then $\limsup |\frac{a_{n+1}}{a_n}| = \liminf |\frac{a_{n+1}}{a_n}| = 1$, and $\sum_{k=1}^{n} \frac{1}{k^2}$ converges.

Theorem 9.2.4. (Root test for series) Let $\{a_n\}$ be a sequence of complex numbers. Let $L = \limsup |a_n|^{1/n}$.

(1) If $L < 1$, then $\sum_k |a_k|$, $\sum_k a_k$ converge.

(2) If $L > 1$, then $\sum_k |a_k|$, $\sum_k a_k$ diverge.

Proof. If $L < 1$, choose $r \in (L, 1)$. Since $L = \inf\{\sup\{|a_n|^{1/n} : n \geq m\} : m \geq 1\}$ and $r > L$, there exists $m \geq 1$ such that $r > \sup\{|a_n|^{1/n} : n \geq m\}$. Thus for all $n \geq m$, $r^n \geq |a_n|$. Thus by the Comparison test (Theorem 9.2.2), since the geometric series $\sum r^k$ converges, we have that $\sum a_k$ and $\sum |a_k|$ converge. The proof of (2) is similar, and is omitted here.

$\square$

Theorem 9.2.5. (Alternating series test) If $\{a_n\}$ is a non-increasing sequence of positive real numbers such that $\lim a_n = 0$. Then $\sum_{k=1}^{\infty} (-1)^k a_k$ converges.

Proof. Let m, n be positive integers. Then

$$0 \leq (a_n - a_{n+1}) + (a_{n+2} - a_{n+3}) + \cdots + (a_{n+2m} - a_{n+2m+1})$$

$$= a_n - a_{n+1} + a_{n+2} - a_{n+3} + \cdots + a_{n+2m} - a_{n+2m+1}$$

$$= a_n - (a_{n+1} - a_{n+2}) - (a_{n+3} - a_{n+4}) - \cdots - (a_{n+2m-1} - a_{n+2m}) - a_{n+2m+1}$$

$$\leq a_n,$$

and similarly

$$0 \le a_n - a_{n+1} + a_{n+2} - a_{n+3} + \cdots + a_{n+2m} \le a_n.$$

Thus by Cauchy's criterion Theorem 9.2.1, $\sum_k (-1)^k a_k$ converges. □

Example 9.2.6. Recall from Example 9.1.7 that the harmonic series $\sum_k 1/k$ diverges. But the alternating series $\sum_k (-1)^k/k$ converges by this theorem. (In fact, $\sum_k (-1)^k/k$ converges to $-\ln 2$, but proving the limit is harder — see the proof after Example 9.7.7.)

We examine this infinite series more carefully:

$$-1 + \frac{1}{2} - \frac{1}{3} + \frac{1}{4} - \frac{1}{5} + \frac{1}{6} - \frac{1}{7} + \frac{1}{8} - \frac{1}{9} + \frac{1}{10} - \cdots$$

We **cannot** rearrange the terms in this series as $\left(\frac{1}{2} + \frac{1}{4} + \frac{1}{6} + \frac{1}{8} + \frac{1}{10} + \cdots\right)$ minus $\left(1 + \frac{1}{3} + \frac{1}{5} + \frac{1}{7} + \frac{1}{9} + \cdots\right)$ because both of these series diverge to infinity. More on changing the order of summation is in Exercise 9.2.16 and in Theorem 9.2.8.

Definition 9.2.7. *A series $\sum_{k=1}^{\infty} a_k$ is called* **absolutely convergent** *if $\sum_{k=1}^{\infty} |a_k|$ converges.*

Theorem 9.2.8. *Let $\sum_{k=1}^{\infty} a_k$ be absolutely convergent. Let $r : \mathbb{N}^+ \to \mathbb{N}^+$ be a bijective function. Then $\sum_{k=1}^{\infty} a_{r(k)}$ converges.*

Proof. By comparison test (Theorem 9.2.2), $\sum_{k=1}^{\infty} a_k$ converges to some number $L \in \mathbb{C}$. We will prove that $\sum_{k=1}^{\infty} a_{r(k)}$ converges to L.

Let $\epsilon > 0$. Since $\sum_{k=1}^{\infty} a_k = L$, there exists $N_0 > 0$ such that for all integers $n > N_0$, $|\sum_{k=1}^{n} a_k - L| < \epsilon/2$. Since $\sum_{k=1}^{\infty} |a_k|$ converges, by Cauchy's criterion for sequences there exists $N_1 > 0$ such that for all integers $n \ge m > N_1$, $\sum_{k=m+1}^{n} |a_k| < \epsilon/2$. Pick an integer $N > \max\{N_0, N_1\}$. Since r is a bijective and hence an invertible function, we can define $M = \max\{r^{-1}(1), r^{-1}(2), \ldots, r^{-1}(N)\}$. Let n be an integer strictly bigger than M. Then by definition the set $\{r(1), r(2), \ldots, r(n)\}$ contains $1, 2, \ldots, N$. Let $K = \{r(k) : k \le n\} \setminus \{1, 2, \ldots, N\}$. Then

$$\left| \sum_{k=1}^{n} a_{r(k)} - L \right| = \left| \sum_{k=1}^{N} a_k + \sum_{k \in K} a_k - L \right| \le \left| \sum_{k=1}^{N} a_k - L \right| + \sum_{k \in K} |a_k| < \epsilon$$

since all the finitely many indices in K are strictly bigger than N_0. □

Theorem 9.2.9. (Integral test for series convergence) *Let* $f :$ $[1, \infty) \to [0, \infty)$ *be a decreasing function. Suppose that for all* $n \in \mathbb{N}^+$, $\int_1^n f$ *exists. Then* $\sum_{k=1}^{\infty} f(k)$ *converges if and only if* $\lim_{n \to \infty} \int_1^n f$ *exists and is a real number.*

(It is not necessarily the case that $\sum_{k=1}^{\infty} f(k)$ *equals* $\lim_{n \to \infty} \int_1^n f$.*)*

Proof. Since f is decreasing, for all $x \in [n, n+1]$, $f(n) \geq f(x) \geq f(n+1)$. Thus

$$f(n+1) = \int_n^{n+1} f(n+1)dx \leq \int_n^{n+1} f(x)dx \leq \int_n^{n+1} f(n)dx = f(n).$$

Suppose that $\sum_k f(k)$ converges. Then by the definition this means that $\lim_{n \to \infty}(f(1) + f(2) + \cdots + f(n))$ exists. By the displayed inequalities, $\int_1^{n+1} f = \int_1^2 f + \int_2^3 f + \cdots + \int_n^{n+1} f \leq f(1) + f(2) + \cdots + f(n)$, so that $\{\int_1^{n+1} f\}_n$ is a bounded increasing sequence of real numbers, so that $\lim_{n \to \infty} \int_1^{n+1} f$ exists, and hence that $\lim_{n \to \infty} \int_1^n f$ exists.

Conversely, suppose that $\lim_{n \to \infty} \int_1^n f$ exists. Let $L \in \mathbb{R}$ be this limit. By the displayed inequalities, $f(2) + f(3) + \cdots + f(n+1) \leq \int_2^3 f + \cdots + \int_n^{n+1} f = \int_2^{n+1} f$. Since f takes on only non-negative values, this says that $f(2) + f(3) + \cdots + f(n+1) \leq L$. Thus $\{f(2) + \cdots + f(n + 1)\}_n$ is a non-decreasing sequence that is bounded above by L. Thus by Theorem 8.6.4, this sequence converges. By adding the constant $f(1)$, the sequence $\{f(1) + \cdots + f(n+1)\}_n$ converges, so that by the definition of series, $\sum_k f(k)$ converges. $\square$

Theorem 9.2.10. (The p-series convergence test) *Let* p *be a real number. The series* $\sum_k k^p$ *converges if* $p < -1$ *and diverges if* $p \geq -1$.

Proof. If $p = -1$, then the series is the harmonic series and hence diverges. If $p \geq -1$, then $n^p \geq n^{-1}$ for all n by Theorem 7.6.5. Thus by the comparison test (Theorem 9.2.2), $\sum_k k^p$ diverges.

Now suppose that $p < -1$. The function $f : [1, \infty) \to \mathbb{R}$ given by $f(x) = x^p$ is differentiable, continuous, and decreasing. Since f is continuous, for all positive integers n, $\int_1^n f$ exists. By the Fundamental theorem of calculus, $\int_1^n f = \int_1^n x^p \, dx = \frac{n^{p+1}-1}{p+1}$. By the composite rule for sequences (either Theorem 8.5.8 or Theorem 8.4.7), since the function that exponentiates by the positive $-(p+1)$ is continuous at all real numbers and $\lim \frac{1}{n} = 0$, it follows that $\lim n^{p+1} = \lim \left(\frac{1}{n}\right)^{-(p+1)} = 0^{-(p+1)} = 0$, so that $\lim \int_1^n f$

exists and equals $\frac{-1}{p+1}$. Thus by the Integral test (Theorem 9.2.9), $\sum_k k^p$ converges.   □

Exercises for Section 9.2

9.2.1. Prove that $\sum_{k=0}^{\infty} \frac{1}{(2k+1)^2}$, $\sum_{k=0}^{\infty} \frac{(-1)^k}{(2k+1)^2}$, $\sum_{k=1}^{\infty} \frac{1}{k^2}$ and $\sum_{k=1}^{\infty} \frac{(-1)^k}{k^2}$ all converge.

9.2.2. Prove that the following statements are equivalent. (They are also all true, but we do not yet have enough methods to prove them.)

(1) $\sum_{k=0}^{\infty} \frac{1}{(2k+1)^2} = \frac{\pi^2}{8}$.

(2) $\sum_{k=1}^{\infty} \frac{1}{k^2} = \frac{\pi^2}{6}$.

(3) $\sum_{k=0}^{\infty} \frac{(-1)^k}{(2k+1)^2} = \frac{\pi^2}{16}$.

(4) $\sum_{k=1}^{\infty} \frac{(-1)^k}{k^2} = \frac{\pi^2}{12}$.

(Hint for a part: Write out the first few summands of $\sum_{k=1}^{\infty} \frac{(-1)^k}{k^2} + \frac{1}{2} \sum_{k=1}^{\infty} \frac{1}{k^2} = \sum_{k=1}^{\infty} \frac{(-1)^k}{k^2} + 2 \sum_{k=1}^{\infty} \frac{1}{(2k)^2}$.)

9.2.3. For each of the following series, determine with proof whether they converge or diverge. You may need to use Examples 8.2.9 and 8.3.6.

i) $\sum_{k=1}^{\infty} \frac{3i}{\sqrt{k^2 + k^4}}$.

ii) $\sum_{k=1}^{\infty} \frac{1}{\sqrt{k}}$.

iii) $\sum_{k=1}^{\infty} \frac{1}{k^3}$.

iv) $\sum_{k=1}^{\infty} \frac{2^k}{k!}$.

v) $\sum_{k=1}^{\infty} \frac{2^k}{k^3}$.

vi) $\sum_{k=1}^{\infty} \frac{1}{k^{\sqrt{2}}}$.

9.2.4. Find a convergent series $\sum_k a_k$ and a divergent series $\sum_k b_k$ with $\limsup |a_n|^{1/n} = \limsup |b_n|^{1/n} = 1$.

9.2.5. Make a list of all encountered criteria of convergence for series.

9.2.6. The goal of this exercise is to show that if the ratio test (Theorem 9.2.3) determines the convergence/divergence of a series, then the root test (Theorem 9.2.4) determines it as well. Let $\{a_n\}$ be a sequence of non-zero complex numbers.

    i) Suppose that $\limsup \left|\frac{a_{n+1}}{a_n}\right| < 1$. Prove that $\limsup |a_n|^{1/n} < 1$.

    ii) Suppose that $\liminf \left|\frac{a_{n+1}}{a_n}\right| > 1$. Prove that $\limsup |a_n|^{1/n} > 1$.

9.2.7. Apply the ratio test (Theorem 9.2.3) and the root test (Theorem 9.2.4) to $\sum_{k=1}^{\infty} \frac{5^k}{k!}$. Was one test easier? Repeat for $\sum_{k=1}^{\infty} \frac{1}{k^k}$.

9.2.8. Let $\{a_n\}$ be a complex sequence, and let $c \in \mathbb{C}$. Is it true that $\sum_{k=1}^{\infty} a_k$ converges if and only if $\sum_{k=1}^{\infty} ca_k$ converges? If true, prove; if false, give a counterexample.

9.2.9. Let $a_n = (-1)^n/n$, $b_n = 2(-1)^n$. Prove that $\sum_{k=1}^{\infty} a_k$ converges, that for all n, $|b_n| = 2$, and that $\sum_{k=1}^{\infty} \frac{a_k}{b_k}$ diverges.

9.2.10. (Compare with Exercise 9.2.9.) Suppose that $\sum_{k=1}^{\infty} |a_k|$ converges. Let $\{b_n\}$ be a sequence of complex numbers such that for all n, $|b_n| > 1$. Prove that $\sum_{k=1}^{\infty} \frac{a_k}{b_k}$ converges.

9.2.11. (Summation by parts) Let $\{a_n\}, \{b_n\}$ be complex sequences. Prove that

$$\sum_{k=1}^{n} a_k b_k = a_n \sum_{k=1}^{n} b_k - \sum_{k=1}^{n} (a_{k+1} - a_k) \sum_{j=1}^{k} b_j.$$

(Hint: Set $a_0 = b_0 = 0$. Let $f, g : [1, \infty) \to \mathbb{C}$ be defined as follows: for each $n \in \mathbb{N}^+$, on the interval $[n, n+1)$, f is the constant function $a_n - a_{n-1}$ and g is the constant function b_n. Both f and g are piecewise continuous, $F(n) = \int_1^{n+1} f = a_n$, $G(n) = \int_1^{n+1} g = \sum_{k=1}^{n} b_k$. The problem should remind you of integration by parts $\int (Fg) = FG - \int (fG)$.)

9.2.12. Let $x \in \mathbb{C}$ with $|x| \leq 1$ and $x \neq 1$. Prove that $\sum_{k=1}^{\infty} \frac{x^k}{k}$ converges. (Hint: Exercise 9.2.11, Example 1.6.4.)

9.2.13. Prove that if $\sum_{k=1}^{\infty} a_k$ converges then $\sum_{k=1}^{\infty} \frac{a_k}{k}$ converges. (Hint: Exercise 9.2.11.)

9.2.14. (The Dirichlet test) Let $\{a_n\}$ be a decreasing real sequence such that $\lim a_n = 0$, and let $\{b_n\}$ be a sequence of complex numbers whose sequence of partial sums is bounded.

    i) Prove that the series $\sum_{k=1}^{\infty} a_k b_k$ converges. (Hint: Exercise 9.2.11.)

    ii) Prove the alternating series test using part i).

9.2.15. The following guides through another proof of the p-series convergence test for $p < -1$. (Confer Theorem 9.2.10 for the first proof).

i) Prove that for each positive integer n there exists $c \in (n, n+1)$ such that $(p + 1)c^p = (n + 1)^{p+1} - n^{p+1}$. (Hint: Mean value theorem (Theorem 6.3.4).)

ii) Prove that for all positive integers n,

$$(n + 1)^p < \frac{(n + 1)^{p+1} - n^{p+1}}{p + 1} \text{ and } n^p < \frac{n^{p+1} - (n - 1)^{p+1}}{p + 1}.$$

iii) Prove by induction on $n \geq 1$ that

$$\sum_{k=1}^{n} k^p \leq \frac{1}{p + 1} n^{p+1} + \frac{p}{p + 1}.$$

iv) Prove that the positive sequence $\{\sum_{k=1}^{n} k^p\}_n$ of partial sums is an increasing sequence bounded above.

v) Prove that the sequence $\{\sum_{k=1}^{n} k^p\}_n$ converges, and so that $\sum_{k=1}^{\infty} k^p$ converges.

9.2.16. (Order of summation in infinite sums is important.)

i) Prove that $\sum_{k=1}^{\infty} \frac{1}{2k}$ and $\sum_{k=1}^{\infty} \frac{1}{2k+1}$ diverge. Refer to Example 9.2.6.

ii) Observe that $\sum_{k=1}^{n} \frac{1}{2k} - \sum_{k=1}^{n} \frac{1}{2k+1} = \sum_{k=1}^{2n+1} \frac{(-1)^k}{k}$.

iii) Argue that $\sum_{k=1}^{\infty} \frac{(-1)^k}{k} \neq \sum_{k=1}^{\infty} \frac{1}{2k} - \sum_{k=1}^{\infty} \frac{1}{2k+1}$. Why does this not contradict the "expected" summation and difference rules?

9.2.17. (**Raabe's test**) Let $a_1, a_2, \ldots$ be positive real numbers such that for some $\alpha > 1$ and for some $N \in \mathbb{N}$, $\frac{a_{n+1}}{a_n} \leq 1 - \frac{\alpha}{n}$ for all $n \geq N$. Prove that $\sum_n a_n$ converges. (Hint: Let $f(x) = 1 - x^\alpha$. Use the Mean value theorem to get $c \in (x, 1)$ such that $f'(c)(1 - x) = f(1) - f(x)$. Conclude that $1 - x^\alpha \leq (1 - x)\alpha$. Apply this to $x = 1 - \frac{1}{n}$. Use that $\sum n^{-\alpha}$ converges.)

9.3   Power series

In this section we deal with sums where the index varies through $\mathbb{N}_0$, and furthermore, the terms of the sequence are special functions rather than constants:

Definition 9.3.1. *A* **power series** *is an infinite series of the form*

$$\sum_{k=0}^{\infty} a_k x^k = \sum_{k \geq 0} a_k x^k = a_0 + a_1 x + a_2 x^2 + a_3 x^3 + \cdots,$$

where $a_0, a_1, a_2, \ldots$ *are fixed complex numbers, and* x *is a variable that can be replaced by any complex number. (By convention as on page 27, $0^0 = 1$.)*

The table below identifies the coefficients a_n of x^n in several powers series.

power series	a_n
$\displaystyle\sum_{k=0}^{\infty} x^k$	1
$\displaystyle\sum_{k=0}^{\infty} x^{2k+1} = x + x^3 + x^5 + x^7 + \cdots$	$\begin{cases} 1, & \text{if } n \text{ is odd;} \\ 0, & \text{if } n \text{ is even.} \end{cases}$
$\displaystyle\sum_{k=0}^{\infty} x^{2k+12} = x^{12} + x^{14} + x^{16} + x^{18} + \cdots$	$\begin{cases} 0, & \text{if } n \text{ is odd;} \\ 0, & \text{if } n \text{ is even and } n < 12; \\ 1, & \text{if } n \geq 12 \text{ is even.} \end{cases}$
$\displaystyle\sum_{k=0}^{\infty} k x^k$	n
$\displaystyle\sum_{k=0}^{\infty} \frac{x^k}{k!}$	$\frac{1}{n!}$
$\displaystyle\sum_{k=0}^{\infty} (kx)^k$	n^n

The partial sums of power series are polynomials, so a power series is a limit of polynomials.

A power series is a function of x, and the domain is to be determined. Clearly, 0 is in the domain of every power series: plugging in $x = 0$ returns $\sum_{k=0}^{\infty} a_k 0^k = a_0$. If all except finitely many a_n are 0, then the power series is actually a polynomial, and is thus defined on all of $\mathbb{C}$. When 1 is in the domain, evaluation of the power series $\sum_k a_k x^k$ at $x = 1$ is the (ordinary) series $\sum_k a_k$.

The most important question that we address in this section is: which x are in the domain of the power series, i.e., for which x does such an infinite series converge. We prove that for every power series whose domain is not all of $\mathbb{C}$ there exists a non-negative real number R such that the series converges for all $x \in \mathbb{C}$ with $|x| < R$ and the series diverges for all $x \in \mathbb{C}$ with $|x| > R$. What happens at x with $|x| = R$ depends on the series.

Example 9.3.2. Let $f(x) = \sum_{k=0}^{\infty} x^k$. By Theorem 9.1.6, the domain of f contains all complex numbers with absolute value strictly smaller than 1, and by Theorem 9.1.10, the domain of f contains no other numbers, so that the domain equals $\{x \in \mathbb{C} : |x| < 1\}$. Moreover, for all x in the domain of f, by Theorem 9.1.6, $\sum_{k=0}^{\infty} x^k = \frac{1}{1-x}$. Note that the domain of $\frac{1}{1-x}$ is strictly larger than the domain of f.

Whereas for general power series it is impossible to get a true numerical infinite sum, for geometric series this is easy: $f(\frac{1}{2}) = \sum_{k \geq 0} \frac{1}{2^k} = \frac{1}{1 - \frac{1}{2}} = 2$, $f(\frac{1}{3}) = \sum_{k \geq 0} \frac{1}{3^k} = \frac{1}{1 - \frac{1}{3}} = \frac{3}{2}$, $f(0.6) = \sum_{k \geq 0} 0.6^k = \frac{1}{1 - 0.6} = 2.5$.

Theorem 9.3.3. (Root test for the convergence of power series) Let $\sum a_k x^k$ be a power series, and let $\alpha = \limsup |a_n|^{1/n}$. Define R by

$$R = \begin{cases} 1/\alpha, & \text{if } 0 < \alpha < \infty; \\ 0, & \text{if } \alpha = \infty; \\ \infty, & \text{if } \alpha = 0. \end{cases}$$

Then for all $x \in \mathbb{C}$ with $|x| < R$, $\sum |a_k||x|^k$ and $\sum a_k x^k$ converge in $\mathbb{C}$, and for all $x \in \mathbb{C}$ with $|x| > R$, $\sum |a_k||x|^k$ and $\sum a_k x^k$ diverge.

Proof. By the definition of limits, α is either a non-negative real number or ∞. We apply the Root test for series (Theorem 9.2.4): $\limsup |a_n x^n|^{1/n} = |x| \limsup |a_n|^{1/n} = |x|\alpha$. If $|x|\alpha < 1$, then both of the series converge, and if $|x|\alpha > 1$, then the two series diverge. If $\alpha = 0$, then $|x|\alpha < 1$ is true for all $x \in \mathbb{C}$, so $R = \infty$ has the stated property. If $\alpha = \infty$, then $|x|\alpha < 1$ is true only for $x = 0$, so $R = 0$ has the stated property. If $0 < \alpha < \infty$, then $|x|\alpha < 1$ is true only for all $x \in \mathbb{C}$ with $|x| < 1/\alpha = R$. $\qquad \square$

Definition 9.3.4. The R from Theorem 9.3.3 is called **the radius of convergence** of the series $\sum a_k x^k$.

This is really a radius of convergence because inside the circle $B(0, R)$ the series converges and outside of the circle the series diverges. Whether the power series converges at points on the circle depends on the series; see Example 9.3.6.

Theorem 9.3.5. *The series* $\sum a_k x^k$, $\sum |a_k| x^k$, $\sum |a_k||x|^k$, $\sum k a_k |x|^k$, $\sum k a_k x^{k-1}$, $\sum k^2 a_k x^k$, $\sum k(k-1) a_k x^k$, *all have the same radius of convergence.*

Proof. By Example 8.2.9, $\lim \sqrt[n]{|n|} = 1$. For any integer $n \geq 2$, $1 \leq \sqrt[n]{n-1} \leq \sqrt[n]{n}$, so that by the squeeze theorem, $\lim \sqrt[n]{n-1} = 1$. Thus by Theorem 8.9.6, $\limsup \sqrt[n]{|n(n-1)a_n|} = \lim \sqrt[n]{n(n-1)} \cdot \limsup \sqrt[n]{|a_n|} = \limsup \sqrt[n]{|a_n|}$. This proves that the α as in Theorem 9.3.3 for $\sum a_k x^k$ is the same as the α for $\sum k(k-1)a_k x^k$, which proves that these two power series have the same radius of convergence. The proofs of the other parts are similar. $\qquad\square$

Example 9.3.6. We have seen that $\sum x^k$ has radius of convergence 1. By Theorem 9.1.10, this series does not converge at any point on the unit circle. By the previous theorem, the radius of convergence of $\sum_{k=1}^{\infty} \frac{1}{k} x^k$ and of $\sum_{k=1}^{\infty} \frac{1}{k^2} x^k$ is also 1. By the p-series test (Theorem 9.2.10) or by the harmonic series fact, $\sum_{k=1}^{\infty} \frac{1}{k} x^k$ diverges at $x = 1$, and by the alternating series test (Theorem 9.2.5), $\sum_{k=1}^{\infty} \frac{1}{k} x^k$ converges at $x = -1$. By the p-series test (Theorem 9.2.10), $\sum_{k=1}^{\infty} \frac{1}{k^2}$ converges, so that by the comparison test Theorem 9.2.2, $\sum_{k=1}^{\infty} \frac{1}{k^2} x^k$ converges on the unit circle.

Theorem 9.3.7. *Let the radius of convergence for $\sum a_k x^k$ be R. Then the radius of convergence for $\sum a_k x^{2k}$ is ∞ if $R = \infty$ and it is $\sqrt{R}$ otherwise.*

Proof. Let $\alpha = \limsup \sqrt[n]{|a_n|}$. In the second power series, the nth coefficient is 0 if n is odd, and it is $a_{n/2} x^n$ if n is even. Then applying the root test to this power series gives

$$\limsup \left\{ \sqrt[1]{0}, \sqrt[2]{|a_1|}, \sqrt[3]{0}, \sqrt[4]{|a_2|}, \sqrt[5]{0}, \sqrt[6]{|a_3|}, \sqrt[7]{0}, \sqrt[8]{|a_4|}, \ldots \right\}$$

$$= \limsup \left\{ \sqrt[2]{|a_1|}, \sqrt[4]{|a_2|}, \sqrt[6]{|a_3|}, \sqrt[8]{|a_4|}, \ldots \right\}$$

(0s do not contribute to limsup of non-negative numbers)

$$= \limsup \left\{ \sqrt{\sqrt[1]{|a_1|}}, \sqrt{\sqrt[2]{|a_2|}}, \sqrt{\sqrt[3]{|a_3|}}, \sqrt{\sqrt[4]{|a_4|}}, \ldots \right\}$$

$$= \sqrt{\limsup \left\{ \sqrt[1]{|a_1|}, \sqrt[2]{|a_2|}, \sqrt[3]{|a_3|}, \sqrt[4]{|a_4|}, \ldots \right\}}$$

$$= \sqrt{\alpha},$$

and the conclusion follows. $\qquad\square$

Example 9.3.8. Similarly to the last example and by Theorem 9.3.5, $\sum x^k$, $\sum kx^k$, $\sum k^2 x^k$, $\sum k(k-1)x^k$, $\sum x^{2k+1}$ all have radius of convergence 1.

Theorem 9.3.9. (Ratio test for the convergence of power series)
Suppose that all a_n are non-zero complex numbers.

(1) If $|x| < \liminf \left| \frac{a_n}{a_{n+1}} \right|$, then $\sum |a_k||x|^k$ and $\sum a_k x^k$ converge.

(2) If $|x| > \limsup \left| \frac{a_n}{a_{n+1}} \right|$, then $\sum |a_k||x|^k$ and $\sum a_k x^k$ diverge.

Thus if $\lim \left| \frac{a_n}{a_{n+1}} \right|$ exists, it equals the radius of convergence of $\sum a_k x^k$.

Warning: Compare with the Ratio test for convergence of series (Theorem 9.2.3) where fractions are different. Explain to yourself why that is necessarily so, possibly after going through the proof below.

Proof. The two series converge in case $x = 0$, so that we may assume that $x \neq 0$. We may then apply the Ratio test for convergence of series (Theorem 9.2.3):

$$\limsup \left| \frac{a_{n+1} x^{n+1}}{a_n x^n} \right| = |x| \limsup \left| \frac{a_{n+1}}{a_n} \right|$$

$$= |x| \inf \left\{ \sup \left\{ \left| \frac{a_{n+1}}{a_n} \right| : n \geq m \right\} : m \geq 1 \right\}$$

$$= |x| \inf \left\{ \frac{1}{\inf \left\{ \left| \frac{a_n}{a_{n+1}} \right| : n \geq m \right\}} : m \geq 1 \right\}$$

$$= \frac{|x|}{\sup \left\{ \inf \left\{ \left| \frac{a_n}{a_{n+1}} \right| : n \geq m \right\} : m \geq 1 \right\}}$$

$$= \frac{|x|}{\liminf \left| \frac{a_n}{a_{n+1}} \right|}.$$

If this is strictly smaller than 1, then the two series converge, which proves (1). Similarly,

$$\liminf \left| \frac{a_{n+1} x^{n+1}}{a_n x^n} \right| = \frac{|x|}{\limsup \left| \frac{a_n}{a_{n+1}} \right|},$$

and if this is strictly larger than 1, then the two series diverge. The last part is then immediate by the definition of radius of convergence. $\square$

Examples 9.3.10.

(1) In Example 9.3.8 we established via the root test that $\sum x^{2k+1}$ has radius of convergence 1. The ratio test is inapplicable for this power series. However, note that $\sum x^{2k+1} = x \sum (x^2)^k$, and by the ratio test for series (not power series), this series converges for non-zero x if $\limsup |(x^2)^{k+1}/(x^2)^k| < 1$, i.e., if $|x^2| < 1$, i.e., if $|x| < 1$, and it diverges if $|x| > 1$.

(2) The radius of convergence of $\sum_{k=1}^{\infty} \left(\frac{x}{k}\right)^k$ is ∞. For this we apply the root test: $\alpha = \limsup \left|\frac{1}{n^n}\right|^{1/n} = \limsup \frac{1}{n} = 0.$

(3) By the ratio test, the radius of convergence of $\sum_{k=1}^{\infty} \frac{x^k}{k!}$ is $\lim \left|\frac{\frac{1}{n!}}{\frac{1}{(n+1)!}}\right| = \lim \frac{(n+1)!}{n!} = \lim (n+1) = \infty$. The root test gives $\alpha = \limsup |1/n!|^{1/n} = \limsup (1/n!)^{1/n}$, and by Example 8.3.6 this is 0. Thus the radius of convergence is ∞ also by the root test.

Exercises for Section 9.3

9.3.1. Let the radius of convergence of $\sum_{k=0}^{\infty} a_k x^k$ be R. Let $c \in \mathbb{C}$.

  i) Prove that $c \cdot \sum_{k=0}^{\infty} a_k x^k = \sum_{k=0}^{\infty} (c \cdot a_k) x^k$ is convergent with radius of convergence equal to R if $c \neq 0$, and with radius of convergence ∞ otherwise.

  ii) Prove that $\sum_{k=0}^{\infty} a_k (cx)^k$ is convergent with radius of convergence equal to ∞ if $c = 0$ and radius $R/|c|$ otherwise.

9.3.2. Suppose that $\sum_k a_k$ converges for some $a_k \in \mathbb{C}$.

  i) Prove that the function defined as the power series $\sum_k a_k x^k$ has radius of convergence at least 1.

  ii) Give an example of $a_k \in \mathbb{C}$ for which the radius of convergence of $\sum_{k=0}^{\infty} a_k x^k$ is strictly greater than 1. (Bonus points for easiest example.)

  iii) Give an example of $a_k \in \mathbb{C}$ for which the radius of convergence of $\sum_{k=0}^{\infty} a_k x^k$ is equal to 1.

9.3.3. Compare with the previous exercise: find the radii of convergence of the power series $\sum_{k=0}^{\infty} x^k$, $\sum_{k=0}^{\infty} (-1) x^k$, $\sum_{k=0}^{\infty} (2x)^k$, $\sum_{k=0}^{\infty} x^k + \sum_{k=0}^{\infty} (-1) x^k$, $\sum_{k=0}^{\infty} x^k + \sum_{k=0}^{\infty} (2x)^k$.

9.3.4. Compute and justify the radius of convergence for the following series:

i) $\sum 3x^k$.

ii) $\sum (3x)^k$.

iii) $\sum 3kx^k$.

iv) $\sum k(3x)^k$.

v) $\sum \frac{x^k}{k^3}$.

vi) $\sum \frac{3x^k}{k^3}$.

vii) $\sum \frac{(3x)^k}{k^3}$.

9.3.5. Compute and justify the radius of convergence for the following series:

i) $\sum \frac{x^k}{k^k}$.

ii) $\sum \frac{x^k}{k^{2k}}$.

iii) $\sum \frac{x^k}{(2k)^k}$.

9.3.6. Let $\sum_{k=0}^{\infty} a_k x^k$ and $\sum_{k=0}^{\infty} b_k x^k$ be convergent power series with radii of convergence R_1 and R_2, respectively. Let $R = \min\{R_1, R_2\}$. Prove that $\sum_{k=0}^{\infty} a_k x^k \pm \sum_{k=0}^{\infty} b_k x^k = \sum_{k=0}^{\infty}(a_k \pm b_k)x^k$ is convergent with radius of convergence at least R. (Hint: Theorem 9.1.4.)

9.3.7. Let R be the radius of convergence of $\sum_{k=0}^{\infty} a_k x^k$. Let p be a positive integer.

i) Determine the radius of convergence of $\sum_{k=p}^{\infty} a_k x^k$.

ii) Determine the radius of convergence of $\sum_{k=0}^{\infty} a_k x^{pk}$.

9.3.8. What would be a sensible definition for generalized **power series** $\sum_{k=0}^{\infty} a_k(x - a)^k$? What would be a sensible definition of the radius of convergence of $\sum_{k=0}^{\infty} a_k(x - a)^k$? Draw a relevant picture in $\mathbb{C}$.

9.4   Differentiation of power series

Power series are functions. In this section we prove that they are differentiable at all x inside the circle of convergence. Since a differentiable function is continuous, it follows that a power series is continuous inside the circle of convergence.

Recall that for any differentiable function f,

$$f'(x) = \lim_{h \to 0} \frac{f(x + h) - f(x)}{h},$$

and any power series $\sum_{k=0}^{\infty} a_k x^k$ is the limit of a sequence: $\sum_{k=0}^{\infty} a_k x^k = \lim\{\sum_{k=0}^{n} a_k x^k\}_n$. Thus

$$\left(\sum_{k=0}^{\infty} a_k x^k\right)' = \lim_{h\to 0} \frac{\lim\{\sum_{k=0}^{n} a_k(x+h)^k\}_n - \lim\{\sum_{k=0}^{n} a_k x^k\}_n}{h}.$$

Certainly by the sum rule for convergent series, $\lim\{\sum_{k=0}^{n} a_k(x+h)^k\}_n - \lim\{\sum_{k=0}^{n} a_k x^k\}_n = \lim\{\sum_{k=0}^{n} a_k((x+h)^k - x^k)\}_n$, and by the constant rule we get that

$$\left(\sum_{k=0}^{\infty} a_k x^k\right)' = \lim_{h\to 0} \lim_{n\to\infty} \sum_{k=0}^{n} a_k \frac{(x+h)^k - x^k}{h}.$$

If we could change the order of limits, then we would get by the polynomial rule for derivatives that

$$\left(\sum_{k=0}^{\infty} a_k x^k\right)' = \lim_{n\to\infty} \lim_{h\to 0} \sum_{k=0}^{n} k a_k x^{k-1}.$$

In fact, it turns out that this is the correct derivative, but our reasoning above was based on an unproven (and generally false) switch of the two limits.

We give a correct proof of derivatives in the rest of the section. By Theorem 9.3.5 we already know that the series $\sum_{k=0}^{\infty} a_k x^k$ and $\sum_{k=0}^{\infty} k a_k x^{k-1} = \sum_{k=1}^{\infty} k a_k x^{k-1}$ have the same radius of convergence.

The following theorem is not necessarily interesting in its own right, but it is a stepping stone in the proof of derivatives of power series.

Theorem 9.4.1. *Let $\sum_k a_k x^k$ have radius convergence R. Let $c \in \mathbb{C}$ satisfy $|c| < R$. Then the function $g(x) = \sum_{k=1}^{\infty} a_k(x^{k-1} + cx^{k-2} + c^2 x^{k-3} + \cdots + c^{k-1})$ is defined on $B(0, R)$ and is continuous at c.*

Proof. There is nothing to prove if $R = 0$, so we may assume that $R > 0$.

First of all, $\sum_{k=1}^{\infty} k a_k x^{k-1}$ converges on $B(0, R)$ by Theorem 9.3.5. In particular, $\sum_{k=1}^{\infty} k a_k c^{k-1} = \sum_{k=1}^{\infty} a_k(c^{k-1} + cc^{k-2} + c^2 c^{k-3} + \cdots + c^{k-1})$ is well-defined, but this is simply $g(c)$, so that c is in the domain of g.

Let $\epsilon > 0$ and let $d \in \mathbb{R}$ satisfy $|c| < d < R$. Then again by Theorem 9.3.5, d is in the domain of $\sum_k a_k x^k$, $\sum_k k a_k x^{k-1}$, and also $\sum_k k(k-1)a_k x^{k-2}$ and $\sum_k k(k-1)|a_k||x|^{k-2}$. Set $D = \frac{1}{2}\sum_k k(k-1)|a_k|d^{k-2}$, and $\delta = \min\{d - |c|, \frac{\epsilon}{D+1}\}$. Then δ is positive. Let $c \in \mathbb{C}$ with $0 < |x - c| < \delta$.

Then by the triangle inequality, $|x| = |x - c + c| \le |x - c| + |c| < \delta + |c| \le d - |c| + |c| = d < R$, so that x is in the domain of the power series $\sum_k a_k x^k$.

But is x in the domain of g? Since the radius of convergence of $\sum_k a_k x^k$ is R, by Theorem 9.3.5 also the radius of convergence of $\sum_k k a_k x^{k-1}$ is R, so that $\sum_k k a_k d^{k-1}$ converges. Then from the Comparison theorem (Theorem 9.2.2) and from

$$|a_k(x^{k-1} + cx^{k-2} + c^2 x^{k-3} + \cdots + c^{k-1})|$$

$$\le |a_k|(|x|^{k-1} + |c||x|^{k-2} + |c|^2|x|^{k-3} + \cdots + |c|^{k-1})$$

$$\text{(by the triangle inequality)}$$

$$\le |a_k|(d^{k-1} + dd^{k-2} + d^2 d^{k-3} + \cdots + d^{k-1})$$

$$= k|a_k|d^{k-1}$$

we deduce that x is in the domain of g. Furthermore,

$$|g(x) - g(c)|$$

$$= |\sum a_k(x^{k-1} + cx^{k-2} + c^2 x^{k-3} + \cdots + c^{k-1}) - \sum ka_k c^{k-1}|$$

$$= |\sum a_k(x^{k-1} + cx^{k-2} + c^2 x^{k-3} + \cdots + c^{k-1} - kc^{k-1})|$$

$$= |\sum a_k((x^{k-1} - c^{k-1}) + (cx^{k-2} - c^{k-1})$$

$$+ (c^2 x^{k-3} - c^{k-1}) + \cdots + (c^{k-1} - c^{k-1})|$$

$$= |\sum a_k((x^{k-1} - c^{k-1}) + c(x^{k-2} - c^{k-2})$$

$$+ c^2(x^{k-3} - c^{k-3}) + \cdots + c^{k-2}(x - c))|$$

$$= |\sum a_k(x - c)((x^{k-2} + x^{k-3}c + x^{k-4}c^2 + \cdots + c^{k-2})$$

$$+ c(x^{k-3} + x^{k-4}c + x^{k-5}c^2 + \cdots + c^{k-3})$$

$$+ c^2(x^{k-4} + x^{k-5}c + x^{k-6}c^2 + \cdots + c^{k-4}) + \cdots + c^{k-2})|$$

$$\text{(by Exercise 1.6.21)}$$

$$\le \sum |a_k||x - c|((|x|^{k-2} + |x|^{k-3}|c| + |x|^{k-4}|c|^2 + \cdots + |c|^{k-2})$$

$$+ |c|(|x|^{k-3} + |x|^{k-4}|c| + |x|^{k-5}|c|^2 + \cdots + |c|^{k-3})$$

$$+ |c|^2(|x|^{k-4} + |x|^{k-5}|c| + |x|^{k-6}|c|^2 + \cdots + |c|^{k-4}) + \cdots + |c|^{k-2})|$$

$$\text{(by the triangle inequality)}$$

$$\le |x - c| \sum |a_k| ((d^{k-2} + d^{k-3}d + d^{k-4}d^2 + \cdots + d^{k-2})$$
$$+ d(d^{k-3} + d^{k-4}d + d^{k-5}d^2 + \cdots + d^{k-3})$$
$$+ d^2(d^{k-4} + d^{k-5}d + d^{k-6}d^2 + \cdots + d^{k-4}) + \cdots + d^{k-2})$$
$$= |x - c| \sum |a_k| ((k - 1)d^{k-2} + (k - 2)d^{k-2} + (k - 3)d^{k-2} + \cdots + d^{k-2})$$
$$= |x - c| \sum |a_k| \frac{k(k - 1)}{2} d^{k-2} \quad \text{(by Example 1.6.1)}$$
$$= |x - c| D$$
$$< \delta D$$
$$< \epsilon.$$

This proves that g is continuous at c. $\qquad\square$

Theorem 9.4.2. *Let* $f(x) = \sum_{k=0}^{\infty} a_k x^k$ *have radius of convergence* R. *Then* f *is differentiable on* $B(0, R)$ *and* $f'(x) = \sum_{k=0}^{\infty} k a_k x^{k-1} = \sum_{k=1}^{\infty} k a_k x^{k-1}$. *The radius of convergence of* f' *equals* R.

Proof. Let $c, x \in B(0, R)$. Then

$$\frac{f(x) - f(c)}{x - c} = \frac{\sum_k a_k x^k - \sum_k a_k c^k}{x - c}$$
$$= \frac{\sum_k a_k (x^k - c^k)}{x - c}$$
$$= \frac{\sum_k a_k (x - c)(x^{k-1} + x^{k-2}c + x^{k-3}c^2 + \cdots + c^{k-1})}{x - c}$$
$$\text{(by Exercise 1.6.21)}$$
$$= \sum_k a_k (x^{k-1} + x^{k-2}c + x^{k-3}c^2 + \cdots + c^{k-1}),$$

which is the function g from the previous theorem. In that theorem we proved that g is continuous at c, so that

$$f'(c) = \lim_{x \to c} \frac{f(x) - f(c)}{x - c} = \lim_{x \to c} g(x) = g(c) = \sum_{k=1}^{\infty} k a_k c^{k-1}.$$

Then the theorem follows from Theorem 9.3.5. $\qquad\square$

Theorem 9.4.3. *Suppose that the power series* $\sum_{k=0}^{\infty} a_k x^k$ *has radius of convergence* R *and that the power series* $\sum_{k=0}^{\infty} b_k x^k$ *has radius of convergence* S. *Suppose that for some positive* $r \le R, S$ *the two power series take on the same values at all* $z \in B_r(0)$. *Then for all* $n \ge 0$, $a_n = b_n$.

In other words, if for all $x \in B_r(0)$, $\sum_{k=0}^{\infty} a_k x^k = \sum_{k=0}^{\infty} b_k x^k$, then the two series are identical.

Proof. Let $f(x) = \sum_{k=0}^{\infty} a_k x^k$ and $g(x) = \sum_{k=0}^{\infty} b_k x^k$. By assumption $a_0 = f(0) = g(0) = b_0$. Let $n \in \mathbb{N}^+$. By Theorem 9.4.2, f and g are differentiable on $B_r(0)$, and since they are identical, their nth derivatives agree on $B_r(0)$ as well. By an n-fold application of the derivative,

$$f^{(n)}(x) = \sum_{k=n}^{\infty} a_k k(k-1)(k-2) \cdots (k-n+1) x^{k-n},$$

and similarly for $g^{(n)}$. But then

$$a_n n(n-1)(n-2) \cdots (n-n+1) = f^{(n)}(0)$$
$$= g^{(n)}(0)$$
$$= b_n n(n-1)(n-2) \cdots (n-n+1),$$

so that $a_n = b_n$. $\qquad\square$

Exercises for Section 9.4

9.4.1. Consider the function $f(x) = \frac{1}{1-x}$ and its corresponding geometric series on $B(0,1)$.

   i) Compute a derivative of f as a rational function and as a power series.

   ii) Compute an antiderivative of f as a rational function and as a power series.

9.4.2. Let $f(x) = \sum_{k=0}^{\infty} \frac{x^{2k+1}}{(2k+1)k!}$.

   i) Find the radius of convergence of the series.

   ii) Compute the derivative of f.

   iii) By Example 9.7.5 we know that for every real x, $e^x = \sum_{k=0}^{\infty} \frac{x^k}{k!}$. Find the series for $e^{(x^2)}$.

   iv) Find a power series whose derivative is $e^{(x^2)}$. (Comment: a^{b^c} might stand for $(a^b)^c$ or for $a^{(b^c)}$. But the first form is simply a^{bc}, so we would not write a^{b^c} to stand for that. So, it is standard notation to write a^{b^c} when we mean $a^{(b^c)}$, In particular, $e^{(x^2)}$ can be written more plainly as e^{x^2}.)

   v) Address the no-closed-form discussion on page 292.

†**9.4.3.** (Invoked in Theorem 9.7.3.) Let $f(x) = \sum_{k=0}^{\infty} a_k x^k$ be a power series. Prove that for all $n \in \mathbb{N}_0$,

$$f^{(n)}(x) = \sum_{k=n}^{\infty} a_k k(k-1)(k-2) \cdots (k-n+1) x^{k-n}$$

$$= \sum_{k=n}^{\infty} a_k \frac{k!}{(k-n)!} x^{k-n}.$$

9.5   Numerical evaluations of some series

Differentiation of power series is a powerful tool. For all complex numbers $x \in B(0,1)$ the geometric series $\sum_{k=0}^{\infty} x^k$ converges to $\frac{1}{1-x}$. Certainly it is easier to compute $\frac{1}{1-x}$ than the infinite sum. We can exploit geometric series and derivatives of power series to compute many other infinite sums. Below we provide a few illustrations of the method.

Example 9.5.1. $\sum_{k=1}^{\infty} \frac{k}{2^{k-1}} = 4$.

Proof. Let $f(x) = \sum_{k=0}^{\infty} x^k$. This is the geometric series with radius of convergence 1 that converges to $\frac{1}{1-x}$ (Example 9.3.2). By Theorem 9.4.2, $f'(x) = \sum_{k=0}^{\infty} kx^{k-1} = \sum_{k=1}^{\infty} kx^{k-1}$, and by Theorem 9.3.5, the radius of convergence of f' is also 1. Thus $\frac{1}{2}$ is in the domain of f'. Since we have two ways of expressing f (as power series and as a rational function), there is similarly also a second form for f':

$$f'(x) = \left(\frac{1}{1-x} \right)' = \frac{1}{(1-x)^2}.$$

From the two forms we deduce that $\sum_{k=0}^{\infty} \frac{k}{2^{k-1}} = \sum_{k=0}^{\infty} k \left(\frac{1}{2} \right)^{k-1} = f'(1/2) = \frac{1}{(1-1/2)^2} = 4$.   □

Example 9.5.2. $\sum_{k=0}^{\infty} \frac{k^2}{2^k} = 6$, and $\sum_{k=0}^{\infty} \frac{k^2}{2^{k-1}} = 12$.

Proof. As in the previous example we start with the geometric series $f(x) = \sum_{k=0}^{\infty} x^k$ that converges on $B(0,1)$. Its derivative $f'(x) = \sum_{k=0}^{\infty} kx^{k-1} = \frac{1}{(1-x)^2}$ also converges on $B(0,1)$. Then $xf'(x) = \sum_{k=0}^{\infty} kx^k$ and its derivative $(xf'(x))' = \sum_{k=0}^{\infty} k^2 x^{k-1}$ also converge on $B(0,1)$. From

$$\sum_{k=0}^{\infty} k^2 x^{k-1} = (xf'(x))' = \left(\frac{x}{(1-x)^2}\right)' = \frac{1+x}{(1-x)^3}$$

we deduce that $\sum_{k=0}^{\infty} \frac{k^2}{2^{k-1}} = \left(\frac{1+\frac{1}{2}}{(1-1/2)^3}\right) = \frac{3/2}{1/8} = 12$, and $\sum_{k=0}^{\infty} \frac{k^2}{2^k} = 6.$ $\square$

Example 9.5.3. From the previous example we know that $\sum_{k=0}^{\infty} k^2 x^{k-1} = \frac{1+x}{(1-x)^3}$. Multiplying both sides by x gives $\sum_{k=0}^{\infty} k^2 x^k = \frac{x+x^2}{(1-x)^3}$, and differentiation gives

$$\sum_{k=0}^{\infty} k^3 x^{k-1} = \frac{d}{dx}\left(\frac{x+x^2}{(1-x)^3}\right) = \frac{1+4x+x^2}{(1-x)^4}.$$

It follows that $\sum_{k=0}^{\infty} \frac{k^3}{2^{k-1}} = \frac{1+4\cdot0.5+0.5^2}{(1-0.5)^4} = 52$ and $\sum_{k=0}^{\infty} \frac{k^3}{2^k} = 26$.

Summary: $\sum_{k=0}^{\infty} \frac{1}{2^{k-1}} = 2$, $\sum_{k=0}^{\infty} \frac{k}{2^{k-1}} = 4$, $\sum_{k=0}^{\infty} \frac{k^2}{2^{k-1}} = 12$, $\sum_{k=0}^{\infty} \frac{k^3}{2^{k-1}} = 26$. Is it possible to predict $\sum_{k=0}^{\infty} \frac{k^4}{2^{k-1}}$?

Example 9.5.4. $\sum_{k=1}^{\infty} \frac{1}{k2^k} = \ln 2$.

Proof. Let $f(x) = \sum_{k=1}^{\infty} \frac{x^k}{k}$. The radius of convergence of f is 1, so $1/2$ is in its domain. Also, $f'(x) = \sum_{k=1}^{\infty} x^{k-1} = \frac{1}{1-x}$, so that $f(x) = -\ln(1-x)+C$ for some constant C. In particular, $C = -0 + C = -\ln(1-0)+C = f(0) = 0$, so that $\sum_{k=1}^{\infty} \frac{1}{k2^k} = \sum_{k=1}^{\infty} \frac{1}{k}\left(\frac{1}{2}\right)^k = f(\frac{1}{2}) = -\ln(1-\frac{1}{2}) = -\ln(\frac{1}{2}) = \ln(2)$. $\square$

Exercises for Section 9.5

9.5.1. Compute and justify $\sum_{k=1}^{\infty} \frac{1}{k3^k}$.

9.5.2. Consider $\sum_{k=1}^{\infty} \frac{k}{k+2\cdot3^k}$. According to my computer the partial sum of the first 1000 terms is a rational number whose numerator and denominator take several screen pages, so this sum as a rational number is hard to comprehend. So instead I computed curtailed decimal expansions for this and for a few other sums: according to my computer, the partial sum of the first 10 terms is about 0.719474635555091, the partial sum of the first 100 terms is about 0.719487336054311, the partial sum of the first 1000 terms is about 0.719487336054311. What can you suspect? How would you go about proving it?

9.6   Some technical aspects of power series

The first theorem in this section is about products of power series and the second is about the convergence of a power series at the points on the boundary of the circle of the radius of convergence. This section is meant as a reference and should be skipped in a first class on power series.

Theorem 9.6.1. Let $\sum_{k=0}^{\infty} a_k x^k$ and $\sum_{k=0}^{\infty} b_k x^k$ be convergent power series with radii of convergence R_1 and R_2, respectively. Let $R = \min\{R_1, R_2\}$. Then on $B(0, R)$ the product sequence $\{(\sum_{k=0}^{n} a_k x^k) \cdot (\sum_{k=0}^{n} b_k x^k)\}_n$ converges to the power series $\sum_{k=0}^{\infty} (\sum_{j=0}^{k} a_j b_{k-j}) x^k$.

We write this as $(\sum_{k=0}^{\infty} a_k x^k) \cdot (\sum_{k=0}^{\infty} b_k x^k) = \sum_{k=0}^{\infty} (\sum_{j=0}^{k} a_j b_{k-j}) x^k$ on $B(0, R)$.

Proof. If $R = 0$, there is nothing to prove. So we may assume that R is a positive real number or it is ∞.

Fix $x \in B(0, R)$. Set $L = \sum_{k=0}^{\infty} a_k x^k$ and $K = \sum_{k=0}^{\infty} b_k x^k$. These limits exist because $|x| < R_1, R_2$. Also set $s_n = \sum_{k=0}^{n} a_k x^k$, $t_n = \sum_{k=0}^{n} b_k x^k$, and $c_n = \sum_{j=0}^{n} a_j b_{n-j}$. By the theorem on the convergence of products of sequences (Theorem 8.4.3), $\lim(s_n t_n) = LK$. Thus we need to prove that $\sum_{k=0}^{\infty} c_k x^k = LK$.

Let $\epsilon > 0$. By convergence of products, there exists $N_1 > 0$ such that for all integers $n > N_1$, $|s_n t_n - LK| < \epsilon/2$. By expanding (repeated "foiling") and by grouping,

$$s_n t_n = \left(\sum_{k=0}^{n} a_k x^k \right) \left(\sum_{j=0}^{n} b_j x^j \right)$$

$$= \sum_{k=0}^{n} \sum_{m=0}^{k} a_m b_{k-m} x^k + \sum_{k=1}^{n} \sum_{m=k}^{n} a_m b_{n+k-m} x^{n+k}$$

$$= \sum_{k=0}^{n} c_k x^k + \sum_{k=1}^{n} \sum_{m=k}^{n} a_m b_{n+k-m} x^{n+k}.$$

Since $|x| < R \leq R_1, R_2$, the series $\sum_{k=0}^{\infty} |a_k x^k|$ and $\sum_{k=0}^{\infty} |b_k x^k|$ converge, to some real numbers $\widetilde{L}, \widetilde{K}$, respectively. Thus there exists $N_2 > 0$ such that for all integers $n > N_2$,

$$\sum_{k=n+1}^{\infty} |a_k x^k| = \left| \sum_{k=0}^{\infty} |a_k x^k| - \sum_{k=0}^{n} |a_k x^k| \right| < \frac{\epsilon}{4\widetilde{K}+1},$$

and similarly there exists $N_3 > 0$ such that for all integers $n > N_3$, $\sum_{k=n+1}^{\infty} |b_k x^k| < \frac{\epsilon}{4\widetilde{L}+1}$. Now let $n > \max\{2N_1, N_2, N_3\}$. Then

$$\sum_{k=1}^{n} \sum_{m=k}^{n} |a_m b_{n+k-m} x^{n+k}|$$

$$= \sum_{k=1}^{n} \sum_{m=k}^{\lfloor n/2 \rfloor} |a_m x^m b_{n+k-m} x^{n+k-m}|$$

$$+ \sum_{k=1}^{n} \sum_{m=\max\{k,\lfloor n/2 \rfloor+1\}}^{n} |a_m x^m b_{n+k-m} x^{n+k-m}|$$

$$\leq \left(\sum_{m=1}^{\lfloor n/2 \rfloor} |a_m x^m| \right) \left(\sum_{k=\lfloor n/2 \rfloor}^{n} |b_k x^k| \right)$$

$$+ \left(\sum_{m=\lfloor n/2 \rfloor}^{n} |a_m x^m| \right) \left(\sum_{k=1}^{\lfloor n/2 \rfloor} |b_k x^k| \right),$$

(this product expands to more non-negative terms)

$$\leq \widetilde{L} \frac{\epsilon}{4\widetilde{L}+1} + \frac{\epsilon}{4\widetilde{K}+1} \widetilde{K}$$

$$< \frac{\epsilon}{2}.$$

Thus

$$\left| \sum_{k=0}^{n} c_k x^k - LK \right| = \left| \sum_{k=0}^{n} c_k x^k - s_n t_n + s_n t_n - LK \right|$$

$$\leq \left| \sum_{k=0}^{n} c_k x^k - s_n t_n \right| + |s_n t_n - LK|$$

$$< \left| \sum_{k=1}^{n} \sum_{m=k}^{n} a_m b_{n+k-m} x^{n+k} \right| + \epsilon/2$$

$$< \epsilon/2 + \epsilon/2$$

$$= \epsilon. \qquad \square$$

Theorem 9.6.2. Let $f(x) = \sum_{k=0}^{\infty} c_k x^k$ have radius of convergence a positive real number R. Let $a \in \mathbb{C}$ with $|a| = R$ such that $\sum_{k=0}^{\infty} c_k a^k$ converges. Let B be an open ball centered at a, and let $g : B \to \mathbb{C}$ be continuous. If $f(x) = g(x)$ for all $x \in B(0, R) \cap B$, then $f(a) = g(a)$.

Proof. Let $\epsilon > 0$. We want to show that $|f(a) - g(a)| < \epsilon$, which via Theorem 2.10.4 then proves that $f(a) = g(a)$. It suffices to prove the inequality $|f(a) - g(a)| < \epsilon$ under the additional assumption that $\epsilon < 1$.

Since $\sum_{k=0}^{\infty} c_k a^k$ converges, by Cauchy's criterion Theorem 9.2.1, there exists a positive integer N such that for all integers $n \geq N$, $|\sum_{k=N}^{n} c_k a^k| < \epsilon/4$. Let $s_m = \sum_{k=N}^{N+m} c_k a^k$. By assumption, for all $m \geq 1$, $|s_m| < \epsilon/4 < 1$. Furthermore, $c_N = s_0/a^N$, and for $n > N$, $c_{N+n} = (s_n - s_{n-1})/a^{N+n}$.

Let r be a real number in the interval $(0, 1)$. By rewriting and by the triangle inequality then

$$
\left| \sum_{k=N}^{N+n} c_k (ra)^k \right| = \left| r^N c_N a^N + c_{N+1} a^{N+1} r^{N+1} + \cdots + c_{N+n} a^{N+n} r^{N+n} \right|
$$

$$
= \left| r^N s_0 + (s_1 - s_0) r^{N+1} + (s_2 - s_1) r^{N+2} + \cdots + (s_n - s_{n-1}) r^{N+n} \right|
$$

$$
= \left| s_0 (r^N - r^{N+1}) + s_1 (r^{N+1} - r^{N+2}) + \cdots \right.
$$

$$
\left. + s_{n-1} (r^{N+n-1} - r^{N+n}) + s_n r^{N+n} \right|
$$

$$
= r^N \left| (1 - r) \left(s_0 + s_1 r + \cdots + s_{n-1} r^{n-1} \right) + s_n r^n \right|
$$

$$
\leq r^N \left((1 - r) \left(|s_0| + |s_1| r + \cdots + |s_{n-1}| r^{n-1} \right) + |s_n| r^n \right)
$$

$$
\leq \frac{\epsilon}{4} r^N \left((1 - r) \left(1 + r + \cdots + r^{n-1} \right) + r^n \right)
$$

$$
\text{(since } |s_m| < \epsilon/4 \text{ for all these } m\text{)}
$$

$$
= \frac{\epsilon}{4} r^N (1 - r^n + r^n) \quad \text{(by Example 1.6.4)}
$$

$$
= \frac{\epsilon}{4} r^N
$$

$$
< \frac{\epsilon}{4}.
$$

Since polynomial functions are continuous, there exists $\delta_1 > 0$ such that for all $x \in B(a, \delta_1)$, $\left| \sum_{k=0}^{N-1} c_k x^k - \sum_{k=0}^{N-1} c_k a^k \right| < \epsilon/4$.

Since g is continuous at a, there exists $\delta_2 > 0$ such that for all $x \in B \cap B(a, \delta_2)$, $|g(x) - g(a)| < \epsilon/4$. Let $r \in (0, 1)$ such that $1 - r < \frac{1}{R}\min\{\delta_1, \delta_2\}$. Then $|a - ra| < \delta_1, \delta_2$, so that

$|g(a) - f(a)|$

$$= \left| g(a) - g(ra) + g(ra) - f(ra) + f(ra) - \sum_{k=0}^{N-1} c_k a^k + \sum_{k=0}^{N-1} c_k a^k - f(a) \right|$$

$$\leq |g(a) - g(ra)| + |g(ra) - f(ra)| + \left| f(ra) - \sum_{k=0}^{N-1} c_k a^k \right| + \left| \sum_{k=0}^{N-1} c_k a^k - f(a) \right|$$

$$= |g(a) - g(ra)| + 0 + \left| \sum_{k=N}^{\infty} c_k (ra)^k + \sum_{k=0}^{N-1} c_k (ra)^k - \sum_{k=0}^{N-1} c_k a^k \right| + \left| \sum_{k=N}^{\infty} c_k a^k \right|$$

$$< \frac{\epsilon}{4} + \left| \sum_{k=N}^{\infty} c_k (ra)^k \right| + \left| \sum_{k=0}^{N-1} c_k (ra)^k - \sum_{k=0}^{N-1} c_k a^k \right| + \left| \sum_{k=N}^{\infty} c_k a^k \right|$$

$$< \frac{\epsilon}{4} + \frac{\epsilon}{4} + \frac{\epsilon}{4} + \frac{\epsilon}{4}$$

$$= \epsilon.$$

Since ϵ is arbitrary, by Theorem 2.10.4, $f(a) = g(a)$. □

Exercises for Section 9.6

9.6.1. Expand $\left(\sum_{k=0}^{\infty} x^k\right)^2$ as a power series, i.e., with proof determine all the coefficients of the series.

9.6.2. Let $\sum_{k=0}^{\infty} a_k x^k$, $\sum_{k=0}^{\infty} b_k x^k$ be power series and R a positive real number such that for all $x \in B(0, R)$, $\left(\sum_{k=0}^{\infty} a_k x^k\right) \cdot \left(\sum_{k=0}^{\infty} b_k x^k\right) = 1$. In other words, at each x in $B(0, R)$, the two infinite series are multiplicative inverses of each other.

   i) Prove that for all $x \in B(0, R)$, $\sum_{k=0}^{\infty} a_k x^k \neq 0$.

   ii) Prove that $a_0 \neq 0$.

   iii) Prove that $b_0 = \frac{1}{a_0}$, and that for all $k > 0$, $b_k = -\frac{1}{a_0}\sum_{j=1}^{k} a_j b_{k-j}$. (Hint: Theorem 9.6.1.)

   iv) Suppose that there exists $M \in \mathbb{R}$ such that for all non-negative integers n, $|a_n x^n| < M$. Prove that for all integers $k \geq 1$, $|b_k x^k| \leq \frac{M}{|a_0|^2}\left(\frac{M + |a_0|}{|a_0|}\right)^{k-1}$. (Hint: induction.)

9.6.3. (Abel's lemma†) Suppose that $a_0, a_1, a_2, \ldots$ are complex numbers and that $\sum_k a_k$ converges. Let $f(x) = \sum_k a_k x^k$. The goal of this exercise is to prove that f is defined on $B(0,1) \cup \{1\}$, that it is continuous on $B(0,1)$, and that when the domain is restricted to $(B(0,1) \cup \{1\}) \cap \mathbb{R}$, then f is continuous also at 1.

i) Prove that the domain of f includes $B(0,1) \cup \{1\}$.

ii) Prove that f is continuous on $B(0,1)$. (Hint: Invoke a theorem.) Set $L = \sum_k a_k$, and for $n \geq 0$, set $s_n = a_0 + a_1 + \cdots + a_n - L$.

iii) Prove that $\sum_{k=0}^n s_k(1-x)x^k = s_n(1 - x^{n+1}) + \sum_{k=0}^n a_k(x^k - 1)$. (Hint: Summation by parts, see Exercise 9.2.11.)

iv) Prove that for every $\epsilon > 0$ there exists $N \in \mathbb{N}$ such that for all integers $n \geq N$, $|s_n| < \epsilon$.

v) Prove that the sequence $\{\sum_{k=0}^n s_k(1-x)x^k\}_n$ converges for all $x \in B(0,1) \cup \{1\}$.

vi) Prove that $f(x) - L = \sum_{k=0}^\infty s_k(1-x)x^k$ for all $x \in B(0,1) \cup \{1\}$.

vii) Let $\epsilon > 0$. Prove that there exists $N \in \mathbb{R}$ such that for all integers $m \geq N$ and for all real numbers x in the interval $[0,1]$, $\left|\sum_{k=m}^\infty s_k(1-x)x^k\right| < \epsilon$.

viii) Let $\epsilon > 0$. Prove that for every positive integer n there exists $\delta > 0$ such that for all complex numbers $x \in (B(0,1) \cup \{1\}) \cap B(1,\delta)$, $\left|\sum_{k=0}^n s_k(1-x)x^k\right| < \epsilon$.

ix) Let $\epsilon > 0$. Prove that there exists $\delta > 0$ such that for every real number x in the interval $[\max\{0, 1-\delta\}, 1]$, $\left|\sum_{k=0}^\infty s_k(1-x)x^k\right| < \epsilon$.

x) Let $g : [0,1] \to \mathbb{C}$ be defined by $g(x) = f(x)$. Prove that g is continuous.

*9.6.4. Let $f(x) = \sum_{k=0}^\infty a_k x^k$ and $g(x) = \sum_{k=0}^\infty b_k x^k$.

i) Express $(f \circ g)(x)$ as a power series in terms of the a_i and b_j.

ii) What is special for the power series of the composition if $a_0 = 0$?

iii) Assuming that $a_0 = 0$, and given the radii of convergence for f and g, what would it take to find the radius of convergence of the composition series?

---

† "Lemma" means a "helpful theorem", possibly not interesting in its own right, but useful later. There are examples of so-called lemmas that have turned out to be very interesting in their own right.

9.7 Taylor series

A common way of generating power series comes from approximations of functions.

Definition 9.7.1. *Let a be in the domain of a function f, and assume that f has derivatives of all orders at a. The* **Taylor series of f** *(centered) at a is the series $\sum_{k=0}^{\infty} \frac{f^{(k)}(a)}{k!}(x-a)^k$.*

Remark 9.7.2. If $a = 0$, the Taylor series is a power series (as defined in this chapter), and for other a this is also a power series but of a more general kind which can nevertheless be easily transformed into a usual power series in the following sense. Let $f : A \to \mathbb{C}$. Set $B = \{x - a : x \in A\}$, and $g : B \to \mathbb{C}$ as $g(x) = f(x + a)$. By straightforward calculus, $\sum_k a_k(x-a)^k$ is a Taylor series for f at a if and only if $\sum_k a_k x^k$ is a Taylor series for g at 0. Furthermore, the radius of convergence of the Taylor series for g is R if and only if for all $x \in \mathbb{C}$, the Taylor series $\sum_k a_k(x-a)^k$ for f at a converges at x whenever $|x-a| < R$ and diverges at x whenever $|x-a| > R$.

Thus any analysis of Taylor series can by a function domain shift be transformed into a Taylor series that is a power series.

Section 6.5 contains many examples of Taylor polynomials and a few more are computed in this section. For some functions this computation is easier than others. The following theorem covers a trivial computation, and it applies in particular to all polynomial functions.

Theorem 9.7.3. *Let $f(x) = \sum_{k=0}^{\infty} a_k x^k$ be a power series with nonzero radius of convergence. Then the nth Taylor polynomial of f centered at 0 is*

$$\sum_{k=0}^{n} a_k x^k,$$

and the Taylor series of f centered at 0 is

$$\sum_{k=0}^{\infty} a_k x^k.$$

Proof. By Exercise 9.4.3, $\frac{f^{(m)}(x)}{m!}$ equals $\sum_{k=m}^{\infty} a_k \frac{k!}{m!(k-m)!} x^{k-m}$. Thus $\frac{f^{(m)}(0)}{m!}$ equals a_m. This proves the form of the Taylor series. $\square$

Remark 9.7.4. Theorems 6.5.5 and 6.5.6 say that some Taylor series are convergent power series at each x near 0 and that they converge to the value of the original function. This is certainly true for functions given as power series (such as in the theorem above). We examine a few further examples in this section with a few left for the exercises. Beware that the Taylor series need not converge to its function at any point other than at a, see Exercise 9.7.3.

Example 9.7.5. Let $f(x) = e^x$, where the domain of f is $\mathbb{R}$. It is easy to compute the Taylor series for f:

$$\sum_{k=0}^{\infty} \frac{x^k}{k!}.$$

By the ratio test or the root test, this series converges for all $x \in \mathbb{C}$, not only for $x \in \mathbb{R}$. (More on this series is in Sections 10.1 and 10.2, where we learn more about exponentiation by complex numbers.) Let $x \in \mathbb{R}$. By Theorem 6.5.5, for every positive integer n there exists d_n between 0 and x such that

$$e^x - \sum_{k=0}^{n} \frac{x^k}{k!} = \frac{e^{d_n}}{(n+1)!} x^{n+1}.$$

Thus

$$\left| e^x - \sum_{k=0}^{n} \frac{x^k}{k!} \right| = \left| \frac{e^{d_n}}{(n+1)!} x^{n+1} \right| \le \left| \frac{e^{|x|}}{(n+1)!} x^{n+1} \right|. \tag{9.7.6}$$

The sequence $\left\{ \frac{e^{|x|}}{(n+1)!} x^{n+1} \right\}$ converges to 0 by the ratio test for sequences. This proves that for each $x \in \mathbb{R}$, the Taylor polynomials for e^x approximate e^x arbitrarily closely, and that the Taylor series at each x equals e^x:

$$e^x = \sum_{k=0}^{\infty} \frac{x^k}{k!}.$$

In practice, this is precisely how one (a human or a computer) computes values e^x to arbitrary precision. For example, to compute $e^{0.1}$ to within 0.0001, we first need to find n such that the remainder as in Equation (9.7.6) is at most 0.0001. We first simplify an upper bound on the remainder to make the finding of n easier:

$$\left| \frac{e^{|0.1|}}{(n+1)!} 0.1^{n+1} \right| \le \frac{3}{(n+1)!} 0.1^{n+1}.$$

We can check that the smallest integer n for which the last expression is at most 0.0001 is $n = 3$, so that to within 0.0001, $e^{0.1}$ is $\sum_{k=0}^{3} \frac{0.1^k}{k!} = 6631/6000 \cong 1.10516666666667$. To compute $e^{-0.1}$, the same n works for the desired precision, giving $e^{-0.1}$ to be within 0.0001 equal to $\sum_{k=0}^{3} \frac{(-0.1)^k}{k!} = \frac{5429}{6000} \cong 0.904833333333333$. Incidentally, $\sum_{k=0}^{16} \frac{0.1^k}{k!} = \frac{210211444629316695849651076651}{190207180800000000000000000000} \cong 1.10517091807565$ and $\sum_{k=0}^{16} \frac{(-0.1)^k}{k!} = \frac{701174932606177917695472088 3}{774918144000000000000000000000} \cong 0.90483741803596$.

To compute e^{π} to within 0.0001, then with some simplication we need to find an integer n such that

$$\left| \frac{e^{\pi}}{(n+1)!} \pi^{n+1} \right| \le \frac{3^4}{(n+1)!} 4^{n+1} < 0.0001.$$

With elementary (but tedious) arithmetic we can compute that the inequality holds for $n = 19$ and bigger but not for $n = 18$. Thus we can compute e^{π} to 0.0001 of precision as $\sum_{k=0}^{19} \frac{\pi^k}{k!} = 23.1406926285462 \pm 0.0001$.

Example 9.7.7. Let $f(x) = \ln(x+1)$, where the domain of f is $(-1, \infty)$. It is straightforward to compute the Taylor series for f centered at 0:

$$\sum_{k=1}^{\infty} \frac{(-1)^{k-1} x^k}{k} = -\sum_{k=1}^{\infty} \frac{(-x)^k}{k}.$$

By the Ratio test for series, Theorem 9.2.3, the radius of convergence for this series is 1. It is worth noting that the domain of the function f is all real numbers strictly bigger than -1, whereas the computed Taylor series converges at all complex numbers in $B(0, 1)$ and diverges at all complex (including real) numbers whose absolute value is strictly bigger than 1. By Example 9.1.7, the series diverges at $x = -1$, and by Theorem 9.2.5, it converges at $x = 1$. Furthermore, by Exercise 9.2.14, the series converges at all complex numbers x with $|x| \le 1$ and $x \ne 1$. You should test and invoke Theorem 6.5.5 to show that for all $x \in (-1, 1)$, $\ln(x+1) = -\sum_{k=1}^{\infty} \frac{(-x)^k}{k}$.

Incidentally, since $\ln(x+1)$ is continuous on its domain and since its Taylor series converges at $x = 1$ by the Alternating series test (Theorem 9.2.5), it follows from Theorem 9.6.2 that

$$\sum_{k=1}^{\infty} \frac{(-1)^{k-1}}{k} = \ln 2.$$

With the help of Taylor series we can similarly get finite-term expressions for other infinite sums. Below is a harder example (and the reader may wish to skip it).

Example 9.7.8. Let $f(x) = \sqrt{1-x}$. The domain of f is the interval $(-\infty, 1]$. On the sub-interval $(-\infty, 1)$ the function has derivatives of all orders:

$$f'(x) = -\frac{1}{2}(1-x)^{-1/2},$$

$$f''(x) = -\frac{1}{2} \cdot \frac{1}{2}(1-x)^{-3/2},$$

$$f'''(x) = -\frac{1}{2} \cdot \frac{1}{2} \cdot \frac{3}{2}(1-x)^{-5/2}, \ldots,$$

$$f^{(n)}(x) = -\frac{1}{2} \cdot \frac{1}{2} \cdot \frac{3}{2} \cdot \ldots \cdot \frac{2n-3}{2}(1-x)^{-(2n-1)/2}.$$

Thus the Taylor series for f centered at 0 is

$$1 - \frac{1}{2}x - \sum_{k=2}^{\infty} \frac{1 \cdot 3 \cdot 5 \cdots (2k-3)}{k! 2^k} x^k.$$

For large n, the quotient of the $(n+1)$st coefficient divided by the nth coefficient equals $\frac{2n-1}{2(n+1)}$, whose limsup equals 1. Thus by the Ratio test for power series (Theorem 9.3.9), the Taylor series converges absolutely on $B(0,1)$, and in particular it converges absolutely on $(-1,1)$. Furthermore, the quotient $\frac{2n-1}{2(n+1)} = \frac{2(n+1)-3}{2(n+1)}$ above is at most $1 - \frac{4/3}{n}$ for all $n \geq 4$. Thus by Raabe's test (Exercise 9.2.17), this Taylor series converges at $x = 1$, and so it converges absolutely on $[-1, 1]$. But what does it converge to? Consider $x \in (-1, 1)$. We use the integral form of the Taylor's remainder theorem (Exercise 7.4.9):

$$T_{n,f,0}(x) - f(x) = \int_0^x \frac{(x-t)^n}{n!} \cdot \frac{1}{2} \cdot \frac{1}{2} \cdot \frac{3}{2} \cdots \frac{2n-1}{2}(1-t)^{-(2n+1)/2} \, dt$$

$$= \int_0^x \left(\frac{x-t}{1-t}\right)^n \frac{1 \cdot 3 \cdot 5 \cdots (2n-1)}{n! 2^{n+1} \sqrt{1-t}} \, dt.$$

As the integrand goes to 0 with n, and as $|x| < 1$, the integral goes to 0 with n, so that the Taylor series converges to f on $(-1, 1)$. (Incidentally, an application of Exercise 9.6.3 shows that the Taylor series is continuous on $[-1, 1]$, and as f is also continuous there, necessarily the Taylor series converges to f on $[-1, 1]$.)

Exercises for Section 9.7

9.7.1. (This exercise has overlap with Exercise 6.5.7.) Let $f(x) = \frac{1}{1-x}$.

  i) Compute $f^{(n)}(x)$ for all integers $n \geq 0$.

  ii) Compute and justify the Taylor series of f centered at 0.

  iii) Determine the radius of convergence.

  iv) What is the domain of f and what is the domain of its Taylor series?

  v) Compute the Taylor series for f centered at 2. Determine its radius of convergence.

  vi) Compute the Taylor series for f centered at 3. Determine its radius of convergence. Compare with the previous parts.

9.7.2. Let $f(x) = \ln(x + 1)$, where the domain of f is $(-1, \infty)$ as in Example 9.7.7. Compute $f(0.001)$ from one of its Taylor polynomials to four digits of precision. Show all work, and in particular invoke Theorem 6.5.5.

9.7.3. Let $f : \mathbb{R} \to \mathbb{R}$ be given by

$$f(x) = \begin{cases} e^{-1/x^2}, & \text{if } x \neq 0; \\ 0, & \text{if } x = 0 \end{cases}$$

as in Exercise 7.6.16.

  i) Compute and justify the Taylor series for f centered at 0.

  ii) Compute the radius of convergence for the series.

  iii) Discuss whether it possible to compute $f(0.001)$ to arbitrary precision from this series.

*9.7.4. In introductory analysis courses we typically handle infinite sums but not infinite products or even longer and longer products. In Example 9.7.8 we proved that the Taylor series for $f(x) = \sqrt{1-x}$ centered at 0 converges on $[-1, 1]$. This in particular means that the sequence $\{\frac{1 \cdot 3 \cdot 5 \cdots (2n-3)}{n! 2^n}\}_n$ converges to 0. Can you prove this directly without involving Raabe's test?

Chapter 10

Exponential and trigonometric functions

The culmination of the chapter and of the course are the exponential and trigonometric functions with their properties. Section 10.4 covers more varied examples of L'Hôpital's rule than what was possible in Section 6.4.

10.1   The exponential function

Define the power series

$$E(x) = \sum_{k=0}^{\infty} \frac{x^k}{k!}.$$

By Examples 9.3.10 (3), the domain of E is $\mathbb{C}$. Thus by Section 9.4, E is differentiable everywhere.

We prove below that this function is the exponential function: the base is e and the exponent is allowed to be any complex number x not just any real number.

Remarks 10.1.1.
(1) $E'(x) = \sum_{k=1}^{\infty} \frac{kx^{k-1}}{k!} = \sum_{k=1}^{\infty} \frac{x^{k-1}}{(k-1)!} = \sum_{k=0}^{\infty} \frac{x^k}{k!} = E(x)$.
(2) $E(0) = 1$.
(3) Let $a \in \mathbb{C}$. Define $g : \mathbb{C} \to \mathbb{C}$ by $g(x) = E(x + a) \cdot E(-x)$. Then g is a product of two differentiable functions, hence differentiable, and

$$g'(x) = E'(x + a) \cdot E(-x) + E(x + a) \cdot E'(-x) \cdot (-1)$$
$$= E(x + a) \cdot E(-x) - E(x + a) \cdot E(-x)$$
$$= 0,$$

so that g is a constant function. This constant has to equal

$$g(0) = E(0 + a) \cdot E(-0) = E(a).$$

Thus for all a and x in $\mathbb{C}$, $g(x) = g(0)$, i.e., that $E(x+a)E(-x) = E(a)$.

(4) Thus for all $c, d \in \mathbb{C}$, $E(c+d) = E(c)E(d)$ (set $c = x+a, d = -x$).

(5) By induction on n and by the previous part, it follows that for all positive integers n and all $a \in \mathbb{C}$, $E(na) = (E(a))^n$.

(6) Part (3) applied to $a = 0$ gives that $1 = E(0) = E(x)E(-x)$. We conclude that $E(x)$ is never 0, and that $E(-x) = \frac{1}{E(x)}$.

(7) By parts (5) and (6), $E(na) = (E(a))^n$ for all integers n and all $a \in \mathbb{C}$.

(8) Let $a, b \in \mathbb{R}$. By part (4,) $E(a + bi) = E(a) \cdot E(bi)$. Thus to understand the function $E : \mathbb{C} \to \mathbb{C}$, it suffices to understand E restricted to real numbers and E restricted to i times real numbers. We accomplish the former in this section and the latter in the next section.

Theorem 10.1.2. *For all $x \in \mathbb{R}$, $E(x) = \ln^{-1}(x) = e^x$, the exponential function from Section 7.6. In particular, $E(1)$ equals the Euler's constant e.*

Proof. Define $f : \mathbb{R} \to \mathbb{R}$ by $f(x) = \frac{E(x)}{e^x}$. Then f is differentiable and

$$f'(x) = \frac{E'(x)e^x - E(x)(e^x)'}{(e^x)^2} = \frac{E(x)e^x - E(x)e^x}{(e^x)^2} = 0.$$

Thus f is a constant function, so that for all $x \in \mathbb{R}$, $\frac{E(x)}{e^x} = f(x) = f(0) = \frac{E(0)}{e^0} = \frac{1}{1} = 1$. Thus $E(x) = e^x$. By Theorem 7.6.7, this is the same as $\ln^{-1}(x)$. In particular, $E(1) = e^1 = e$. $\qquad\square$

The following is now immediate:

Theorem 10.1.3. *The function E restricted to $\mathbb{R}$ has the following properties:*

(1) *The range is all of $\mathbb{R}^+$.*

(2) *E restricted to $\mathbb{R}$ is invertible with $E^{-1}(x) = \ln(x)$.*

The sequence $\{\sum_{k=0}^n \frac{1}{k!}\}_n$ of partial sums for $E(1) = e^1 = e$ is convergent, and starts with $1, 2, 2.5, \frac{8}{3} \cong 2.66667, \frac{65}{24} \cong 2.70833, \frac{163}{60} \cong 2.71667, \frac{1957}{720} \cong 2.71806, \frac{685}{252} \cong 2.71825, \frac{109601}{40320} \cong 2.71828, \frac{98641}{36288} \cong 2.71828$. The last approximation is from the Taylor polynomial of degree 9 and it

is correct to 5 decimal places because by the Taylor remainder theorem
(Theorem 6.5.5) there exists $d \in (0,1)$ such that

$$|E(1) - T_{9,E,0}(1)| = \left| \frac{E^{(10)}(d)}{10!}(1-0)^{10} \right| = \frac{e^d}{10!} < \frac{3}{10!} < 8.3 \times 10^{-7}.$$

Exercises for Section 10.1

10.1.1. Use the definition of ln from Theorem 10.1.3 to prove that for all $x, y \in \mathbb{R}^+$, $\ln(xy) = \ln x + \ln y$, and that for all integers n, $\ln(x^n) = n \ln x$.

10.1.2. Use the definition of ln from Theorem 10.1.3 and derivatives of inverse functions (Theorem 6.2.7) to prove that $(\ln)'(x) = \frac{1}{x}$.

10.1.3. Prove that any function from $\mathbb{R}$ to $\mathbb{R}$ that equals its own derivative equals cE(restricted to $\mathbb{R}$) for some $c \in \mathbb{R}$. Similarly prove that $\{f : \mathbb{C} \to \mathbb{C}$ differentiable $\mid f' = f\} = \{cE : c \in \mathbb{C}\}$.

10.1.4. Find $a_0, a_1, a_2, \ldots \in \mathbb{R}$ such that the power series $\sum_{k=0}^{\infty} a_k x^k$ converges for all $x \in \mathbb{R}$ to e^{x^2}. (Hint: Use the function E to do this easily.) With that, determine a power series whose derivative is $e^{(x^2)}$. (This problem is related to Exercise 9.4.2; it turns out that there is no simpler, finite-term antiderivative of e^{x^2}.)

10.1.5. Determine a power series whose derivative is $2^{(x^2)}$. (It turns out that there is no simpler, finite-term antiderivative of this function — see the no-closed-form discussion on page 292).

10.1.6. Express $\sum_{k=0}^{\infty} \frac{(-1)^k}{4^k k!}$ in terms of e.

10.1.7. Express $\sum_{k=0}^{\infty} \frac{2^k}{3^k (k+1)!}$ in terms of e.

10.1.8. Compute a Taylor polynomial of degree n of the function E centered at 0. (Do as little work as possible, but do explain your reasoning.)

10.1.9. Numerically evaluate e^2 from the power series for $e^x = E(x)$ to 5 significant digits. Prove that you have achieved desired precision. (Hint: Theorem 6.5.5.)

10.1.10. Compute $\lim_{x \to a} E(x)$. (Do as little hard work as possible, but do explain your reasoning.)

10.1.11. With proof, for which real numbers α does the sequence $\{E(n\alpha)\}_n$ converge?

10.2   The exponential function, continued

In this section we restrict E from the previous section to the imaginary axis. Thus we look at $E(ix)$ where x varies over real numbers. Note that

$$E(ix) = \sum_{k=0}^{\infty} \frac{(ix)^k}{k!}$$

$$= 1 + \frac{ix}{1!} + \frac{(ix)^2}{2!} + \frac{(ix)^3}{3!} + \frac{(ix)^4}{4!} + \frac{(ix)^5}{5!} + \frac{(ix)^6}{6!} + \frac{(ix)^7}{7!} + \frac{(ix)^8}{8!} + \cdots$$

$$= 1 + i\frac{x}{1!} - \frac{x^2}{2!} - i\frac{x^3}{3!} + \frac{x^4}{4!} + i\frac{x^5}{5!} - \frac{x^6}{6!} - i\frac{x^7}{7!} + \frac{x^8}{8!} + \cdots.$$

We define two new functions (their names may be purely coincidental, but pronounce them for the time being as "cause" and "sin"):

$$\text{COS}\,(x) = \text{Re}\,E(ix) = 1 - \frac{x^2}{2!} + \frac{x^4}{4!} - \frac{x^6}{6!} + \frac{x^8}{8!} + \cdots,$$

$$\text{SIN}\,(x) = \text{Im}\,E(ix) = \frac{x}{1!} - \frac{x^3}{3!} + \frac{x^5}{5!} - \frac{x^7}{7!} + \frac{x^9}{9!} + \cdots.$$

Thus

$$E(ix) = \text{COS}\,(x) + i\text{SIN}\,(x).$$

Since $E(ix)$ converges for all x, so do its real and imaginary parts. This means that COS and SIN are defined for all x (even for all complex x; but in this section all x are real). Thus the radius of convergence for the power series COS and SIN is infinity.

We can alternatively practice the root test on these functions to determine their radii of convergence: the α for the function COS is $\limsup\left\{1, 0, \sqrt[2]{\left|\frac{1}{2!}\right|}, 0, \sqrt[4]{\left|\frac{1}{4!}\right|}, 0, \sqrt[6]{\left|\frac{1}{6!}\right|}, \ldots\right\}$. Since 0s do not contribute to limsup of a sequence of non-negative numbers, it follows that $\alpha = \limsup\left\{1, \sqrt[2]{\left|\frac{1}{2!}\right|}, \sqrt[4]{\left|\frac{1}{4!}\right|}, \sqrt[6]{\left|\frac{1}{6!}\right|}, \ldots\right\}$. By Example 8.3.6 this is 0. Thus the radius of convergence of COS is ∞. The proof for the infinite radius of convergence of SIN is similar.

Remarks 10.2.1.

(1) $E(ix) = \text{COS}\,(x) + i\,\text{SIN}\,(x)$. This writing is not a random rewriting of the summands in $E(ix)$, but it is the sum of the real and the imaginary parts.

(2) $\mathrm{COS}\,(0) = \mathrm{Re}\,E(i\cdot 0) = \mathrm{Re}\,1 = 1$, $\mathrm{SIN}\,(0) = \mathrm{Im}\,E(i\cdot 0) = \mathrm{Im}\,1 = 0$.

(3) By the powers appearing in the power series for the two functions, for all $x \in \mathbb{R}$,

$$\mathrm{COS}\,(-x) = \mathrm{COS}\,(x), \quad \mathrm{SIN}\,(-x) = -\mathrm{SIN}\,(x).$$

Thus $E(-ix) = \mathrm{COS}\,(x) - i\,\mathrm{SIN}\,(x)$, which is the complex conjugate of $E(ix)$.

(4) For all $x \in \mathbb{R}$,

$$(\mathrm{COS}\,(x))^2 + (\mathrm{SIN}\,(x))^2$$
$$= (\mathrm{COS}\,(x) + i\,\mathrm{SIN}\,(x)) \cdot (\mathrm{COS}\,(x) - i\,\mathrm{SIN}\,(x))$$
$$= (\mathrm{COS}\,(x) + i\,\mathrm{SIN}\,(x)) \cdot (\mathrm{COS}\,(-x) + i\,\mathrm{SIN}\,(-x))$$
$$= E(ix)E(-ix)$$
$$= E(ix - ix)$$
$$= E(0)$$
$$= 1.$$

Thus $|E(ix)| = 1$ for all $x \in \mathbb{R}$.

(5) We conclude that for all $x \in \mathbb{R}$, the real and imaginary parts of $E(ix)$ have absolute value at most 1. In other words,

$$-1 \le \mathrm{COS}\,(x), \ \mathrm{SIN}\,(x) \le 1.$$

(6) Since E is differentiable,

$$(E(ix))' = E'(ix)\,i = E(ix)\,i$$
$$= (\mathrm{COS}\,(x) + i\,\mathrm{SIN}\,(x))\,i = i\,\mathrm{COS}\,(x) - \mathrm{SIN}\,(x).$$

It follows by Theorem 6.2.6 that

$$(\mathrm{COS}\,(x))' = (\mathrm{Re}(E(ix)))' = \mathrm{Re}((E(ix))') = -\mathrm{SIN}\,(x),$$

and

$$(\mathrm{SIN}\,(x))' = (\mathrm{Im}(E(ix)))' = \mathrm{Im}((E(ix))') = \mathrm{COS}\,(x).$$

Theorem 10.2.2. *There exists a unique real number $s \in (0, \sqrt{3})$ such that $E(is) = i$. In other words, there exists a unique real number $s \in (0, \sqrt{3})$ such that $\mathrm{COS}\,(s) = 0$ and $\mathrm{SIN}\,(s) = 1$.*

Proof. The function $t - \text{SIN}(t)$ is differentiable and its derivative is $1 - \text{COS}(t)$, which is always non-negative. Thus $t - \text{SIN}(t)$ is non-decreasing for all real t, so that for all $t \geq 0$, $t - \text{SIN}(t) \geq 0 - \text{SIN}(0) = 0$. Hence the function $\frac{t^2}{2} + \text{COS}(t)$ has a non-negative derivative on $[0, \infty)$, so that $\frac{t^2}{2} + \text{COS}(t)$ is non-decreasing on $[0, \infty)$. Thus for all $t \geq 0$, $\frac{t^2}{2} + \text{COS}(t) \geq \frac{0^2}{2} + \text{COS}(0) = 1$. It follows that the function $\frac{t^3}{6} - t + \text{SIN}(t)$ is non-decreasing on $[0, \infty)$. [HOW LONG WILL WE KEEP GOING LIKE THIS?] Thus for all $t \geq 0$, $\frac{t^3}{6} - t + \text{SIN}(t) \geq \frac{0^3}{6} - 0 + \text{SIN}(0) = 0$. Thus $\frac{t^4}{24} - \frac{t^2}{2} - \text{COS}(t)$ is non-decreasing on $[0, \infty)$, so that for all $t \geq 0$, $\frac{t^4}{24} - \frac{t^2}{2} - \text{COS}(t) \geq \frac{0^4}{24} - \frac{0^2}{2} - \text{COS}(0) = -1$. We conclude that for all $t \geq 0$,

$$\text{COS}(t) \leq \frac{t^4}{24} - \frac{t^2}{2} + 1.$$

In particular, $\text{COS}(\sqrt{3}) \leq \frac{\sqrt{3}^4}{24} - \frac{\sqrt{3}^2}{2} + 1 = -\frac{1}{8} < 0$. We also know that $\text{COS}(0) = 1 > 0$. Since COS is differentiable, it is continuous, so by the Intermediate value theorem (Theorem 5.3.1) there exists $s \in (0, \sqrt{3})$ such that $\text{COS}(s) = 0$.

Now suppose that there exists a different $u \in (0, \sqrt{3})$ such that $\text{COS}(u) = 0$. By possibly switching the names of s and u we may assume that $s < u$. Note that since s and u are both positive, $s^2 < u^2$. Since $\frac{t^4}{24} - \frac{t^2}{2} - \text{COS}(t)$ is non-decreasing on $[0, \infty)$, it follows that $\frac{u^4}{24} - \frac{u^2}{2} - \text{COS}(u) \geq \frac{s^4}{24} - \frac{s^2}{2} - \text{COS}(s)$. Since $\text{COS}(u) = \text{COS}(s) = 0$, this says that $\frac{u^4}{24} - \frac{u^2}{2} \geq \frac{s^4}{24} - \frac{s^2}{2}$. In other words,

$$\frac{1}{4} = \frac{1}{24}(\sqrt{3}^2 + \sqrt{3}^2) \geq \frac{1}{24}(u^2 + s^2) = \frac{1}{24}(u^2 + s^2)\frac{u^2 - s^2}{u^2 - s^2} = \frac{1}{24}\frac{u^4 - s^4}{u^2 - s^2} = \frac{1}{2},$$

which is a contradiction. Thus s is unique.

Since $(\text{COS}(s))^2 + (\text{SIN}(s))^2 = 1$, we have that $\text{SIN}(s) = \pm 1$. Since COS is positive on $(0, \sqrt{3})$, this means that SIN is increasing on $(0, \sqrt{3})$, and so by continuity of SIN, $\text{SIN}(s)$ must be positive, and hence $\text{SIN}(s) = 1$. This proves that $E(is) = i$.   □

Remark 10.2.3. The proof above establishes the following properties for all $t \geq 0$:

$$\text{COS}(t) \leq 1, \quad \text{COS}(t) \geq 1 - \frac{t^2}{2}, \quad \text{COS}(t) \leq 1 - \frac{t^2}{2} + \frac{t^4}{24},$$

$$\text{SIN}(t) \leq t, \quad \text{SIN}(t) \geq t - \frac{t^3}{6}.$$

Observe that the polynomials above alternately over- and under-estimating COS are the Taylor polynomials of COS (by Theorem 9.7.3) and are converging to the Taylor series for COS. Similarly, the polynomials above alternately over- and under-estimating SIN are the Taylor polynomials of SIN converging to the Taylor series for SIN.

Remark 10.2.4. (About the numerical value of s.) What role does $\sqrt{3}$ play in Theorem 10.2.2? We deduced that $s < \sqrt{3}$ from knowing that $COS(t) \leq 1 - t^2/2 + t^4/4!$ at all $t > 0$, from knowing that the latter polynomial evaluates to a negative number at $\sqrt{3}$, and from an application of the Intermediate value theorem. You probably suspect that $s = \pi/2$. One can check that $\pi/2 < \sqrt{3}$, and even that $\pi/2 < 0.91 \cdot \sqrt{3}$. However, $1 - t^2/2 + t^4/4!$ at $0.91 \cdot \sqrt{3}$ is positive, so we cannot conclude that $s < 0.91 \cdot \sqrt{3}$ from **this** argument. But if in the proof of Theorem 10.2.2 we compute a few more upper and lower bounding polynomials of SIN and COS, we obtain that $COS(t) \leq 1 - t^2/2 + t^4/4! - t^6/6! + t^8/8!$ for all $t > 0$, and this latter polynomial is negative at $0.91 \cdot \sqrt{3}$, which would then by the Intermediate value theorem guarantee that $s < 0.91 \cdot \sqrt{3}$. By taking higher and higher degree polynomials as in the proof we could get tighter and tighter upper bounds on s. We can also get lower bounds on s. From $COS(t) \geq 1 - t^2/2$ (as in the proof) we deduce that $COS(t)$ is positive at $0.999999 \cdot \sqrt{2}$ (with an arbitrary finite number of digits 9). Thus the unique s must be in $[\sqrt{2}, 0.91 \cdot \sqrt{3})$. More steps in the proof also show that from $COS(t) \geq 1 - t^2/2 + t^4/4! - t^6/6!$ for all $t > 0$, and this polynomial is negative at $0.9 \cdot \sqrt{3}$. Thus by the Intermediate value theorem,

$$0.9 \cdot \sqrt{3} < s < 0.91 \cdot \sqrt{3}.$$

Incidentally, my computer gives me 1.55884572681199 for $0.9 \cdot \sqrt{3}$ and 1.57616623488768 for $0.91 \cdot \sqrt{3}$. Thus we know the in-between s to a few digits of precision. (Incidentally, my computer gives me 1.5707963267949 for $\pi/2$.) By taking higher-degree polynomials in the more-step version of the proof of Theorem 10.2.2 we can get even more digits of precision of s. However, this numerical approach does not prove that s is equal to $\pi/2$; we need different reasoning to prove that $s = \pi/2$ (see Theorem 10.2.5).

Finally, we connect COS and SIN to trigonometric functions. First we need to specify the trigonometric functions: For any real number t,

$$\cos(t) + i\sin(t)$$

is the unique complex number on the unit circle centered at the origin that is on the half-ray from the origin at angle t radians measured counterclockwise from the positive x-axis. In terms of ratio geometry (see page 10), this says that in a right triangle with one angle t radians, $\cos(t)$ is the ratio of the length of the adjacent edge divided by the length of the hypotenuse, and $\sin(t)$ is the ratio of the length of the opposite edge divided by the length of the hypotenuse. Geometrically it is clear, and we do assume this fact, that cos and sin are continuous functions. In general continuity does not imply differentiability, but for these two functions we use their continuity in the proof of their differentiability.

Theorem 10.2.5. *The trigonometric functions* cos *and* sin *are differentiable. Furthermore,* COS *and* SIN *are the functions* cos *and* sin, $s = \pi/2$, *and for any real number* x, $E(ix)$ *is the point on the unit circle centered at* 0 *at angle* x *radians measured counterclockwise from the positive real axis.*

Proof. We know that $E(ix)$ is a point on the unit circle with coordinates $(\text{COS}(x), \text{SIN}(x))$. What we do not yet know is whether the angle of this point counterclockwise from the positive real axis equals x radians.

Let s be as in Theorem 10.2.2. The angle of $E(is) = 0 + i \cdot 1 = i$ measured in radians counterclockwise from the positive real axis is $\pi/2$. Uniqueness of s guarantees that COS is positive on $[0, s)$. Thus SIN is increasing on $[0, s]$, and it increases from 0 to 1. Thus for all $t \in (0, s)$, $E(it)$ is in the first quadrant.

Let n be an integer strictly greater than 1. Let α be the angle of $E(is/n)$ measured counterclockwise from the positive real axis. For every integer j, $(E(ijs/n)) = E(is)^j$, so by Theorem 3.12.2, the angle of $(E(ijs/n))$ is $j\alpha$, and in particular the angle $n\alpha$ coincides with the angle of $E(is) = E(ins/n)$, i.e., with $\frac{\pi}{2} + 2\pi k$ for all integers k. Thus $\alpha = (\frac{\pi}{2} + 2\pi k)/n$ for some integer k. For all $j = 1, \ldots, n-1$, $j/n \in (0, 1)$, and so $E(ijs/n)$ is in the first quadrant by the previous paragraph. Thus

$$0 < j\alpha = j\frac{\pi}{2n}(1 + 4k) \le \frac{\pi}{2}.$$

The first inequality says that k is not negative, and the second inequality used with $j = n - 1$ says that k is not positive. Thus $k = 0$ and $\alpha = \frac{\pi}{2n}$.

Together with Theorem 3.12.2 we just established that for all positive integers n and all integers j, the angle of $E(isj/n) = (E(is/n))^j$ is $j\alpha = \frac{\pi}{2} \cdot \frac{j}{n}$. In other words, for all rational numbers r,

$$\mathrm{COS}\,(sr) + i\,\mathrm{SIN}\,(sr) = E(isr) = \cos\left(\frac{\pi}{2}r\right) + i\sin\left(\frac{\pi}{2}r\right).$$

By continuity of the functions E, cos and sin we conclude that for any real number x,

$$\mathrm{COS}\,(sx) + i\,\mathrm{SIN}\,(sx) = E(isx) = \cos\left(\frac{\pi}{2}x\right) + i\sin\left(\frac{\pi}{2}x\right).$$

By matching the real and the imaginary parts in this equation we get that for all $x \in \mathbb{R}$,

$$\cos(x) = \mathrm{COS}\left(\frac{2s}{\pi}x\right), \quad \sin(x) = \mathrm{SIN}\left(\frac{2s}{\pi}x\right),$$

and so cos and sin are also differentiable functions. By the chain rule,

$$\cos'(x) = -\frac{2s}{\pi}\mathrm{SIN}\left(\frac{2s}{\pi}x\right) = -\frac{2s}{\pi}\sin(x),$$

$$\sin'(x) = \frac{2s}{\pi}\mathrm{COS}\left(\frac{2s}{\pi}x\right) = \frac{2s}{\pi}\cos(x).$$

By geometry (see Exercise 1.1.19), for small positive real x, $0 < \cos(x) < \frac{x}{\sin x} < \frac{1}{\cos x}$. Since cos is differentiable, it is continuous. Since $\cos(0) = \mathrm{COS}\,(0) = 1$, by the Squeeze theorem (Theorem 4.4.11), $\lim_{x \to 0^+} \frac{x}{\sin x} = 1$. Hence

$$\frac{2s}{\pi} = \frac{2s}{\pi}\cos(0) = \sin'(0) = \lim_{x \to 0}\frac{\sin x - \sin 0}{x} = \lim_{x \to 0^+}\frac{\sin x}{x} = 1,$$

whence $s = \pi/2$. This proves that $\cos = \mathrm{COS}$ and $\sin = \mathrm{SIN}$. $\square$

Theorem 10.2.6. *Every complex number x can be written in the form $rE(i\theta)$, where $r = |x|$ is the length and θ the angle of x counterclockwise from the positive real axis.*

Proof. Let $r = |x|$. Thus x lies on the circle centered at 0 and of radius r. If $r = 0$, then $x = 0$, and the angle is irrelevant. If instead r is non-zero, it is necessarily positive, and x/r is a complex number of length $|x|/r = 1$.

By Theorem 3.12.1, x/r and x have the same angle. Let θ be that angle. Then $x/r = E(i\theta)$, so that $x = rE(i\theta)$. $\square$

Notation 10.2.7. It is common to write $E(x) = e^x$ for any complex number x. We have seen that equality does hold if x is real, but we adopt this notation also for other numbers. With this, if $x, y \in \mathbb{R}$, then

$$e^{x+iy} = e^x e^{iy},$$

and e^x is the length and y is the angle of e^{x+iy} counterclockwise from the positive x axis.

Exercises for Section 10.2

10.2.1. Let α and r be real numbers with $r > 0$. Using the findings from this section prove that for any complex number x, the complex number $re^{i\alpha} \cdot x$ is obtained from x by rotating x counterclockwise around the origin by angle α radians and by stretching the result by a factor of r. (Compare with Theorem 3.12.1.)

10.2.2. Prove that $e^{i\pi} + 1 = 0$. (This has been called one of the most beautiful equalities, because it connects the fundamental numbers 0, 1, e and π.)

10.2.3. Let $x, y \in \mathbb{R}$. Expand both sides of $E(i(x + y)) = E(ix)E(iy)$ in terms of sin and cos to prove that $\cos(x+y) = \cos(x)\cos(y) - \sin(x)\sin(y)$, and $\sin(x + y) = \sin(x)\cos(y) + \cos(x)\sin(y)$. (Cf. Exercise 1.1.16.)

10.2.4. Express the following complex numbers in the form e^{x+iy}: i, $-i$, 1, -1, e, $2 + 2i$. Can 0 be expressed in this way? Justify.

10.2.5. Express $(3 + 4i)^6$ in the form e^{a+bi} for some $a, b \in \mathbb{R}$. (Do not give numeric approximations of a and b.)

10.2.6. Prove that for any integer m, the sequence $\{E(2\pi imn)\}_n$ converges. Prove that the sequence $\{E(\sqrt{2}\pi in)\}_n$ does not converge. Determine all real numbers α for which the sequence $\{E(in\alpha)\}_n$ converges. Prove your conclusion.

10.2.7. For which real numbers α and β does the sequence $\{E(n(\alpha+i\beta))\}_n$ converge? Prove your conclusion.

10.2.8. Let $f(x) = \sin(x^2)$ or $f(x) = \cos(x^2)$.

i) Determine the Taylor series for f centered at 0. (This should not be hard; do not compute derivatives of all orders.)

ii) Determine a power series whose derivative is f. (There is no simpler or finite-term antiderivative of f.)

iii) Determine the radius of convergence of the antiderivative power series.

iv) Use the Taylor remainder theorem (Theorem 6.5.5) to compute $\int_0^1 f$ to within five digits of precision.

*10.2.9. Theorem 10.2.2 proves the existence of a special number s for which Theorem 10.2.5 proves that it equals $\pi/2$. The goal of this exercise is to give another proof that $s = \pi/2$. Let $f : [0, 1] \to \mathbb{R}$ be given by $f(x) = \sqrt{1 - x^2}$. Then the graph of f is the part of the circle of radius 1 that is in the first quadrant.

i) Use lengths of curves and that 2π is the perimeter of the circle of radius 1 to justify that

$$\frac{\pi}{2} = \int_0^1 \sqrt{1 + \left(\frac{-2x}{2\sqrt{1 - x^2}} \right)^2} \, dx.$$

(Caution: This is an improper integral.)

ii) Compute the integral via substitution $x = \mathrm{COS}\,(u)$. The integral should evaluate to s from Theorem 10.2.2.

10.2.10. Prove **de Moivre's formula**: for all $n \in \mathbb{Z}$ and all real numbers x,

$$(\cos(x) + i\sin(x))^n = \cos(nx) + i\sin(nx).$$

10.2.11. Prove that for every real number x that is not an integer multiple of 2π, the series

$$\sum_{k=1}^\infty \frac{\cos(kx)}{k} \quad \text{and} \quad \sum_{k=1}^\infty \frac{\sin(kx)}{k}$$

converge. (Hint: The complex series $\sum_{k=1}^\infty \frac{\cos(kx)+i\sin(kx)}{k}$ converges if and only if its real and imaginary parts converge. Use De Moivre's formula (Exercise 10.2.10) and Exercise 9.2.12.)

10.3   Trigonometry

In this section we review the semester's concepts of limits, continuity, differentiability, and integrability on the newly established functions from the previous section. We use notation is at the end of the last section.

Definition 10.3.1. (Trigonometric functions) *By Theorem 10.2.5, for any real number x, e^{ix} is the complex number that is on the unit circle at angle x radians counterclockwise from the positive horizontal axis.*
  (1) $\sin(x)$ *is the imaginary part of e^{ix}.*
  (2) $\cos(x)$ *is the real part of e^{ix}.*
  (3) $\tan(x) = \frac{\sin(x)}{\cos(x)}$, $\cot(x) = \frac{\cos(x)}{\sin(x)}$, $\sec(x) = \frac{1}{\cos(x)}$, $\csc(x) = \frac{1}{\sin(x)}$.

Remarks 10.3.2. The following is straightforward from Section 10.2.
  (1) The domain of sin and cos is $\mathbb{R}$.
  (2) sin and cos are differentiable and hence continuous functions.
  (3) The Taylor series for sin is

$$\sum_{k=0}^{\infty}(-1)^k \frac{x^{2k+1}}{(2k+1)!}$$

and it converges to $\sin(x)$ for all $x \in \mathbb{R}$. Similarly, the Taylor series for cos is

$$\cos(x) = \sum_{k=0}^{\infty}(-1)^k \frac{x^{2k}}{(2k)!}.$$

  (4) For all $x \in \mathbb{R}$, $\sin(x + 2\pi) = \sin(x)$ and $\cos(x + 2\pi) = \cos(x)$. This is not obvious from the power series definition of $E(ix)$, but if follows from Theorem 10.2.6.
  (5) $\sin(x) = \frac{e^{ix} - e^{-ix}}{2i}$.
  (6) $\cos(x) = \frac{e^{ix} + e^{-ix}}{2}$.
  (7) $(\sin(x))^2 + (\cos(x))^2 = 1$. (Recall from Remark 2.4.8 that for trigonometric functions we also write this as $\sin^2(x) + \cos^2(x) = 1$, but for an arbitrary function f, $f^2(x)$ refers to $f(f(x))$ rather than to $(f(x))^2$.)
  (8) Dividing the equality in the previous part by $(\cos(x))^2$ yields the equality $(\tan(x))^2 + 1 = (\sec(x))^2$.
  (9) $(\cot(x))^2 + 1 = (\csc(x))^2$.
  (10) $\sin'(x) = \cos(x)$.

(11) $\cos'(x) = -\sin(x)$.

(12) For all $x \in \mathbb{R}$, $\sin(x) = -\sin(-x)$ and $\cos(x) = \cos(-x)$.

(13) sin and cos take on non-negative values on $[0, \pi/2]$. By the previous part, cos takes on non-negative values on $[-\pi/2, \pi/2]$.

(14) sin is increasing on $[-\pi/2, \pi/2]$ since its derivative cos is non-negative there and positive on $(-\pi/2, \pi/2)$.

(15) sin, when restricted to $[-\pi/2, \pi/2]$, has an inverse by Theorem 2.9.4. The inverse is called **arcsin**. The domain of arcsin is $[-1, 1]$.

(16) Geometrically, sin takes on non-negative values on $[0, \pi]$ and positive values on $(0, \pi)$, so that cos, when restricted to $[0, \pi]$, has an inverse, called **arccos**. The domain of arccos is $[-1, 1]$.

(17) By the quotient rule for differentiation,
$$\tan'(x) = (\sec(x))^2, \qquad \cot'(x) = -(\csc(x))^2,$$
$$\sec'(x) = \sec(x)\tan(x), \quad \csc'(x) = -\csc(x)\cot(x).$$

(18) The derivative of tan is always non-negative, so that on $(-\pi/2, \pi/2)$, tan is invertible. The inverse of the function tan is called **arctan**, and the domain of arctan is $(-\infty, \infty)$.

(19) cot restricted to $(0, \pi)$ is strictly decreasing and invertible; its inverse function is called **arccot**, and the domain of arccot is $(-\infty, \infty)$.

Theorem 10.3.3. *Refer to Theorem 6.2.7.*

(1) For $x \in (-\pi/2, \pi/2)$, $\arcsin'(x) = \frac{1}{\sqrt{1-x^2}}$.

(2) For $x \in (0, \pi)$, $\arccos'(x) = \frac{-1}{\sqrt{1-x^2}}$.

(3) $\arctan'(x) = \frac{1}{1+x^2}$.

Proof. We prove the first and the third parts; the proof of the second part is similar to the proof of the first.

$$\arcsin'(x) = \frac{1}{\sin'(\arcsin(x))} \quad \text{(by Theorem 6.2.7)}$$
$$= \frac{1}{\cos(\arcsin(x))}$$
$$= \frac{1}{\sqrt{(\cos(\arcsin(x)))^2}}$$
$$\text{(because cos is non-negative on } [-\pi/2, \pi/2])$$

$$= \frac{1}{\sqrt{1 - (\sin(\arcsin(x)))^2}}$$

$$= \frac{1}{\sqrt{1 - x^2}},$$

$$\arctan'(x) = \frac{1}{\tan'(\arctan(x))} \quad \text{(by Theorem 6.2.7)}$$

$$= \frac{1}{(\sec(\arctan(x)))^2}$$

$$= \frac{1}{1 + (\tan(\arctan(x)))^2}$$

$$= \frac{1}{1 + x^2}. \qquad \qquad \square$$

Exercises for Section 10.3

10.3.1. Prove that there exists a real number x such that $\cos x = x$. (Hint: Intermediate value theorem Theorem 5.3.1.)

10.3.2. Prove that for all positive x, $x > \sin(x)$. (Hint: prove that $f(x) = x - \sin(x)$ is an increasing function.)

10.3.3. Fix a constant c and let $f : \mathbb{R} \to \mathbb{R}$ be defined as

$$f(x) = \begin{cases} \sin(\frac{1}{x}), & \text{if } x \neq 0; \\ c, & \text{if } x = 0. \end{cases}$$

Prove that f is not continuous at 0 no matter what c is.

10.3.4. Let $f : \mathbb{R} \to \mathbb{R}$ be defined as

$$f(x) = \begin{cases} x \sin(\frac{1}{x}), & \text{if } x \neq 0; \\ 0, & \text{if } x = 0. \end{cases}$$

  i) Prove that f is differentiable at all non-zero x.

  ii) Prove that f is continuous but not differentiable. (Hint: not differentiable at 0.)

10.3.5. Let $f : \mathbb{R} \to \mathbb{R}$ be defined as

$$f(x) = \begin{cases} x^2 \sin(\frac{1}{x}), & \text{if } x \neq 0; \\ 0, & \text{if } x = 0. \end{cases}$$

  i) Compute f'.

  ii) Prove that f' is not continuous (at 0).

10.3.6. Let $f : \mathbb{R} \to \mathbb{R}$ be defined as

$$f(x) = \begin{cases} x^3 \sin(\frac{1}{x}), & \text{if } x \neq 0; \\ 0, & \text{if } x = 0. \end{cases}$$

i) Compute f'.

ii) Prove that f' is continuous but not differentiable.

10.3.7. Let $f : \mathbb{R} \to \mathbb{R}$ be defined as

$$f(x) = \begin{cases} x^4 \sin(\frac{1}{x}), & \text{if } x \neq 0; \\ 0, & \text{if } x = 0. \end{cases}$$

i) Compute f' and f''.

ii) Is f'' continuous or differentiable? Prove your claims.

10.3.8. Let $f : [0, \pi/2] \to \mathbb{R}$ be defined by $f(x) = (\pi/2) \sin(x) - x$.

i) Prove that there is exactly one $c \in [0, \pi/2]$ such that $f'(c) = 0$.

ii) Prove that $(\pi/2) \sin(x) \geq x$ for all $x \in [0, \pi/2]$. (Compare with Exercise 10.3.2.)

10.3.9. Prove that for all $x \in \mathbb{R}$,

$$(\sin(x))^2 = \frac{1 - \cos(2x)}{2}, \quad (\cos(x))^2 = \frac{1 + \cos(2x)}{2}.$$

(Hint: Exercise 10.2.3.)

10.3.10. Let f be sin or cos.

i) Compute the Taylor polynomial $T_{10,f,0}$ of f of degree 10 centered at 0.

ii) Compute $T_{10,f,0}(0)$ and $T_{10,f,0}(1)$.

iii) Estimate the errors $|f(0) - T_{10,f,0}(0)|$ and $|f(1) - T_{10,f,0}(1)|$ with Theorem 6.5.5.

iv) In fact, calculators and computers use Taylor polynomials to compute values of trigonometric and also exponential functions. Compute $f(1)$ and $f(100)$ to within 0.000001 of their true value. What degree Taylor polynomial suffices for each?

10.3.11. Prove that $\tan^{-1}(x) + \tan^{-1}\left(\frac{1}{x}\right) = \frac{\pi}{2}$ for all $x \in \mathbb{R} \setminus \{0\}$. (Hint: take the derivative of the expression on the right and use a clever x.)

10.3.12. Use integration by substitution (Exercise 7.4.5).

  i) Determine antiderivatives of tan and cot.

  ii) Determine antiderivatives of sec and csc. (Hint: Use first the rewriting trick $\sec(x) = \frac{\sec(x)(\sec(x)+\tan(x))}{\sec(x)+\tan(x)}$.)

10.3.13. Compute $\int (\sin(x))^2 \, dx$ (or $\int (\cos(x))^2 \, dx$) twice, with two different methods:

  i) Use Exercise 10.3.9.

  ii) Use integration by parts twice (Exercise 7.4.6) and rewrite after the second usage.

10.3.14.

  i) Compute $\lim_{N\to\infty} \int_{-N}^{N} \sin(x) \, dx$.

  ii) Compute $\lim_{N\to\infty} \int_{-2N\pi}^{2N\pi+\pi/2} \sin(x) \, dx$.

  iii) Compute $\lim_{N\to\infty} \int_{-2N\pi+\pi/2}^{2N\pi} \sin(x) \, dx$.

  iv) Conclude that sin is not integrable over $(-\infty, \infty)$ in the sense of Exercise 7.4.10.

10.3.15. Let $k, j \in \mathbb{Z}$. Prove the following equalities (possibly by multiple integration by parts (Exercise 7.4.6)):

  i) $\int_{-\pi}^{\pi} \sin(kt) \cos(jt) \, dt = 0$.

  ii) $\int_{-\pi}^{\pi} \sin(kt) \sin(jt) \, dt = \begin{cases} 0, & \text{if } j \neq k \text{ or } jk = 0, \\ \pi, & \text{otherwise.} \end{cases}$

  iii) $\int_{-\pi}^{\pi} \cos(kt) \cos(jt) \, dt = \begin{cases} 0, & \text{if } j \neq k \text{ or } jk = 0, \\ \pi, & \text{otherwise.} \end{cases}$

10.3.16. Prove that for any non-zero integer k,

$$\int x \sin(kx) \, dx = -\frac{1}{k} x \cos(kx) + \frac{1}{k^2} \sin(kx) + C,$$

and that

$$\int x \cos(kx) \, dx = \frac{1}{k} x \sin(kx) + \frac{1}{k^2} \cos(kx) + C.$$

10.3.17. Let $t \in \mathbb{R}$ and $n \in \mathbb{N}^+$.

  i) Prove that $\sin((n + \frac{1}{2})t) = \sin((n - \frac{1}{2})t) + 2\cos(nt)\sin(\frac{1}{2}t)$. (Hint: Exercise 10.2.3, multiple times.)

  ii) Prove that $\sin((n + \frac{1}{2})t) = \sin(\frac{1}{2}t)\big(1 + 2\sum_{k=1}^{n} \cos(kt)\big)$.

  iii) Give a formula or formulas for simplifying $\sum_{k=1}^{n} \cos(kt)$.

10.3.18. (This is from the Reviews section edited by P. J. Campbell, page 159 in *Mathematics Magazine* **91** (2018).) Let P be a polynomial function and m a non-zero constant.

i) Prove that $P + \frac{P'}{m} + \frac{P''}{m^2} + \frac{P'''}{m^3} + \cdots$ is a polynomial function.

ii) Compute the derivative with respect to x of

$$-\frac{e^{-mx}}{m}\left(P(x) + \frac{P'(x)}{m} + \frac{P''(x)}{m^2} + \frac{P'''(x)}{m^3} + \cdots\right).$$

iii) Integrate $\int e^{-mx}P(x)dx$.

10.3.19. (Fourier analysis, I) Let $f : [-\pi, \pi] \to \mathbb{C}$ be a differentiable function (or possibly not quite differentiable, but that would take us too far). For $n \in \mathbb{N}_0$ define $a_n = \int_{-\pi}^{\pi} f(x)\cos(nx)\,dx$, and for $n \in \mathbb{N}^+$ define $b_n = \int_{-\pi}^{\pi} f(x)\sin(nx)\,dx$. The **Fourier series** of f is

$$\frac{a_0}{2} + \sum_{n=1}^{\infty}(a_n\cos(nx) + b_n\sin(nx)).$$

i) Compute the Fourier series for $f(x) = \sin(3x)$. (Hint: Exercise 10.3.15.)

ii) Compute the Fourier series for $f(x) = (\sin(x))^2$. (Hint: Exercises 10.3.9 and 10.3.15.)

iii) Compute a_0, a_1, a_2, b_1, b_2 for $f(x) = x$. (Hint: Exercise 10.3.16.)

10.3.20. (Fourier analysis, II) Let $f : [-\pi, \pi] \to \mathbb{C}$ be a differentiable function (or possibly not quite differentiable, but that would take us too far). For $n \in \mathbb{Z}$ define $a_n = \int_{-\pi}^{\pi} f(x)e^{inx}\,dx$. The **Fourier series** of f is

$$\sum_{n=-\infty}^{\infty} a_n e^{-inx}.$$

Compute the Fourier series in this sense for the functions in the previous exercise.

10.3.21. Use Exercise 10.3.18 to elaborate on the two types of Fourier series from the previous two exercises for any polynomial function. Note that Fourier series are also functions in the form of infinite sums but they are not power series.

*10.3.22. (This exercise is used in Exercise 10.5.4.) Use de Moivre's formula (see Exercise 10.2.10) to prove that for all $x \in \mathbb{R}$ and all positive integers n,

$$\cos(nx) + i\sin(nx) = (\sin(x))^n (\cot(x) + i)^n$$

$$= (\sin(x))^n \sum_{k=0}^{n} \binom{n}{k} i^k (\cot(x))^{n-k}.$$

If $n = 2m + 1$, this says that

$$\cos((2m+1)x) + i\sin((2m+1)x)$$

$$= (\sin(x))^{2m+1} \sum_{k=0}^{2m+1} \binom{2m+1}{k} i^k (\cot(x))^{2m+1-k}.$$

The imaginary part of this is:

$$\sin((2m+1)x) = (\sin(x))^{2m+1} \sum_{k=0}^{m} (-1)^k \binom{2m+1}{2k+1} (\cot(x))^{2m-2k}.$$

Let $P(x) = \sum_{k=0}^{m} (-1)^k \binom{2m+1}{2k+1} x^{m-k}$. Prove that P is a polynomial of degree m with leading coefficient $2m + 1$. Prove that for $k = 1, \ldots, m$, $(\cot(k\pi/(2m+1)))^2$ is a root of P. Prove that P has m distinct roots. Prove that

$$P(x) = (2m+1) \cdot \prod_{k=1}^{m} \left(x - \left(\cot\left(\frac{k\pi}{2m+1} \right) \right)^2 \right).$$

By equating the coefficient of x^{m-1} in the two forms of P prove that $\sum_{k=1}^{m} (\cot(k\pi/(2m+1)))^2 = \frac{m(2m-1)}{3}$.

10.4   Examples of L'Hôpital's rule

Several versions of L'Hôpital's rule were proved in Section 6.4. With an increased repertoire of functions we can now show more interesting examples, including counterexamples if some hypothesis is omitted. All work in this section is in the exercises.

Exercises for Section 10.4

10.4.1. The proof of Theorem 10.2.5 required knowing that $\lim\limits_{x\to 0} \frac{\sin(x)}{x} = 1$. Re-prove this with L'Hôpital's rule. Discuss why this is not an independent confirmation of the fact.

10.4.2. Prove that $\lim\limits_{x\to 0} \frac{1-\cos(x)}{x} = 0$ first by L'Hôpital's rule. Then justify all steps in the following proof:

$$\lim_{x\to 0} \frac{1-\cos(x)}{x} = \lim_{x\to 0} \frac{1-\cos(x)}{x} \cdot \frac{1+\cos(x)}{1+\cos(x)}$$

$$= \lim_{x\to 0} \frac{1-(\cos(x))^2}{x(1+\cos(x))}$$

$$= \lim_{x\to 0} \frac{(\sin(x))^2}{x(1+\cos(x))}$$

$$= \lim_{x\to 0} \frac{\sin(x)}{x} \cdot \frac{\sin(x)}{1+\cos(x)}$$

$$= 1 \cdot \frac{0}{1+1}$$

$$= 0.$$

The next two exercises are taken from R. J. Bumcrot's article "Some subtleties in L'Hôpital's rule" in *Century of Calculus. Part II. 1969–1991,* edited by T. M. Apostol, D. H. Mugler, D. R. Scott, A. Sterrett, Jr., and A. E. Watkins. Raymond W. Brink Selected Mathematical Papers. Mathematical Association of America, Washington, DC, 1992, pages 203–204.

10.4.3. Let $f(x) = x^2 \sin(x^{-1})$ and $g(x) = \sin x$.
   i) Prove that $\lim_{x\to 0} \frac{f(x)}{g(x)} = 0$.
   ii) Prove that $\lim_{x\to 0} \frac{f'(x)}{g'(x)}$ does not exist.
   iii) Why does this not contradict L'Hôpital's rule Theorem 6.4.2?

10.4.4. Let $f(x) = x \sin(x^{-1})$ and $g(x) = \sin x$.
   i) Compute $\lim_{x\to 0} f(x)$ and $\lim_{x\to 0} g(x)$.
   ii) Compute f' and g'.
   iii) Prove that neither $\lim_{x\to 0} \frac{f(x)}{g(x)}$ nor $\lim_{x\to 0} \frac{f'(x)}{g'(x)}$ exist.

*10.4.5. (This is from R. P. Boas's article "Counterexamples to L'Hôpital's rule", *American Mathematical Monthly* **93** (1986), 644–645.) Define $f(x) = \int_0^x \cos^2(t)\, dt$, $g(x) = \frac{f(x)}{2+\sin(x)}$.

i) Prove that $\lim_{x\to\infty} f(x) = \infty = \lim_{x\to\infty} g(x)$.

ii) Prove that $\lim_{x\to\infty} \dfrac{f(x)}{g(x)} \neq 0$.

iii) Compute $f'(x), g'(x)$.

iv) Write $f'(x) = f_1(x)\cdot\cos x$ and $g'(x) = g_1(x)\cdot\cos x$ for some functions f_1, g_1. Prove that $\lim_{x\to\infty} \dfrac{f_1(x)}{g_1(x)} = 0$.

v) Prove that $\lim_{x\to\infty} \dfrac{f'(x)}{g'(x)} \neq 0$.

vi) Why does this not contradict L'Hôpital's rule at infinity (proved in Exercise 6.4.4)?

*10.4.6. (This is from R. C. Buck's "Advanced Calculus", McGraw-Hill Book Company, Inc., 1956, page 48.) Let $f(x) = 2x + \sin(2x)$, $g(x) = (2x + \sin(2x))e^{\sin(x)}$, and $h(x) = e^{-\sin(x)}$.

i) Prove that $\lim_{x\to\infty} h(x)$ does not exist.

ii) Prove that $\lim_{x\to\infty} \frac{f(x)}{g(x)}$ does not exist.

iii) Verify that $f'(x) = 4\cos^2(x)$ and that $g'(x) = (4\cos(x) + 2x + \sin(2x))\cos(x)e^{\sin(x)}$.

iv) Prove that $\lim_{x\to\infty} \frac{4\cos(x)}{(4\cos(x)+2x+\sin(2x))e^{\sin(x)}} = 0$.

v) Why does L'Hôpital's rule not apply here?

*10.4.7. (Modification of the previous exercise.) Let $f(x) = \frac{2}{x} + \sin(\frac{2}{x})$, and $g(x) = (\frac{2}{x} + \sin(\frac{2}{x}))e^{\sin(\frac{1}{x})}$. Prove that $\lim_{x\to 0+} \frac{f(x)}{g(x)}$ does not exist but that $\lim_{x\to 0+} \frac{f'(x)}{g'(x)} = 0$. Why does L'Hôpital's rule not apply here?

10.5   Trigonometry for the computation of some series

The goal of this section is to handle a few more infinite series and power series. Namely, so far we have been able to add up few infinite series numerically, and here we compute for example $\sum_{k=1}^{\infty} \frac{1}{k^2}$ in two different ways. (In Example 9.1.9 we proved that the sum converges but we did not know what it converges to.) All work is in the exercises. As the title suggests, these uses are not covered in a standard first analysis course and it is fine to omit this section.

Exercises for Section 10.5

10.5.1. (This is taken from the Monthly Gem article by D. Velleman in the *American Mathematical Monthly* **123** (2016), page 77. The gem is a simplification of the article by Y. Matsuoka, An elementary proof of the formula $\sum_{k=1}^{\infty} k^{-2} = \pi^2/6$, in *American Mathematical Monthly* **68** (1961), 485–487.)

For every non-negative integer n define

$$I_n = \int_0^{\pi/2} (\cos(x))^{2n}\, dx \quad \text{and} \quad J_n = \int_0^{\pi/2} x^2(\cos(x))^{2n}\, dx.$$

i) Prove that $I_0 = \pi/2$ and $J_0 = \pi^3/24$.

ii) Prove that for all $n \geq 1$, $I_n = (2n-1)(I_{n-1}-I_n)$. (Hint: integration by parts (Exercise 7.4.6) and $dv = \cos(x)\, dx$.)

iii) Prove that $I_n = \frac{2n-1}{2n} \cdot I_{n-1}$.

iv) Use integration by parts twice, first with $dv = dx$ and then with $dv = 2xdx$, to prove that for all $n \geq 1$, $I_n = n(2n-1)J_{n-1} - 2n^2 I_n$.

v) Prove that for all $n \geq 1$, $\frac{1}{n^2} = 2\left(\frac{J_{n-1}}{I_{n-1}} - \frac{J_n}{I_n}\right)$.

vi) Prove that for all integers N,

$$\sum_{k=1}^{N} \frac{1}{k^2} = 2\left(\frac{J_0}{I_0} - \frac{J_N}{I_N}\right) = \frac{\pi^2}{6} - 2\frac{J_N}{I_N}.$$

vii) Use the inequality $x \leq (\pi/2)\sin(x)$ for $x \in [0, \pi/2]$ (see Exercise 10.3.8) to estimate

$$0 \leq J_N \leq \frac{\pi^2}{4} \int_0^{\pi/2} (\sin(x))^2(\cos(x))^{2N}\, dx$$

$$= \frac{\pi^2}{4}(I_N - I_{N+1})$$

$$= \frac{\pi^2}{4} \cdot \frac{1}{2N+2} I_N.$$

viii) Prove that $\sum_{k=1}^{\infty} \frac{1}{k^2} = \frac{\pi^2}{6}$.

10.5.2. Fill in the explanations and any missing steps in the two **double** improper integrals. (While integrating with respect to x, think of y as a constant, and while integrating with respect to y, think of x as a constant.)

$$\int_0^\infty \left(\int_0^\infty \frac{1}{1+y} \cdot \frac{1}{1+x^2 y} dx \right) dy = \int_0^\infty \frac{1}{1+y} \left(\int_0^\infty \frac{1}{1+x^2 y} dx \right) dy$$

$$= \int_0^\infty \frac{1}{1+y} \left(\lim_{N \to \infty} \int_0^N \frac{1}{1+x^2 y} dx \right) dy$$

$$= \int_0^\infty \frac{1}{1+y} \left(\lim_{N \to \infty} \frac{\arctan(\sqrt{y}x)}{\sqrt{y}} \Bigg|_{x=0}^N \right) dy$$

$$= \frac{\pi}{2} \int_0^\infty \frac{1}{\sqrt{y}(1+y)} \, dy$$

$$= \frac{\pi}{2} \int_0^\infty \frac{2u}{u(1+u^2)} \, du$$

$$\text{(by substitution } y = u^2)$$

$$= \frac{\pi^2}{2},$$

$$\int_0^\infty \left(\int_0^\infty \frac{1}{1+y} \cdot \frac{1}{1+x^2 y} dy \right) dx$$

$$= \int_0^\infty \left(\int_0^\infty \frac{1}{1-x^2} \left(\frac{1}{1+y} - \frac{x^2}{1+x^2 y} \right) dy \right) dx$$

$$= \int_0^\infty \frac{1}{1-x^2} \ln \left(\frac{1}{x^2} \right) dx$$

$$= 2 \int_0^\infty \frac{\ln x}{x^2 - 1} \, dx$$

$$= 2 \left(\int_0^1 \frac{\ln x}{x^2 - 1} \, dx + \int_1^\infty \frac{\ln x}{x^2 - 1} \, dx \right)$$

$$= 2 \left(\int_0^1 \frac{\ln x}{x^2 - 1} \, dx + \int_0^1 \frac{\ln u}{u^2 - 1} \, du \right)$$

$$= 4 \int_0^1 \frac{\ln x}{x^2 - 1} \, dx. \text{ (Stop here.)}$$

10.5.3. (This is from the article D. Ritelli, Another proof of $\zeta(2) = \frac{\pi^2}{6}$, *American Mathematical Monthly* **120** (2013), 642–645.) We proved in Section 9.4 that derivatives and (definite) integrals commute with infinite sums for power series. There are other cases where integrals commute with infinite sums, but the proofs in greater generality are harder. Accept that $\int_0^1 \frac{-\ln x}{1-x^2} dx = \int_0^1 (-\ln x) \sum_{k=0}^{\infty} x^{2k} dx = \sum_{k=0}^{\infty} \int_0^1 (-\ln x) x^{2k} dx$. Also accept that the two integrals in Exercise 10.5.2 are the same (order of integration matters sometimes). Use Exercises 7.6.7 and 9.2.1 to prove that $\sum_{k=1}^{\infty} \frac{1}{k^2} = \frac{\pi^2}{6}$.

10.5.4. (See also Exercises 10.5.1 and 10.5.3.) (This is from the article I. Papadimitriou, A simple proof of the formula $\sum_{k=1}^{\infty} k^{-2} = \pi^2/6$, *American Mathematical Monthly* **80** (1973), 424–425.) Let x be an angle measured in radians strictly between 0 and $\pi/2$. Draw the circular wedge with angle x on a circle of radius 1. The largest right triangle in this wedge whose hypotenuse is one of the wedge sides of length 1 has area $\frac{1}{2}\sin(x)\cos(x)$, and the smallest right triangle containing this wedge whose side is one of the wedge sides (of length 1) has area $\frac{1}{2}\tan(x)$.

   i) Prove that $\sin(x)\cos(x) < x < \tan(x)$, and that $(\cot(x))^2 < \frac{1}{x^2}$.

   ii) Use Exercise 10.3.2 to prove that $\frac{1}{x^2} < 1 + (\cot(x))^2$.

   iii) Prove that for all integers k, m with $1 \le k \le m$,

$$\sum_{k=1}^{m}\left(\cot\left(\frac{k\pi}{2m+1}\right)\right)^2 < \frac{(2m+1)^2}{\pi^2}\sum_{k=1}^{m}\frac{1}{k^2}$$

$$< m + \sum_{k=1}^{m}\left(\cot\left(\frac{k\pi}{2m+1}\right)\right)^2.$$

   iv) Use Exercise 10.3.22 to prove that $\frac{m(2m-1)}{3} < \frac{(2m+1)^2}{\pi^2}\sum_{k=1}^{m}\frac{1}{k^2} < m + \frac{m(2m-1)}{3}$.

   v) Prove that $\sum_{k=1}^{\infty}\frac{1}{k^2} = \frac{\pi^2}{6}$. (Hint: multiply the previous part by $\pi^2/4m^2$.)

Appendices

Appendix A

Advice on writing mathematics

Make your arguments succinct and straightforward to read

Write preliminary arguments for yourself on scratch paper: your first attempt may yield some dead ends which definitely should not be on the final write-up. In the final write-up, write succinctly and clearly; write what you mean and mean what you write; write with the goal of not being misunderstood. Use good English grammar, punctuate properly. And above all, use correct logical reasoning.

Do not allow yourself to turn in work that is half-thought out or that is produced in a hurry. Use your best writing, in correct logical order, with good spatial organization on paper, with only occasional crossing out of words or sections, on neat paper. Represent your reasoning and yourself well. Take pride in your good work.

Process is important

Perhaps the final answer to the question is 42. It is not sufficient to simply write "42", "The answer is 42," or similar, without the process that led to that answer. While it is extremely beneficial to have the intuition, the smarts, the mental calculating and reasoning capacity, the inspiration, or what-not, to conclude "42", a huge part of learning and understanding is to be able to explain clearly the reasoning that lead to your answer.

I encourage you to discuss the homework with others before, during or after completing it: the explanations back-and-forth make you a better thinker and expositor.

Write your solutions in your own words on your own, and for full disclosure write the names of all of your collaborators on the work that you turn in for credit. I do not take points off, but you should practice full honesty.

Sometimes you may want to consult a book or the internet. Again, on the work that you turn in disclose the help that you got from outside sources.

Keep in mind that the more you have to consult outside sources, the more fragile your stand-alone knowledge is, the less well you understand the material, and the less likely you are to be able to do satisfactory work on closed-book or limited-time projects.

Do not divide by 0

Never write "1/0", "0/0", "$0^2/0$", "$\infty/0$". (Erase from your mind that you ever saw this in print! It cannot exist.)

Sometimes division by zero creeps in in subtler ways. For example, to find solutions to $x^2 = 3x$, it is wrong to simply cancel the x on both sides to get only one solution $x = 3$. Yes, $x = 3$ is one of the solutions, but $x = 0$ is another one. Cancellation of x in $x^2 = 3x$ amounts to dividing by 0 in case the solution is $x = 0$.

Never plug numbers into a function that are not in the domain of the function

By design, the only numbers you can plug into a function are those that are in the domain of the function. What else is there to say?

I will say more, by way of examples. Never plug 0 into the function $f(x) = \frac{1}{x}$ (see previous admonition). Even never plug 0 into the function $f(x) = \frac{x}{x}$: the latter function is undefined at $x = 0$ and is constant 1 at all other x.

Never plug -1 into $\sqrt{}$ or into ln.

Do not plug $x = 0$ or $x = 1.12$ into the function f that is defined by
$$f(x) = \begin{cases} 3x - 4, & \text{if } x > 3; \\ 2x + 1, & \text{if } x < -1. \end{cases}$$

Order of writing is important

The meanings of "Everybody is loved by somebody" and "Somebody loves everybody" are very different. Another way of phrasing these two statements is as follows: "For every x there exists y such that y loves x," and "There exists y such that for every x, y loves x." In crisp symbols, without distracting "such that", "for", and commas, these are written as "$\forall x \; \exists y, y$ loves x" and "$\exists y \; \forall x, y$ loves x."

Conclusion: order of quantifiers matters.

The order also matters in implications; simply consider the truth values of "If $x > 2$ then $x > 0$" and "If $x > 0$ then $x > 2$."

The statement "if A then B" can be written also as "B if A", but in general it is better to avoid the latter usage. In particular, when writing a long proof, do not write out a very long B and only at the very end add that you were assuming A throughout; you could have lost the doubting-your-statements reader before the end came in sight.

Here is a fairly short example where "B if A" form is not as elegant. With the statement "$\frac{2}{x}$ is ... if $x \neq 0$" you might have gone over the abyss of dividing by 0 and no ifs can make you whole again. It is thus better to first obtain proper assurances, and write "if $x \neq 0$ then $\frac{2}{x}$ is ... ."

Write parentheses

$\cdot -$ is not a recognized binary operator. Do not write "$5 \cdot -2$"; instead write "$5 \cdot (-2)$".

$\lim\limits_{x \to -1} 4 - 3x = 4 - 3x$, whereas $\lim\limits_{x \to -1} (4 - 3x) = 7$.

"$\int 4 - 3x \, dx$" is terrible grammar; instead write "$\int (4 - 3x) \, dx$".

π does not equal 3.14159

If the answer to a problem is $\pi\sqrt{17}/59$, leave it at that. This is an exact number from which one can get an approximation to arbitrary precision, but from 0.21954437870195 one cannot recover further digits. Never write "$\pi\sqrt{17}/59 = .21954437870195$", but it is fine to write "$\pi\sqrt{17}/59 \cong .21954437870195$". Usually it is not necessary to write out the numerical approximation, but sometimes the approximation helps us get a sense of the size of the answer and to check our derivation with any intuition about the problem. The answer to how far a person can run in one minute certainly should not exceed a kilometer or a mile.

To prove an (in)equality, manipulate one side in steps to get to the other side

If you have to prove that $\sum_{k=1}^{n} k^2 = \frac{n(n+1)(2n+1)}{6}$ for $n = 1$, do **NOT** do the following:

$$\sum_{k=1}^{1} k^2 = \frac{1(1+1)(2 \cdot 1 + 1)}{6}$$

$$1^2 = \frac{1 \cdot 2 \cdot 3}{6}$$

$$1 = 1$$

ON THE LEFT!   WRITE AS   DO NOT

The reasoning above is wrong-headed because in the first line you are asserting the equality that you are expected to prove, and in subsequent lines you are simply repeating your assumptions more succinctly. If you add question marks over the three equal sums and a check mark on the last line, then you are at least acknowledging that you are not yet sure of the equality. However, even writing with question marks over equal signs is inelegant and long-winded. That kind of writing is what we do on scratch paper to get our bearings on how to tackle the problem. But a cleaned-up version of the proof would be better as follows:

$$\sum_{k=1}^{1} k^2 = 1^2 = 1 = \frac{1 \cdot 2 \cdot 3}{6} = \frac{1(1+1)(2 \cdot 1 + 1)}{6}.$$

Do you see how this is shorter and proves succinctly the desired equality by transitivity of equality, with each step on the way sure-footed?

Another reason why the three-line reasoning above is bad is because it can lead to the following nonsense:

$$1 \overset{?}{=} 0$$

add 1 to both sides: $2 \overset{?}{=} 1$

multiply both sides by 0: $0 \overset{?}{=} 0$ $\checkmark$

But we certainly cannot conclude that the first line "$1 = 0$" is correct.

Do not assume what is to be proved

If you have to prove that a is positive, do not assume that a is positive. (For a proof by contradiction suppose that $a \le 0$, do correct logical reasoning until you get a contradiction, from which you can conclude that $a \leq 0$ is impossible, so that a must have been positive. See page 11 for proofs by contradiction.)

Limit versus function value

Asking for $\lim_{x \to a} f(x)$ is in general different from asking for $f(a)$. (If the latter is always meant, would there be a point to developing the theory of limits?)

Do not start a sentence with a mathematical symbol

This admonition is in a sense an arbitrary stylistic point, but it helps avoid certain confusions, such as in "Let $\epsilon > 0$. x can be taken to be negative." Here one could read or confuse the part "0. x" as "$0 \cdot x = 0$", but then ϵ would have to be both positive and negative. Do not force a reader to have to do a double-take: write unambiguous and correct sentences.

Appendix B

What one should never forget

Logic

We should remember the basic truth tables, correct usage of "or" and of implications, how to justify/prove a statement, and how to negate a statement.

If A implies B and if A is true, we may conclude B.

If A implies B and if B is false, we may conclude that A is false.

If A implies B and if A is false, we may not conclude anything.

If A implies B and if B is true, we may not conclude anything.

Truth table:

P	Q	not P	P and Q	P or Q	P xor Q	$P \Rightarrow Q$	$P \Leftrightarrow Q$
T	T	F	T	T	F	T	T
T	F	F	F	T	T	F	F
F	T	T	F	T	T	T	F
F	F	T	F	F	F	T	T

Negation chart:

Statement	Negation
not P	P
P and Q	$(\text{not } P)$ or $(\text{not } Q)$
P or Q	$(\text{not } P)$ and $(\text{not } Q)$
$P \Rightarrow Q$	P and $(\text{not } Q)$
$P \Leftrightarrow Q$	$P \Leftrightarrow (\text{not } Q) = (\text{not } P) \Leftrightarrow Q$
For all x of the specified type, property P holds for x.	There exists x of the specified type such that P is false for x.
There exists x of the specified type such that property P holds for x.	For all x of the specified type, P is false for x.

Statement	How to prove it
P (via contradiction).	Suppose not P. Establish some nonsense that makes not P impossible so that P must hold.
P and Q.	Prove P. Prove Q.
P or Q.	Suppose that P is false. Then prove Q. Alternatively: Suppose that Q is false and then prove P. (It may even be the case that P is always true. Then simply prove P. Or simply prove Q.)
If P then Q.	Suppose that P is true. Then prove Q. Contrapositively: Suppose that Q is false. Prove that P is false.
$P \Leftrightarrow Q$.	Prove $P \Rightarrow Q$. Prove $Q \Rightarrow P$.
For all x of a specified type, property P holds for x.	Let x be arbitrary of the specified type. Prove that property P holds for x. (Possibly break up into a few subcases.)
There exists x of a specified type such that property P holds for x.	Find/construct an x of the specified type. Prove that property P holds for x. Alternatively, invoke a theorem guaranteeing that such x exists.
An element x of a specified type with property P is unique.	Suppose that x and x' are both of specified type and satisfy property P. Prove that $x = x'$. Alternatively, show that x is the only solution to an equation, or the only element on a list, or ... .
x with property P is unique.	Suppose that x and y have property P. Prove that $x = y$.

Mathematical induction

The goal is to prove a property for all integers $n \geq n_0$. First prove the **base case**, namely that the property holds for n_0. For the **inductive step**, assume that for some $n-1 \geq n_0$, the property holds for $n-1$ (alternatively, for $n_0, n_0 + 1, \ldots, n - 1$), and then prove the property for n.

The limit definition of derivative

$$f'(a) = \lim_{h \to 0} \frac{f(a + h) - f(a)}{h} = \lim_{x \to a} \frac{f(x) - f(a)}{x - a}.$$

The limit-partition definition of integrals

A **partition** of $[a, b]$ is a finite set $P = \{x_0, x_1, \ldots, x_n\}$ such that $x_0 = a < x_1 < x_2 < \cdots < x_{n-1} < x_n = b$. Let $f : [a, b] \to \mathbb{R}$ be a bounded function. For each $j = 1, \ldots, n$, let

$$m_j = \inf\{f(x) : x \in [x_{j-1}, x_j]\}, \ M_j = \sup\{f(x) : x \in [x_{j-1}, x_j]\}.$$

The **lower sum of f with respect to P** is

$$L(f, P) = \sum_{j=1}^{n} m_j(x_j - x_{j-1}).$$

The **upper sum of f with respect to P** is

$$U(f, P) = \sum_{j=1}^{n} M_j(x_j - x_{j-1}).$$

The **lower integral of f over [a, b]** is

$$L(f) = \sup\{L(f, P) : \text{as } P \text{ varies over partitions of } [a, b]\},$$

and the **upper integral of f over [a, b]** is

$$U(f) = \inf\{U(f, P) : \ P \text{ varies over partitions of } [a, b]\}.$$

We say that f is **integrable over** $[a, b]$ when $L(f) = U(f)$. We call this common value **the integral of f over [a, b]**, and we write it as

$$\int_a^b f = \int_a^b f(x)\, dx = \int_a^b f(t)\, dt.$$

The Fundamental theorems of calculus

I: Let $f, g : [a, b] \to \mathbb{R}$ such that f is continuous and g is differentiable with $g' = f$. Then

$$\int_a^b f = g(b) - g(a).$$

II: Let $f : [a, b] \to \mathbb{R}$ be continuous. Then for all $x \in [a, b]$, f is integrable over $[a, x]$, and the function $g : [a, b] \to \mathbb{R}$ given by $g(x) = \int_a^x f$ is differentiable on (a, x) with

$$\frac{d}{dx} \int_a^x f = f(x).$$

Geometric series

$\sum_{k=1}^{\infty} r^k$ diverges if $|r| \geq 1$ and converges to $\frac{r}{1-r}$ if $|r| < 1$.

$\sum_{k=0}^{\infty} r^k$ diverges if $|r| \geq 1$ and converges to $\frac{1}{1-r}$ if $|r| < 1$.

For all $r \in \mathbb{C} \setminus \{1\}$, $\sum_{k=0}^{n} r^k = \frac{r^{n+1}-1}{r-1}$.

Never divide by 0

It bears repeating. Similarly do not plug 0 or negative numbers into ln, do not plug negative numbers into the square root function, do not ascribe a function (or a person) a task that makes no sense.

Index